Modern Approaches in Solid Earth Sciences

Volume 24

Series Editors

Background and motivation

Earth Sciences are going through an interesting phase as the traditional disciplinary boundaries are collapsing. Disciplines or sub-disciplines that have been traditionally separated in the past have started interacting more closely, and some new fields have emerged at their interfaces. Disciplinary boundaries between geology, geophysics and geochemistry have become more transparent during the last ten years. Geodesy has developed close interactions with geophysics and geology (tectonics). Specialized research fields, which have been important in development of fundamental expertise, are being interfaced in solving common problems.

In Earth Sciences the term System Earth and, correspondingly, Earth System Science have become overall common denominators. Of this full System Earth, Solid Earth Sciences – predominantly addressing the Inner Earth - constitute a major component, whereas others focus on the Oceans, the Atmosphere, and their interaction. This integrated nature in Solid Earth Sciences can be recognized clearly in the field of Geodynamics. The broad research field of Geodynamics builds on contributions from a wide variety of Earth Science disciplines, encompassing geophysics, geology, geochemistry, and geodesy. Continuing theoretical and numerical advances in seismological methods, new developments in computational science, inverse modelling, and space geodetic methods directed to solid Earth problems, new analytical and experimental methods in geochemistry, geology and materials science have contributed to the investigation of challenging problems in geodynamics. Among these problems are the high-resolution 3D structure and composition of the Earth's interior, the thermal evolution of the Earth on a planetary scale, mantle convection, deformation and dynamics of the lithosphere (including orogeny and basin formation), and landscape evolution through tectonic and surface processes. A characteristic aspect of geodynamic processes is the wide range of spatial and temporal scales involved. An integrated approach to the investigation of geodynamic problems is required to link these scales by incorporating their interactions. Scope and aims of the new series

The book series "Modern Approaches in Solid Earth Sciences" provides an integrated publication outlet for innovative and interdisciplinary approaches to problems and processes in Solid Earth Sciences, including Geodynamics.

It acknowledges the fact that traditionally separate disciplines or sub-disciplines have started interacting more closely, and some new fields have emerged at their interfaces. Disciplinary boundaries between geology, geophysics and geochemistry have become more transparent during the last ten years. Geodesy has developed close interactions with geophysics and geology (tectonics). Specialized research fields (seismic tomography, double difference techniques etc), which have been important in development of fundamental expertise, are being interfaced in solving common problems.

Accepted for inclusion in Scopus.

Prospective authors and/or editors should consult one of the Series Editors or the Springer Contact for more details. Any comments or suggestions for future volumes are welcomed.

Arthur H. Hickman

Archean Evolution of the Pilbara Craton and Fortescue Basin

Springer

Arthur H. Hickman
Geological Survey of Western Australia
East Perth, WA, Australia

ISSN 1876-1682 ISSN 1876-1690 (electronic)
Modern Approaches in Solid Earth Sciences
ISBN 978-3-031-18005-7 ISBN 978-3-031-18007-1 (eBook)
https://doi.org/10.1007/978-3-031-18007-1

This Springer imprint is published by the registered company Springer Nature Switzerland AG
The registered company address is: Gewerbestrasse 11, 6330 Cham, Switzerland

To the Pilbara Craton mapping team, 1994–2005, and to my wife Caroline for her understanding and support during my work on the book.

Preface

Based on information gathered over almost half a century of investigations, this book reviews and interprets the Eoarchean to Neoarchean crustal evolution of the northern, best exposed part of the Pilbara Craton. The investigations included two periods of systematic geological mapping by the Geological Survey of Western Australia (GSWA), government airborne geophysical surveys, regional geochemical and geochronological investigations, and detailed local studies by numerous researchers and government organizations. Through full-time employment with the GSWA since 1972, the author has been heavily involved in much of the work, including geological mapping, management of a major mapping project, and extensive research and publication. With the completion of mapping and follow-up geological interpretation, it is timely to review the results.

The Pilbara Craton is one of two Archean cratons in Western Australia, the other being the Yilgarn Craton, which together have provided important contributions to our understanding of Earth's Archean crustal evolution. Since 1970, Perth has been selected to host six successive international symposia on Archean geology, which testifies to the international recognition of the work conducted on Western Australia's cratons, and the significance of the results to our understanding of Archean geology. The Pilbara Craton contains Earth's most complete geological record of Paleoarchean crustal evolution and provides the best evidence on a transition from vertical to horizontal tectonics between the Paleoarchean and the Mesoarchean. From the advent of plate tectonic theory in the 1970s, a long-running controversial issue has been whether or not plate tectonic processes operated throughout the Archean or if these processes evolved during or after the Archean. As documented in this book, evidence from the Pilbara Craton has resolved this issue, at least as far as this craton is concerned.

The Pilbara Craton has also revealed the processes and timing involved in the formation of its dome-and-keel crustal architecture, a feature common to many Archean cratons worldwide. Another important finding in the Pilbara is that geochemical, geochronological, and isotopic data support early suggestions that the oldest well-preserved greenstone succession of the craton, the 3530–3235 Ma

Pilbara Supergroup, was deposited on older sialic crust. Formation of this pre-3530 Ma crust occurred over at least 250 million years in the Eoarchean and early Paleoarchean. This negates some previous assumptions that the Paleoarchean 'greenstones' of the Pilbara Craton are the oldest rocks preserved, and that they originated as oceanic crust.

Any review of the regional geology and crustal evolution of the Pilbara Craton would not be complete without mention of the strong geological similarities to the Barberton region in the eastern Kaapvaal Craton of South Africa. Almost all previous comparisons have focussed on the Neoarchean to Paleoproterozoic successions of these areas. However, as explained in this book, there are even greater similarities between the Paleoarchean successions of the cratons. The concept that both cratons are fragments of the same Archean supercontinent, 'Vaalbara' (Cheney et al., 1988), has been slow to gain general acceptance. However, stratigraphic, tectonic, geochronological, and paleomagnetic data have now combined to support the interpretation that the Pilbara and Kaapvaal Cratons shared a common, billion-year-long, evolutionary history on the same continent. The Pilbara Craton is now interpreted to have experienced two episodes of continental breakup, and recognition that the present Pilbara Craton is merely a fragment of a once far more extensive Archean continental plate provides a far better appreciation of the original scale of its stratigraphy and tectonic units.

In summary, the main aims of this book are to review all currently available data on the Archean geology of the northern half of the Pilbara Craton. The southern half of the craton is largely concealed by Neoarchean and Paleoproterozoic cover. The review cites previous sources of information on all the subjects covered and provides up-to-date interpretations of the craton's stratigraphy, evolving tectonic and depositional environments, deformation history, geochronology, and overall crustal evolution. In the process, several new concepts on Paleoarchean crustal evolution are introduced or expanded upon, including the existence of mantle plume-related large igneous provinces from ~3530 Ma onwards, and that from ~3335 Ma onwards the individual granite–greenstone domes of the east Pilbara evolved into separate tectonic domains. Additionally, geochronology, geochemistry, and Sm–Nd isotope data have been combined to confirm a long-held interpretation that the Mesoarchean evolution of the northwest Pilbara Craton was controlled by plate convergence and subduction. Finally, the Neoarchean Fortescue Basin is reviewed because its evolution marks the second continental breakup in the history of the craton.

The book fills a gap in the present literature by providing students, researchers, and geoscientific organizations with a detailed and comprehensive modern account of the Archean geology of the Pilbara Craton, and of important previous investigations on which present interpretations are based. As such, it will facilitate access to background information for future geoscientific studies. It should also help identify remaining questions requiring more work. It is hoped that the book will lead to the Archean geology of the Pilbara Craton featuring more prominently in future global reviews of Archean granite–greenstone terranes, and of Archean crustal evolution generally.

The author has drawn heavily on previously published data and interpretations, in particular in publications by GSWA and Geoscience Australia. However, he takes full responsibility for any new interpretations using previously published data, and for summaries and comments relating to the evidence and interpretations of previous workers.

East Perth, Australia Arthur H. Hickman

Acknowledgements

The author thanks Franco Pirajno for his suggestion to write this review of geological information on the Pilbara Craton and Fortescue Basin gained from investigations by GSWA staff and others over the past 50 years, and in particular from the 1994–2005 Pilbara Craton Mapping Project. This project, conducted jointly by GSWA and Geoscience Australia, provided a wealth of new geoscientific data that resulted in numerous key publications, both during and after the project. The present book would not have been possible without extensive use of observations and interpretations made within the project, and from related research over the past 15 years. Special acknowledgment is made to major geoscientific contributions to the project from Martin Van Kranendonk, Hugh Smithies, Ian Williams, Leon Bagas, and Terry Farrell of GSWA, and from Andrew Glikson, David Huston, David Champion, and Richard Blewett of Geoscience Australia. The author publishes with permission of the Executive Director of the Geological Survey of Western Australia and with his permission to include many figures drafted for previous GSWA publications.

About This Book

The responsible series editor of this book is Dr. F. Pirajno.

Contents

About the Author

Arthur Hugh Hickman joined the Geological Survey of Western Australia (GSWA) early in 1972 shortly after being awarded a PhD (Geology) at Birmingham University, England. His interpretations of the stratigraphy, structure, and geochemistry of 800 km^2 of the Southwest Highlands of Scotland were subsequently published in UK journals between 1975 and 1983. Parts of his geological mapping were incorporated into maps produced by the British Geological Survey. Arthur's main aim during his first 20 years at GSWA was to contribute to increasing the geological understanding of parts of the vast state of Western Australia. His initial investigations covered the 100,000 km^2 Pilbara region, and after a few years Arthur wrote a book (GSWA Bulletin 127) providing the first comprehensive interpretation of this area's stratigraphy, crustal evolution, and mineralization. GSWA priorities then moved his investigations to other parts of Western Australia, but in 1993 a project was set up for a more detailed geological study of the Pilbara. The Pilbara Craton Mapping Project was conducted jointly between GSWA and Geoscience Australia between 1994 and 2005. The project was focused on more detailed geological mapping, with the GSWA Pilbara team quickly expanding to eight geoscientists under Arthur's management. After 2005, members of the Pilbara team were moved to other projects and Arthur commenced follow-up compilation of maps and reports, geological interpretations, and working to make the large volume of data from the project more accessible in digital format. During his 50-year career with GSWA, Arthur has led many local and international geological excursions to the Pilbara and is recognized as a leading authority on the geology of this scientifically important area. He has published widely, and has given presentations to international audiences in Australia, South Africa, Japan, and USA.

List of Figures

List of Tables

Chapter 1
Outline of the Pilbara Craton

Abstract Previous investigations of the northern Pilbara Craton are briefly summarized, followed by an outline of the region's lithostratigraphy and major tectonic units. Previous interpretations of its tectonic evolution have not taken account evidence that the presently preserved 500,000 km^2 Pilbara Craton is composed of fragments of much larger Paleoarchean and Mesoarchean continents. This consideration provides important new insights on the original scales of the processes and tectonic units that existed before two major events of continental breakup.

Keywords Previous investigations · Lithostratigraphy · Tectonic units · Continental breakup

1.1 Introduction

The Pilbara Craton is a 500,000 km^2 segment of Eo–Mesoarchean granite–greenstone crust underlying the Pilbara region of northwestern Australia (Figs. 1.1 and 1.2). Although large sections of the craton are concealed by Neoarchean–Paleoproterozoic successions of the Fortescue, Hamersley, and Turee Creek Basins, there are areas of erosion that provide large windows onto the granite–greenstones. The largest exposure is a 60,000 km^2 inlier that extends 500 km east from the northwest Pilbara coast (Fig. 1.3). Once described as the 'Pilbara Block' (Ryan 1965; Blockley 1975; Hickman 1983), the eastern part of this inlier contains the 40,000 km^2 East Pilbara Terrane (EPT, Fig. 1.3). This is arguably the world's best preserved Paleoarchean terrane, providing unique insights on Paleoarchean crustal evolution and the earliest life on Earth. Equally informative is the western area of the inlier where Mesoarchean terranes and basins record the beginning of plate tectonic processes in the Pilbara.

Trendall (1990) argued that the Fortescue, Hamersley, and Turee Creek Basins should be included in the Pilbara Craton because crustal stability was not attained until about 2400 Ma. However, more recent geological investigations have established that cratonization was completed during orogenic episodes between 2955 and 2890 Ma, after which there was over 100 Ma of crustal stability (Hickman

A. H. Hickman, *Archean Evolution of the Pilbara Craton and Fortescue Basin*, Modern Approaches in Solid Earth Sciences 24,
https://doi.org/10.1007/978-3-031-18007-1_1

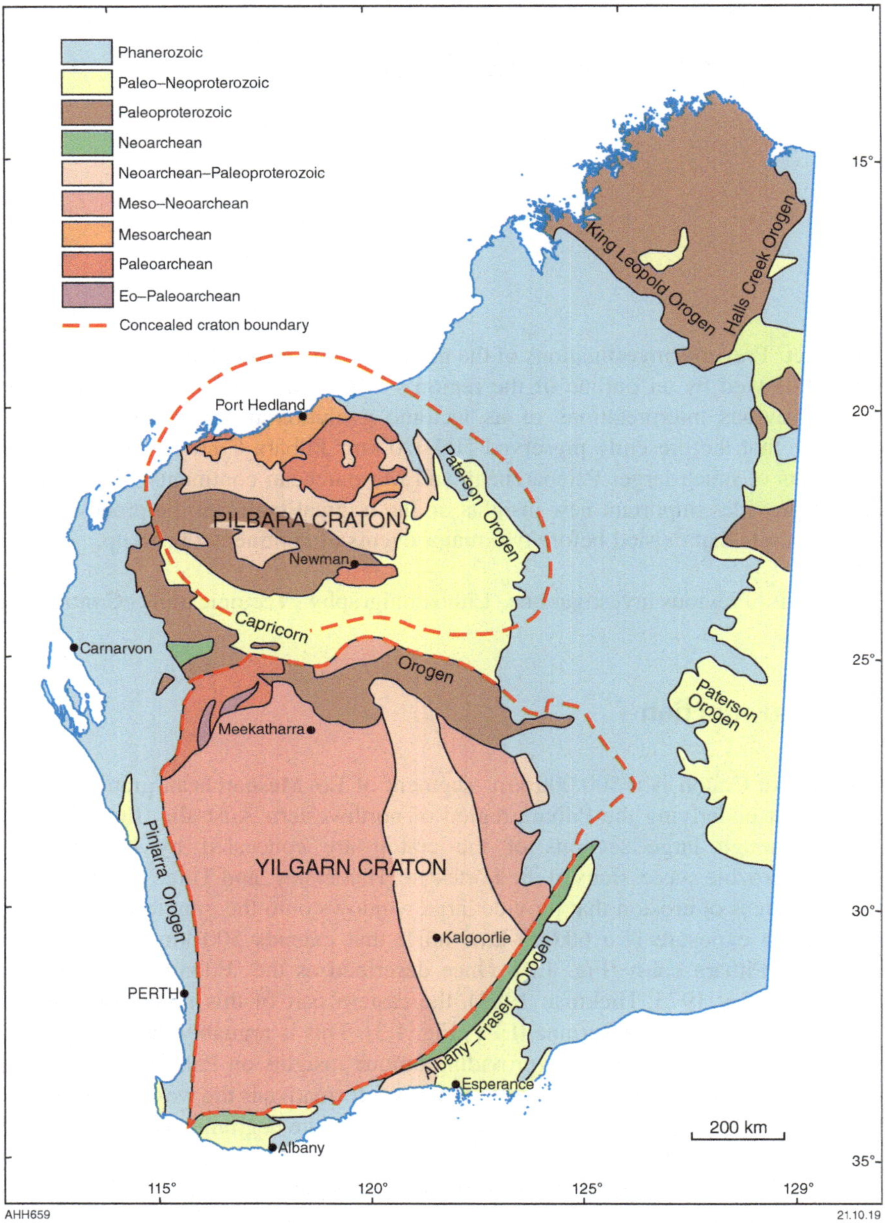

Fig. 1.1 Simplified Archean and Proterozoic chronological divisions of Western Australia, showing an interpretation of the concealed extents of the Pilbara and Yilgarn Cratons (From Hickman 2016; with Geological Survey of Western Australia permission)

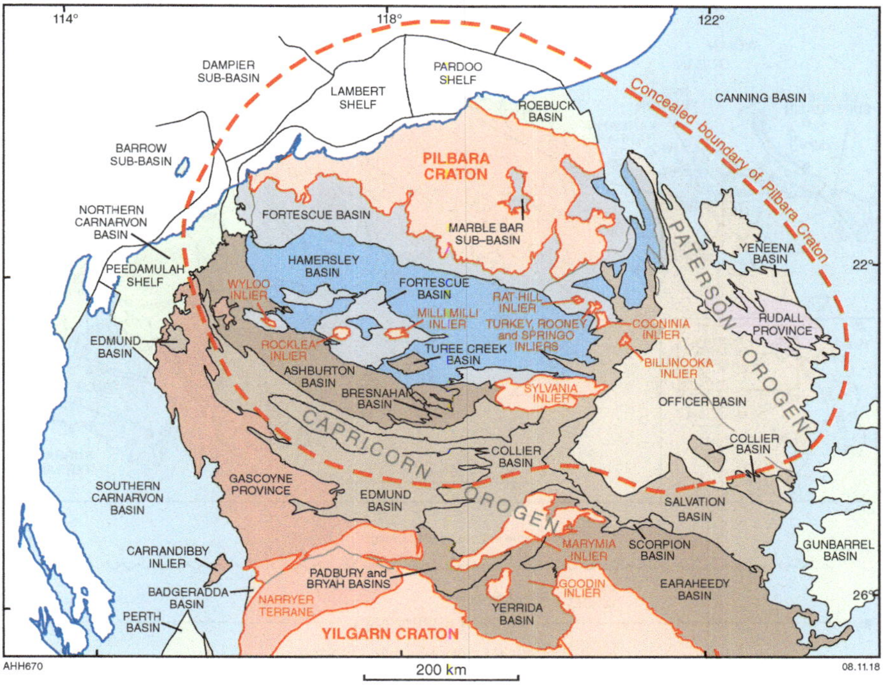

Fig. 1.2 Tectonic units of northwestern Western Australia, showing the setting of the Pilbara Craton. The southern half of the craton is concealed by Neoarchean and Proterozoic rocks except for rare exposures within inliers (From Hickman 2016; with Geological Survey of Western Australia permission)

et al. 2006; Van Kranendonk et al. 2006; Hickman and Van Kranendonk 2012). A c. 2775 Ma regional unconformity at the base of the Fortescue Basin marks the end of this stability and the commencement of rifting and breakup of the Pilbara Craton (Blake and Barley 1992; Blake 1993; Martin et al. 1998a, b; Thorne and Trendall 2001; Barley et al. 2005; Hickman et al. 2010; Pirajno and Santosh 2015).

1.1.1 Investigations of the Pilbara Craton

Early geological investigations of the Pilbara Craton were reviewed by Noldart and Wyatt (1962) and Hickman (1983). Present geological interpretations are based on information obtained during three periods of investigation:

- *1972 to 1975:* the Geological Survey of Western Australia (GSWA) undertook systematic 1: 50,000 geological mapping of the East Pilbara for the compilation of four 1: 250,000 maps. Results included stratigraphic interpretations of greenstone stratigraphy (Hickman and Lipple 1975; Lipple 1975; Hickman 1977,

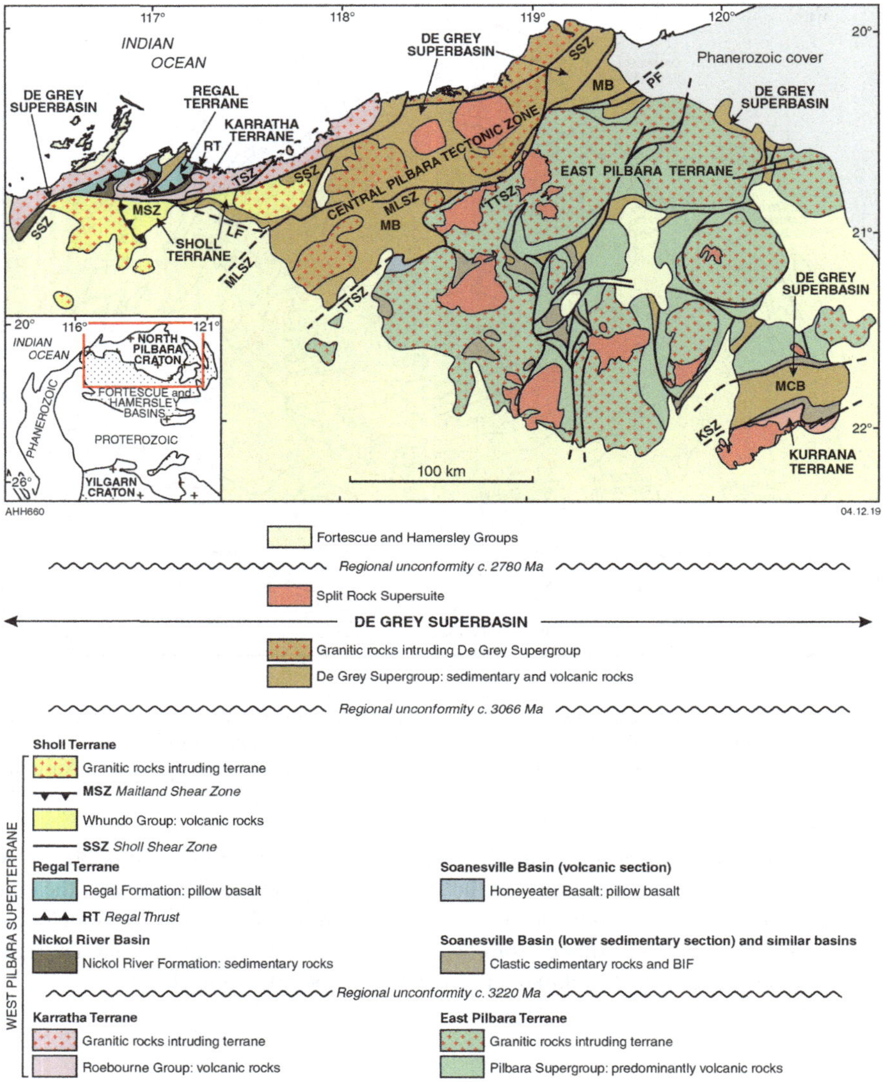

Fig. 1.3 Major tectonic units of the northern Pilbara Craton. The mainly Paleoarchean East Pilbara Terrane is separated from Mesoarchean terranes and basins of the northwest Pilbara by the Tabba Tabba Shear Zone. The Central Pilbara Tectonic Zone is a Mesoarchean zone of deformation and magmatic intrusion formed by 3165 to 2900 Ma plate convergence between the East Pilbara and Karratha Terranes. Abbreviations: *KSZ* Kurrana Shear Zone; *LF* Loudens Fault; *MB* Mallina Basin; *MCB* Mosquito Creek Basin; *MLSZ* Mallina Shear Zone; *MSZ* Maitland Shear Zone; *PF* Pardoo Fault (part of TTSZ); *TSZ* Terenar Shear Zone; *TTSZ* Tabba Tabba Shear Zone (From Hickman 2016; with Geological Survey of Western Australia permission)

1980a, b, c) and an interpretation of the area's crustal evolution (Hickman 1981, 1983). In 1976, the northwest Pilbara was briefly examined to clarify local stratigraphy and to assess if correlations could be made with the geology of the East Pilbara; however, limited time did not permit a mapping program.

- *1994 to 2005:* a collaborative project between GSWA and Geoscience Australia (GA) investigated the entire 60,000 km^2 granite–greenstone inlier, plus all surrounding outcrops of the Fortescue and Hamersley Groups north of the Fortescue River. Work involved systematic 1:25,000-scale geological mapping (for publication of 1:100,000 scale geological maps), airborne aeromagnetic and radiometric surveys, extensive U–Pb zircon geochronology, zircon Lu–Hf and whole-rock Sm–Nd isotope studies, geochemistry, and local studies of structural geology and mineralization. For most of its duration, this 'Pilbara Craton Mapping Project' (PCMP) employed between 10 and 12 geoscientists. In addition to a new series of geological maps over almost 100,000 km^2, the PCMP resulted in a major revision of the stratigraphy, structural geology, crustal evolution, and mineralization of the northern Pilbara Craton (Van Kranendonk et al. 2002, 2006, 2007a, b; Huston et al. 2002; Hickman 2004; Smithies et al. 2004, 2005a, b, 2007; Champion and Smithies 2007).
- *2006 onwards:* the geological evidence provided by the PCMP laid the foundation for numerous subsequent studies by Australian and international researchers, all of which have contributed valuable additional information.

This book reviews and reinterprets the data and conclusions from all these investigations. Over 300 U–Pb zircon dates on intrusive and volcanic rocks proved essential in better defining the stratigraphy of the Pilbara Craton. In combination with new geochemical data, the geochronology led to the recognition of discrete granitic supersuites (Van Kranendonk et al. 2006) and the correlation of these between all eleven granite–greenstone domes of the EPT (Fig. 1.4).

1.2 Stratigraphy of the Northern Pilbara Craton

The first detailed stratigraphic correlations between the east and northwest Pilbara assumed that the northern Pilbara Craton was essentially a single terrane (Hickman 1980a, b, c, 1981, 1983). Contrary evidence appeared in the early 1980s when Landsat imagery revealed a series of northeast-trending lineaments cutting across the northern Pilbara Craton. These lineaments were first interpreted to coincide with late-stage faults that had no significance to the early crustal evolution of the craton (Krapež and Barley 1987). However, subsequent geochronological evidence (Horwitz and Pidgeon 1993) cast doubt on existing east–west correlations between the older sections of the Pilbara stratigraphy. This led to interpretations that the lineaments marked either boundaries between separate tectonostratigraphic domains

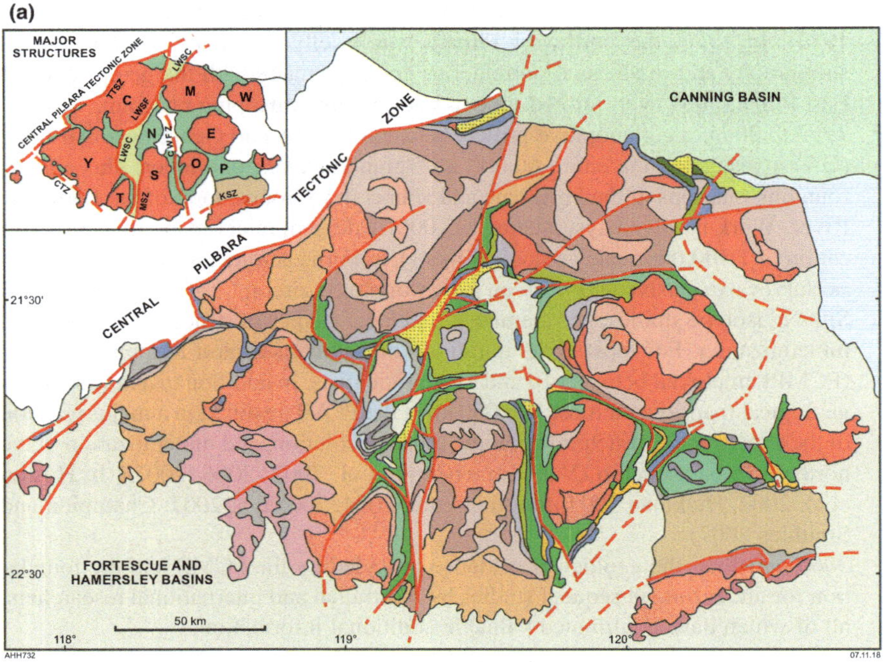

Fig. 1.4 Simplified geological map of the eastern section of the northern Pilbara Craton. Mainly volcanic groups and subgroups of the Paleoarchean East Pilbara Terrane are unconformably overlain by mainly sedimentary Mesoarchean groups. The Paleoarchean stratigraphy is continuous across the East Pilbara Terrane but shows deformation into a dome–and–keel crustal architecture. Paleoarchean granitic intrusions were emplaced into the cores of the domes during diapiric deformation, whereas Mesoarchean granitic intrusions were emplaced in zones controlled by plate-tectonic processes and are therefore unrelated to the dome–and–keel structure. Paleoarchean granitic intrusions have contemporaneous felsic volcanic equivalents in the Paleoarchean succession, whereas Mesoarchean granitic intrusions have no volcanic equivalents in the East Pilbara. Inset figure: shows the East Pilbara Terrane separated into east and west sections by the Lalla Rookh–Western Shaw Structural Corridor (LWSC) and separated from the Mesoarchean Central Pilbara Tectonic Zone of the northwest Pilbara Craton by the Tabba Tabba Shear Zone (TTSZ). The Coongan–Warralong Fault Zone (CWFZ) defines the western limit of the 3324–3290 Ma Emu Pool Supersuite, whereas the Kurrana Shear Zone (KSZ) is the southeast limit of the East Pilbara Terrane and overlying Mosquito Creek Basin. The Chichester Tectonic Zone (CTZ) is a broad east–southeast trending zone of 3070–2920 Ma deformation and metamorphism. Dome abbreviations: *C* Carlindi; *E*, Mount Edgar; *I* Yilgalong; *M*, Muccan; *N* North Pole; *O* Corunna Downs; *P* McPhee; *S* Shaw; *T* Tambourah; *W* Warrawagine; *Y* Yule (From Hickman 2021; with Geological Survey of Western Australia permission)

(Krapež 1993; Krapež and Eisenlohr 1998) or sutures between accreted terranes (Barley 1997).

Mid-way through the PCMP, it was recognized that the Pilbara Craton had evolved through stages of rifting and continental breakup between 3280 and 3165 Ma (Hickman 2001a, b, 2004, 2016, 2021; Hickman et al. 2001; Van

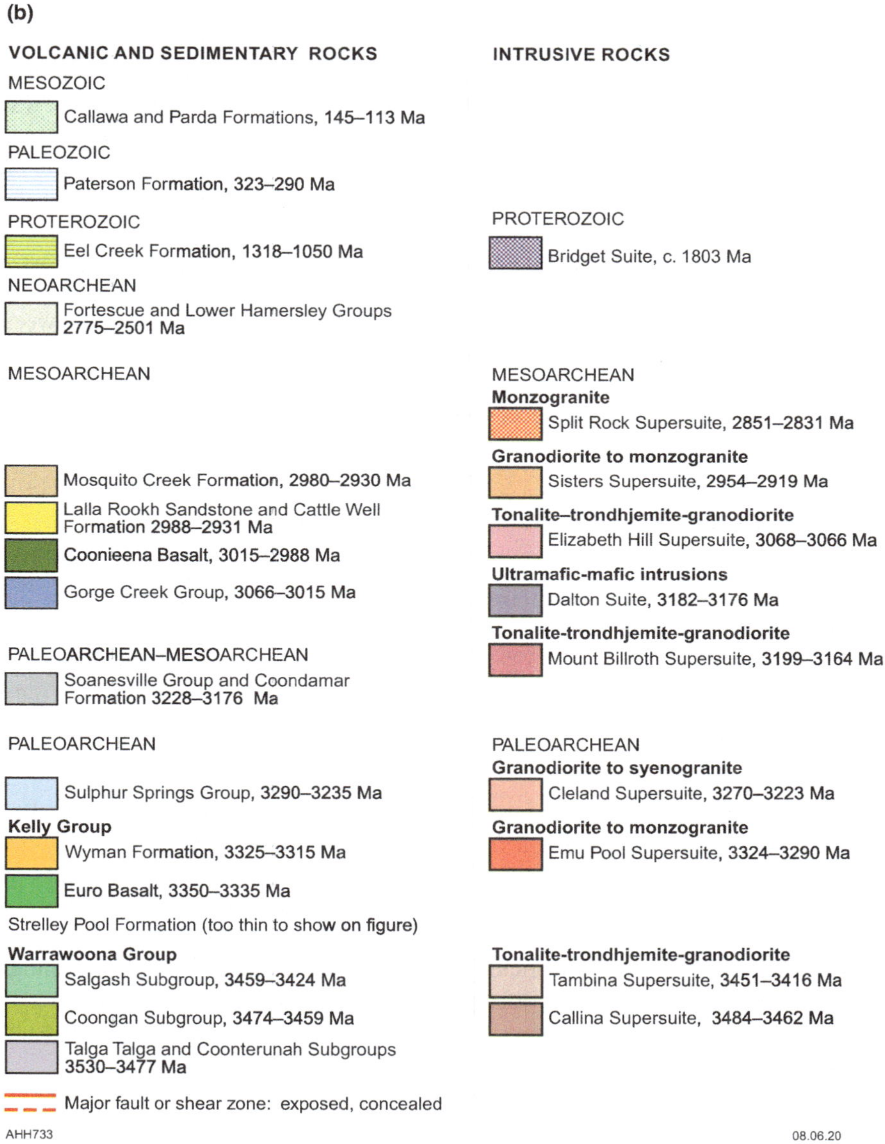

Fig. 1.4 (continued)

Kranendonk et al. 2002, 2006, 2010; Smithies et al. 2005b). From breakup and plate separation that occurred at 3220 Ma, the east and northwest parts of the Pilbara Craton evolved independently until 3070 Ma when, following re-convergence and plate collision, the east and northwest Pilbara were recombined. Therefore, the northwest Pilbara stratigraphy that evolved between 3220 and 3070 Ma is unrelated to the stratigraphy of the East Pilbara. Owing to separate evolution of the east and northwest Pilbara, no single stratigraphic column can be applied across the northern part of the craton. Following the PCMP, a major revision of the stratigraphy was provided by Van Kranendonk et al. (2006). This included stratigraphic comparisons between the east and northwest parts of the Pilbara Craton (Van Kranendonk et al. 2006) and a diagrammatic interpretation of the tectonic processes involved in the evolution of the stratigraphy (Fig. 1.5) (Van Kranendonk et al. 2006).

Figure 1.6 makes a stratigraphic comparison between the east and northwest Pilbara Craton based on a statistical analysis of over 300 U–Pb zircon dates on volcanic and intrusive units and highlights the differences and similarities between the two areas. However, it should be noted that there is relatively limited exposure of Paleoarchean rocks in the northwest Pilbara. Table 1.1 reviews the relative ages of formations and groups in both successions.

1.3 Tectonic Units

Tectonic units provide the geological framework used in most descriptions of the Pilbara Craton.

1.3.1 Terminology

The Pilbara Craton is divided into three main types of tectonic unit, terranes, basins, and tectonic zones. A terrane is a fault-bounded body of rock of regional extent, characterized by a geological history different from that of contiguous bodies of rock. A basin is an area underlain by a substantial thickness of sedimentary or volcanic rocks, which has unifying characteristics of stratigraphy and structure due to deposition in a regionally restricted area. Basins are bounded by unconformities except where major faults have juxtaposed a basin with another tectonic unit. A superbasin is a connected series of basins. Tectonic zones are linear belts of deformation crosscutting the terranes and basins.

Additionally, the craton is composed of greenstone belts and granitic complexes. These are principally lithological divisions restricted to individual terranes. Greenstone belts contain successions of volcanic and sedimentary formations and groups, typically including mafic intrusive rocks but lacking large granitic intrusions. Granitic complexes are large accumulations of granitic intrusions, in many cases amalgamated over hundreds of millions of years. In the EPT, the granitic complexes are composed of Paleoarchean intrusions forming the central cores of granite–greenstone domes with total diameters between 30 and 120 km. Within each dome, the outer boundary of the granitic core is a tectonic or intrusive contact with one or more greenstone belts (Fig. 1.7). The age ranges of the granites and

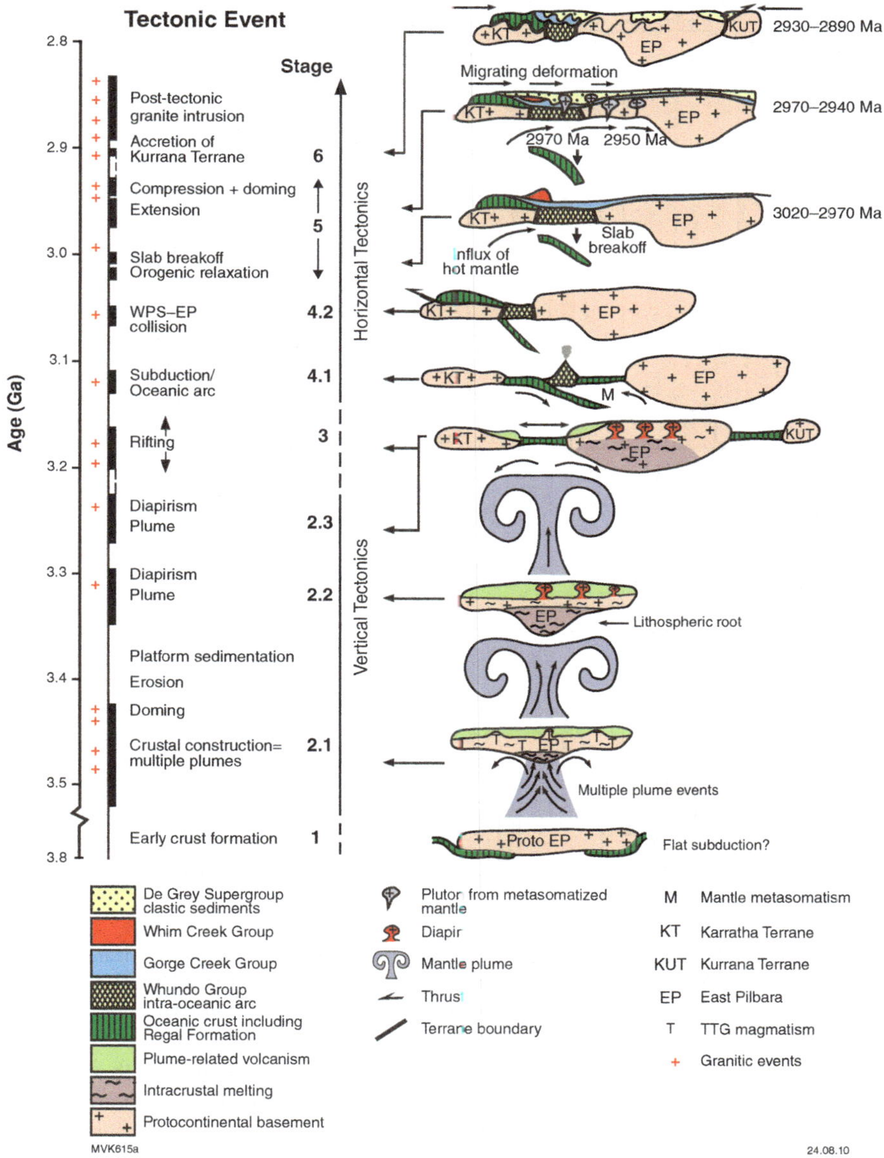

Fig. 1.5 Diagrammatic illustration of the main events in the evolution of the Pilbara Craton. Following the formation of 3800–3530 Ma continental crust, a series of Paleoarchean mantle plume events resulted in the eruption of the mafic volcanic Warrawoona, Kelly, and Sulphur Springs Groups. Resulting gravitational instability led to phases of diapiric doming between 3460 and 3223 Ma. The Sulphur Springs plume uplifted and extended the crust causing rifting and the first breakup of the craton at 3220 Ma. Plate separation developed basins of oceanic-like basaltic crust between the newly formed continental microplates (KT, EPT, and KUT). Compression from c. 3160 to 2920 Ma led to Mesoarchean plate tectonic processes including subduction, obduction, evolution of magmatic arcs, terrane accretion, and orogenic deformation (Modified from Van Kranendonk et al. 2006; with Geological Survey of Western Australia permission)

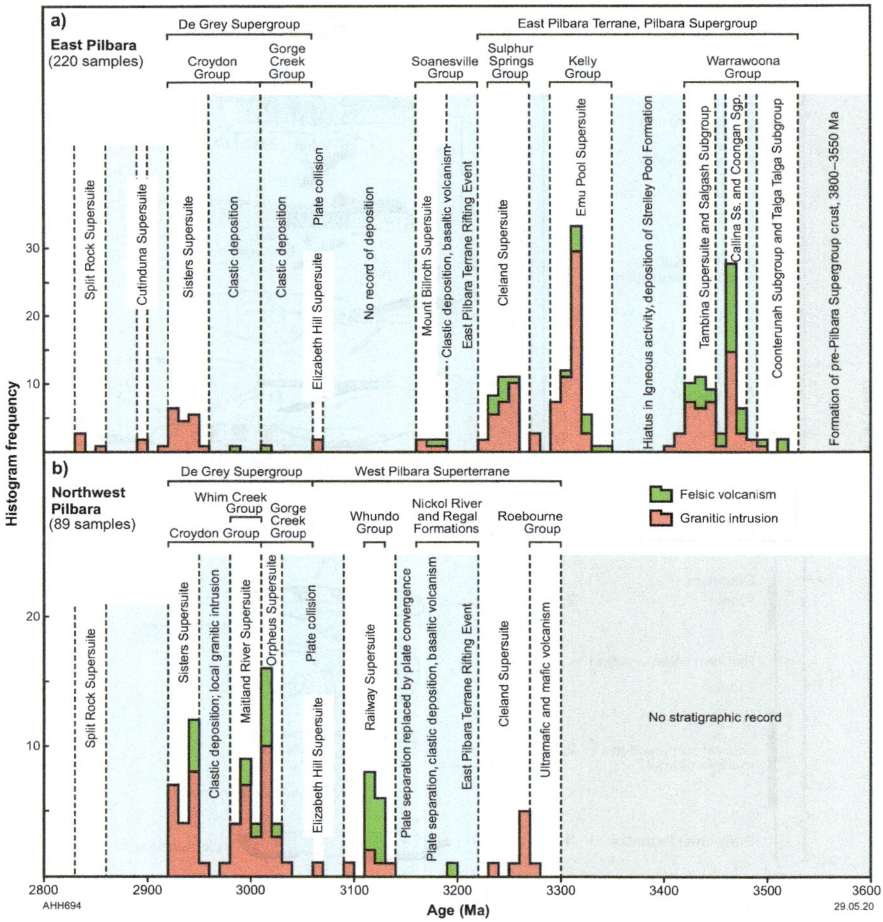

Fig. 1.6 Comparison of the timing of episodes of granitic intrusion and felsic volcanism in the Northern Pilbara Craton, as indicated by published U–Pb zircon geochronology (309 samples) (From Hickman 2021; with Geological Survey of Western Australia permission)

greenstones within each dome are typically the same, consistent with a genetic relationship. In many instances, this relationship is confirmed where granites pass upwards into subvolcanic felsic intrusions, and these feed into overlying felsic volcanic formations. This is an important feature of the EPT, although not previously been described in detail. Genetic relationships between granitic complexes and adjacent greenstones are also evident in the northwest Pilbara, but that part of the craton contains no granite–greenstone domes comparable to those in the East Pilbara.

Table 1.1 Generalized Paleoarchean to Mesoarchean lithostratigraphy (excluding intrusive units) of the northern Pilbara Craton

Age (Ma)	Terrane/Superbasin	Basin	Group	Formation
2980–2930		Mosquito Creek Basin (upper part)	Nullagine Group (upper part)	Mosquito Creek Formation
	NO CONTACT			
3067–2931	De Grey Superbasin			
3015–2931		Mallina Basin (central trough) (WP)	Croydon Group	Mallina Formation
3015–2955				Constantine Sandstone
		TECTONIC CONTACT		
2988–2931		Mallina Basin (southeast shelf)	Croydon Group	Lalla Rookh Sandstone
c. 2988				Cattle Well Formation
3015–2988				Coonieena Basalt
		TECTONIC CONTACT		
2948–2941		Mallina Basin (north shelf) (WP)	Bookingarra Group (WP)	Kialrah Rhyolite
2950–2948				Mount Negri Volcanics
2950–2948				Louden Volcanics
2955–2950				Rushall Slate
c. 2955				Cistern Formation
		UNCONFORMITY		
		Whim Creek Basin (WP)	Whim Creek Group (WP)	Red Hill Volcanics

(continued)

Table 1.1 (continued)

Age (Ma)	Terrane/Superbasin	Basin	Group	Formation
3000–2990				
c.3000				Warambie Basalt
		UNCONFORMITY		
c. 3015		Gorge Creek Basin	Gorge Creek Group	Cundaline Formation
3022–3015				Cleaverville Formation
3067–3022				Farrel Quartzite
	REGIONAL UNCONFORMITY			
3130–3110	Sholl Terrane (WP)			
3115–3110		Whundo Basin (WP)	Whundo Group (WP)	Woodbrook Formation
c. 3115				Bradley Basalt
c. 3120				Tozer Formation
3130–3120				Nallana Formation
	TECTONIC CONTACT			
3200–3160	Regal Terrane (WP)			
3195–3160		Regal Basin (WP)		Port Robinson Basalt
c. 3195				Dixon Island Formation
3200–3160				Regal Formation

	TECTONIC CONTACT			
3223–3165	Rift-related basins			
3176–3165		Soanesville Basin	Soanesville Group	Empress Formation
3176–3165				Hong Kong Chert
3176–3165				Pyramid Hill Formation
3185–3176				Honeyeater Basalt
3190–3185				Paddy Market Formation
c. 3190				Corboy Formation
3223–3190				Cardinal Formation
		NO CONTACT		
3223–3165				Budjan Creek Formation
		NO CONTACT		
3220–3165		Mosquito Creek Basin (lower part)	Nullagine Group (lower part)	Coondamar Formation
		NO CONTACT		
3220–3165		Nickol River Basin (WP)		Nickol River Formation
	UNCONFORMITY			
c. 3280	Karratha Terrane (WP)			
c. 3280			Roebourne Group (WP)	Weerianna Basalt
				Ruth Well Formation

(continued)

Table 1.1 (continued)

Age (Ma)	Terrane/Superbasin	Basin	Group	Formation
	NO CONTACT			
3530–3235	East Pilbara Terrane (Pilbara Supergroup)			
3252–3235		Sulphur Springs Basin	Sulphur Springs Group	Kangaroo Caves Formation
3275–3253				Kunagunarrina Formation
3290–3255				Leilira Formation
		UNCONFORMITY		
c. 3315		Kelly LIP	Kelly Group	Charteris Basalt
3325–3315				Wyman Formation
3350–3335				Euro Basalt
		DISCONFORMITY		
3426–3350				Strelley Pool Formation
		EROSIONAL UNCONFORMITY to PARACONFORMITY		
3448–3427		Warrawoona LIP	Warrawoona Group	Panorama Formation
3455–3441				Apex Basalt
				DISCONFORMITY
3474–3459				Duffer Formation
3474–3469				Mount Ada Basalt

				DISCONFORMITY
3481–3477				Dresser/McPhee Formations
3530–3490				North Star Basalt
				NO CONTACT
3498–3490				Double Bar Formation
3515–3498				Coucal Formation
3530–3515				Table Top Formation

(WP), restricted to the West Pilbara

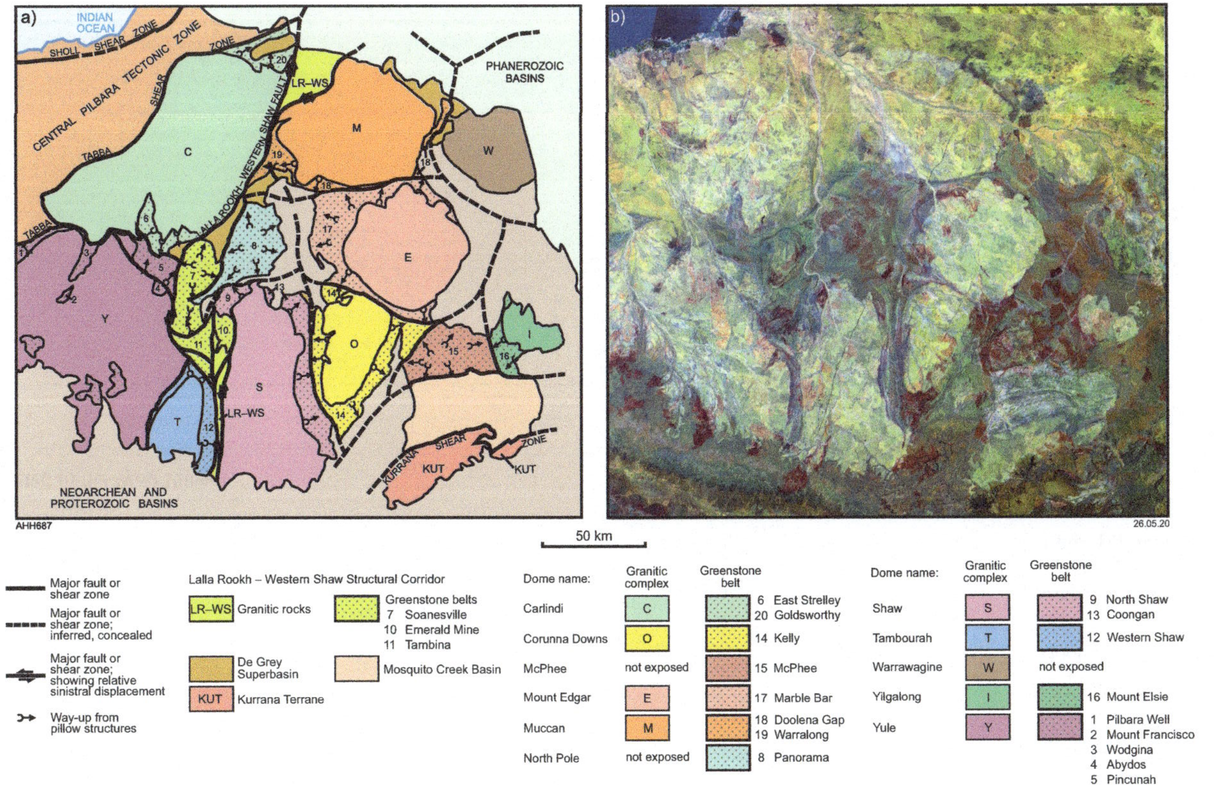

Fig. 1.7 Granite–Greenstone Domes of the Eastern Pilbara Craton: (**a**) simplified structural map showing separation of domes by major faults; (**b**) Landsat Thematic Mapper image (Bands 7, 4, and 1). Domes composed of coupled granitic cores and greenstone belts are separated by major faults. Colours have no age significance (Modified from Gardiner et al. 2018; with Geological Survey of Western Australia permission)

1.3.2 Summary of Tectonic Units

The northern Pilbara Craton comprises three Paleoarchean terranes (East Pilbara, Karratha and Kurrana), two Mesoarchean terranes (Regal and Sholl), and six Mesoarchean basins (Soanesville, Nickol River, Gorge Creek, Whim Creek, Mallina and Mosquito Creek). Additionally, the craton contains remnants of pre-3530 Ma sialic crust. Windows to the largely concealed southern part of the craton include the Sylvania, Milli Milli, Rocklea, and Wyloo Inliers (Fig. 1.2). The northern part of the Pilbara Craton comprises the following tectonic units:

1. *Early crust, 3800–3530 Ma:* the existence of continental crust pre-dating the Paleoarchean greenstone succession (Pilbara Supergroup) has been established from several sources, including U–Pb zircon geochronology, whole-rock Sm–Nd model ages, and Lu–Hf isotope data from zircons. Enclaves of tonalite gneiss and metagabbro within Paleoarchean granitic rocks have been dated between 3650 and 3580 Ma (McNaughton et al. 1988; Nelson et al. 1999; Williams and Hickman 2000; Wingate et al. 2010; Petersson et al. 2019a). Published $^{207}Pb/^{206}Pb$ ages of inherited zircons in igneous rocks and of detrital zircons in sedimentary rocks commonly exceed 3530 Ma (Thorpe et al. 1992; Van Kranendonk et al. 2002, 2007a, b; Bagas et al. 2004, 2008; Hickman 2012; Kemp et al. 2015a, b; Sheppard et al. 2017; Gardiner et al. 2019; Wiemer et al. 2018; Petersson et al. 2019b). Additionally, whole-rock Sm–Nd isotope data (Jahn et al. 1981; Collerson and McCulloch 1983; Gruau et al. 1987; McCulloch 1987; Bickle et al. 1989, 1993; Tyler et al. 1992; Van Kranendonk et al. 2002, 2007a, b; Smithies et al. 2007; Champion 2013; Champion and Huston 2016; Gardiner et al. 2017, 2018) and zircon Lu–Hf isotope data (Amelin et al. 2000; Guitreau et al. 2012; Nebel et al. 2014; Kemp et al. 2015a, b; Gardiner et al. 2017, 2018, 2019; Petersson et al. 2019a, b) have indicated pre-3530 Ma sources for many rocks. The conclusion is that 3800–3530 Ma crust formed an extensive 'basement' during deposition of the Pilbara Supergroup. This constrains possible tectonic processes for the Paleoarchean crustal evolution of the craton; in particular, it precludes some previous concepts that the Pilbara Supergroup formed as oceanic crust.
2. *East Pilbara Terrane (EPT), 3530–3223 Ma:* the principal Paleoarchean granite-greenstone terrane of the Pilbara Craton is exposed across 40,000 km^2 of the east Pilbara (Fig. 1.3). Other Paleoarchean terranes are the Karratha Terrane (KT) of northwest Pilbara and the Kurrana Terrane (KUT) in the southeast Pilbara. All three terranes are interpreted to be rifted fragments of the Pilbara crust that had evolved up to ~3220 Ma (Sun and Hickman 1998; Hickman 2016). Evidence suggests that an additional terrane of the Pilbara Craton might underlie the Fortescue and Hamersley Basins in the South Pilbara. Onshore exposures of the KT occupy only 3000 km^2 but geophysical data indicate that the terrane underlies 20,000 km^2 of the Lambert Shelf and Dampier Sub-basin of the Northern Carnarvon Basin (Fig. 1.2). Based on interpretation of gravity data (Hickman 2004), the KUT is largely covered by the Neoarchean

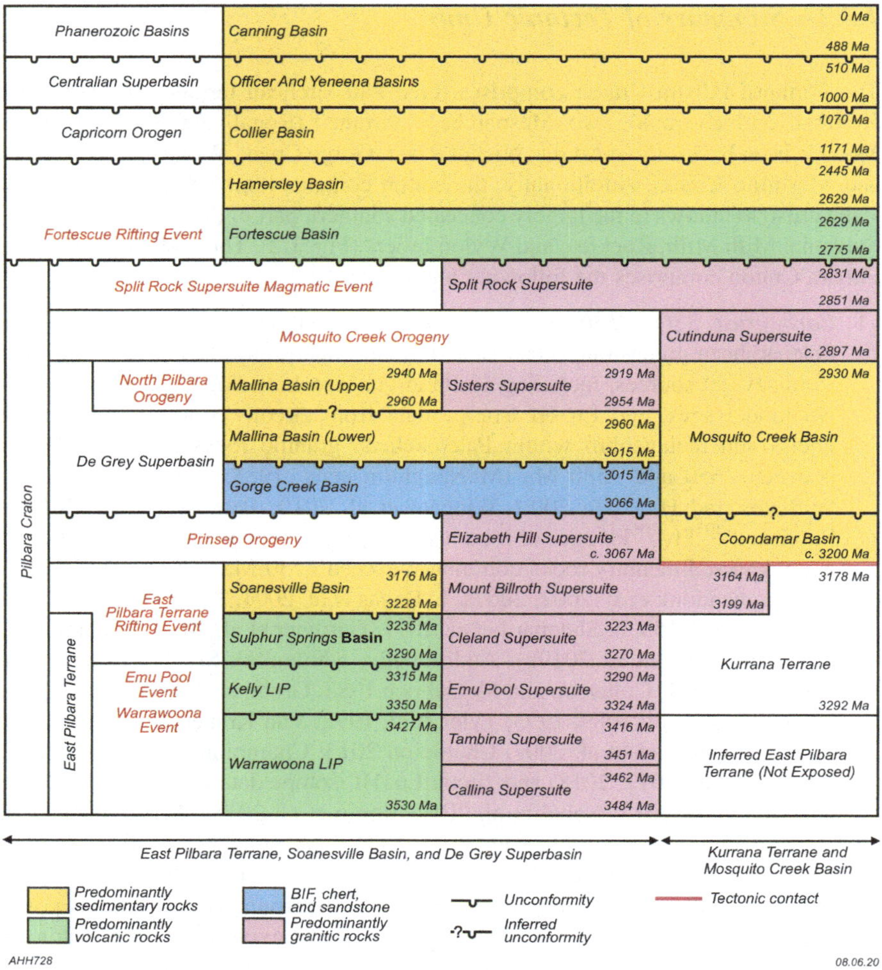

Fig. 1.8 Diagrammatic illustration of the ages and contact relationships of terranes, basins, supersuites, and events in the East Pilbara Craton. The East Pilbara Terrane Rifting Event separates the Paleoarchean East Pilbara Terrane from Mesoarchean units commencing with the Soanesville Basin and Mount Billroth Supersuite. The Mosquito Creek Basin and Kurrana Terrane have uncertain stratigraphic relationships to the successions overlying the East Pilbara Terrane, although the Coondamar Basin (not shown), underlying the Mosquito Creek Basin, is about the same age as the Soanesville Basin (From Hickman 2021; with Geological Survey of Western Australia permission)

Fortescue and Hamersley Basins. The EPT comprises three volcanic groups (Warrawoona, Kelly, and Sulphur Springs) and a sedimentary formation (Strelley Pool Formation) that make up the Pilbara Supergroup (Fig. 1.8, Table 1.1) and five granitic supersuites (Mulgundoona, Callina, Tambina,

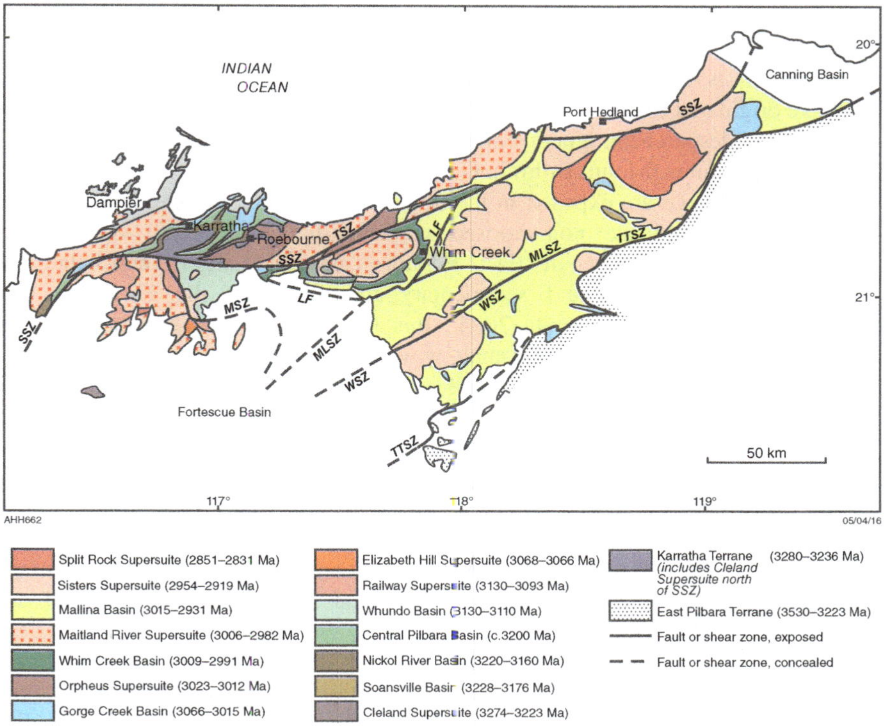

Fig. 1.9 Geological map showing basins and supersuites of the northwest Pilbara Craton. Granitic supersuites are distributed in east-northeast trending linear zones, with decreasing intrusive ages toward the southeast. *LF* Loudens Fault; *MLSZ* Mallina Shear Zone; *SSZ* Sholl Shear Zone; *TSZ* Terenar Shear Zone; *TTSZ* Tabba Tabba Shear Zone; *WSZ* Wohler Shear Zone (From Hickman 2016; with Geological Survey of Western Australia permission)

Emu Pool, and Cleland). Spanning 300 Ma, the Pilbara Supergroup was deposited on a volcanic plateau (Van Kranendonk et al. 2002), the dimensions of which far exceeded the present area of the EPT.

3. *Karratha Terrane (KT), 3530–3236 Ma:* mainly located north of the Sholl Shear Zone (SSZ, Figs. 1.3 and 1.9), the KT is extensively concealed by Phanerozoic rocks of the Northern Carnarvon Basin (Fig. 1.2) and is widely intruded by three Mesoarchean granitic supersuites (Orpheus, Maitland River, and Sisters). Exposures of the Paleoarchean terrane are consequently limited to komatiite and tholeiite outcrops of the 3280–3270 Ma Roebourne Group and granodiorite and tonalite of the 3270–3236 Ma Cleland Supersuite (age range in Karratha Terrane). Evidence that the KT once included older Paleoarchean crust is provided by Sm–Nd two-stage depleted mantle model ages ranging from ~3520 to ~3360 Ma (Smith et al. 1998; Arndt et al. 2001; Smithies et al. 2007; Hickman 2016). The KT separated from the EPT at ~3220 Ma (Sun and Hickman 1998; Van Kranendonk et al. 2006), accreted Mesoarchean terranes

and basins between 3160 and 3070 Ma, and re-amalgamated with the EPT at 3070 Ma.

4. *Kurrana Terrane (KUT), 3530–3178 Ma:* like the KT, the KUT is interpreted to have originated from late Paleoarchean breakup of the EPT. The KUT (Fig. 1.3) exposes Mesoarchean igneous and metasedimentary units including three granitic supersuites (Mount Billroth, Cutinduna, and Split Rock). Greenstone units of uncertain age are exposed in the northern part of the terrane, and inherited zircon ages indicate non-exposed units of the 3270–3223 Ma Cleland Supersuite (age range in East Pilbara Terrane). Largely concealed by Neoarchean volcanic and sedimentary formations of the Fortescue and Hamersley Basins (Hickman 2004), this large terrane still has an uncertain history of crustal evolution. By analogy with exposed areas of the EPT, Bouguer gravity data in the concealed southeast part of the Pilbara Craton suggest a concealed 'dome-and-keel' crustal architecture. The southwestern extent of the Kurrana Terrane is unknown.
5. *Soanesville Basin, 3223–3165 Ma:* ~3220 Ma breakup of the Paleoarchean Pilbara Craton (Van Kranendonk et al. 2010), and plate separation from the KT, resulted in deposition of a passive margin succession along the northwest margin of the EPT. Initially the basin was composed of clastic sedimentary rocks, BIF and chert, but between 3185 and 3165 Ma deep rifting of the extended EPT crust led to basaltic volcanism.
6. *Nickol River Basin, 3220–3160 Ma:* entirely composed of metamorphosed sedimentary rocks of the Nickol River Formation, the basin is a passive margin succession that was deposited over the southeast margin of the KT. Conglomerate, sandstone, siltstone, carbonaceous shale, carbonate rocks and BIF form a succession up to 1 km thick. Due to the formation immediately underlying a major thrust (Regal Thrust), large sections of the stratigraphy are tectonically attenuated or replaced by fault breccia and mylonite. The preserved 100 km strike length of the Nickol River Basin (Fig. 1.9) is interpreted to greatly under-represent its depositional extent.
7. *Regal Basin, 3200–3160 Ma:* rifting and plate separation of the EPT and KT formed a rift basin containing MORB-like basaltic crust, the Regal Formation. Plate separation ceased at ~3160 Ma when a regional metamorphic event (Karratha Event) was recorded in the KT. The Karratha Event is attributed to a collision between the KT and another plate converging from the northwest (Hickman 2016). Following this collision, ongoing convergence of the northwest plate pushed the KT back towards the EPT. Resulting northwest–southeast compression of the Regal Basin led to northwesterly thrusting (obduction) of the Regal Formation onto the KT above the Regal Thrust (Hickman et al. 2000, 2001, 2010). The 3-km-thick slab of obducted ophiolites forms the Regal Terrane (RT) across 3000 km^2 of the northwest Pilbara Craton. Based on evidence that the Regal Formation originated in the Regal Basin south of the SSZ, and on balanced sections through the KT and RT north of the Sholl Shear Zone (Hickman et al. 2006), the Regal Thrust 'transported' the ophiolites at least 50 km northwest across the KT.

8. *Sholl Terrane (ST), 3130–3093 Ma:* compression related to convergence of the EPT and KT between 3160 and 3070 Ma led to the development of a subduction zone within the thin basaltic crust of the Regal Basin. Above this zone, the ST evolved as a magmatic arc, with mafic and felsic volcanism forming the 10-km-thick, 3130–3110 Ma Whundo Group above large mafic–felsic intrusions of the 3130–3093 Ma Railway Supersuite. The terrane outcrops over an area of 4000 km^2 south of the SSZ (Fig. 1.3) and is interpreted to underlie most of the Mallina Basin.
9. *De Grey Superbasin, 3066–2930 Ma:* following a ~ 3070 Ma collision between terranes of the northwest Pilbara Craton and the EPT, closure of the Regal Basin and ongoing northwest–southeast compression resulted in the northwest plate being subducted under the northwest margin of the Pilbara Craton. The oldest exposed magmatic arc related to this subduction is that of the 3024–3007 Ma Orpheus Supersuite, although an older arc might be preserved offshore within the ~150 km-wide part of the craton concealed by the Northern Carnarvon Basin (Fig. 1.2). Successively younger magmatic arcs are represented by the 3006–2982 Ma Maitland River Supersuite and the 2954–2919 Ma Sisters Supersuite. Evolution of the De Grey Superbasin between c. 3066 and 2930 Ma is directly related to that of the northwest subduction zone. The superbasin comprises three back-arc basins, two (Gorge Creek and Mallina) being mainly sedimentary, and the third (Whim Creek), on the rifted northwest margin of the Mallina basin, being mainly volcanic.
10. *Gorge Creek Basin, 3066–3015 Ma:* southeast of the Orpheus Supersuite arc, the sedimentary succession of the Gorge Creek Group was deposited in a large retro-arc basin overlying thick continental crust. In the East Pilbara, early sediment input was coarsely clastic with conglomerate and sandstone derived exclusively from erosion of the EPT and older crust. Overlying this clastic succession, represented by the Farrell Quartzite, BIF and shale of the Cleaverville Formation were deposited in deeper water environments that extended hundreds of kilometres into the northwest Pilbara Craton. Sediment provenance in the northwest Pilbara included the KT, ST, and the Orpheus magmatic arc; and felsic volcaniclastic rocks related to the arc and interbedded with BIF, sandstone, and shale.
11. *Mallina Basin, 3015–2930 Ma:* the central part of the Mallina Basin is a deep-water, fault-controlled rift basin in which sandstone, wacke, and shale of the Croydon Group conformably to disconformably overlie the Gorge Creek Group. The two groups are separated by a thin transitional succession of interbedded conglomerate, shale, ferruginous chert, and BIF. The upper part of the Croydon Group in this central section of the basin is mainly composed of deep-water turbidites, although local abrupt lateral facies changes to sandstone are attributed to syn-depositional rifting, strike-slip faulting, and transpressional folding. Granitic intrusion (Sisters Supersuite) between 2954 and 2919 Ma was confined to the central part of the basin between the Tabba Tabba Shear Zone (TTSZ) in the southeast and the SSZ and Loudens Fault (LF) in the northwest (Fig. 1.3). Southeast of the TTSZ, the upper Croydon Group is represented by the Lalla

Rookh Sandstone, a 150-km-wide, shallow-water clastic succession unconformably overlying the EPT. Due to major folding and subsequent deep erosion, exposures of the Lalla Rookh Sandstone are mainly restricted to the cores of synclines. The northwest margin of the Mallina Basin is faulted along the SSZ and the LF, and major stratigraphic mismatches between the central basin and the 2955–2943 Ma Bookingarra Group northwest of these faults testify to large strike-slip movements. Geochemical evidence suggests that the Bookingarra Group is related to very much thinner volcanic units in the Mallina Basin (Van Kranendonk et al. 2006), although the Bookingarra and Croydon Groups were tectonically juxtaposed after ~2945 Ma.

12. *Whim Creek Basin, 3010–2990 Ma:* unconformably overlying the Gorge Creek Basin, and contemporaneous with the lower stratigraphy of the Croydon Group in the Mallina basin, the Whim Creek Basin is composed of mafic–felsic volcaniclastic rocks and subvolcanic intrusions of the Whim Creek Group. Deposition of the volcanic and volcaniclastic rocks was contemporaneous with intrusion of tonalite, granodiorite, and monzogranite of the 3006–2982 Ma Maitland River Supersuite (Fig. 1.9). Volcanic and sedimentary facies of the Whim Creek Group (Pike and Cas 2002; Pike et al. 2006) suggest transport of detritus from the northwest where the Maitland River Supersuite outcrops along a > 200 km-long, >50 km-wide belt along the northwest Pilbara coast.
13. *Central Pilbara Tectonic Zone (CPTZ), 3160–2900 Ma:* the CPTZ (Fig. 1.3) evolved as a zone of deformation, metamorphism, and magmatic intrusion during oblique convergence of the Pilbara Craton and the northwest plate. Due to the existence of the Regal Basin, the EPT remained isolated from the effects of this convergence until 3070 Ma when it collided with the ST, RT, and KT of the northwest Pilbara Craton.
14. *Mosquito Creek Basin, 3200–2930 Ma:* originating during the late Paleoarchean breakup of the EPT, the Mosquito Creek Basin is a rift basin separating the EPT from the KUT (Fig. 1.3). The basin contains mafic volcanic and clastic sedimentary units of the 3200–3176 Ma Coondamar Formation overlain by various sedimentary facies of the 2980–2930 Ma Mosquito Creek Formation. Ultramafic–mafic dykes and sills of the ~3185 Ma Dalton Suite intruded the Coondamar Formation during rifting, but it is likely that the basin remained relatively inactive between ~3160 and 2980 Ma. Doming of the EPT after deposition of the Gorge Creek Group led to erosion of uplifted areas and deposition of the Mosquito Creek Formation.
15. *Split Rock Supersuite, 2851–2831 Ma:* cratonization of the Pilbara Craton during the 2955–2900 Ma North Pilbara and Mosquito Creek Orogenies was followed by post-orogenic intrusion of the Split Rock Supersuite (Van Kranendonk et al. 2006). The supersuite comprises intrusions of highly fractionated, Sn–Ta–Li-bearing monzogranites emplaced along a northwest-trending belt from the KUT to the northeastern Mallina Basin. This distribution might be evidence that the Pilbara Craton drifted southeast to northwest across a hot spot (Hickman 2016).

1.4 Fragment of an Archean Continent

Evidence from the Pilbara Craton, and from the overlying Fortescue, Hamersley, and Turee Creek Groups, establishes that the present boundaries of the craton (Figs. 1.1 and 1.2) are tectonic and that it represents a fragment of a much larger mass of Archean continental crust (Martin et al. 1998a; Krapež and Eisenlohr 1998; Zegers et al. 1998; Thorne and Trendall 2001; Eriksson et al. 2002; Barley et al. 2005; de Kock et al. 2009; Hickman 2016). PCMP data indicate two Archean continental breakups: the first occurred between 3280 and 3165 Ma and is referred to as the East Pilbara Terrane Rifting Event (EPTRE, Hickman 2016, 2021); the second occurred between 2775 and 2501 Ma and was the Neoarchean breakup of the late Mesoarchean craton (Blake and Barley 1992; Blake 1993; Martin et al. 1998a, b; Thorne and Trendall 2001; Eriksson et al. 2002; Barley et al. 2005; Hickman et al. 2010; Pirajno and Santosh 2015).

The EPTRE involved the rifting and breakup of Eoarchean–Paleoarchean continental crust in the Pilbara and was initiated by uplift and crustal extension above a c. 3280 Ma mantle plume (Van Kranendonk et al. 2002, 2006, 2007b; Hickman 2004, 2012, 2016, 2021; Hickman and Van Kranendonk 2008, 2012). The Neoarchean breakup resulted in the evolution of the Fortescue Basin at c. 2775 Ma which defines the minimum age of the Pilbara Craton.

From the regional geology of the northern Pilbara Craton, it is evident that many depositional basins and large-scale structures are abruptly truncated at the present boundaries of the craton. Recognition that these boundaries were produced by the breakup events, and do not define the original sizes of Paleoarchean and Mesoarchean terranes, igneous provinces, and sedimentary basins within the craton, is important to understanding the evolution of the craton. For example, evidence described in this book indicates that the present boundaries of the EPT (Fig. 1.3) are tectonic and are very different from the boundaries of this terrane during the Paleoarchean. Correlations between the 3530–3223 Ma Pilbara Supergroup and the 3530–3250 Ma Onverwacht Group of South Africa (Zegers et al. 1998; Hickman 2016, 2021; Pirajno and Huston 2019) suggest that the EPT and Kaapvaal Craton were parts of the same continental-scale body of Paleoarchean crust.

As another example, the northeast-trending CPTZ (Fig. 1.3), which developed between the EPT and the Karratha Terrane over 300 million years (3220–2920 Ma), was originally far longer than its currently exposed length of 250 km. Geophysical and stratigraphic evidence establishes that the northeast extent of the CPTZ was truncated by the Proterozoic Paterson Orogen (Hickman 2021). Its southwest extent is obscured by Neoarchean and Proterozoic cover but is also likely to have been cut short by Neoarchean and Proterozoic deformation. Within the present boundaries of the Pilbara Craton, the maximum possible length (exposed and concealed) of the CPTZ is c. 500 km, whereas by analogy with Phanerozoic convergent margins, the CPTZ would initially have been at least 1000 km long (Hickman 2021).

1.5 Vaalbara Continent?

Early stratigraphic comparisons between the Neoarchean–Paleoproterozoic successions of the Pilbara and South Africa (Trendall 1968; Button 1976) revealed strong similarities. This led to an interpretation that the two areas originated on the same Neoarchean–Paleoproterozoic supercontinent (Cheney et al. 1988; Cheney 1996). Based on the locations of the correlated areas (Kaapvaal and Pilbara), Cheney et al. (1988) named the inferred supercontinent 'Vaalbara'. Several workers have reported evidence to support the existence of Vaalbara, although almost all only consider its existence from ~2700 Ma onwards (Bleeker 2003; Barley et al. 2005; De Kock et al. 2009, 2012; Huston et al. 2012; Smirnov et al. 2013). Paleomagnetic data suggest that the Pilbara and Kaapvaal Cratons had similar paleolatitudes in the Neoarchean (Zegers et al. 1998; Strik et al. 2003). De Kock et al. (2009) argued that paleolongitudes and polar wander paths were also similar. Kumar et al. (2017) reported paleomagnetic data from c. 2765 Ma dolerite dykes in the Singhbhum Craton of eastern India that are similar to data from the c. 2770 Ma Black Range Dolerite of the East Pilbara, suggesting that the Singhbhum Craton was also part of Vaalbara.

Some workers have argued that similarities in the Neoarchean–Paleoproterozoic stratigraphy of the Pilbara and Kaapvaal areas are merely fortuitous, resulting from contemporaneous similar depositional settings in different areas of the Earth (Nelson et al. 1992, 1999; Eriksson et al. 2009). Button (1976) suggested that the Neoarchean–Paleoproterozoic successions of South Africa and the Pilbara were deposited in separate basins with similar geologic histories.

Evidence from the Pilbara indicates that from 2775 to 2501 Ma the Pilbara Craton was in the process of breaking up, and the same might apply to the Kaapvaal Craton (Martin et al. 1998a; de Kock et al. 2009). Therefore, the Neoarchean–Paleoproterozoic correlations between the Pilbara and Kaapvaal areas relate to stages of continental breakup, and not to processes in the evolution of a stable continent. With reference to Vaalbara, a more important question relates to the time when this continent was first formed. It might be argued that Vaalbara was formed by plate tectonic processes commencing after the ~3220 Ma breakup. The observation that stratigraphic similarities between the Pilbara and Kaapvaal Cratons are stronger in the Paleoarchean than in the Mesoarchean might be evidence that they separated at ~3220 Ma and through much of the Mesoarchean they evolved far apart. However, based on the Pilbara evidence (Hickman 2016, 2021), the Mesoarchean between 3160 and 2900 Ma was a period of plate accretion during which plates that separated at ~3220 Ma progressively re-converged and collided. Earlier in this Introduction, it was noted that convergence of a northwest plate with the Karratha Terrane was a dominant factor in the plate-tectonic evolution of the Pilbara Craton from 3160 Ma onwards. Because the northwest plate was subducted, it is likely to have been basaltic, possibly formed in an oceanic-like basin between the Karratha Terrane and the Paleoarchean section of the Kaapvaal Craton.

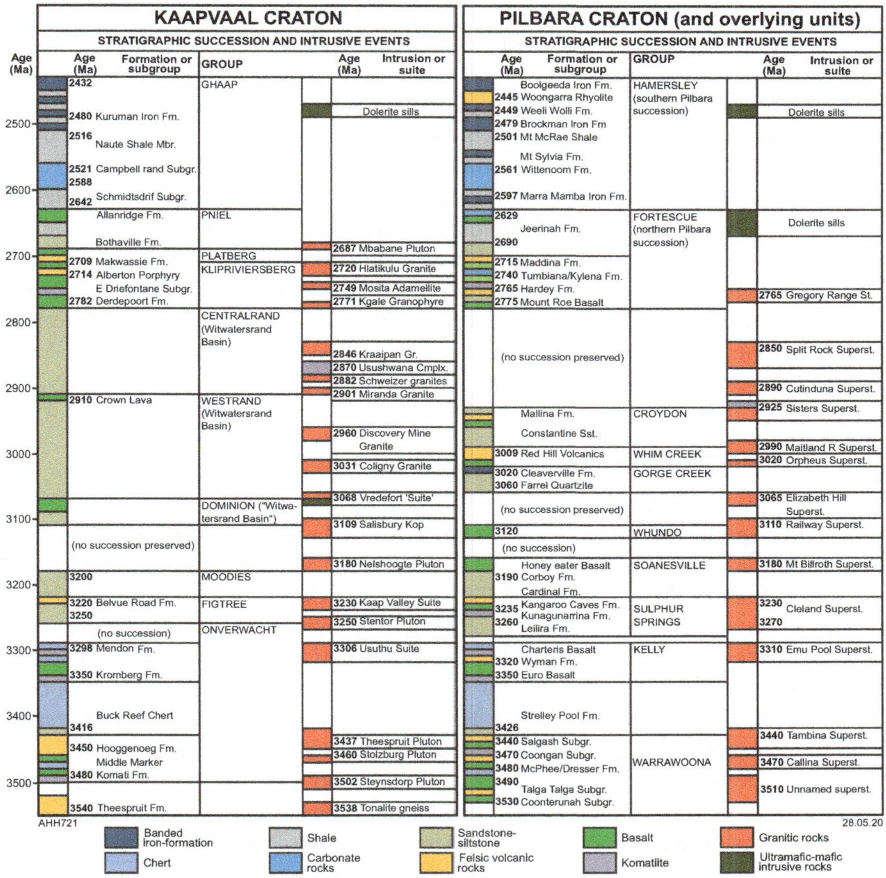

Fig. 1.10 Stratigraphic comparison of the Pilbara and Kaapvaal successions between 3550 and 2450 Ma. Major similarities include <2780 Ma, successions of the Fortescue and Hamersley Groups compared to the successions of the Klipriviersberg, Platberg, Pniel, and Ghaap Groups; >3180 Ma, successions of the Warrawoona, Kelly, Sulphur Springs, and Soanesville Groups compared to the successions of the Onverwacht, Fig Tree, and Moodies Groups; 3426–3350 Ma, sedimentary deposition of the Strelley Pool Formation and Buck Reef Chert during a contemporaneous c. 75 Ma break in volcanic activity; 3500–3065 Ma, contemporaneous granitic intrusion at c. 3500, 3470–3460, 3440, 3310, 3250–3220, 3180, 3110, and 3068–3065 Ma (From Hickman 2021; with Geological Survey of Western Australia permission)

Figure 1.10 illustrates the remarkable stratigraphic and geochronological matches between the Pilbara and Kaapvaal successions, both with respect to volcanic and sedimentary units and with regard to episodes of granitic intrusion. The high degree of correlation between (a), the Pilbara Supergroup (Warrawoona and Kelly Groups) and the Onverwacht Group, and (b), between the Fortescue Group–Hamersley Group succession and the Ventersdorp Supergroup–Ghaap Group succession, strongly suggests that the Paleoarchean and Neoarchean sections of the Pilbara and Kaapvaal Cratons evolved in close proximity. Geochronological evidence from the Barberton Greenstone Belt and the adjacent Ancient Gneiss Complex of the eastern Kaapvaal Craton (Kröner and Tegtmeyer 1994; Zegers et al. 1998; Kröner 2007; Zeh et al. 2009, 2011; Kröner et al. 2016; Hoffmann and Kroner 2019) reveals matching Paleoarchean magmatic events in the two cratons. The most likely interpretation is that the cratons were in sufficient proximity (within about 2000 km) to be affected by the same Paleoarchean mantle plumes. An alternative explanation that these events were parts of global mantle disturbances is not supported by similar geochronological data from other continents, nor from comparisons between later supercontinents (Bleeker 2003). Paleomagnetic data (de Kock et al. 2009, 2012) and extremely similar ~3400 Ma microfossils (Oehler et al. 2017) (Chap. 4) are also consistent with proximity of the Pilbara and Kaapvaal Cratons in the Paleoarchean.

1.6 Concept of an 'Ancient Nucleus'

Some descriptions of the Pilbara Craton have included a concept that the EPT represents an 'ancient nucleus' (Barley 1997; Van Kranendonk 2000; Van Kranendonk et al. 2002, 2006, 2007a, b, 2019). Ancient nuclei have also been proposed for several other Archean cratons (De Wit et al. 1992; Wilson et al. 1995; Kröner et al. 1996; Zeh et al. 2009; Percival et al. 2012; Santosh 2013). However, in the case of the Pilbara Craton, the concept has lacked supportive evidence. If the EPT and the early Kaapvaal Craton formed parts of the same Paleoarchean continent, as suggested by strong stratigraphic similarities (Fig. 1.10) and large igneous provinces apparently common to both cratons (Zegers et al. 1998; Hickman 2016, 2021), the area of Paleoarchean Vaalbara is likely to have exceeded 1,000,000 km^2. Rather than being a nucleus, such a large unit of continental crust would be better regarded as a basement. Any suggestion that the present EPT represents the first and only continental nucleus of the Pilbara Craton also takes no account of the widespread existence of Eoarchean crust (older than the EPT) which existed during the Mesoarchean, as explained in Chap. 2. For example, Petersson et al. (2019b) used inherited zircon ages to suggest that c. 3750 Ma Eoarchean crust was a nucleus in one part of the Pilbara Craton.

It might be interpreted that the EPT became a nucleus after the 3220 Ma continental breakup of the Pilbara Craton. However, this breakup also produced several other potential nuclei for subsequent Mesoarchean accretion, for example, the Karratha Terrane, Kurrana Terrane, and outside Australia the eastern Kaapvaal Craton of Africa and the Singhbhum and Bastar Cratons of eastern India. During plate convergence from 3160 Ma to 3070 Ma, new basins and terranes were formed in the northwest Pilbara, but these Mesoarchean units were first accreted onto the Karratha Terrane to form the West Pilbara Superterrane (WPS) (Van Kranendonk et al. 2006, 2007b, 2010; Hickman and Van Kranendonk 2008; Hickman et al. 2010). There is no evidence of any accretion onto the EPT between 3160 and

3070 Ma. The WPS amalgamated with the EPT only after the Regal Basin between the WPS and the EPT had been completely closed.

1.7 Concealed Pilbara Craton

A seismic survey between the southern Pilbara Craton and the Gascoyne Province (Fig. 1.2) indicates that a concealed section of the Pilbara Craton extends up to 200 km beyond the most southerly exposures of the craton and forms an Archean basement to the Proterozoic Ashburton, Edmund, and Collier Basins (Johnson et al. 2011, 2013; Korsch et al. 2011; Thorne et al. 2011). Farther east, to the southeast from the Sylvania Inlier (Fig. 1.2), gravity data suggest that the craton is concealed by Proterozoic cover extending at least 200 km southeast. Geophysical data from the Canning Basin (Frogtech Geoscience 2017) indicate that the Pilbara Craton extends at least 150 km east of the most easterly outcrops in the Gregory Range and underlies the Paterson Orogen, including the Rudall Province. These geophysical data support previous structural evidence that the Rudall Province was thrust southwestwards across the eastern margin of the Pilbara Craton (Maidment 2017). Isotopic data from the Rudall Province (Kirkland et al. 2013) also support this interpretation. Farther north, Reading et al. (2012) interpreted data from a passive seismic transect to propose a deepening of the Moho across the northeastern margin of the Pilbara Craton. North and northwest from the Pilbara coast, gravity data, and numerous seismic lines undertaken during petroleum exploration, indicate that the Pilbara Craton, possibly with overlying Neoarchean–Proterozoic units, extends between 150 and 200 km offshore beneath Phanerozoic sedimentary successions of the Northern Carnarvon Basin. Within 30–150 km of the Pilbara coast, the successions of the Lambert Shelf and Peedamullah Shelf are less than 1 km thick, whereas farther offshore the Pilbara Craton is covered by thicker Phanerozoic successions in the Dampier Sub-basin (Fig. 1.2). Neoarchean and Paleoproterozoic rocks of the Fortescue and Hamersley Basins were thrust onto the western margin of the Pilbara Craton (Hickman and Strong 1998).

References

Amelin Y, Lee D-C, Halliday AN (2000) Early-middle Archaean crustal evolution deduced from Lu–Hf and U–Pb isotopic studies of single zircon grains. Geochim Cosmochim Acta 62:4205–4225

Arndt N, Bruzak G, Reischmann T (2001) The oldest continental and oceanic plateaus: geochemistry of basalts and komatiites of the Pilbara Craton, Australia. In: Ernst RE, Buchan K (eds) Mantle plumes: their identification through time. Geological Society of America, pp 359–387

Bagas L, Farrell TR, Nelson DR (2004) The age and provenance of the Mosquito Creek Formation. In: Geological Survey of Western Australia annual review 2003–04, pp. 62–70

Bagas L, Bierlein FP, Bodorkos S, Nelson DR (2008) Tectonic setting, evolution, and orogenic gold potential of the late Mesoarchaean Mosquito Creek Basin, North Pilbara Craton, Western Australia. Precambrian Res 160:237–244

Barley M, Bekker A, Krapež B (2005) Late Archean to Early Paleoproterozoic global tectonics, environmental change and the rise of atmospheric oxygen. Earth Planet Sci Lett 238:156–171. https://doi.org/10.1016/j.epsl.2005.06.062

Barley ME (1997) The Pilbara Craton. In: de Wit MJ, Ashwel L (eds) Greenstone belts. Oxford University Press, Oxford, UK, pp 657–664

Bickle MJ, Bettenay LF, Chapman HJ, Groves DI, McNaughton NJ, Campbell IH, de Laeter JR (1989) The age and origin of younger granitic plutons of the Shaw batholith in the Archean Pilbara Block, Western Australia. Contrib Mineral Petrol 101:361–376

Bickle MJ, Bettenay LF, Chapman HJ, Groves DI, McNaughton NJ, Campbell IH, de Laeter JR (1993) Origin of the 3500–3300 Ma calc-alkaline rocks in the Pilbara Archaean: isotopic and geochemical constraints from the Shaw Batholith. Precambrian Res 60:117–149

Blake TS, Barley ME (1992) Tectonic evolution of the late Archaean to early Proterozoic Mount Bruce megasequence set, Western Australia. Tectonics 11:1415–1425

Blake TS (1993) Late Archaean crustal extension, sedimentary basin formation, flood basalt volcanism, and continental rifting. The Nullagine and Mount Jope Supersequences. Western Australia Precambr Res 60:185–241

Bleeker W (2003) The late Archean record: a puzzle in ca. 35 pieces. Lithos 71:99–134

Blockley JG (1975) Pilbara Block. In: The geology of Western Australia. Geological Survey of Western Australia, pp 81–93

Button A (1976) Transvaal and Hamersley Basins–review of basin development and mineral deposits. Mineral Science Engineering 8:262–293

Champion DC (2013) Neodymium depleted mantle model age map of Australia: explanatory notes and user guide. Record, 2013/44. Geoscience Australia

Champion DC, Huston DL (2016) Radiogenic isotopes, ore deposits and metallogenic terranes: novel approaches based on regional isotopic maps and the mineral systems concept. Ore Geol Rev 76:229–256. https://doi.org/10.1016/j.oregeorev.2015.09.025

Champion DC, Smithies RH (2007) Geochemistry of Paleoarchean granites of the East Pilbara Terrane, Pilbara Craton, Western Australia: implications for early Archaean crustal growth. In: Van Kranendonk MJ, Smithies RH, Bennett VC (eds) Earth's oldest rocks, developments in Precambrian geology 15. Elsevier, Amsterdam, pp 369–409

Cheney ES (1996) Sequence stratigraphy and plate tectonic significance of the Transvaal succession of southern Africa and its equivalent in Western Australia. Precambrian Res 79:3–24

Cheney ES, Roering C, Stettler E (1988) Valbara. In: Geocongress '88: extended abstracts. Geological Society of South Africa, pp 85–88

Collerson KD, McCulloch MT (1983) Field and Sr–Nd isotopic constraints on Archaean crust and mantle evolution in the East Pilbara Block, Western Australia. In: Australian National University, Research School of Earth Sciences, Annual Report 1982. Canberra, ACT, pp 167–168

de Kock MO, Beukes NJ, Armstrong RA (2012) New SHRIMP U–Pb zircon ages from the Hartswater Group, South Africa: implications for correlations of the Neoarchean Ventersdorp Supergroup on the Kaapvaal Craton and with the Fortescue group on the Pilbara Craton. Precambrian Res 204–205:66–74

de Kock MO, Evans DAD, Beukes NJ (2009) Validating the existence of Vaalbara in the Neoarchean. Precambrian Res 174:145–154

de Wit MJ, de Ronde CEJ, Tredoux M, Roering C, Hart RJ, Armstrong RA, Green RWE, Peberdy E, Hart RA (1992) Formation of an Archaean continent. Nature 357:553–562

Eriksson PG, Condie KC, van der Westhuizen W, van der Merwe R, de Bruiyn H, Nelson DR, Altermann W, Catuneanu O, Bumby AJ, Lindsay J (2002) Late Archaean superplume events: a Kaapvaal-Pilbara perspective. J Geodyn 34:207–247

Eriksson PG, Banerjee S, Nelson DR, Rigby MJ, Catuneanu O, Sarkar S, Roberts J, Ruban D, Mtimkulu N, Raju S (2009) A Kaapvaal Craton debate: nucleus of an early small supercontinent or affected by an enhanced accretion event? Gondwana Res 15:354–372

Frogtech Geoscience (2017) 2017 Canning Basin SEEBASE study and GIS data package. Report 182. Geological Survey of Western Australia, 297p

Gardiner NJ, Hickman AH, Kirkland CL, Lu Y, Johnson T, Zhao J-X (2017) Processes of crust formation in the early Earth imaged through Hf isotopes from the East Pilbara Terrane. Precambrian Res 297:56–76. https://doi.org/10.1016/j.precamres.2017.05.004

Gardiner NJ, Hickman AH, Kirkland CL, Lu Y, Johnson TE, Wingate MTD (2018) New Hf isotope insights into the Paleoarchean magmatic evolution of the Mount Edgar Dome, Pilbara Craton: Implications for early Earth and crust formation processes. Report, vol 181. Geological Survey of Western Australia

Gardiner NJ, Wacey D, Kirkland CL, Johnson TE, Jeon H (2019) Zircon U–Pb, Lu–Hf and O isotopes from the 3414 Ma Strelley Pool Formation, East Pilbara Terrane, and the Palaeoarchaean emergence of a cryptic cratonic core. Precambrian Res 321:64–84. https://doi.org/10.1016/j.precamres.2018.11.023

Gruau G, Jahn BM, Glikson AY, Davy R, Hickman AH, Chauvel C (1987) Age of the Archean Talga-Talga subgroup, Pilbara Block, Western Australia, and the early evolution of the mantle: new Sm–Nd isotopic evidence. Earth Planet Sci Lett 85:105–116

Guitreau M, Blichert-Toft J, Martin H, Mojzsis SJ, Albarède F (2012) Hafnium isotope evidence from Archean granitic rocks for deep-mantle origin of continental crust. Earth Planet Sci Lett 337:211–223

Hickman AH (1977) New and revised definitions of rock units in the Warrawoona Group, Pilbara Block. In: Annual report for the year 1976. Geological Survey of Western Australia, p 53

Hickman AH (1980a) Excursion guide–Archaean Geology of the Pilbara Block: second International Archaean Symposium, Perth 1980. Geological Society of Australia, W.A. Division, Perth, Western Australia

Hickman AH (1980b) Crustal evolution of the Pilbara Block. In: Glover JE, Groves DI (eds) Second International Archaean Symposium, Perth 1980, extended abstracts. Geological Society of Australia and IGCP Archaean Geochemistry Project, Perth

Hickman AH (1980c) Lithological map and stratigraphic interpretation of the Pilbara Block. Geology if the Pilbara Block and its environs, Geological Survey of Western Australia

Hickman AH (1981) Crustal evolution of the Pilbara Block. In: Glover JE, Groves DI (eds) Archaean Geology: second International Symposium, Perth 1980. Geological Society of Australia, pp 57–69

Hickman AH (1983) Geology of the Pilbara Block and its environs. Bulletin, vol 127. Geological Survey of Western Australia

Hickman AH (2001a) The West Pilbara granite–greenstone terrane and its place in the Pilbara Craton. In: Cassidy KF, Dunphy JM, Van Kranendonk MJ (eds) Extended abstracts–4th International Archaean Symposium, Perth, 24–28 September 2001. AGSO (Geoscience Australia), pp 319–321

Hickman AH (2001b) Geology of the Dampier 1:100 000 sheet. 1:100 000 Geological Series Explanatory Notes. Geological Survey of Western Australia

Hickman AH (2004) Two contrasting granite–greenstones terranes in the Pilbara Craton, Australia: evidence for vertical and horizontal tectonic regimes prior to 2900 Ma. Precambrian Res 131: 153–172

Hickman AH (2012) Review of the Pilbara Craton and Fortescue Basin, Western Australia: crustal evolution providing environments for early life. Island Arc 21:1–31

Hickman AH (2016) Northwest Pilbara Craton: a record of 450 million years in the growth of Archean continental crust. Report 160, Geological Survey of Western Australia

Hickman AH (2021) East Pilbara Craton: a record of one billion years in the growth of Archean continental crust. Report 143. Geological Survey of Western Australia

Hickman AH, Lipple SL (1975) Explanatory notes on the Marble Bar 1:250 000 geological sheet, Western Australia. Record, 1974/20. Geological Survey of Western Australia

Hickman AH, Strong CA (1998) Structures in the Cape Preston area, northwest Pilbara Craton–new evidence for a convergent margin. In: Geological Survey of Western Australia annual review 1997–98. Geological Survey of Western Australia, pp 77–84

Hickman AH, Van Kranendonk MJ (2008) Archean crustal evolution and mineralization of the northern Pilbara Craton–a field guide. Record, 2008/13. Geological Survey of Western Australia

Hickman AH, Van Kranendonk MJ (2012) Early earth evolution: evidence from the 3.5–1.8 Ga geological history of the Pilbara region of Western Australia. Episodes 35:283–297. https://doi.org/10.18814/epiiugs/2012/v35i1/028

Hickman AH, Smithies RH, Huston DL (2000) Archaean geology of the West Pilbara granite–greenstone terrane and the Mallina Basin, Western Australia–a field guide. Record, 2000/9. Geological Survey of Western Australia

Hickman AH, Smithies RH, Pike G, Farrell TR, Beintema KA (2001) Evolution of the West Pilbara granite–greenstone terrane and Mallina Basin, Western Australia–a field guide. Record, 2001/16. Geological Survey of Western Australia

Hickman AH, Smithies RH, Strong CA (2006) Interpreted bedrock geology of the northwestern Pilbara Craton. Geological Survey of Western Australia

Hickman AH, Smithies RH, Tyler IM (2010) Evolution of active plate margins: West Pilbara Superterrane, De Grey Superbasin, and the Fortescue and Hamersley Basins–a field guide. Record, 2010/3. Geological Survey of Western Australia

Hoffmann JE, Kroner A (2019) Chapter 23, early Archean crustal evolution in southern Africa–an updated record of the ancient gneiss complex of Swaziland. In: Van Kranendonk MJ, Bennett VC, Hoffmann JE (eds) Earth's oldest rocks, 2nd edn. Elsevier B.V, pp 437–462. https://doi.org/10.1016/B978-0-444-63901-1.00023-X

Horwitz R, Pidgeon RT (1993) 3.1 Ga tuff from the Sholl Belt in the West Pilbara: further evidence for diachronous volcanism in the Pilbara Craton. Precambrian Res 60:175–183

Huston DL, Blewett RS, Champion DC (2012) Australia through time: a summary of its tectonic and metallogenic evolution. Episodes 35:23–43. https://doi.org/10.18814/epiiugs/2012/v35i1/004

Huston DL, Sun S-S, Blewett R, Hickman AH, Van Kranendonk MJ, Phillips D, Baker D, Brauhart C (2002) The timing of mineralization in the Archean North Pilbara Terrain, Western Australia. Econ Geol 97:733–755

Jahn B, Glikson AY, Peucat JJ, Hickman AH (1981) REE geochemistry and isotopic data of Archean silicic volcanics and granitoids from the Pilbara Block, Western Australia: implications for the early crustal evolution. Geochim Cosmochim Acta 45:1633–1652

Johnson SP, Cutten HN, Tyler IM, Korsch RJ, Thorne AM, Blay OA, Kennett BLN, Blewett RS, Joly A, Dentith MC, Aitken ARA, Goodwin JA, Salmon M, Reading A, Boren G, Ross J, Costelloe RD, Fomin T (2011) Preliminary interpretation of deep seismic reflection lines 10GA-CP2 and 10GA-CP3: crustal architecture of the Gascoyne Province, and Edmund and Collier Basins. In: Johnson SP, Thorne A, Tyler IM (eds) Capricorn Orogen seismic and magnetotelluric (MT) workshop 2011: extended abstracts. Geological Survey of Western Australia, pp. 49–60

Johnson SP, Thorne AM, Tyler IM, Korsch RJ, Kennett BLN, Cutten HN, Goodwin J, Blay OA, Blewett RS, Joly A, Dentith MC, Aitken ARA, Holzschuh J, Salmon M, Reading A, Heinson G, Boren G, Ross J, Costelloe RD, Fomin T (2013) Crustal architecture of the Capricorn Orogen, Western Australia and associated metallogeny. Aust J Earth Sci 60:681–705. https://doi.org/10.1080/08120099.2013.826735

Kemp AIS, Hickman AH, Kirkland CL (2015a) Early evolution of the Pilbara Craton, Western Australia, from hafnium isotopes in detrital and inherited zircons. Report, vol 151. Geological Survey of Western Australia

Kemp AIS, Hickman AH, Kirkland CL, Vervoort JD (2015b) Hf isotopes in detrital and inherited zircons of the Pilbara Craton provide no evidence for hadean continents. Precambrian Res 261: 112–126

Kirkland CL, Johnson SP, Smithies RH, Hollis JA, Wingate MTD, Tyler IM, Hickman AH, Cliff JB, Belousova EA, Murphy RC, Tessalina S (2013) The crustal evolution of the Rudall Province from an isotopic perspective. Report 122. Geological Survey of Western Australia, 30p

Korsch RJ, Johnson SP, Tyler IM, Thorne AM, Blewett RS, Cutten HN, Joly A, Dentith MC, Aitken ARA, Goodwin JA, Kennett BLN (2011) Geodynamic implications of the Capricorn deep seismic survey: from the Pilbara Craton to the Yilgarn Craton. In: Johnson SP, Thorne A, Tyler IM (eds) Capricorn Orogen seismic and magnetotelluric (MT) workshop 2011: extended abstracts. Geological Survey of Western Australia, pp. 107–114

Krapež B (1993) Sequence stratigraphy of the Archaean supracrustal belts of the Pilbara Block, Western Australia. Precambrian Res 60:1–45

Krapež B, Barley ME (1987) Archaean strike-slip faulting and related ensialic basins: evidence from the Pilbara Block. Australia Geol Mag 124:555–567. https://doi.org/10.1017/S0016756800017386

Krapež B, Eisenlohr B (1998) Tectonic settings of Archaean (3325–2775 Ma) crustal–supracrustal belts in the West Pilbara Block. Precambrian Res 88:173–205

Kröner A, Tegtmeyer A (1994) Gneiss–greenstone relationships in the ancient gneiss complex of southwestern Swaziland, southern Africa, and implications for early crustal evolution. Precambrian Res 67:109–139

Kröner A, Hegner E, Wendt JI, Byerly GR (1996) The oldest part of the Barberton granitoid-greenstone terrain, South Africa: evidence for crust formation between 3.5 and 3.7 Ga. Precambrian Res 78:105–124

Kröner A, Anhaeusser CR, Hoffmann JE, Wong J, Geng H, Hegner E, Xie H, Yang J, Liu D (2016) Chronology of the oldest supracrustal sequences in the Palaeoarchaean Barberton Greenstone Belt, South Africa and Swaziland. Precambrian Res 279:123–143

Kröner A (2007) The ancient gneiss complex of Swaziland and environs: record of early Archaean crustal evolution in Southern Africa. In: Van Kranendonk MJ, Smithies RH, Bennett VC (eds) Earth's oldest rocks. Elsevier, Amsterdam, pp 465–480

Kumar A, Parashuramula V, Shankar R, Besse J (2017) Evidence for a Neoarchean LIP in the Singhbhum Craton, Eastern India: implications to Vaalbara supercontinent. Precambrian Res 292:163–174

Lipple SL (1975) Definitions of new and revised stratigraphic units of the Eastern Pilbara Region. In: Annual report for the year 1974. Geological Survey of Western Australia, pp 58–63

Maidment DW (2017) Geochronology from the Rudall Province, Western Australia: implications for the amalgamation of the West and North Australian Cratons. Geological Survey of Western Australia, Report, p 161

Martin DMB, Clendenin CW, Krapež B, McNaughton NJ (1998a) Tectonic and geochronological constraints on late Archaean and Palaeoproterozoic stratigraphic correlation within and between the Kaapvaal and Pilbara Cratons. J Geol Soc 155:311–322

Martin DMB, Li ZX, Nemchin AA, Powell CMA (1998b) A pre-2.2 Ga age for giant hematite ores of the Hamersley Province, Australia? Econ Geol 93:1084–1090

McCulloch MT (1987) Sm–Nd isotopic constraints on the evolution of Precambrian crust in the Australian continent. In: Kroener A (ed) Proterozoic lithosphere evolution. American Geophysical Union, pp 111–120

McNaughton NJ, Green MD, Compston W, Williams IS (1988) Are anorthositic rocks basement to the Pilbara Craton? Geol Soc Aust Abstr 21:272–273

Nebel O, Campbell IH, Sossi PA, Van Kranendonk MJ (2014) Hafnium and iron isotopes in early Archean komatiites record a plume-driven convection cycle in the Hadean Earth. Earth Planet Sci Lett 397:111–120

Nelson DR, Trendall AF, Altermann W (1999) Chronological correlations between the Pilbara and Kaapvaal Cratons. Precambrian Res 97:165–189

Nelson DR, Trendall AF, de Laeter JR, Grobler NJ, Fletcher IR (1992) A comparative study of the geochemical and isotopic systematics of late Archaean flood basalts from the Pilbara and Kaapvaal Cratons. Precambrian Res 54:231–256

Noldart AJ, Wyatt JD (1962) The geology of portion of the Pilbara Goldfield, covering the marble Bar and Nullagine 4 mile map sheets. Bulletin, vol 115. Geological Survey of Western Australia

Oehler DZ, Walsh MM, Sugitani K, Liu M-C, House CH (2017) Large and robust lenticular microorganisms on the young Earth. Precambrian Res 296:112–119

Percival JA, Skulski T, Sanborn-Barrie M, Stott GM, Leclair AD, Corkery MT, Boily M (2012) Geology and tectonic evolution of the Superior Province, Canada. Chapter 6. In: Percival JA, Cook FA, Clowes RM (eds) Tectonic styles in Canada: the LITHOPROBE perspective, vol 49. Geological Association of Canada, Special Paper, pp 321–378

Petersson A, Kemp AIS, Hickman AH, Van Kranendonk MJ, Whitehouse MJ, Martin L, Gray CM (2019a) A new 3.59 Ga magmatic suite and a chondritic source to the East Pilbara Craton. Chem Geol 511:51–70. https://doi.org/10.1016/j.chemgeo.2019.01.021

Petersson A, Kemp AIS, Whitehouse MJ (2019b) A Yilgarn seed to the Pilbara Craton (Australia)? Evidence from inherited zircons. Geology 47:1098–1102. https://doi.org/10.1130/G46696.1

Pike G, Cas R (2002) Stratigraphic evolution of Archaean volcanic rock-dominated rift basins from the Whim Creek Belt, West Pilbara Craton, Western Australia. In: Altermann W, Corcoran P (eds) Precambrian sedimentary environments: a modern approach to ancient depositional systems. Blackwell Science, Oxford, UK, pp 213–234

Pike G, Cas RAF, Hickman AH (2006) Archean volcanic and sedimentary rocks of the Whim Creek greenstone belt, Pilbara Craton, Western Australia. Report, vol 101. Geological Survey of Western Australia

Pirajno F, Huston DL (2019) Paleoarchean (3.6–3.2 Ga) mineral systems in the context of continental crust building and the role of Mantle Plumes, In: Van Kranendonk MJ, Bennett VC, Hoffmann JE (eds) Earth's oldest rocks, 2nd Elsevier B.V. pp. 187–209, doi:https://doi.org/10.1016/B978-0-444-63901-1.00019-9

Pirajno F, Santosh M (2015) Mantle plumes, supercontinents, intracontinental rifting and mineral systems. Precambrian Res 259:243–261

Reading AM, Tkalčić H, Kennett BLN, Johnson SP, Sheppard S (2012) Seismic structure of the crust and uppermost mantle of the Capricorn and Paterson Orogens and adjacent cratons, Western Australia, from passive seismic transects. Precambrian Res 196–197:295–308. https://doi.org/10.1016/j.precamres.2011.07.001

Ryan GR (1965) The geology of the Pilbara Block. Australasian Institute of Mining and Metallurgy Proceedings 214:61–94

Santosh M (2013) Evolution of continents, cratons and supercontinents: building the habitable Earth. Curr Sci 104:871–879

Sheppard S, Krapež B, Zi J-W, Rasmussen B, Fletcher I (2017) SHRIMP U-Pb zircon geochronology establishes that banded iron formations are not chronostratigraphic markers across Archean greenstone belts of the Pilbara Craton. Precambrian Res 292:290–304

Smirnov AV, Evans ADA, Ernst RE, Söderlund U, Li ZX (2013) Trading partners: tectonic ancestry of southern Africa and western Australia, in Archean supercratons Vaalbara and Zimgarn. Precambrian Res 224:11–12

Smith JB, Barley ME, Groves DI, Krapež B, McNaughton NJ, Bickle MJ, Chapman HJ (1998) The Sholl Shear Zone, West Pilbara; evidence for a domain boundary structure from integrated tectonostratigraphic analysis, SHRIMP U–Pb dating and isotopic and geochemical data of granitoids. Precambrian Res 88:143–172

Smithies RH, Champion DC, Sun S-S (2004) Evidence for early LREE-enriched mantle source regions: diverse magmas from the c. 3.0 Ga Mallina Basin, Pilbara Craton, NW Australia. J Petrol 45:1515–1537

Smithies RH, Champion DC, Van Kranendonk MJ, Hickman AH (2007) Geochemistry of volcanic rocks of the northern Pilbara Craton, Western Australia. Report, vol 104. Geological Survey of Western Australia

Smithies RH, Champion DC, Van Kranendonk MJ, Howard HM, Hickman AH (2005a) Modern-style subduction processes in the Mesoarchean: geochemical evidence from the 3.12 Ga Whundo intra-oceanic arc. Earth Planet Sci Lett 231:221–237

Smithies RH, Van Kranendonk MJ, Champion DC (2005b) It started with a plume–early Archaean basaltic proto-continental crust. Earth Planet Sci Lett 238:284–297

Strik G, Blake TS, Zegers TE, White SH, Langereis CG (2003) Palaeomagnetism of flood basalts in the Pilbara Craton, Western Australia: late Archaean continental drift and the oldest known reversal of the geomagnetic field. J Geophys Res 108. https://doi.org/10.1029/2003JB002475

Sun S-S, Hickman AH (1998) New Nd-isotopic and geochemical data from the West Pilbara–implications for Archaean crustal accretion and shear zone development. Australian Geological Survey Organisation, Research Newsletter:25–29

Thorne AM, Trendall AF (2001) Geology of the Fortescue Group, Pilbara Craton, Western Australia. Bulletin 144. Geological Survey of Western Australia

Thorne AM, Tyler IM, Korsch RJ, Johnson SP, Brett JW, Cutten HN, Blay OA, Kennett BLN, Blewett RS, Joly A, Dentith MC, Aitken ARA, Holzschuh J, Goodwin JA, Salmon M, Reading A, Boren G (2011) Preliminary interpretation of deep seismic reflection line 10GA-CP1: crustal architecture of the northern Capricorn Orogen. In: Johnson SP, Thorne A, Tyler IM (eds) Capricorn Orogen seismic and magnetotelluric (MT) workshop 2011: extended abstracts. Geological Survey of Western Australia, pp 19–26

Thorpe R, Hickman AH, Davis D, Morton JGG, Trendall AF (1992) U–Pb zircon geochronology of Archaean felsic units in the Marble Bar region, Pilbara Craton, Western Australia. Precambrian Res 56:169–189

Trendall AF (1968) Three great basins of Precambrian banded iron-formation deposition: a systematic comparison. Geol Soc Am Bull 79:1527–1544

Trendall AF (1990) Pilbara Craton–Introduction. In: Geological Survey of Western Australia (ed) Geology and mineral resources of Western Australia. Geological Survey of Western Australia, p 128

Tyler IM, Fletcher IR, de Laeter JR, Williams IR, Libby WG (1992) Isotope and rare earth element evidence for a late Archaean terrane boundary in the southeastern Pilbara Craton, Western Australia. Precambrian Res 54:211–229

Van Kranendonk MJ (2000) Geology of the North Shaw 1:100 000 sheet. 1:100 000 Geological series explanatory notes. Geological Survey of Western Australia

Van Kranendonk MJ, Hickman AH, Smithies RH, Champion DC (2007a) Paleoarchean development of a continental nucleus: the East Pilbara Terrane of the Pilbara Craton, Western Australia. In: Van Kranendonk MJ, Bennett VC, Smithies RH (eds) Earth's oldest rocks. Elsevier B.V, Burlington, MA, USA, pp 307–337

Van Kranendonk MJ, Hickman AH, Smithies RH, Nelson DN, Pike G (2002) Geology and tectonic evolution of the Archaean North Pilbara Terrain, Pilbara Craton, Western Australia. Econ Geol 97:695–732. https://doi.org/10.2113/gsecongeo.97.4.695

Van Kranendonk MJ, Hickman AH, Smithies RH, Williams IR, Bagas L, Farrell TR (2006) Revised lithostratigraphy of Archean supracrustal and intrusive rocks in the northern Pilbara Craton, Western Australia. Record, 2006/15. Geological Survey of Western Australia

Van Kranendonk MJ, Smithies RH, Champion DC (2019) Chapter 19–Paleoarchean development of a continental nucleus: the East Pilbara Terrane of the Pilbara Craton, Western Australia. In: Van Kranendonk MJ, Bennett VC, Hoffmann JE (eds) Earth's oldest rocks, 2nd edn. Elsevier B. V, pp 437–462

Van Kranendonk MJ, Smithies RH, Hickman AH, Champion DC (2007b) Review: secular tectonic evolution of Archean continental crust: interplay between horizontal and vertical processes in the formation of the Pilbara Craton, Australia. Terra Nova 19:1–38

Van Kranendonk MJ, Smithies RH, Hickman AH, Wingate MTD, Bodorkos S (2010) Evidence for Mesoarchean (~3.2 Ga) rifting of the Pilbara Craton: the missing link in an early Precambrian Wilson cycle. Precambrian Res 177:145–161

Wiemer D, Schrank CE, Murphy DT, Wenham L, Allen CM (2018) Earth's oldest stable crust in the Pilbara Craton formed by cyclic gravitational overturns. Nat Geosci 11:357–361. https://doi.org/10.1038/s41561-018-0105-9

Williams IR, Hickman AH (2000) Archaean geology of the Muccan region, East Pilbara granite–greenstone terrane, Western Australia–a field guide. Record, 2000/4. Geological Survey of Western Australia

Wilson JF, Nesbitt RW, Fanning CM (1995) Zircon geochronology of Archaean felsic sequences in the Zimbabwe Craton: a revision of greenstone stratigraphy and a model for crustal growth. In: Coward MP, Ries AC (eds) Early Precambrian processes, Geological Society of London special publication, vol 95. Geological Society, London, pp 109–126

Wingate MTD, Kirkland CL, Bodorkos S, Hickman AH (2010) 160258: felsic metavolcanic rock. Orchard Well

Zegers TE, de Wit MJ, Dann J, White SH (1998) Vaalbara, Earth's oldest assembled continent? A combined structural, geochronological, and palaeomagnetic test. Terra Nova 10:250–259

Zeh A, Gerdes A, Barton JM Jr (2009) Archean accretion and crustal evolution of the Kalahari Craton – the Zircon Age and Hf isotope record of granitic rocks from Barberton/Swaziland to the Francistown Arc. J Petrol 50(5):933–966. https://doi.org/10.1093/petrology/egp027

Zeh A, Gerdes A, Millonig L (2011) Hafnium isotope record of the ancient gneiss complex, Swaziland, southern Africa: evidence for Archaean crust–mantle formation and crust reworking between 3.66 and 2.73 Ga. J Geol Soc Lond 168:953–963

Chapter 2
Eoarchean and Early Paleoarchean Crust of the Pilbara Craton

Abstract U–Pb zircon geochronology and Sm–Nd and Lu–Hf isotope data indicate that the Pilbara Craton includes Eoarchean sialic crust formed up to 270 million years before deposition of the greenstone succession (3530–3235 Ma Pilbara Supergroup). Dates on pre-3530 Ma xenocrystic zircons in felsic igneous rocks and detrital zircons in Paleoarchean and Mesoarchean metasedimentary rocks suggest that this early crust evolved from about 3800 Ma. Zircon ages suggest episodic magmatic events at 3760–3700 Ma, c. 3650 Ma and 3590–3570 Ma. Negative ε_{Hf} values from zircons that crystallized during these events, and Hf two-stage depleted mantle model ages, indicate derivation of melts from 4000 to 3800 Ma sources.

Zircon Lu–Hf isotope evidence indicates that 3530–3460 volcanics and granitic rocks were derived from more juvenile sources than applied to pre-3530 Ma felsic igneous rocks. This is interpreted to be evidence for an abrupt change in magma sources between 3550 and 3530 Ma, marking a major event in the early crustal evolution of the Pilbara Craton. The precise nature of this event is unclear, although one likely explanation is that uplift, crustal extension, and rifting of the 3800–3530 Ma crust was due to the arrival of the mantle plume that triggered eruption of the Pilbara Supergroup.

Keywords Eoarchean crust · Detrital zircons · Sm–Nd isotopes · Lu–Hf isotopes · Model ages

2.1 Introduction

One of the most contentious questions in the interpretation of Archean granite–greenstone terranes has been whether the oldest greenstones were deposited on even older continental crust or if the greenstones are the oldest rocks and originated as oceanic crust. There are known to be two areas of the world where a Paleoarchean granite–greenstone stratigraphy is sufficiently well preserved to resolve this question: the East Pilbara Terrane (EPT) of the Pilbara Craton and the Barberton Greenstone Belt (BGB) of the eastern Kaapvaal Craton in southern Africa.

A. H. Hickman, *Archean Evolution of the Pilbara Craton and Fortescue Basin*, Modern Approaches in Solid Earth Sciences 24,
https://doi.org/10.1007/978-3-031-18007-1_2

Geological mapping of the EPT in the 1970s led to a conclusion that the earliest greenstones were deposited on older sialic crust (Hickman 1975, 1981, 1983, 1984; Hickman and Lipple 1975, 1978). Geochemical, geochronological, and Sm–Nd isotope studies over the next 20 years added evidence in support of this interpretation (Hamilton et al. 1981; Jahn et al. 1981; Collerson and McCulloch 1983; Gruau et al. 1987; McCulloch 1987; Bickle et al. 1989, 1993; Green et al. 2000). However, before results from the Pilbara were widely known, the BGB had become a model for the evolution of Paleoarchean greenstone belts (Anhaeusser 1971a, b; Viljoen and Viljoen 1971; Glikson 1972, 1979; Glikson and Lambert 1976; Windley 1976; de Wit 1983, 1998; de Wit et al. 1992, 2011; de Wit and Ashwal 1997; Kamo and Davis 1994). As it evolved from the 1970s through the 1980s, this model incorporated plate-tectonic processes based on tectonic interpretations of Phanerozoic terranes. However, the evidence from the Pilbara Craton was not consistent with plate-tectonic processes (Hickman 1981, 1983, 1984).

The origin of the BGB and the relationship of this belt to the adjacent granitic terrane of the Ancient Gneiss Complex have been controversial. As in the Pilbara, Sm–Nd model ages of 3700–3600 Ma from the Kaapvaal Craton (Kröner et al. 1996) suggested evolved crustal sources for felsic igneous rocks. Later, Kröner et al. (2016) interpreted zircon Hf isotope data from c. 3530 Ma felsic volcanic rocks at the base of the Barberton greenstone succession to indicate derivation from melting of significantly older felsic crust. They concluded that the Barberton greenstones were deposited on a Paleoarchean volcanic plateau underlain by older continental crust. The interpretation that the BGB evolved by plate-tectonic processes deforming oceanic crust (de Wit 1983, 1998; de Wit et al. 1992, 2011; Kamo and Davis 1994; Moyen et al. 2006; Schoene et al. 2008; Kisters et al. 2010; Schoene and Bowring 2010; Furnes et al. 2012; Taylor et al. 2012) has been contested by Van Kranendonk (2011), Van Kranendonk et al. (2015a, b), Kröner et al. (2016), and Byerly et al. (2019).

Gneisses from the Ancient Gneiss Complex have been dated between c. 3660 and 3550 Ma (Compston and Kröner 1988; Kröner et al. 1989, 2014; Zeh et al. 2011; Suhr et al. 2015; Hoffmann and Kroner 2019), although the bulk of the complex is composed of Paleoarchean granitic rocks. It has been proposed that the older gneisses of the Ancient Gneiss Complex formed a basement to the Barberton greenstones, but the contact is tectonic. In a review of the evolution of the BGB, Byerly et al. (2019) concluded that multiple lines of evidence demonstrate that continental crust had already formed in the Barberton area by around 3600 Ma and probably substantially earlier. They attributed the almost complete absence of pre-3550 Ma detrital zircons in Barberton Paleoarchean sedimentary rocks, a feature noted by Drabon et al. (2017), to non-exposure of this basement until the Mesoarchean.

Throughout the latter part of the twentieth century, most tectonic models for early Archean granite–greenstone evolution have favoured the oceanic interpretations applied to the BGB. The opposing interpretation reached from studies of the Pilbara Craton has rarely been mentioned. However, investigations over the past 25 years have established that the Paleoarchean greenstone successions and granitic rocks of the Pilbara and Kaapvaal Cratons are virtually identical, now making different tectonic models unsustainable (Van Kranendonk et al. 2015a, b). This chapter examines the various lines of evidence for Eoarchean and early Paleoarchean continental crust in the Pilbara Craton.

The existence of continental crust pre-dating the 3530–3235 Ma Pilbara Supergroup has been indicated by several lines of evidence: rare exposures of remnants of the early crust within the East Pilbara Terrane (McNaughton et al. 1988; Nelson 1999; Wingate et al. 2009a, b; Petersson et al. 2019a); U–Pb dating of 3800–3530 Ma xenocrystic zircons in felsic igneous rocks (Thorpe et al. 1992; Nelson 1998, 1999, 2000, 2006; Beintema 2003; Kabashima et al. 2003; Kitajima et al. 2008; Wingate et al. 2009a, b; Petersson et al. 2019b); U–Pb dating of detrital zircons in sedimentary rocks (Van Kranendonk et al. 2002, 2007a, b; Bagas et al. 2004, 2008; Hickman 2012, 2021); whole-rock Sm–Nd isotopic data (Hamilton et al. 1981; Jahn et al. 1981; Collerson and McCulloch 1983; Gruau et al. 1987; McCulloch 1987; Bickle et al. 1989, 1993; Tyler et al. 1992; Van Kranendonk et al. 2002, 2007a, b; Smithies et al. 2007b; Champion 2013; Champion and Huston 2016; Hasenstab et al. 2021); and zircon Lu–Hf isotopic data (Amelin et al. 2000; Guitreau et al. 2012; Nebel et al. 2014; Kemp et al. 2015a, b; Gardiner et al. 2017, 2018; Hasenstab et al. 2021; Hickman 2021).

This chapter commonly refers to formations and groups within the Pilbara Supergroup. Table 1.1 summarizes the stratigraphy of the East Pilbara Craton, and Fig. 1.6 shows the ages and relationships of the igneous supersuites.

2.2 Eoarchean to Early Paleoarchean Crust (3800–3530 Ma)

Rocks dated at 3590–3580 Ma in the centre of the East Pilbara Terrane (McNaughton et al. 1988; Wiemer et al. 2018; Petersson et al. 2019a) are consistent with the interpretation that the 3530–3235 Ma Pilbara Supergroup was constructed on a substrate of older crust (Hickman 1981; Van Kranendonk et al. 2002). Concordia plots of U–Pb zircon data from the two localities in the Warrawagine Dome (Fig. 1.7) show several inherited zircons with crystallization ages >3570 Ma (Fig. 2.1). Additionally, as seen in Table 2.1, xenocrystic zircons in various Paleoarchean rhyolite outcrops in the North Pole area have been dated between 3760 and 3700 Ma (Thorpe et al. 1992; Petersson et al. 2019b) and c. 3590 Ma (Kitajima et al. 2008). Other evidence for pre-Pilbara Supergroup felsic crust include the existence of sandstone units low in the stratigraphy of the Warrawoona Group (Hickman 1981); geochemical evidence that the stratigraphically lowest basalts were contaminated by felsic crust (Glikson and Hickman 1981; Green et al. 2000; Smithies et al. 2007b); Sm–Nd isotope evidence of pre-3530 Ma sources for the volcanic and intrusive rocks (Jahn et al. 1981; Hamilton et al. 1981; Gruau et al. 1987; McCulloch 1987; Bickle et al. 1989, 1993; Tyler et al. 1992; Smithies et al.

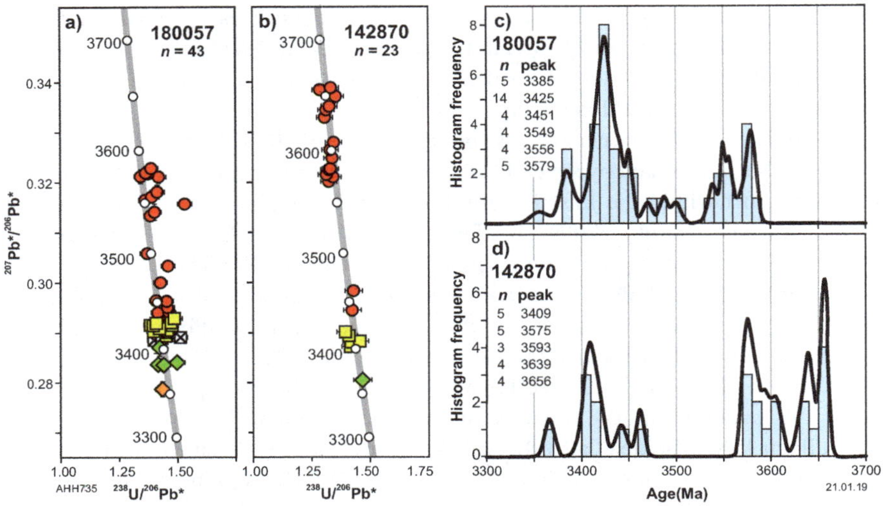

Fig. 2.1 U–Pb analytical data for samples Geological Survey of Western Australia 180057 and 142870 from the Warrawagine Dome, revealing inheritance from Eoarchean to early Paleoarchean crust. In concordia diagrams (**a**, **b**), n = number of analyses: yellow squares indicate magmatic zircons; green diamonds indicate metamorphic zircon rims; orange diamond indicates a younger metamorphic rim; red circles indicate xenocrystic zircons; crossed squares indicate possible core–rim mixtures. In probability diagrams (c, d), n = number of ages in each significant age component (Ma, based on three or more ages) (From Hickman 2021; with Geological Survey of Western Australia permission)

2004, 2007b; Van Kranendonk et al. 2006, 2007a, b; Champion 2013; Champion and Huston 2016; Gardiner et al. 2017, 2018; Hasenstab et al. 2021); and zircon Lu–Hf isotope evidence for crustal sources at old as 4000 Ma (this Chapter).

Early evidence for the existence of pre-Pilbara Supergroup crust was obtained through the recognition that essentially the same stratigraphy of the 15-km-thick Warrawoona Group was present in most greenstone belts of the EPT (Hickman 1980a, b, 1981). This effectively precluded the possibility that each individual greenstone belt represented a separate depositional basin, as in a tectonic model derived from the Barberton Greenstone Belt of southern Africa (Anhaeusser et al. 1969). The lateral continuity of the thick Warrawoona Group implied the existence of a similarly continuous older substrate or 'basement' onto which it had been deposited. Information on the thickness and composition of the Warrawoona Group was inconsistent with oceanic crust, and it was concluded that the basement was continental crust. An additional consideration was that oceanic crust, unless obducted, is destroyed by subduction within about 250 million years, even on modern Earth which has large oceanic plates. Some workers consider that Archean Earth had much smaller plates (Komiya and Maruyama 2007).

The nature of the pre-Pilbara Supergroup crust, especially its thickness, extent, composition, and age range, will help to constrain viable tectonic models for the crustal evolution of the EPT. Some of these features have been considered in

Table 2.1 Summary of >3530 Ma U–Pb zircon ages in igneous rocks of the East Pilbara Craton (Extract of data from Hickman 2021; with Geological Survey of Western Australia permission)

Age (Ma)	Disc. %	Method	Sample	Rock type	Location	Reference
3760 ± 3	5	SHRIMP U-Pb zircon	18APB12	Rhyolite	North Pole	Petersson et al. (2019b)
3753 ± 3	8	SHRIMP U-Pb zircon	18APB12	Rhyolite	North Pole	Petersson et al. (2019b)
3746 ± 4	0	SHRIMP U-Pb zircon	15TKPB17	Rhyolite	North Pole	Petersson et al. (2019b)
3746 ± 3	1	SIMS U-Pb zircon	15TKPB17	Rhyolite	North Pole	Petersson et al. (2019b)
3735 ± 2	1	SIMS U-Pb zircon	15TKPB17	Rhyolite	North Pole	Petersson et al. (2019b)
3733 ± 5	1	SHRIMP U-Pb zircon	15TKPB17	Rhyolite	North Pole	Petersson et al. (2019b)
3732 ± 3	1	SIMS U-Pb zircon	15TKPB17	Rhyolite	North Pole	Petersson et al. (2019b)
3731 ± 2	9	SHRIMP U-Pb zircon	15TKPB17	Rhyolite	North Pole	Petersson et al. (2019b)
3729 ± 2	1	SIMS U-Pb zircon	15TKPB17	Rhyolite	North Pole	Petersson et al. (2019b)
3729 ± 4	2	SHRIMP U-Pb zircon	18APB13	Rhyolite	North Pole	Petersson et al. (2019b)
3724 ± 1	1	U-Pb zircon, conventional	100507	Rhyolite	North Pole	Thorpe et al. (1992)
3718 ± 5	6	SHRIMP U-Pb zircon	15TKPB17	Rhyolite	North Pole	Petersson et al. (2019b)
3718 ± 3	0	SHRIMP U-Pb zircon	18APB13	Rhyolite	North Pole	Petersson et al. (2019b)
3715 ± 4	1	SHRIMP U-Pb zircon	15TKPB17	Rhyolite	North Pole	Petersson et al. (2019b)
3713 ± 5	4	SIMS U-Pb zircon	15TKPB17	Rhyolite	North Pole	Petersson et al. (2019b)
3710 ± 2	1	SIMS U-Pb zircon	15TKPB17	Rhyolite	North Pole	Petersson et al. (2019b)
3704 ± 3	30	SIMS U-Pb zircon	15TKPB17	Rhyolite	North Pole	Petersson et al. (2019b)
3699 ± 1	2	SIMS U-Pb zircon	15TKPB17	Rhyolite	North Pole	Petersson et al. (2019b)
3696 ± 4	9	SHRIMP U-Pb zircon	15TKPB17	Rhyolite	North Pole	Petersson et al. (2019b)
3696 ± 3	17	SHRIMP U-Pb zircon	18APB16	Rhyolite	North Pole	Petersson et al. (2019b)
3689 ± 5	25	SHRIMP U-Pb zircon	15TKPB17	Rhyolite	North Pole	Petersson et al. (2019b)
3681 ± 5	44	SHRIMP U-Pb zircon	15TKPB17	Rhyolite	North Pole	Petersson et al. (2019b)

(continued)

Table 2.1 (continued)

Age (Ma)	Disc. %	Method	Sample	Rock type	Location	Reference
3678 ± 3	61	SHRIMP U-Pb zircon	18APB13	Rhyolite	North Pole	Petersson et al. (2019b)
3676 ± 4	1	SHRIMP U-Pb zircon	15TKPB17	Rhyolite	North Pole	Petersson et al. (2019b)
3675 ± 4	48	SHRIMP U-Pb zircon	18APB13	Rhyolite	North Pole	Petersson et al. (2019b)
3660 ± 62	0	LA-ICPMS U-Pb zircon	96NS-410	Rhyolite	North Pole	Kitajima et al. (2008)
3658 ± 3	1	SHRIMP U-Pb zircon	142870	Tonalite gneiss	Warrawagine	Nelson (1999)
3656 ± 4	−1	SHRIMP U-Pb zircon	142870	Tonalite gneiss	Warrawagine	Nelson (1999)
3655 ± 3	0	SHRIMP U-Pb zircon	142870	Tonalite gneiss	Warrawagine	Nelson (1999)
3650 ± 6	2	SHRIMP U-Pb zircon	142870	Tonalite gneiss	Warrawagine	Nelson (1999)
3650 ± 3	38	SHRIMP U-Pb zircon	18APB13	Rhyolite	North Pole	Petersson et al. (2019b)
3641 ± 4	1	SHRIMP U-Pb zircon	142870	Tonalite gneiss	Warrawagine	Nelson (1999)
3641 ± 3	69	SHRIMP U-Pb zircon	18APB12	Rhyolite	North Pole	Petersson et al. (2019b)
3640 ± 3	11	SIMS U-Pb zircon	15TKPB17	Rhyolite	North Pole	Petersson et al. (2019b)
3638 ± 4	0	SHRIMP U-Pb zircon	142,870	Tonalite gneiss	Warrawagine	Nelson (1999)
3632 ± 8	6	SIMS U-Pb zircon	15TKPB17	Rhyolite	North Pole	Petersson et al. (2019b)
3631 ± 5	−1	SHRIMP U-Pb zircon	142870	Tonalite gneiss	Warrawagine	Nelson (1999)
3629 ± 35	N/A	LA-ICPMS U-Pb zircon	KB312	Granodiorite	Carlindi	Beintema et al. (2003)
3626 ± 4	32	SHRIMP U-Pb zircon	18APB13	Rhyolite	North Pole	Petersson et al. (2019b)
3621 ± 3	0	SHRIMP U-Pb zircon	178031	Granodiorite	Muccan	Nelson (2006)
3612 ± 3	1	SHRIMP U-Pb zircon	18APB16	Rhyolite	North Pole	Petersson et al. (2019b)
3609 ± 31	19	LA-ICPMS U-Pb zircon	95NS – 281	Rhyolite	North Pole	Kabashima et al. (2003)
3608 ± 5	1	SHRIMP U-Pb zircon	142870	Tonalite gneiss	Warrawagine	Nelson (1999)
3605 ± 3	45	SHRIMP U-Pb zircon	18APB12	Rhyolite	North Pole	Petersson et al. (2019b)

(continued)

Table 2.1 (continued)

Age (Ma)	Disc. %	Method	Sample	Rock type	Location	Reference
3603 ± 3	1	SHRIMP U-Pb zircon	18APB13	Rhyolite	North Pole	Petersson et al. (2019b)
3601 ± 5	0	SHRIMP U-Pb zircon	142870	Tonalite gneiss	Warrawagine	Nelson (1999)
3601 ± 5	24	SHRIMP U-Pb zircon	18APB13	Rhyolite	North Pole	Petersson et al. (2019b)
3600 ± 3	5	SHRIMP U-Pb zircon	142884	Syenogranite	Tambourah	Nelson (1998)
3593 ± 4	0	SHRIMP U-Pb zircon	142870	Tonalite gneiss	Warrawagine	Nelson (1999)
3590 ± 3	6	SHRIMP U-Pb zircon	18APB13	Rhyolite	North Pole	Petersson et al. (2019b)
3589 ± 10	1	SHRIMP U-Pb zircon	15TKPB17	Rhyolite	North Pole	Petersson et al. (2019b)
3589 ± 2	15	SHRIMP U-Pb zircon	15TKPB17	Rhyolite	North Pole	Petersson et al. (2019b)
3583 ± 5	−1	SHRIMP U-Pb zircon	142870	Tonalite gneiss	Warrawagine	Nelson (1999)
3583 ± 3	3	SHRIMP U-Pb zircon	180057	Tonalite gneiss	Warrawagine	Wingate et al. (2009b)
3582 ± 5	−1	SHRIMP U-Pb zircon	142870	Tonalite gneiss	Warrawagine	Nelson (1999)
3582 ± 9	12	SHRIMP U-Pb zircon	178203	Granodiorite	Mosquito Creek	Nelson (2000)
3580 ± 5	3	SHRIMP U-Pb zircon	142967	Monzogranite	Shaw	Nelson (2005)
3580 ± 3	25	SHRIMP U-Pb zircon	15TKPB17	Rhyolite	North Pole	Petersson et al. (2019b)
3579 ± 6	1	SHRIMP U-Pb zircon	180057	Tonalite gneiss	Warrawagine	Wingate et al. (2009b)
3578 ± 3	2	SHRIMP U-Pb zircon	180057	Tonalite gneiss	Warrawagine	Wingate et al. (2009b)
3577 ± 5	−2	SHRIMP U-Pb zircon	142870	Tonalite gneiss	Warrawagine	Nelson (1999)
3577 ± 4	-1	SHRIMP U-Pb zircon	178203	Biotite	Mosquito Creek	Nelson (2000)
3576 ± 44	-1	LA-ICPMS U-Pb zircon	96NS-430	Felsic tuff	North Pole	Kabashima et al. (2003)
3575 ± 3	0	SHRIMP U-Pb zircon	142870	Tonalite gneiss	Warrawagine	Nelson (1999)
3575 ± 7	0	SHRIMP U-Pb zircon	180057	Tonalite gneiss	Warrawagine	Wingate et al. (2009b)
3575 ± 6	4	SHRIMP U-Pb zircon	180057	Tonalite gneiss	Warrawagine	Wingate et al. (2009b)

(continued)

Table 2.1 (continued)

Age (Ma)	Disc. %	Method	Sample	Rock type	Location	Reference
3574 ± 3	2	SHRIMP U-Pb zircon	142828	Granodiorite	Muccan	Nelson (1998)
3570 ± 3	−2	SHRIMP U-Pb zircon	142870	Tonalite gneiss	Warrawagine	Nelson (1999)
3566 ± 5	3	SIMS U-Pb zircon	15TKPB17	Rhyolite	North Pole	Petersson et al. (2019b)
3565 ± 3	8	SHRIMP U-Pb zircon	15TKPB17	Rhyolite	North Pole	Petersson et al. (2019b)
3561 ± 7	3	SHRIMP U-Pb zircon	180057	Tonalite gneiss	Warrawagine	Wingate et al. (2009b)
3558 ± 45	6	U-Pb zircon (LA-ICPMS)	96NS-430	Felsic tuff	North Pole	Kabashima et al. (2003)
3556 ± 3	2	SHRIMP U-Pb zircon	180057	Tonalite gneiss	Warrawagine	Wingate et al. (2009b)
3556 ± 6	9	SHRIMP U-Pb zircon	15TKPB17	Rhyolite	North Pole	Petersson et al. (2019b)
3554 ± 2	9	SHRIMP U-Pb zircon	18APB12	Rhyolite	North Pole	Petersson et al. (2019b)
3553 ± 5	0	SHRIMP U-Pb zircon	180057	Tonalite gneiss	Warrawagine	Wingate et al. (2009b)
3552 ± 5	4	SHRIMP U-Pb zircon	142936	Monzogranite	Yule	Nelson (2000)
3552 ± 3	3	SHRIMP U-Pb zircon	180056	Granophyre	Muccan	Wingate et al. (2009a)
3549 ± 2	9	SHRIMP U-Pb zircon	180057	Tonalite gneiss	Warrawagine	Wingate et al. (2009b)
3544 ± 5	42	SIMS U-Pb zircon	15TKPB17	Rhyolite	North Pole	Petersson et al. (2019b)
3544 ± 30	20	LA-ICPMS U-Pb zircon	95NS-281	Rhyolite	North Pole	Kabashima et al. (2003)
3541 ± 8	2	SHRIMP U-Pb zircon	180057	Tonalite gneiss	Warrawagine	Wingate et al. (2009b)
3538 ± 3	7	SHRIMP U-Pb zircon	142828	Granodiorite	Muccan	Nelson (1998)
3538 ± 4	-2	SHRIMP U-Pb zircon	178022	Granodiorite	Muccan	Nelson (2006)
3538 ± 4	1	SHRIMP U-Pb zircon	180057	Tonalite gneiss	Warrawagine	Wingate et al. (2009b)
3535 ± 8	18	SHRIMP U-Pb zircon	15TKPB17	Rhyolite	North Pole	Petersson et al. (2019b)

previous studies (Van Kranendonk et al. 2006, 2007a, b; Kemp et al. 2015a, b; Gardiner et al. 2017, 2018).

2.2.1 U–Pb Zircon Geochronology

Research within the Pilbara Craton Mapping Project (PCMP, Chap. 1), and from other investigations, has provided a large body of evidence indicating the widespread existence of Eoarchean to early Paleoarchean continental crust in the Pilbara Craton. This evidence includes Sm–Nd and Lu–Hf isotope data in addition to pre-Pilbara Supergroup U–Pb zircon dates on xenocrystic and detrital zircons. Until recently, some workers questioned the widespread existence of pre-Pilbara Supergroup crust (pre-3530 Ma) because it has rarely been identified in outcrop. However, an understanding of the tectonic evolution of the Pilbara Craton explains the present scarcity of such outcrops as being a consequence of crustal reworking during the Paleoarchean and the Mesoarchean. In terms of the evolution of the craton, it is now firmly established that such pre-Pilbara Supergroup was present, and exposed, in the Archean. Principal lines of evidence for Eoarchean to early Paleoarchean crust include:

- *Ages of xenocrystic zircons in felsic igneous rocks:* zircons significantly older than the crystallization age of a host igneous rock were, with few exceptions, inherited from an older felsic igneous rock during intrusion of one igneous rock by another. An exception might occur in cases where an igneous intrusion incorporates old detrital zircons from an intruded sedimentary rock. Most xenocrystic zircons in igneous rocks of the East Pilbara are post-3530 Ma, but early Paleoarchean and Eoarchean zircons are also present (Table 2.1).
- *Age spectra of detrital zircons in Paleoarchean and Mesoarchean metasedimentary formations*: the zircon age groups in a clastic sedimentary rock potentially provide a representative sample of the ages of felsic igneous rock exposed in the provenance regions during deposition (Cawood et al. 2012). Recycling of xenocrystic and detrital zircons from older igneous and sedimentary rocks is also possible.
- *Whole-rock Nd model ages of igneous rocks*: these ages indicate the average mantle extraction ages of the sources of the melts for currently exposed igneous rocks. Epsilon Nd (ε_{Nd}) values provide another measure of the difference between a rock's crystallization age and the time at which its sources were extracted from the mantle. Negative ε_{Nd} values are typical of continental crust in which the Nd system has been separated from the Sm-enriched mantle for many millions of years, whereas positive ε_{Nd} values indicate juvenile mantle-derived melts.
- *Lu–Hf isotope data from zircons in felsic igneous and sedimentary rocks*: these data convey information on the mantle extraction ages of sources and on the relative importance of melts derived by crustal recycling versus additions of juvenile magma directly from the mantle or from recently extracted crust. Hf

isotope ratios, between ^{176}Hf (produced by decay of ^{176}Lu) and ^{177}Hf, are expressed using the same ε-notation used for Nd. Strongly negative ε_{Hf} values imply reworking of significantly older crustal material, whereas positive ε_{Hf} values suggest juvenile sources. Hf isotope ratios are therefore useful indicators of tectonic processes and settings (Gardiner et al. 2017, 2018).

2.2.1.1 Pre-3530 Ma Xenocrystic Zircons in Felsic Igneous Rocks

For both igneous and sedimentary rocks in the East Pilbara, it is interpreted that zircons with ^{207}Pb/^{206}Pb ages older than 3530 Ma were derived from the crust on which the Pilbara Supergroup was deposited. There is no alternative source of pre-3530 Ma zircons in Australia because the Yilgarn Craton was too remote from the Pilbara Craton in the Archean (e.g. Smirnov et al. 2013). Many xenocrystic zircons older than 3530 Ma have been recorded in the East Pilbara, and the oldest, dated at >3700 Ma, were reported from 3450 Ma rhyolite in the North Pole Dome (Petersson et al. 2019b). These indicate the existence of Eoarchean crust in the area of the North Pole Dome at the time when the rhyolite crystallized. Although xenocrystic zircons in igneous rocks might be derived during intrusion of older sedimentary rocks, this is a very unlikely explanation for the >3700 Ma North Pole zircons. Sedimentary rocks older than 3450 Ma in the North Pole Dome are restricted to the c. 3481 Ma Dresser Formation (Wingate et al. 2009a, GSWA 180070) and a thin c. 3470 Ma sandstone within the Mount Ada Basalt (Byerly et al. 2002), and these units contain no detrital zircons older than 3534 Ma.

Table 2.1 lists xenocrystic zircons from igneous rocks of the EPT that have been dated at >3530 Ma. The 3451–3416 Ma Tambina Supersuite has yielded 80 percent of the zircons derived from 3760–3530 Ma crust, probably because the first widespread major doming and crustal melting occurred during its intrusion. The Tambina Supersuite is mainly composed of banded gneiss and includes numerous enclaves metamorphosed to at least amphibolite facies. TTG magmas for the Tambina Supersuite were derived by partial melting of older felsic crust, as in the case of leucogranite in the Shaw Dome (Pawley et al. 2004) or, more generally, from infracrustal melting of enriched basaltic crust (Champion and Smithies 2007; Smithies et al. 2009; Van Kranendonk et al. 2014). Clusters of zircon ages between 3760 and 3690 Ma, and from 3610 to 3570 Ma, suggest separate episodes of increased magmatic activity. Until now, 3760–3690 Ma xenocrystic zircons have all been obtained from a single area of the North Pole Dome, but the 3610–3570 Ma zircons have been recorded from five domes and from the Kurrana Terrane.

Gabbroic rocks dated at 3590–3580 Ma are exposed in the western Shaw Dome and are surrounded and intruded by c. 3440 Ma granitic rocks. One of these gabbroic units was dated by the U–Pb zircon method at 3578 ± 4 Ma (McNaughton et al. 1988), although supporting analytical data were not published. Subsequently, a more detailed geochronological investigation of several large gabbroic bodies in the same area provided consistent U–Pb zircon dates of c. 3580 Ma (Petersson et al. 2019a).

Wiemer et al. (2018) reported U–Pb zircon dates of 3591 ± 36 Ma and 3576 ± 22 Ma for samples of trondhjemite and tonalite gneiss from the southern margin of the Muccan Dome. Similar results were obtained from xenoliths of banded tonalite gneiss in the Warrawagine Dome where zircons have crystallization ages between 3593 and 3575 Ma (Fig. 2.1) (GSWA 142870, Nelson 1999; GSWA 180057, Wingate et al. 2009b; sample 12TKPB06, Petersson et al. 2020). The conclusion is that a supersuite of mafic and felsic igneous rocks intruded a large area of the Pilbara Craton between 3610 and 3570 Ma.

2.2.1.2 Evidence from $^{207}Pb/^{206}Pb$ Ages of Detrital Zircons

PCMP dating of Paleoarchean and Mesoarchean metasedimentary rocks in the northern Pilbara Craton identified many detrital zircons much older than the depositional ages of the rocks. The significance of the old zircons was uncertain until the chronology of magmatic events in the crustal evolution of the craton had been established. This occurred when U–Pb zircon data from more than 200 samples were used to recognize, define, and name eight granitic supersuites and seven granitic suites (Van Kranendonk et al. 2004a, b, 2006). Figure 1.6 summarizes the published crystallization ages of igneous rocks in the northern Pilbara Craton.

Approximately 15% of the detrital zircons in the sedimentary rocks analysed are older than 3530 Ma (Fig. 2.2a), the maximum depositional age of the Pilbara Supergroup and the intrusive suites represented in Fig. 1.6. These pre-3530 Ma zircons were obtained from samples of the Pilbara Supergroup and from formations of overlying Mesoarchean sedimentary basins (Figs. 2.2b-f). On the basis that these old detrital zircons were eroded from the Pilbara crust, they contribute evidence on its age, distribution, and composition.

The scarcity of known outcrops of pre-3530 Ma rocks in the northern Pilbara Craton, in contrast to the common occurrence of pre-3530 Ma detrital zircons in sedimentary formations, led some workers to suggest derivation from terranes outside the craton (Bagas et al. 2004, 2008; Kemp et al. 2015a). However, this would be inconsistent with the fact that there are no potential alternative sources of Eoarchean to early Paleoarchean zircons within Australia. Although the northwest Yilgarn Craton contains >3530 Ma rocks, the Pilbara and Yilgarn Cratons were widely separated in the Paleoarchean and Mesoarchean and did not collide until the 2005–1950 Ma Glenburgh Orogeny (Sheppard et al. 2005; Spaggiari et al. 2008; Johnson et al. 2011, 2013; Smirnov et al. 2013).

As described by Kemp et al. (2015b), the age spectra of the detrital zircons from particular samples are consistent with provenances within the Pilbara Craton. The evidence falls into two main categories: specific, based on consideration of individual samples, and general, based on data from large numbers of samples.

2.2.1.2.1 Specific Evidence on Provenance

In some instances, a direct relationship is indicated between the detrital zircons in sandstone and magmatic zircons in adjacent granites. On the northern margin of the Muccan Dome, the 3060–3022 Ma Farrel Quartzite (Gorge Creek Group) contains a dominant age group of 3434 ± 6 Ma detrital zircons (Nelson 1998, GSWA 143996), and, within error, this is consistent with the 3443 ± 6 Ma crystallization age of underlying granodiorite of the Tambina Supersuite (Nelson 1996, GSWA 124755). A major erosional unconformity separating the sandstone from the granodiorite was described by Dawes et al. (1995a, b). Other workers in the East Pilbara have also interpreted local provenance for sandstones overlying unconformities (Buick et al. 1995; Green et al. 2000; Allwood et al. 2006, 2007a, b; Wacey et al. 2010).

Sedimentary facies commonly demonstrate local provenance, as in the example of the Duffer Formation in the Warralong greenstone belt (Fig. 2.3). Sample GSWA 168996 (Nelson 2002) was collected from a feldspathic sandstone of the Duffer Formation that overlies polymictic conglomerate containing boulders of chert up to

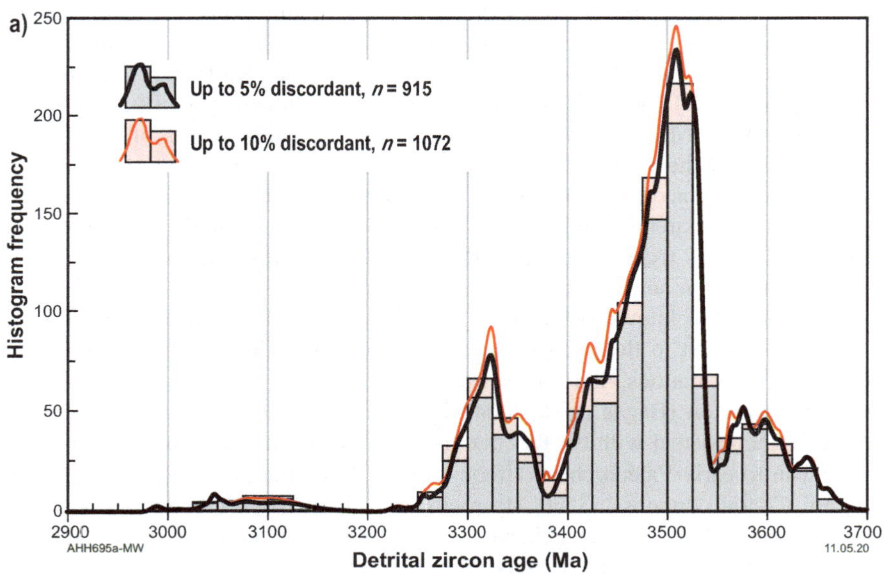

Fig. 2.2 Histograms showing the frequency of detrital zircon ages in sedimentary formations of the East Pilbara Craton, excluding the Mosquito Creek and Mallina Formations. The total dataset (**a**) is distinguished by dome and formation (**b** and **c**). The total dataset shows three well-defined peaks at 3660–3560 Ma (early crust), 3540–3400 Ma (Warrawoona Group and EPT granitic intrusions of the same age), and 3360–3280 Ma (Kelly Group and Emu Pool Supersuite). Pre-3530 Ma detrital zircons are concentrated in the Corboy and Strelley Pool Formations (large sedimentary basins), although the Apex Basalt and parts of the Duffer and Wyman Formations also contain old zircons, presumably from more proximal sources. Differences between the zircon age spectra of different domes support the interpretation of relatively local derivation of detritus in most formations (From Hickman 2021; with Geological Survey of Western Australia permission)

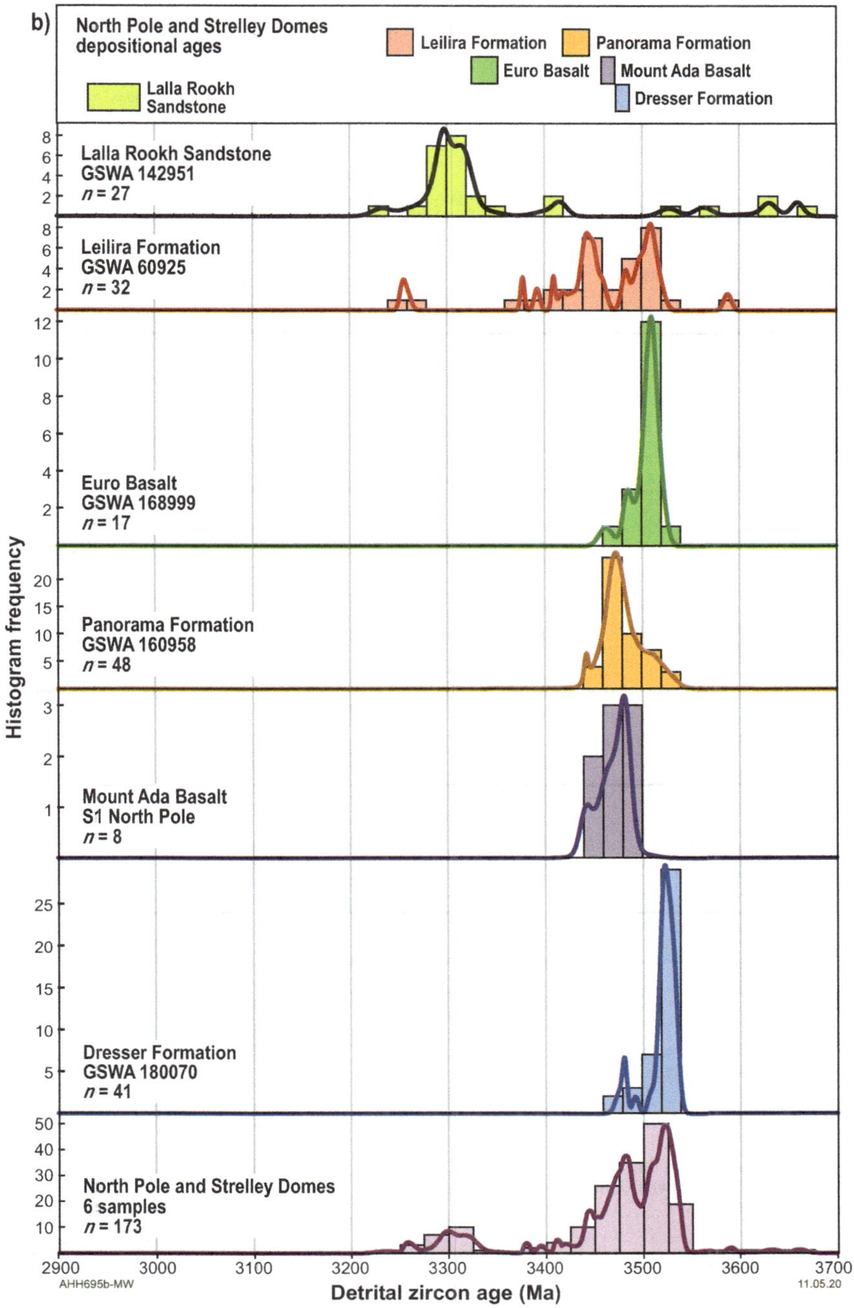

Fig. 2.2 (continued)

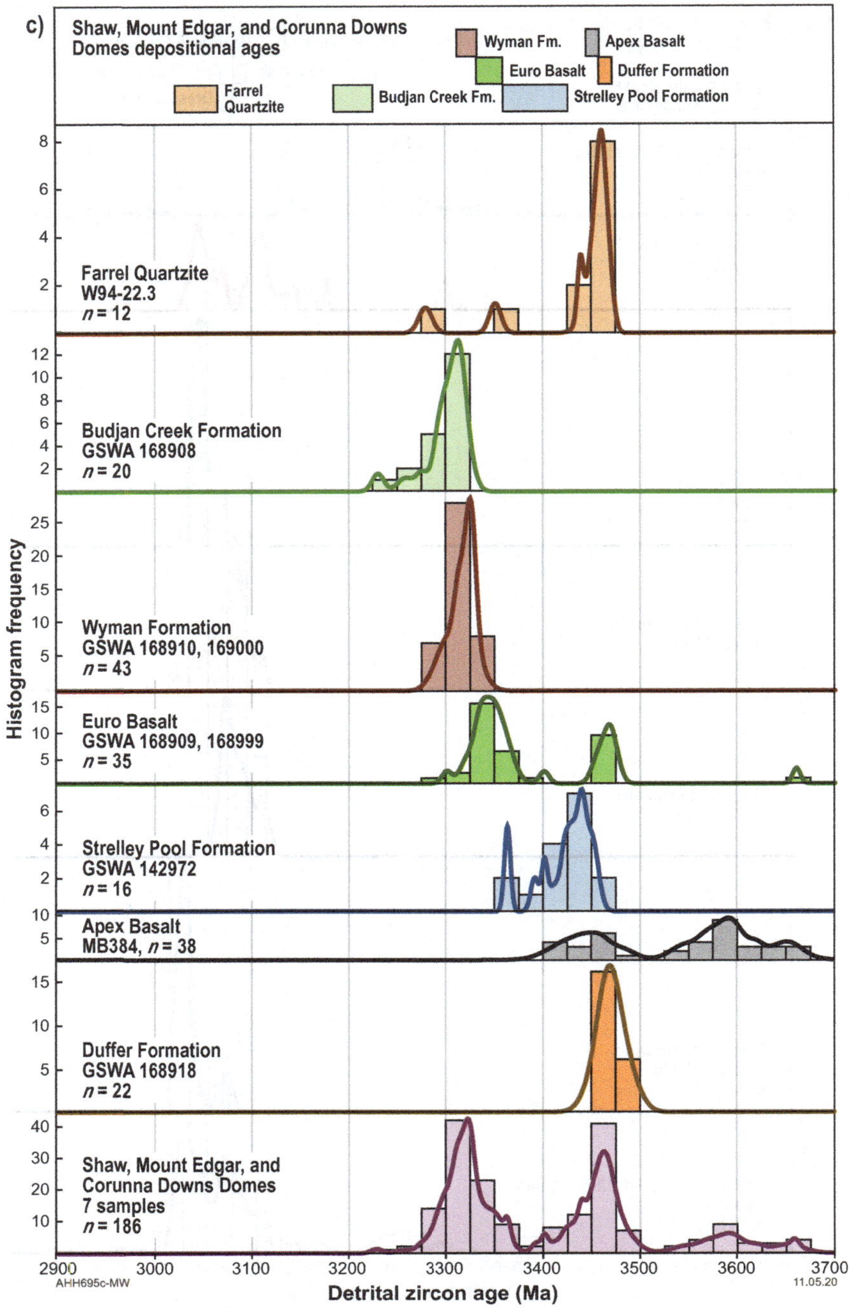

Fig. 2.2 (continued)

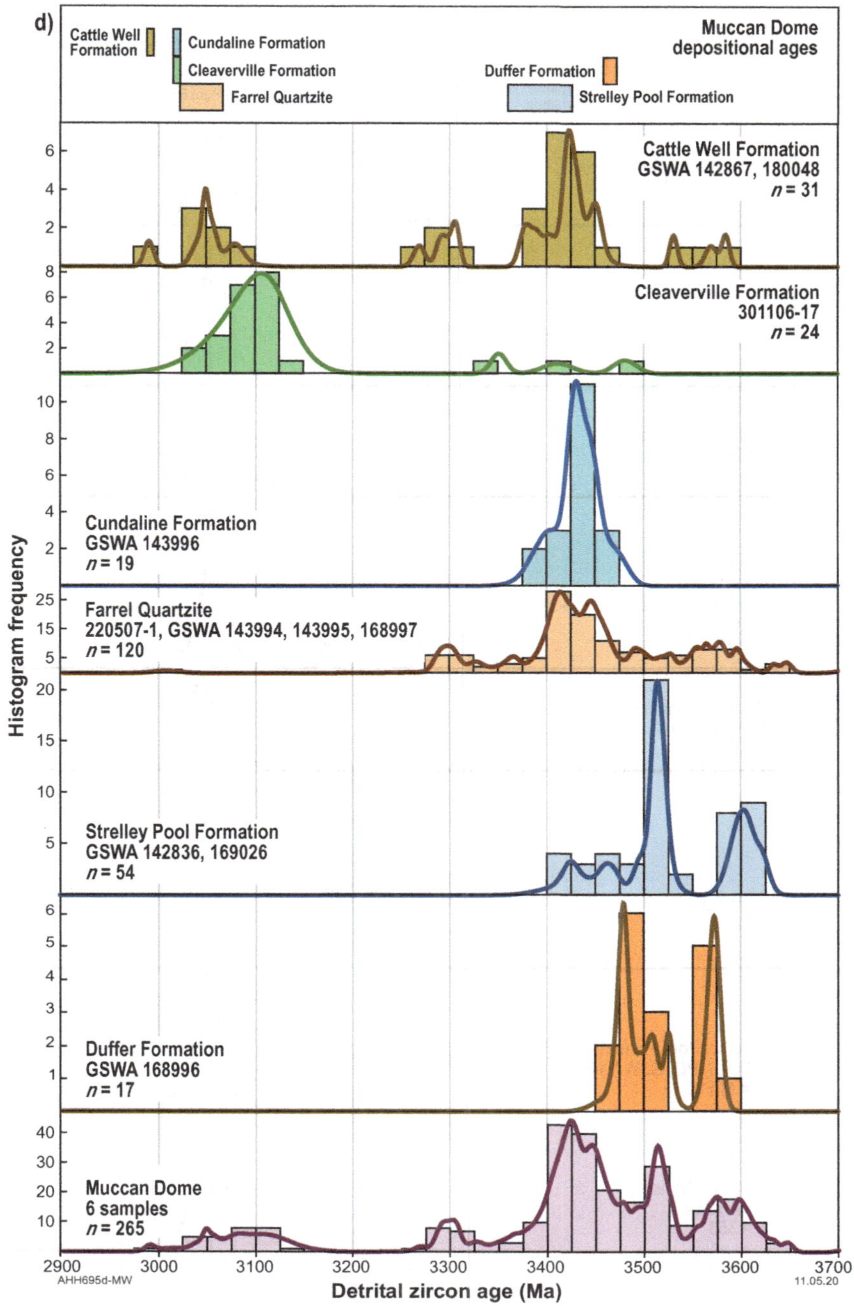

Fig. 2.2 (continued)

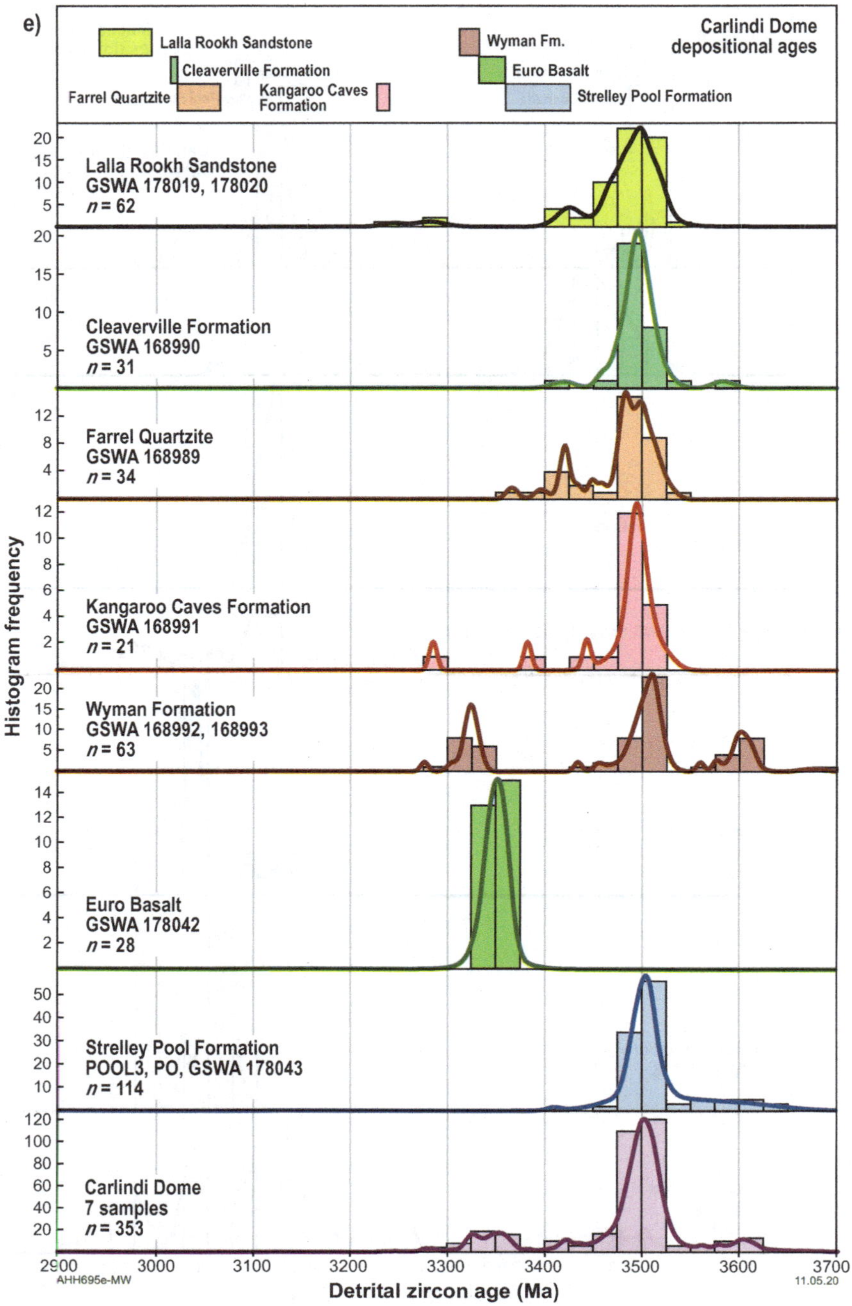

Fig. 2.2 (continued)

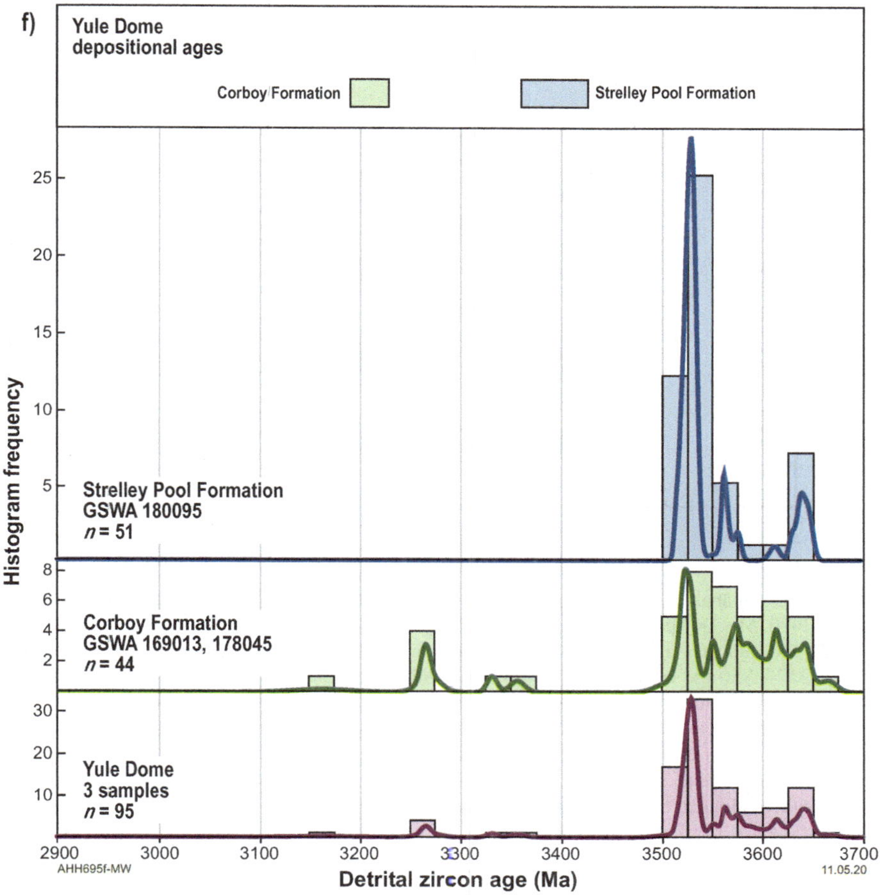

Fig. 2.2 (continued)

20 cm in diameter. This facies provides strong evidence for local derivation of the clastic material in the sandstone, and this is particularly significant because eight detrital zircons were dated between 3575 and 3566 Ma (Fig. 2.3). Other zircon age groups in the sample are 3524 ± 4 Ma, indicating erosion of the 3530–3490 Ma Counterunah Subgroup which outcrops to the east, and 3477 ± 4 Ma, consistent with erosion of either the 3484–3462 Ma Callina Supersuite, which forms part of adjacent granitic complexes (Muccan and Carlindi Domes), or older volcanic rocks of the 3474–3459 Ma Duffer Formation. Granitic rocks of the Muccan Dome contain inherited zircons dated at 3621 ± 3 Ma (Nelson 2006, GSWA 178031) and 3574 ± 3 Ma and 3538–3499 Ma (Nelson 1998, GSWA 142828), suggesting that this dome was a probable local source.

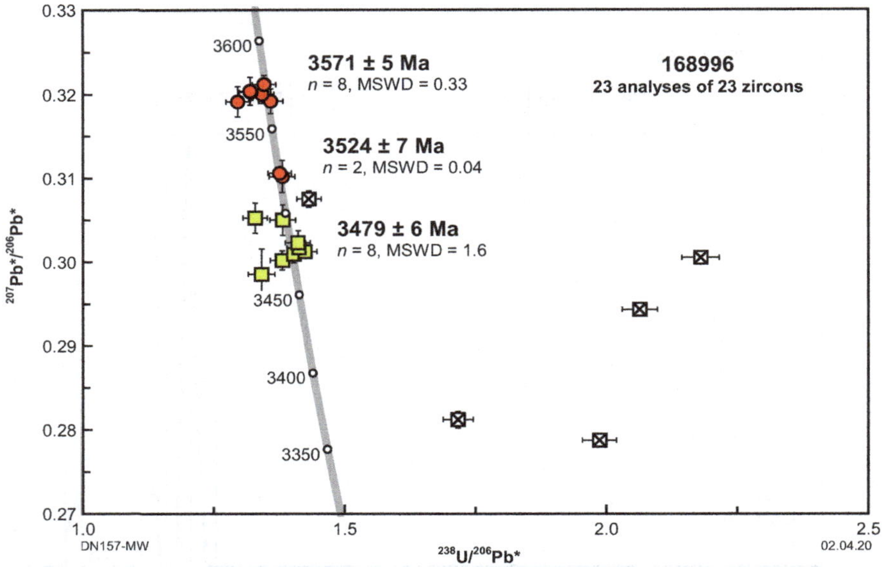

Fig. 2.3 U–Pb analytical data for sample GSWA 168996, an altered felsic volcaniclastic rock of the c. 3470 Ma Duffer Formation, Warralong greenstone belt. The zircon age component at c. 3571 Ma indicates a felsic source of this age in either the Muccan Dome or Carlindi Dome. The c. 3524 Ma age component is consistent with derivation from the Coonterunah Subgroup currently exposed in the Carlindi Dome, or from granitic rocks of similar age. Age components in this sample are recalculated from Nelson (2002). Yellow squares indicate magmatic (or detrital) zircons; red circles indicate inherited or detrital zircons; crossed squares indicate analyses >5% discordant. n, number of analyses; MSWD, mean square of weighted deviates (From Hickman 2021; with Geological Survey of Western Australia permission)

Other types of specific evidence include:

- Depositional setting: for example, sandstones deposited along growth faults, in grabens, adjacent to transpressional faults, or within intracratonic basins are evidently derived from relatively local erosion (Eriksson 1981; Buick and Barnes 1984; Krapež 1984; Wilhelmij and Dunlop 1984; Krapež and Barley 1987; DiMarco and Lowe 1989b; De Vries et al. 2006).
- Paleocurrent data: current direction indicators, combined with geological evidence of depositional setting of a clastic unit, can suggest local provenance. Examples include evidence of deposition on the outer slopes of Paleoarchean volcanoes in the North Pole and Mount Edgar Domes (DiMarco and Lowe 1989a; Olivier et al. 2012).

2.2.1.2.2 General Evidence on Provenance

The close match between detrital zircon age peaks and the timing of felsic magmatic events in the same terrane is compelling evidence of a provenance within that terrane because, by definition, each terrane has a geological history different from the histories of adjacent terranes. Figure 1.6 uses all published dates on intrusive and extrusive felsic rocks in the northern Pilbara Craton to illustrate a well-defined record of felsic magmatic events. When detrital zircon ages from Paleoarchean and Mesoarchean sandstones of the East Pilbara are plotted on frequency–age histograms, it is evident that detrital zircons have ages matching the ages of the magmatic events.

Another line of general evidence is the strong agreement between the previously interpreted depositional settings of formations and groups and the detrital zircon age profiles on Figs. 2.4 and 2.5. Cawood et al. (2012) pointed out that detrital zircon spectra reflect the tectonic setting of the basin in which they are deposited. They observed that convergent plate margins are characterized by a large proportion of zircon ages close to the depositional age of the sediment, whereas sediments in collisional, extensional, and intracratonic settings contain greater proportions with older ages that reflect the history of the underlying basement. For East Pilbara Mesoarchean intracratonic or passive margin basins, the zircon ages are very much older than sediment depositional ages (Fig. 2.5). By contrast, in the northwest Pilbara, where Mesoarchean basins were within the convergent margin setting of the Central Pilbara Tectonic Zone (Hickman 2016), detrital zircon ages are much closer to depositional ages. These close links between depositional tectonic setting and detrital zircon ages support proximal sources.

Based on the 1970s mapping of the East Pilbara (Hickman 1984), it was interpreted that sedimentary deposition during the evolution of the dome-and-keel structure of the EPT was related to the erosion of uplifted domes. That scenario implies that the zircon provenance for any particular formation would mainly have been within the dome hosting that formation (Hickman 2012; Wiemer et al. 2016). Exceptions would have occurred immediately after events of regional peneplanation when sedimentary deposition extended across several eroded domes; formations deposited in that environment include the Strelley Pool Formation and Leilira Formation which overlie regional unconformities (Buick et al. 1995, 2002; Van Kranendonk 2000; Van Kranendonk et al. 2007a, b; Hickman 2008, 2012).

The c. 3450 Ma Apex Basalt (Table 1.1), which overlies an older erosional interval represented by the Marble Bar Chert Member of the Duffer Formation, might also fall into this category. In the Salgash area, 15 km south of Marble Bar, a thin sandstone unit about 600 m above the base of the Apex Basalt provides abundant evidence of pre-Pilbara Supergroup continental crust. U–Pb zircon dating of this sandstone (WJ Collins, written communication, 2016) revealed detrital zircon age groups with ages of c. 3650 Ma and c. 3592 Ma, in addition to zircon ages consistent with erosion of underlying felsic volcanic rocks of the Duffer Formation (Fig. 2.6). The pre-3530 Ma zircons were evidently derived from the erosion of uplifted Eoarchean and early Paleoarchean crust.

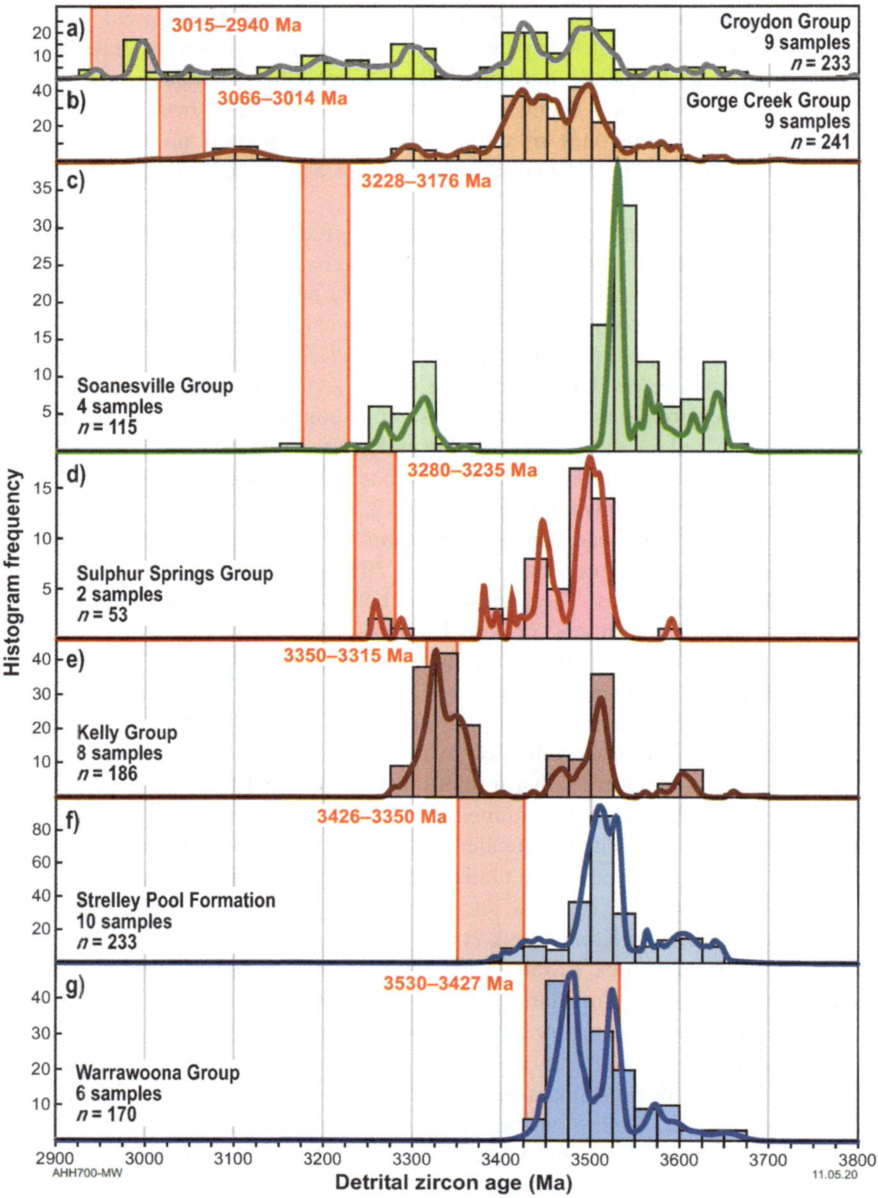

Fig. 2.4 Histograms showing the frequency of detrital zircon ages in sedimentary formations of the Croydon, Gorge Creek, Soanesville, and Sulphur Springs Groups. Gaps between depositional age and average detrital zircon age increase with decreasing depositional age, and the dominant source for all groups except the Soanesville Group was 3520–3400 Ma, indicating derivation of detritus from the Warrawoona Group and granites of the same age (From Hickman 2021; with Geological Survey of Western Australia permission)

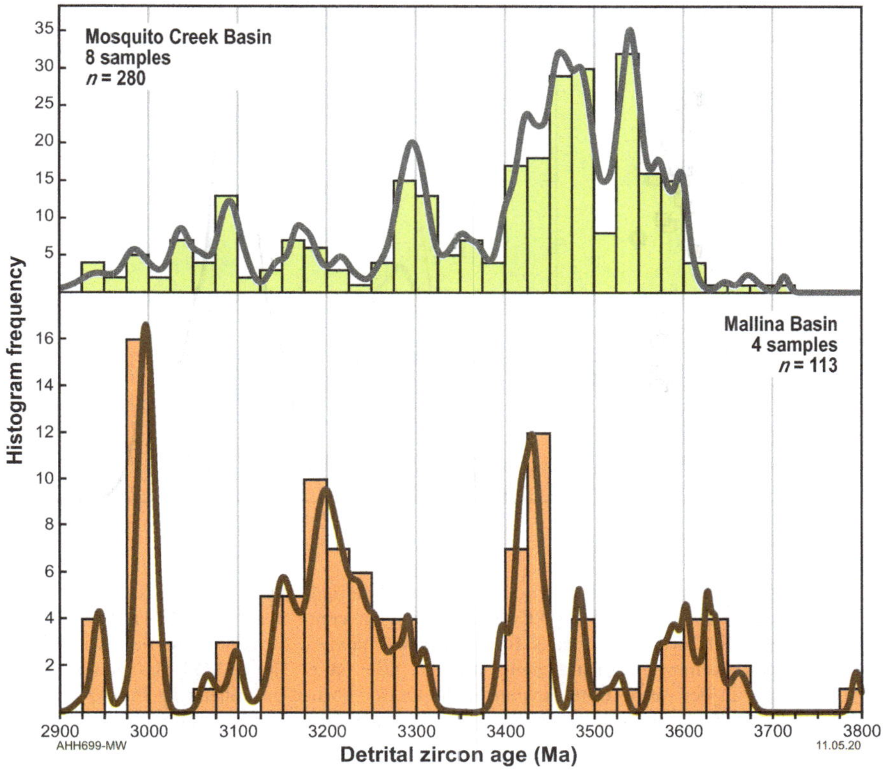

Fig. 2.5 Histogram showing the frequency of detrital zircon ages in the 3015–2930 Ma Mallina and Mosquito Creek Basins. Both these large Late Mesoarchean basins contain far more diverse zircon age components than are present in the Paleoarchean and rifting-related early Mesoarchean formations. The Mosquito Creek Basin contains the higher percentage of 3600–3400 Ma zircons (From Hickman 2021; with Geological Survey of Western Australia permission)

2.2.1.2.3 Statistical Analysis of Data

Prior to a recent study by the author (Hickman 2021), statistical processing of Pilbara detrital zircon data was limited to data from the Mesoarchean Mosquito Creek Basin (Bagas et al. 2004, 2008).

Figure 2.2a shows the relative frequency of detrital zircon ages (1072 grains) from different sandstone formations and domes of the East Pilbara. Data from sandstones of the Mesoarchean Mosquito Creek and Mallina Basins (Fig. 2.5) are excluded because these basins derived some sediment from outside the East Pilbara. Several important observations are made from Figure 2.2a:

1. The peak in detrital zircon ages between 3660 and 3560 Ma is consistent with xenocrystic zircon ages (Table 2.1) indicating extensive pre-3530 Ma crust.

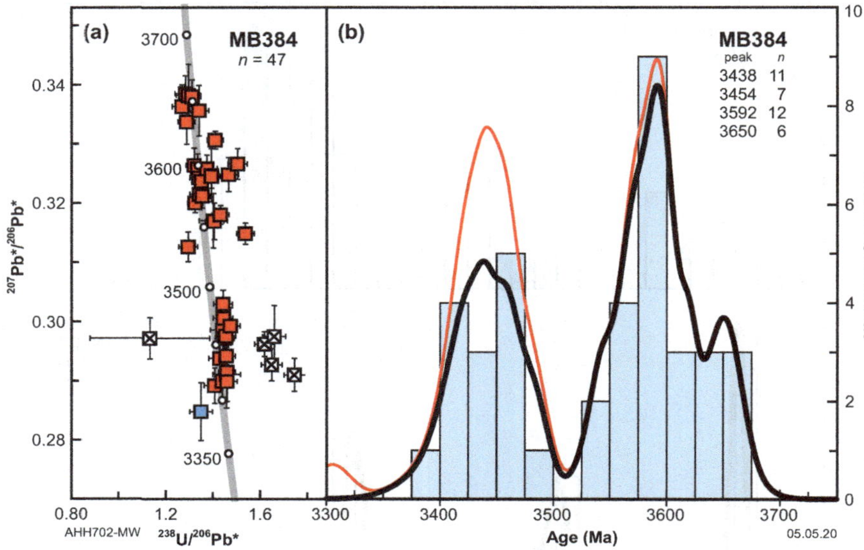

Fig. 2.6 Concordia plot for detrital zircons in a sandstone unit of the Apex Basalt, Warrawoona Group (data from sample MB384, WJ Collins). Zircon age components at c. 3650 and c. 3592 Ma are consistent with zircon age peaks from other sedimentary formations in the East Pilbara, supporting crust-forming magmatic events at about these times. The c. 3454 Ma age component is consistent with derivation from erosion of the Duffer Formation underlying the Apex Basalt (From Hickman 2021; with Geological Survey of Western Australia permission)

2. The major peak of detrital zircon ages between 3540 and 3400 Ma is partly explained by erosion of 3484–3416 Ma granitic rocks of the Callina and Tambina Supersuites and felsic volcanic rocks of the Duffer and Panorama Formations. However, the large number of zircons dated between 3530 and 3480 Ma indicates an additional older source of detritus. Until recently, only the c. 3515 Ma Coucal Formation of the Coonterunah Subgroup (lower Warrawoona Group) was known to contain felsic igneous rocks of this age, and there are very few outcrops of this formation in the EPT. The discovery of a 3530–3490 Ma granitic supersuite in the EPT (Petersson et al. 2020) explains the likely source of these zircons, and it suggests that it was exposed in the Paleoarchean and Mesoarchean.
3. The youngest of the three well-defined zircon age peaks coincides with the combined age range of the 3324–3290 Ma Emu Pool Supersuite and 3350–3315 Ma Kelly Group. Importantly, this peak is largely composed of zircon ages from volcaniclastic units in the Kelly Group itself (Fig 2.2c, e) so that most of the zircons represented in this peak were derived from reworking of igneous rocks of the same age.
4. There is a pronounced scarcity of detrital zircon ages between 3400 and 3360 Ma. This interval coincides with sedimentary deposition of the Strelley Pool

Formation and supports there being little or no igneous activity at this time (Hickman 2008, 2012).

5. The ages of the detrital zircons reveal that Mesoarchean felsic igneous rocks of the northwest Pilbara (e.g. granitic rocks of the Railway, Orpheus, Maitland River, or Sisters Supersuites, and felsic volcanic rocks of the Whundo and Whim Creek Groups) contributed no detrital zircon to the Mesoarchean sandstones of the East Pilbara. This supports the conclusion that detrital zircons in the East Pilbara sandstones were derived from proximal sources in the East Pilbara crust.

Figure 2.2b and f examine the detrital zircon ages according to dome and stratigraphic host formation. Two additional conclusions are reached from these plots:

1. The various domes present different 'profiles' of zircon ages. For example, in the Yule Dome (Fig. 2.2f), almost all detrital zircons are older than 3500 Ma, whereas in the Shaw, Mount Edgar, and Corunna Downs Domes (Fig. 2.2c), almost all formations except the Apex Basalt (Fig. 2.6) contain only zircons derived from the presently exposed Paleoarchean felsic igneous rocks. The Carlindi Dome (Fig. 2.2e) is dominated by 3520–3480 Ma detrital zircons irrespective of the host formation whereas sandstones of the Muccan Dome (Fig. 2.2d) contain a spread of detrital zircon ages. Sandstones of the North Pole Dome and Strelley area of the Soanesville greenstone belt (Fig. 2.2b) contain detrital zircons of Warrawoona Group age, including a peak age at c. 3520 Ma.
2. The depositional ages of the formations from which the detrital zircons were extracted reveal that formations deposited after 3420 Ma (i.e. younger than the Warrawoona Group) contain most of the oldest zircons. The likely explanation for this is that the first major doming in the EPT between 3440 and 3420 Ma uplifted and exposed old sections of the Pilbara crust. Evidence for the magnitude of vertical uplift during 3440–3420 Ma doming comes from metamorphic evidence (Délor et al. 1991; Hickman and Van Kranendonk 2004; Van Kranendonk et al. 2004a; François et al. 2014) and regional variations in the nature of the c. 3420 Ma unconformity at the base of the Strelley Pool Formation (Hickman 2008).

In summary, detrital zircon ages indicate that 3660–3540 Ma felsic crust was exposed to erosion during the Paleoarchean and Mesoarchean and that the exposures were relatively widespread across the northern Pilbara Craton.

2.2.2 Sm–Nd Isotope Data

Whole-rock Sm–Nd isotope data (i.e. $^{143}Nd/^{144}Nd$ in combination with Sm and Nd concentrations) are commonly used to indicate ages of crustal sources. The isotope ^{147}Sm decays radioactively to ^{143}Nd, and thus the $^{143}Nd/^{144}Nd$ ratio of a rock or mineral increases over geological time at a rate proportional to the Sm/Nd ratio in the rock. The Nd model ages quoted in this chapter are Sm–Nd two-stage depleted mantle model ages (T_{DM}^2) calculated assuming that the first fractionation event occurred during mantle melting and a second event occurred in the crust during partial melting. The method of two-stage model age calculation was described by Champion and Huston (2016). The model ages were calculated assuming a depleted mantle (see below) and using the linear depletion model which assumes linear depletion from $\varepsilon_{Nd} = 0$ at 4560 Ma to +10 today (Champion and Huston 2016). The resulting Nd model ages are geological useful because they suggest the time at which a sample was separated from its source in the mantle (McCulloch and Wasserburg 1978). However, crystallizing magmas in continental settings were rarely derived from a single mantle source. In a situation of multiple sources, Nd model ages indicate only the average times from which material in samples had been resident in the continental crust (Arndt and Goldstein 1987), and the ages of the oldest sources cannot be determined.

Except for six new Sm–Nd analyses (Gardiner et al. 2017, 2018) and Nd data on two samples from the Kurrana Terrane reported by Tyler et al. (1992), all the Nd model ages quoted in this chapter were calculated and tabulated by Smithies et al. (2007b) and Champion (2013). Another geologically useful source of information on the source of magmas is provided by epsilon Nd values (ε_{Nd}). The ε_{Nd} value indicates the relative deviation of the $^{143}Nd/^{144}Nd$ ratio from the chondritic ratio (0.512638) in parts per 10^4 (White 2013). Epsilon Nd values are useful in crustal evolution studies by providing a measure of the difference between a rock's crystallization age and the time at which its sources were extracted from the mantle. A negative ε_{Nd} value implies that, over time, the average Sm/Nd ratio of the rock or its precursors has been lower than the chondritic ratio, indicating light rare earth element (LREE) enrichment. Negative ε_{Nd} values are typical of continental crust in which the Nd system has been separated from the Sm-enriched mantle for many millions of years and indicate crustal reworking. Positive ε_{Nd} values indicate juvenile mantle-derived melts, as in MORB (mid-ocean ridge basalt) and magmatic arcs produced by subduction and melting of oceanic crust. Table 2.2 summarizes published Nd T_{DM}^2 model ages and ε_{Nd} values obtained from Paleoarchean and Mesoarchean igneous rocks of the East Pilbara. All Sm–Nd data from the northern Pilbara Craton prior to 2021 are listed by Hickman (2021).

Sm–Nd isotope data indicate extensive pre-3530 Ma crust in the EPT (Van Kranendonk et al. 2006, 2007a, b; Gardiner et al. 2017). The data in Table 2.2 suggest that the ages of the crustal sources for Paleoarchean igneous rocks were predominantly 3700–3550 Ma. However, evidence for considerably older crust in the Pilbara Craton is provided by Hf model ages obtained from 3700–3550 Ma zircons, which are effectively proxies for the older crust.

Table 2.2 Sm–Nd model ages ($T_{DM}2$) and ε_{Nd} values from igneous stratigraphic units of the East Pilbara Craton. (From Hickman 2021; with GSWA permission)

Stratigraphic unit, age (Ga)	Lithology	Samples	ε_{Nd}	T_{DM}^2 (Ga)	Data Source
Pilbara Supergroup					
Warrawoona Group					
Coonterunah Subgroup:					
Table Top Formation, 3.53–3.52	Basalt	2	1.40 to 0.89	3.63 to 3.59	Smithies et al. (2007b)
Coucal Formation, c. 3.52–3.50	Rhyolite	1	0.95	3.62	Smithies et al. (2007b)
"	Andesite	2	1.71 to 0.64	3.64 to 3.56	Smithies et al. (2007b)
"	Basalt	1	0.93	3.62	Smithies et al. (2007b)
Talga Talga Subgroup:					
North Star Basalt, 3.53–3.49	Basalt	5	1.71 to 0.27	3.66 to 3.55	Gruau et al. (1987)
"	Komatiitic basalt	1	0.53	3.64	Gruau et al. (1987)
"	Dacite	2	1.40 to 0.94	3.61 to 3.57	Hamilton et al. (1981)
"	Andesite	1	0.98	3.61	Hamilton et al. (1981)
"	Komatiitic basalt	1	1.34	3.58	Hamilton et al. (1981)
"	Komatiite	1	2.31	3.51	Hamilton et al. (1981)
"	Dacite	3	1.60 to 0.15	3.67 to 3.56	Jahn et al. (1981)
"	Andesite	1	0.35	3.65	Jahn et al. (1981)
Dresser Formation, c. 3.48	Altered basalt	8	−3.30	4.46 to 3.71	Tessalina et al. (2010)
Coongan Subgroup:					
Mount Ada Basalt, c. 3.47	Basalt	3	1.57 to 0.55	3.61 to 3.54	Gruau et al. (1987)
"	Komatiitic basalt	2	1.95 to −0.41		Gruau et al. (1987)

(continued)

Table 2.2 (continued)

Stratigraphic unit, age (Ga)	Lithology	Samples	ε_{Nd}	T_{DM}^{2} (Ga)	Data Source
				3.69 to 3.51	
"	Basalt	1	0.77	3.60	Smithies et al. (2007b)
Duffer Formation, 3.47–3.46	Dacite	3	1.73 to 0.76	3.60 to 3.52	Smithies et al. (2007b)
"	Andesite	1	1.05	3.58	Smithies et al. (2007b)
"	Dacite	1	0.45	3.62	McCulloch (1987)
Salgash Subgroup:					
Apex Basalt, c. 3.45	Basalt	1	0.86	3.58	Smithies et al. (2007b)
Panorama Formation, 3.45–3.43	Rhyolite	3	1.21 to 0.72	3.57 to 3.53	Smithies et al. (2007b)
Kelly Group					
Euro Basalt, 3.35–3.33	Basalt	5	1.07 to 0.48	3.51 to 3.47	Arndt et al. (2001)
Sulphur Springs Group					
Kangaroo Caves Formation, c. 3.24	Rhyolite	1	−0.48	3.50	Brauhart et al. (2000)
MESOARCHEAN FORMATIONS					
Soanesville Group					
Honeyeater Basalt, 3.18–3.17	Basalt	5	2.56 to −1.52	3.54 to 3.24	Smithies et al. (2007b)
Empress Formation, c. 3.17	Basalt	1	1.32	3.33	Smithies et al. (2007b)
DE GREY SUPERGROUP					
Coonieena Basalt, c. 3.00	Basalt	2	−2.04 to −2.13	3.42 to 3.41	Smithies et al. (2007b)
GRANITIC ROCKS					

(continued)

Table 2.2 (continued)

Stratigraphic unit, age (Ga)	Lithology	Samples	ε_{Nd}	T_{DM}^2 (Ga)	Data Source
Pre-3.53 Ga gneiss, < 3.66					
Warrawagine Dome	Tonalite gneiss	2	1.49 to 1.24	3.65 to 3.63	Smithies et al. (2007b)
Callina Supersuite, 3.48–3.46					
Shaw Dome	Tonalite–granodiorite	6	1.69 to −1.49	3.77 to 3.53	Bickle et al. (1993)
"	Granite	1	−0.53	3.69	Bickle et al. (1993)
"	Granite	1	0.70	3.62	McCulloch (1987)
Mount Edgar Dome	Quartz diorite	1	0.75	3.59	Champion (2013)
Carlindi Dome	Granodiorite	1	1.25	3.57	Smithies et al. (2007b)
Muccan Dome	Granodiorite	1	−0.26	3.67	Gardiner et al. (2017)
Tambina Supersuite, 3.45–3.42					
Yule Dome	Monzogranite	1	−1.34	3.71	Smithies et al. (2007b)
Mount Edgar Dome	Tonalite–granodiorite	2	0.40 to −0.03	3.63 to 3.59	Gardiner et al. (2017)
Corunna Downs Dome	Tonalite	2	2.11 to 1.84	3.47 to 3.45	Smithies et al. (2007b)
Carlindi Dome	Monzogranite	1	−0.34	3.54	Smithies et al. (2007b)
Emu Pool Supersuite, 3.32–3.28					
Corunna Downs Dome	Monzogranite	3	1.49 to 0.25	3.50 to 3.42	Smithies et al. (2007b)
Mount Edgar Dome	Granodiorite	1	−1.44	3.63	Gardiner et al. (2017)
Cleland Supersuite, 3.27–3.22					

(continued)

Table 2.2 (continued)

Stratigraphic unit, age (Ga)	Lithology	Samples	ε_{Nd}	T_{DM}^2 (Ga)	Data Source
Strelley Monzogranite	Diorite–monzogranite	3	−0.41 to −1.00	3.54 to 3.49	Brauhart et al. (2000)
Warrawagine Dome	Granodiorite	2	0.12 to −0.28	3.49 to 3.46	Smithies et al. (2007b)
Muccan Dome	Monzogranite	2	−0.81 to −1.27	3.56 to 3.53	Smithies et al. (2007b)
Mount Billroth Supersuite, 3.20–3.16					
Yule Dome	Tonalite	2	1.13 to −0.33	3.42 to 3.31	Smithies et al. (2007b)
Elizabeth Hill Supersuite, c. 3.07					
Yule Dome	Tonalite	2	−4.09 to −4.73	3.56 to 3.52	Smithies et al. (2007b)
Sisters Supersuite, 2.95–2.92					
Shaw Dome	Monzogranite	7	−3.60 to −3.19	3.70 to 3.42	Smithies et al. (2007b)
Yule Dome	Monzogranite	7	−3.27 to 0.20	3.45 to 3.18	Smithies et al. (2007b)
Carlindi Dome	Monzogranite	3	3.19 to 3.10	3.12 to 2.97	Smithies et al. (2007b)
Split Rock Supersuite, 2.85–2.83					
Shaw Dome	Monzogranite	6	−8.03 to −4.13	3.74 to 3.44	Smithies et al. (2007b)
Mount Edgar Dome	Monzogranite	1	−7.08	3.65	Gardiner et al. (2018)
Carlindi Dome	Monzogranite	4	−2.02 to 1.64	3.28 to 3.01	Smithies et al. (2007b)
Yule Dome	Monzogranite	1	−0.81	3.19	Smithies et al. (2007b)

2.2.2.1 Depleted or Chondritic Paleoarchean Mantle?

Interpretation of the ages of crustal sources depends on the models used, in particular in relation to the assumed composition of the Paleoarchean mantle (i.e. depleted or chondritic). Most workers have assumed that from the Hadean onwards, there was progressive depletion of incompatible elements in the mantle due to repeated crustal extractions over time (McCulloch and Wasserburg 1978; DePaolo 1988; Hofmann 1997; Griffin et al. 2000). If Archean mantle depletion occurred at a rate similar to Phanerozoic depletion, there would have been a significantly depleted mantle by c. 3500 Ma (Bennett et al. 1993; McCulloch and Bennett 1993; Amelin et al. 1999, 2000). An alternative interpretation adopted by some workers is that Earth's mantle was still essentially chondritic c. 3500 Ma (Vervoort and Blichert-Toft 1999; Vervoort 2011; Vervoort and Kemp 2016; Fisher and Vervoort 2018). However, strong evidence for a Paleoarchean depleted mantle is provided by the global database of Roberts and Spencer (2015) which contains hundreds of >3500 Ma zircons with very positive ε_{Hf} values.

Assuming a depleted mantle during the evolution of the EPT, the Lu–Hf data from zircons younger than 3530 Ma indicate average source ages (i.e. the age of the underlying crust) between 3700 and 3600 Ma. In contrast, the average Hf $T_{DM}{}^2$ age of 126 zircons that crystallized before 3530 Ma is 3976 Ma (data in Hickman 2021), implying derivation of the pre-3530 Ma crust from early Eoarchean sources.

2.2.2.2 Sm–Nd Isotope Data from the East Pilbara

The earliest Sm–Nd studies on rocks of the East Pilbara were conducted in the 1980s (Hamilton et al. 1981; Jahn et al. 1981; Collerson and McCulloch 1983; Gruau et al. 1987; McCulloch 1987; Bickle et al. 1989, 1993). Hamilton et al. (1980) plotted data from six samples of the Talga Talga Subgroup on a Sm–Nd evolution diagram. This indicated a depositional age of 3556 ± 32 Ma, which was consistent with then available U–Pb zircon data that the depositional age of the overlying Duffer Formation was c. 3450 Ma (Pidgeon 1978). The 3556 ± 32 Ma age is within error of the currently understood maximum age of the Talga Talga Subgroup at c. 3530 Ma (Hickman 2021). Jahn et al. (1981) used the Sm–Nd isochron method to date the North Star Basalt at 3556 ± 542 Ma; the large age uncertainty reflected a limited range of $^{143}Nd/^{144}Nd$ and $^{147}Sm/^{144}Nd$ in the four samples analysed. Collerson and McCulloch (1983) reported Nd T_{CHUR} model ages (assuming the Archean mantle had a chondritic composition) from the Shaw and Mount Edgar Domes that were < 3500 Ma and similar to previously reported U–Pb zircon ages. Considerably older Nd model ages would have been obtained using a depleted Archean mantle model. Gruau et al. (1987) combined basaltic samples from the North Star Basalt and Mount Ada Basalt and used the Sm–Nd isochron method to arrive at a 3712 ± 98 Ma depositional age for the lower part of the Warrawoona Group. They also calculated the isotopic data from the two basalt formations separately and obtained ages of

3737 ± 117 Ma for the North Star Basalt and 3715 ± 170 Ma for the Mount Ada Basalt. Noting that these Sm–Nd ages were considerably older than current estimates for the depositional age of the North Star Basalt, Gruau et al. (1987) interpreted that, rather than indicating depositional ages, the isochrons represented mixing lines and the c. 3720 Ma ages dated the time of extraction of sources from the mantle. McCulloch (1987) analysed nine felsic igneous rocks from the East Pilbara as part of a larger study. The $^{143}Nd/^{144}$ Nd and $^{147}Sm/^{144}Nd$ results from four of these samples were interpreted by Champion (2013) to indicate Nd $T_{DM}{}^2$ model ages ranging from 3700 to 3440 Ma and ε_{Nd} values of +0.45 to −6.67. Bickle et al. (1989) reported Sm–Nd data for eight samples from the Shaw Granitic Complex. As recalculated by Champion (2013), the oldest Nd $T_{DM}{}^2$ model ages, 3740 Ma and 3590 Ma, were obtained on two monzogranites of the 2851–2831 Ma Split Rock Supersuite. The ε_{Nd} values of these rocks were − 8.01 and − 6.1 suggesting partial melting of much older crust. Data from the Garden Creek Monzogranite (not dated by the U–Pb zircon method), which is possibly an intrusion of the Split Rock Supersuite (associated tin mineralization), was recalculated by Champion (2013) to give a Nd $T_{DM}{}^2$ model age of 3590 Ma and ε_{Nd} value of −5.23 (assuming a crystallization age of 2925 Ma). Data from five samples from the 2954–2919 Ma Sisters Supersuite were recalculated by Champion (2013) to give Nd $T_{DM}{}^2$ model ages from 3530 to 3420 Ma and ε_{Nd} values of −4.47 to–3.31. Bickle et al. (1989) interpreted the negative ε_{Nd} values as evidence for derivation of melts from older crust. Bickle et al. (1993) provided Nd model ages (T_{CHUR} and T_{DM}) and ε_{Nd} values for seven samples from Paleoarchean granitic rocks in the Shaw Granitic Complex. Assuming a crystallization age of c. 3470 Ma (Callina Supersuite) for all the samples, Champion (2013) recalculated the Nd $T_{DM}{}^2$ model ages for these older granites to range between 3770 and 3530 Ma and ε_{Nd} values to be between +1.65 and − 1.53. Tyler et al. (1992) used the Sm–Nd method to investigate granitic rocks of the Sylvania Inlier (Fig. 1.2), Kurrana Terrane, and Shaw Granitic Complex. Four of the five samples analysed from the Kurrana Terrane, and Shaw Granitic Complex were from intrusions of the Split Rock Supersuite, and Champion (2013) recalculated the Nd $T_{DM}{}^2$ model ages as between 3490 and 3410 Ma, with ε_{Nd} values of −4.77 to −3.75. Champion (2013) and Champion and Huston (2016) used Nd data from the EPT to suggest that the average crustal source ages of the Paleoarchean rocks were c. 3600 Ma, with maximum average source ages for individual samples being up to c. 3800 Ma. Gardiner et al. (2017, 2018) reached the same conclusion as Champion and Huston (2016) regarding EPT Nd model ages and used new Hf isotope data from the Mount Edgar Dome to interpret average Hf $T_{DM}2$ model ages of 3700–3600 Ma.

Tessalina et al. (2010) analysed metabasalt from the c. 3481 Ma Dresser Formation to derive a Sm–Nd isochron that indicated a depositional age of 3490 ± 100 Ma. Their calculated ε_{Nd} value of −3.3 ± 1.0 was interpreted to indicate a much older source. Based on trace element and isotopic modelling, they interpreted this source to have differentiated from the mantle at >4300 Ma. Hasenstab et al. (2021) used Hf–Nd–Ce isotope data to interpret that mantle-crust differentiation in the Pilbara commenced as early as c. 4200 Ma.

2.2.2.3 Model Ages

Figure 2.7 examines the spatial distribution of Nd T_{DM}^2 model ages in the northern Pilbara Craton. As observed by previous workers (Van Kranendonk et al. 2006, 2007a, b; Champion 2013; Champion and Huston 2016), Nd model ages from samples collected in the Paleoarchean East Pilbara Terrane (EPT) are almost invariably older than Nd model ages from Mesoarchean rocks of the northwest Pilbara. Eoarchean ages (>3600 Ma, Fig. 2.7a) are recorded in the Warrawagine, Muccan, Mount Edgar, Shaw, Carlindi, and Yule Domes (Fig. 1.7). By contrast, model ages younger than 3310 Ma (Fig. 2.7c) are concentrated in the Mallina Basin and Sholl Terrane. Significantly, the differences between Nd T_{DM}^2 model ages and rock crystallization ages are, in most instances, far less in the northwest Pilbara than in the EPT. This is consistent with most Mesoarchean units in the northwest Pilbara having formed as juvenile crust or from juvenile sources, whereas most Paleoarchean units in the EPT were derived by recycling of significantly older crust (Bickle et al. 1993; Collins 1993; Green et al. 2000; Van Kranendonk et al. 2007a, b; Smithies et al. 2009; Champion 2013; Champion and Huston 2016; Gardiner et al. 2017, 2018). Nd model ages in the range 3530–3220 Ma are recorded northwest of the Mallina Basin, mostly in the Paleoarchean Karratha Terrane (Fig. 2.7b). The current lack of >3530 Ma Nd T_{DM}^2 model ages from the Karratha Terrane is attributed to the fact that the few rocks that have been analysed from that terrane are late Paleoarchean (Cleland Supersuite and Roebourne Group). Even in the EPT, rocks <3300 Ma (as in the Karratha Terrane) do not have Nd T_{DM}^2 model ages >3560 Ma (Table 2.2). This is possibly due to Sm–Nd analyses of rocks <3300 Ma giving average mantle extraction ages that included Paleoarchean EPT sources.

Nd T_{DM}^2 model ages are presented differently in Fig. 2.8 and Table 2.2. Figure 2.8a shows that Nd model ages do not significantly decrease with decreasing crystallization age until about 3300 Ma. This is interpreted to indicate no significant change in the ages of crustal sources with time (Eoarchean to early Paleoarchean). Moreover, except for the Carlindi Dome (CA in Fig. 2.8) and the northern Yule Dome (NY), similarly old sources are indicated for East Pilbara rocks that crystallized between 3000 and 2830 Ma.

At one stage in the research, the relatively young Nd model ages from the Carlindi and Yule Domes raised a question about the northwest extent of the EPT (Van Kranendonk et al. 2007b; Smithies et al. 2007c; Champion 2013; Champion and Huston 2016). Smithies et al. (2007c) suggested that these young ages indicate a western limit of the EPT extending northeast–southwest through the Carlindi and Yule Domes. However, this is not the situation because there are c. 3250 Ma granites, felsic volcanics, and gabbro along the Tabba Tabba Shear Zone (TTSZ) (Beintema et al. 2001, 2003; Beintema 2003; GSWA 160258, Wingate et al. 2010) establishing that the EPT extends west to this major crustal structure. These c. 3250 Ma units are assigned to the Cleland Supersuite and Sulphur Springs Group. Additionally, inherited zircons in the c. 3250 granites within the TTSZ

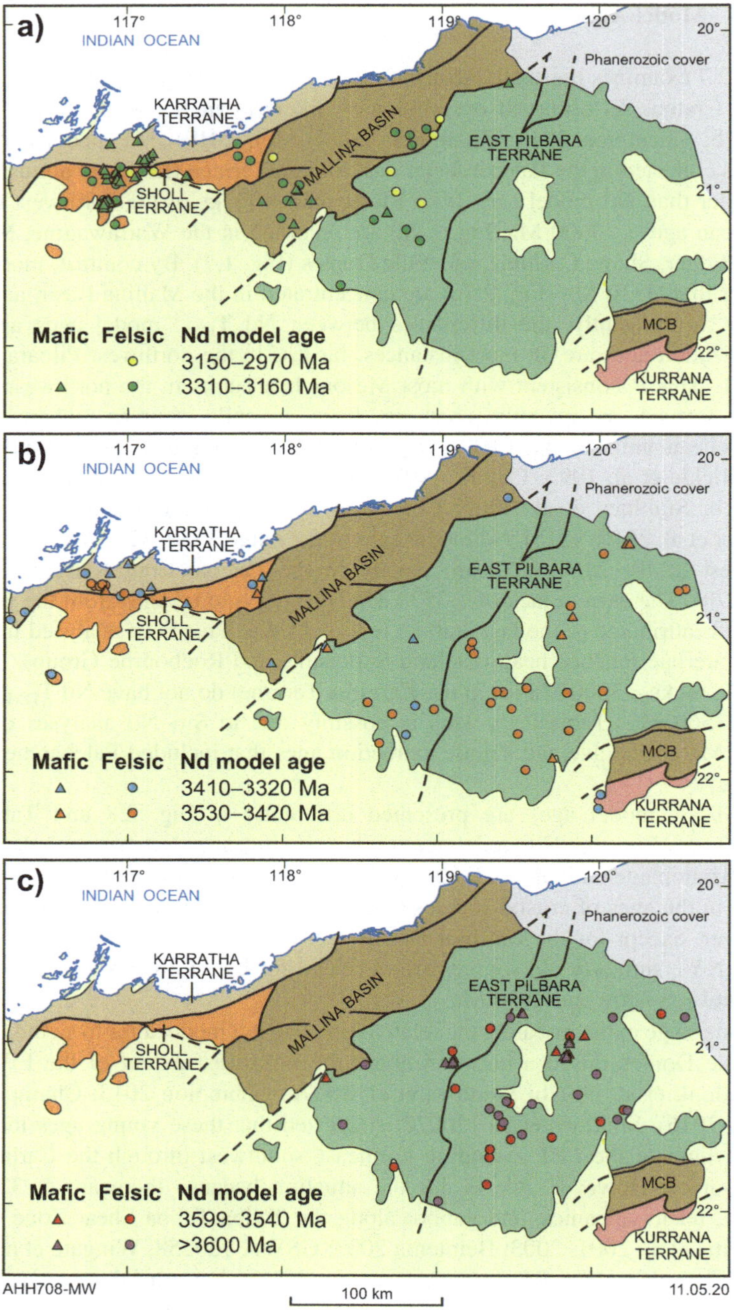

Fig. 2.7 Nd two-stage model ages from the Northern Pilbara Craton, distinguishing data from mafic and felsic igneous rocks. Nd model ages older than the maximum depositional age of the Pilbara Supergroup (3530 Ma) have been obtained only from the East Pilbara Terrane, (**a**) whereas model ages <3310 Ma are mainly confined to the northwest area of the Pilbara Craton (**c**).

have ages consistent with derivation from much older Paleoarchean sources (Nelson 2001, GSWA 160745; Beintema 2003, samples KB 263, 312, 770, 779). Therefore, the northwest boundary of the EPT, as previously defined by geological mapping and stratigraphy (Van Kranendonk et al. 2002), coincides with the TTSZ (Hickman 2016). The explanation for young Nd model ages within the Carlindi and Yule Domes is that late Mesoarchean granites were intruded into these Paleoarchean domes during convergence and collision between the EPT and the Mallina Basin. Relatively juvenile magma was derived from melting of <3200 Ma Mallina Basin crust subducted beneath the EPT, resulting in the late Mesoarchean granites of the Carlindi and Yule Domes giving relatively young Nd model ages.

Table 2.2 summarizes the ranges of Nd $T_{DM}{}^2$ model ages and ε_{Nd} values (see below) of individual stratigraphic units in the East Pilbara. It is notable that extrusive and intrusive units older than c. 3200 Ma have consistently old Nd model ages. For Mesoarchean units, Nd model ages east of the Lalla Rookh–Western Shaw Structural Corridor (LWSC) range between 3700 and 3400 Ma, whereas Nd model ages in the Carlindi and northern Yule Domes are generally younger than c. 3300 Ma. Limited data from the Yule Dome indicate that the southern part of this dome, which includes tonalites of the Elizabeth Hill and Mount Billroth Supersuites, had crustal sources similar to those in the EPT east of the LWSC.

2.2.2.4 Epsilon Nd Values

Sm–Nd isotope data from the northern Pilbara Craton reveal substantial differences in ε_{Nd} values between the East Pilbara and most units of the northwest Pilbara (Smith et al. 1998; Sun and Hickman 1998; Arndt et al. 2001; Smithies et al. 2004, 2007c; Van Kranendonk et al. 2007a, b; Champion 2013; Champion and Huston 2016; Gardiner et al. 2017, 2018). In the northwest Pilbara, most stratigraphic units within the Regal and Sholl Terranes have positive ε_{Nd} values up to +3.5 indicating juvenile crust (Smithies et al. 2007b: Champion 2013; Champion and Huston 2016; Hickman 2016). By contrast, many stratigraphic units of the East Pilbara have moderately to strongly negative ε_{Nd} values consistent with crustal reworking (Champion 2013; Champion and Huston 2016; Gardiner et al. 2017, 2018).

In the East Pilbara, two significant Nd trends are revealed by comparing rocks that vary in crystallization age from 3570 Ma to c. 3220 Ma (Fig. 2.10). Firstly, Nd model ages do not decrease with decreasing rock crystallization age (Fig. 2.10a), indicating no significant change in the ages of crustal sources (Eoarchean to early Paleoarchean) with time. Only in the Carlindi Dome, and to a lesser degree the Yule

Fig. 2.7 (continued) Paleoarchean Nd model ages in the Northwest Pilbara (**b**) were obtained from units in the Karratha Terrane and tectonic units overlying that terrane. These data indicate that, apart from the Karratha Terrane, igneous formations and intrusions in the Northwest Pilbara Craton were derived from juvenile Mesoarchean crust (From Hickman 2021; with Geological Survey of Western Australia permission)

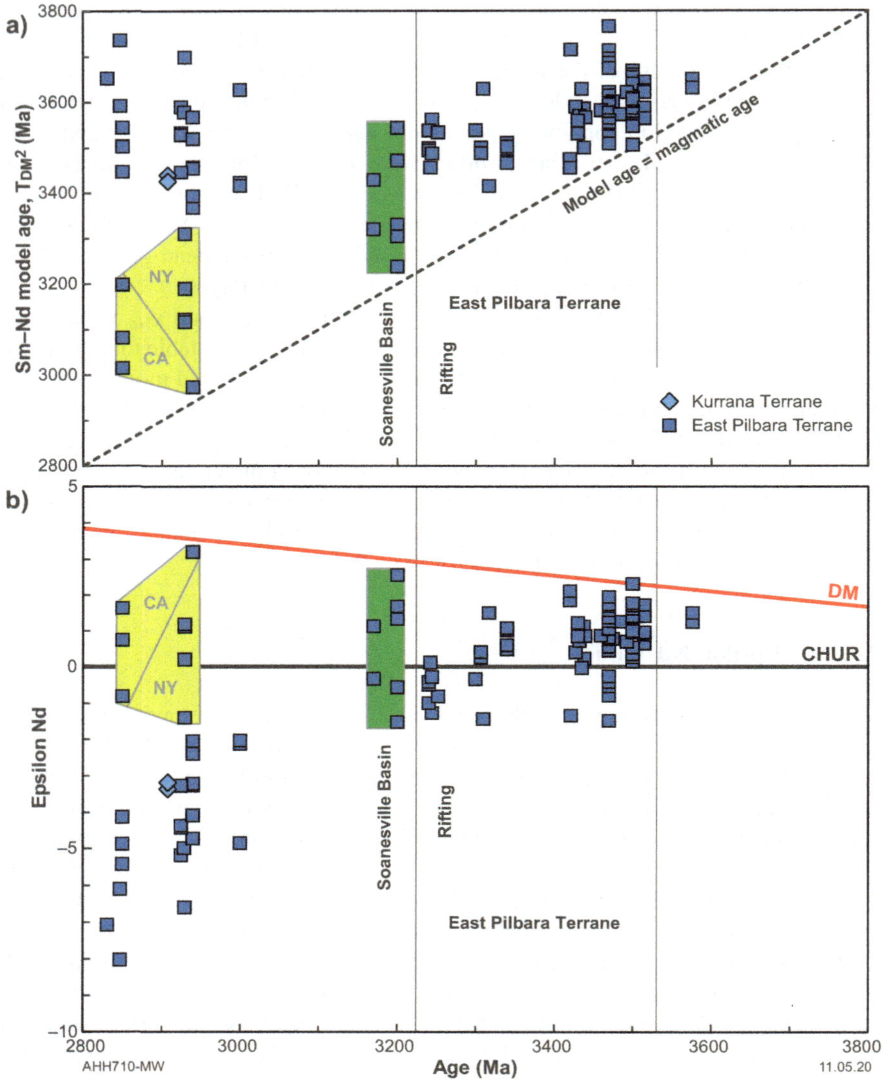

Fig. 2.8 Two-stage depleted mantle Nd model ages ($T_{DM}{}^2$) (**a**) and ε_{Nd} (**b**) versus magmatic ages for the Northern Pilbara Craton. (**a**) Between 3530 and 3220 Ma, the East Pilbara Terrane shows only gradually decreasing model ages indicating reworking of 3700–3500 Ma crust with only minor additions of juvenile material. At c. 3200 Ma, some rocks of the Soanesville Basin (green shaded area) show derivation from juvenile crust. Mesoarchean granitic rocks that intruded the East Pilbara Terrane between 2950 and 2830 Ma show a spread of Nd model ages that is resolved into samples from the Carlindi (CA) and northern Yule (NY) Domes (yellow shaded area) with juvenile sources and those from farther east in the terrane still showing evidence of derivation from Eoarchean to early Paleoarchean sources; (**b**) Apart from samples from the Soanesville Basin and the Yule and Carlindi Domes, rocks from the East Pilbara Terrane show steadily decreasing ε_{Nd} with decreasing magmatic age. This indicates ongoing reworking of Eoarchean to early Paleoarchean sources with time, except in western areas of the East Pilbara Terrane where granitic magmas were derived from

Dome, is there evidence of substantial addition of juvenile magma. Secondly, there is a strong trend for increasingly negative ε_{Nd} values in younger rocks (Fig. 2.10b), consistent with repeated derivation of melts from similar old sources. There are few ε_{Nd} values from the EPT to suggest sources much older than 3700 Ma, although because Nd model ages indicate the average age of sources, this does not preclude contributions from sources >3700 Ma. The maximum age of crustal extraction for any particular rock is likely to be greater than the Nd $T_{DM}{}^{2}$ model age.

The widespread distribution of samples providing Nd model ages >3530 Ma (Fig. 2.7a) indicates that pre-3530 Ma crust was present in all parts of the EPT during deposition of the Pilbara Supergroup and intrusion of the Paleoarchean granitic supersuites. The predominant age range of this older crust was between 3680 and 3540 Ma (Table 2.2). Pilbara Sm–Nd isotopic data indicate widespread crustal recycling until c. 3220 Ma.

2.2.3 Lu–Hf Isotopes in Zircon

The lutetium–hafnium (Lu–Hf) system in zircon can provide important isotopic evidence on crustal and mantle evolution. As for the Sm–Nd system, isotopic values in the Lu–Hf system indicate mantle extraction ages. The Lu–Hf system in zircon has an advantage over the whole-rock Sm–Nd systematics in that it is less prone to disturbance by metamorphism and weathering. Hf isotope ratios, between ^{176}Hf (produced by decay of ^{176}Lu) and ^{177}Hf, are expressed using the ε-notation that was first developed for interpretation of Sm–Nd data. Zircon is particularly useful for measuring the initial ^{176}Hf/^{177}Hf ratio inherited from the melt in which it crystallized because Hf is greatly concentrated relative to Lu; this minimizes post-crystallization changes in ^{176}Hf/^{177}Hf by decay of ^{176}Lu. High values of ^{176}Hf/^{177}Hf, and therefore also high (positive) ε_{Hf} values, indicate a juvenile or mantle-derived origin for a magma, whereas low ε_{Hf} values imply the reworking of old crustal material. The crystallization of zircon preserves the ^{176}Hf/^{177}Hf ratio of the parent magma.

2.2.3.1 Previous Investigations

As part of a wider study, Amelin et al. (2000) provided new U–Pb and Lu–Hf data on zircons from three previously dated samples from the East Pilbara (Thorpe et al. 1992): sample 94,770, a c. 3448 Ma porphyritic felsic volcanic rock from the Panorama Formation at West Bamboo, Mount Edgar Dome; sample 103283, a c. 3441 Ma porphyritic subvolcanic dacite of the Tambina Supersuite at Miralga

Fig. 2.8 (continued) melting of subducted juvenile crust of the Mallina Basin (From Hickman 2021; with Geological Survey of Western Australia permission)

Creek, North Pole Dome; and sample 94754, a c. 3320 Ma rhyolite from the Wyman Formation at Emu Creek, Corunna Downs Dome. Eight zircon grains were analysed from these three Pilbara samples, with positive ε_{Hf} values ranging between +4.56 (sample 94770) and + 1.75 (sample 94754). These early results suggested derivation of juvenile magma from a depleted mantle at c. 3448 Ma changing to derivation from a mantle of near-chondritic composition at c. 3320 Ma. Data in Hickman (2021) suggest that the strongly positive ε_{Hf} values that were obtained from zircons in the Panorama Formation and Tambina Supersuite are atypical of most of the EPT, although they are considered significant in providing evidence of a depleted Pilbara mantle at c. 3450 Ma.

Kemp et al. (2011) briefly summarized unpublished Nd and Hf isotopic data from a large number of Paleoarchean to Neoarchean Pilbara rocks that varied in composition from komatiite to rhyolite and granite and included many samples from the northwest Pilbara. They interpreted a strong temporal trend to increasingly negative ε_{Hf} values in zircons from felsic igneous rocks after 3200 Ma and interpreted this to indicate Mesoarchean reworking and recycling of >3500 Ma crust. They reported approximately chondritic initial $^{176}Hf/^{177}Hf$ in most pre-3500 Ma zircons and, assuming an approximately chondritic mantle composition, interpreted this as evidence against reworking of >3800 Ma crust. Kemp et al. (2014) used a > 3500 Ma chondritic mantle model to argue that crustal evolution in the Pilbara Craton commenced at c. 3650 Ma and that their data provided no evidence for significantly older crust.

Guitreau et al. (2012) included new whole-rock and zircon Lu–Hf isotopic data from the East Pilbara as part of a global study of Archean TTGs. They concluded that, on a global scale, all continental crust was initially derived from unfractionated material from the deep mantle via mantle plumes and that this left a depleted mantle residue in the upper mantle. The large dataset (over 12,000 samples) supported previous findings that suprachondritic ε_{Hf} values were present in some Archean TTGs, thus arguing for a depleted upper mantle in the Paleoarchean. They interpreted the deep mantle to have retained its primitive relative element abundances over most of Earth history because little continental material has been recycled to great depth. Nebel et al. (2014) obtained whole-rock Hf isotope data from 21 East Pilbara komatiite samples ranging in age from c. 3515 to c. 3200 Ma; 8 komatiite samples from the northwest Pilbara were also analysed. Several strongly positive ε_{Hf} values (above +6.0) and initial $^{176}Hf/^{177}Hf$ above the age-corrected depleted mantle were interpreted to provide evidence of mantle depletion from the Hadean onwards. Negative ε_{Hf} values were partly attributed to crustal contamination of the hot komatiitic magmas. In contrast to the interpretation of Nebel et al. (2014), Kemp et al. (2014) interpreted the Hf–Nd systematics to indicate that the oldest Pilbara komatiites and basalts separated from mildly superchondritic mantle, not from mantle that had been significantly depleted by earlier crust extractions. They argued that the isotopic evidence did not support the view that 3520–3120 Ma Pilbara granites were formed by melting of significantly older crust: instead, they proposed that the granites were sourced from precursors with only short crustal residence times and were juvenile continental additions. Nebel-Jacobsen et al.

(2010) presented whole-rock Hf isotope data from 3450–2760 Ma black shales of the Pilbara (Archean Biosphere Drilling Project cores 2, 3, 5, and 6). Results of most relevance to the East Pilbara were obtained from shale in ABDP 2 (Apex Basalt, not the Duffer Formation as stated in the paper) and the Mosquito Creek Formation (ABDP 5). Strongly positive ε_{Hf} values for shales of the Apex Basalt, plotting above the assumed depleted mantle array at c. 3450 Ma, were interpreted to indicate deposition during weathering of rocks from strongly depleted sources. Nebel-Jacobsen et al. (2010) commented that only komatiites from the Pilbara Craton and Barberton Greenstone Belt have comparable Hf isotope compositions, which suggests that detritus in the shales was derived from erosion and weathering of exposed juvenile komatiites rather than more evolved reworked crustal sources. In contrast, all shale samples from the c. 2940 Ma Mosquito Creek Formation, and from the Mount Roe Basalt (ABDP 6) and Hardey Formation (ABDP 3) of the Neoarchean Fortescue Group, have negative ε_{Hf} values indicating evolved crustal sources in their areas of prevenance. Hf model ages for these younger shales average c 3250 Ma. Nebel-Jacobsen et al. (2010) concluded that the Hf isotope compositions of the Pilbara black shales are consistent with a change from juvenile to evolved crust through the Paleoarchean to the Mesoarchean, as previously interpreted from other evidence by Smithies et al. (2007c) and Van Kranendonk et al. (2007b).

2.2.3.2 Current Zircon Lu–Hf Data

Published zircon Lu–Hf isotope data used in this review were derived from Kemp et al. (2015a, b), Gardiner et al. (2017, 2018), and from previously unpublished Lu–Hf data obtained by GSWA between 2010 and 2014. All these data were derived from Lu–Hf analysis of zircons on GSWA mounts previously used for U–Pb zircon dating of rock samples by the SHRIMP method. The $^{207}Pb/^{206}Pb$ ages of all the zircons have been provided in GSWA geochronology records that were released between 1995 and 2016. The $^{207}Pb/^{206}Pb$ data were used to select those grains most suitable for further study and constrained the time at which initial $^{176}Hf/^{177}Hf$ ratios were set. To minimize any effects of zircon heterogeneity (e.g. crystal zonation), Lu–Hf analyses were made on the same grain spots previously dated by SHRIMP. Most >3530 Ma zircons analysed are detrital, whereas most zircons younger than 3500 Ma were extracted from felsic igneous rocks.

2.2.3.3 Lu–Hf Analysis of Detrital Zircons

To investigate the possibility that the Pilbara Craton contains Hf isotopic evidence of Hadean crust (>4000 Ma), Kemp et al. (2015a) analysed 108 Pilbara zircons (xenocrystic and detrital) dated at >3530 Ma (representing approximately 50% of the pre-3530 Ma zircons previously identified by GSWA U–Pb dating). The 108 zircons came from 14 samples collected by GSWA geologists and geochronologists over a wide area of the northern Pilbara. Of the 84 detrital zircons analysed, the

3 oldest (3795–3681 Ma) had ε_{Hf} values between −3 and − 2, suggesting magmatic sources with very old mantle extraction ages (4000–3800 Ma). These detrital grains came from sandstones in the two largest Mesoarchean basins of the northern Pilbara (Mallina and Mosquito Creek). Clearly, the significance of these data to the age of the early Pilbara crust depends on the origin of the zircon grains. Since data from the 3 oldest grains indicated significantly older sources than the other 105 zircons analysed in their study, Kemp et al. (2015a) raised the possibility that these 3 grains might not have originated from the Pilbara Craton. However, there are no plausible alternative Australian sources for >3600 Ma Pilbara zircons. Although Kemp et al. (2015a) observed that sufficiently old felsic igneous rocks exist in the Narryer Terrane of the Yilgarn Craton, all available evidence is that the Yilgarn Craton was too remote from the Pilbara Craton during the c. 2950 Ma deposition of the host rocks (Smirnov et al. 2013). The Yilgarn and Pilbara Cratons did not collide until the 2005–1950 Ma Glenburgh Orogeny (Sheppard et al. 2005; Spaggiari et al. 2008; Johnson et al. 2011, 2013), which precludes a source in the Narryer Terrane. As seen on Fig. 2.9, subsequent work has since revealed many additional zircon grains older than 3650 Ma with Hf two-stage model ages >4000 Ma.

2.2.3.4 Lu–Hf Analysis of Cognate and Inherited Zircons

Figure 2.10 summarizes the Hf two-stage model ages of xenocrystic zircons from the Callina, Tambina, Emu Pool, and Cleland Supersuites. Hf model ages for 18 >3650 Ma xenocrystic zircons extracted from rhyolite of the Tambina Supersuite (Petersson et al. 2019b) are not included in this figure, but the ε_{Hf} values (Fig. 2.9) indicate Hf two-stage model ages >4000 Ma. For the other data, the combined model ages (Fig. 2.10d) suggest that the zircons were derived from sources with mantle extraction ages between 3800 and 3550 Ma.

Some interpretations of Lu–Hf isotope data (Kemp et al. 2015a, b, 2017; Fisher and Vervoort 2018) have not accepted the existence of a depleted mantle prior to 3500 Ma. If the Pilbara 3500 Ma mantle was not significantly depleted, the calculated Hf model ages summarized in Fig. 2.10 are between 100 and 200 million years too old. However, most workers use depleted mantle models proposed by McCulloch and Wasserburg (1978) and DePaolo (1988).

As might be expected, zircons from the Callina and Tambina Supersuites (Fig. 2.10) had, on average, the oldest sources although the gap between zircon crystallization age and Hf two-stage model age increases in the younger supersuites. This indicates that the younger supersuites were substantially derived from recycling of much the same 3800–3550 Ma sources as the Callina and Tambina Supersuites and that the introduction of juvenile material to the EPT diminished after intrusion of the Tambina Supersuite. In the Emu Pool Supersuite, Fig. 2.9 reveals a scatter of positive ε_{Hf} values between CHUR and DM which suggests juvenile input from the mantle in addition to melting of older felsic crust. Mean ε_{Hf} values in the Callina and Tambina Supersuites show no trend with time (Callina +0.6, $n = 149$; Tambina +0.7, $n = 120$), whereas between the Emu Pool and Cleland Supersuites, there is trend

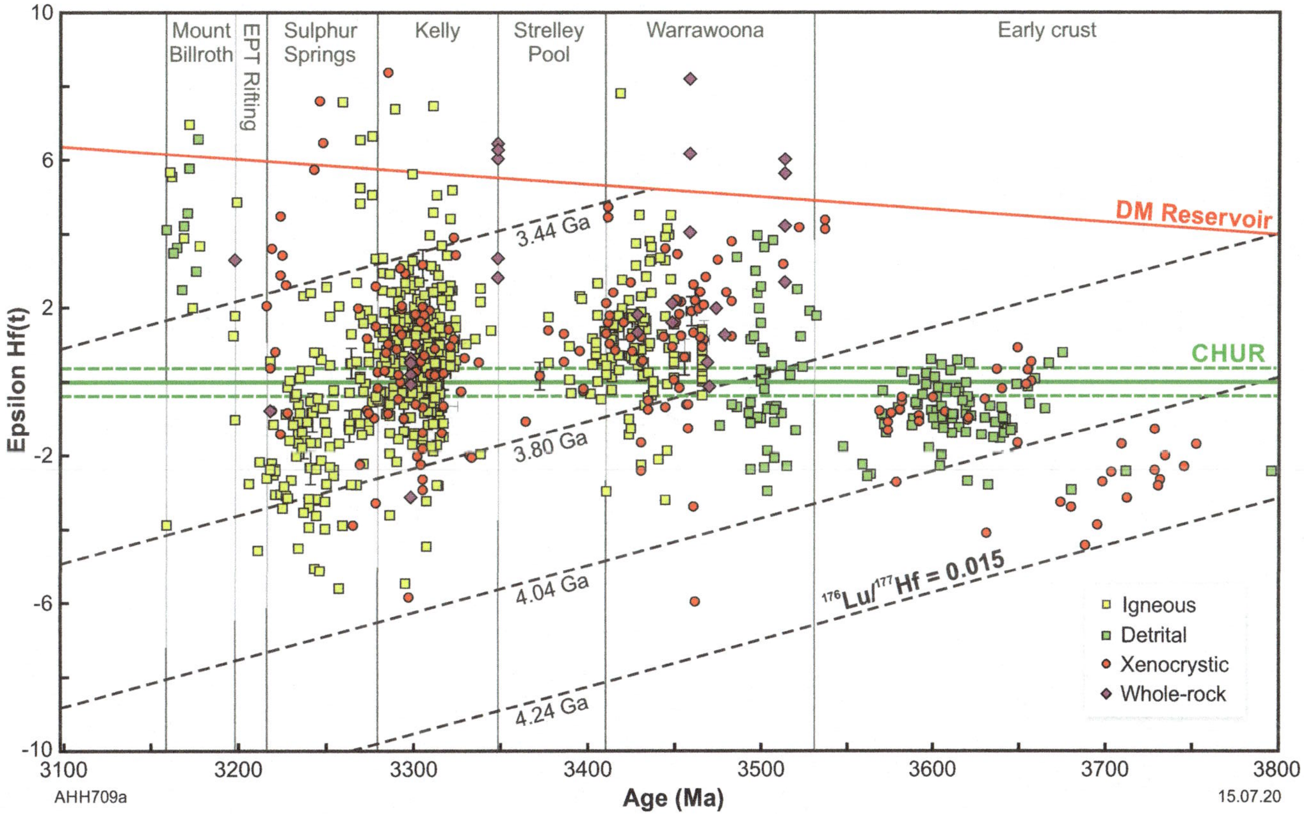

Fig. 2.9 ε_{Hf} evolution diagram for analyses of cognate and inherited zircons, detrital zircons with U–Pb ages older than 3550 Ma or younger than 3200 Ma, and whole-rock samples (mostly komatiites) from the East Pilbara. Blue shading shows Hf evolution line defined by a $^{176}Lu/^{177}Hf$ ratio of 0.010 (after Gardiner et al. 2017). Data for detrital and inherited zircons older than 3550 Ma are from Kemp et al. (2015a, b). Two abrupt changes in ε_{Hf} values (negative to positive)

from chondritic to negative values with time (Emu Pool +0.0, $n = 161$; Cleland −2.2, $n = 64$). The mean Callina Supersuite ε_{Hf} values are mainly from data in Kemp et al. (2017, Owens Gully Diorite, $n = 87$) and unpublished data from three other samples ($n = 42$) (AIS Kemp, written communication, 2018).

Figure 2.11 (**a**) summarizes all available zircon Hf two-stage model ages (from igneous and sedimentary rocks) from the East Pilbara, including some zircons <3200 Ma. The minor peak between 4000 and 3900 Ma is produced by data from zircons that crystallized earlier than 3540 Ma. This is highlighted by comparing Figure (**b** and **c**). Assuming the 3800–3550 Ma zircons were mainly derived from 3800–3550 Ma granitic rocks in the early crust, it is evident that these granites were derived from entirely different sources from those of almost all the <3530 Ma granites emplaced in the EPT. This is consistent with the pre-3530 Ma crust representing an entirely separate terrane (Eoarchean to early Paleoarchean) from the EPT.

Figure 2.9 presents ε_{Hf} values for all East Pilbara zircon grains and whole-rock samples that have crystallization ages older than 3160 Ma. Similar data are available for zircons and rocks younger than 3160 Ma, but these are mainly from the northwest Pilbara where Mesoarchean crustal evolution was dominated by plate-tectonic processes. Plate-tectonic processes included the subduction and melting of relatively juvenile crust. Data used in Fig. 2.9 are listed in Hickman (2021). Observations include:

1. Only 14% of zircons older than 3530 Ma ($n = 106$), representing pre-Pilbara Supergroup crust, have positive ε_{Hf} values, the highest being +0.9. Although the >3530 Ma zircons were obtained from widely separated areas in the East Pilbara (Kemp et al. 2015a, b), they are remarkably uniform with respect to ε_{Hf}, the average value being −0.8. If, as many geochronologists interpret, Earth had a depleted mantle between 3800 and 3530 Ma, the data suggest crustal sources with extraction ages between 4050 and 3850 Ma (Fig. 2.9). Even if a chondritic early Paleoarchean mantle is assumed, the crustal sources of about half the pre-3530 Ma zircons would have been between 3800 and 3700 Ma.
2. Zircon grains with ages between 3530 and 3514 Ma, representing the Counterunah Subgroup (Warrawoona Group) and related granites, have ε_{Hf} values ranging from +6.0 to +2.6, consistent with juvenile depleted mantle sources. This suggests that magma of the Counterunah Subgroup (oldest part of the Warrawoona Group) was at least partly sourced from Paleoarchean or late Eoarchean mantle material. This interpretation is consistent with geochemical evidence from the Coonterunah Subgroup (Smithies et al. 2007a, 2009) indicating that pre-3530 Ma East Pilbara crust was probably mainly composed of

---

Fig. 2.9 (continued) at c. 3530 Ma (initial deposition of Pilbara Supergroup) and c. 3200 Ma (intrusion of Mount Billroth Supersuite) indicate major influxes of magma from juvenile sources at these times. Increasingly negative ε_{Hf} values from 3530 to 3220 Ma indicate progressively more evolved sources with time (recycling of older crust) (From Hickman 2021; with Geological Survey of Western Australia permission)

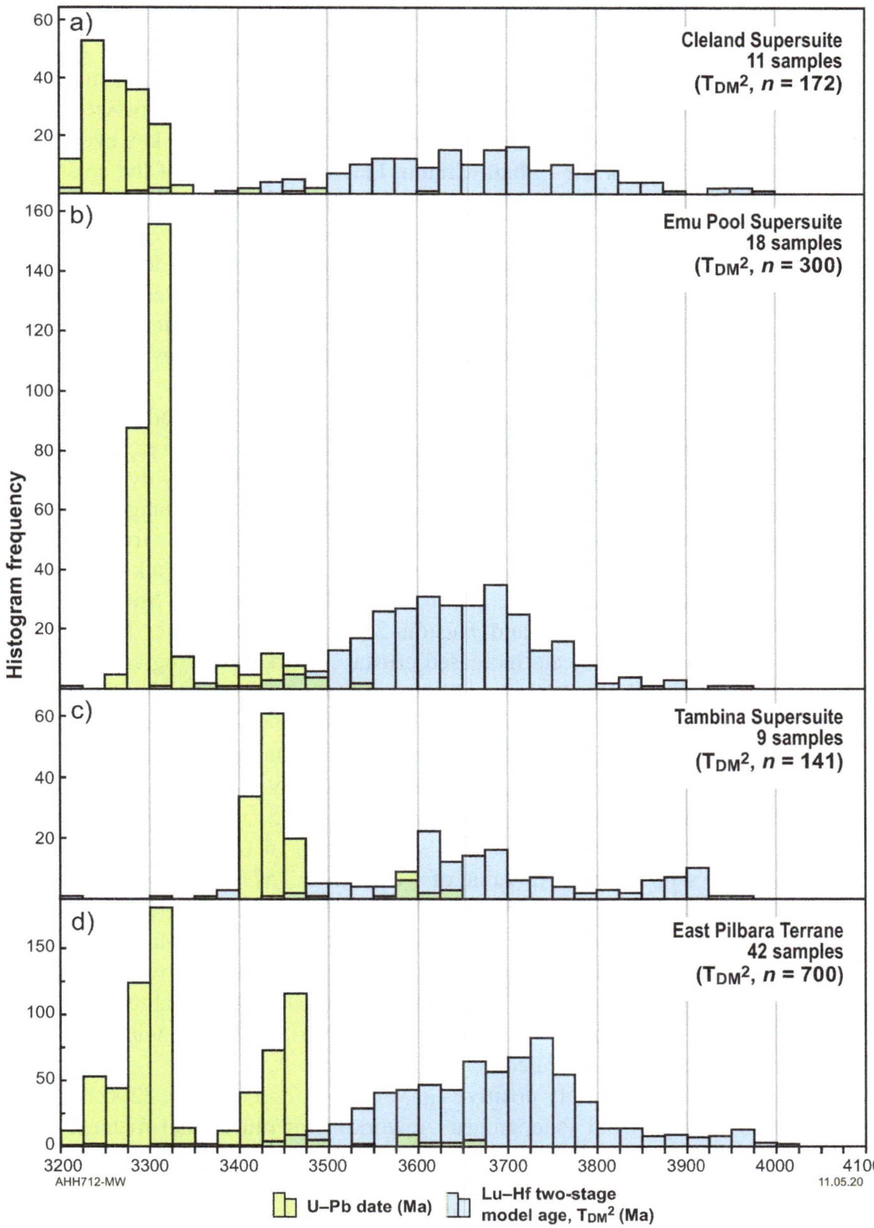

Fig. 2.10 Histograms of two-stage Hf model (T_{DM}^2) ages for cognate and inherited zircons from igneous rocks of the East Pilbara Terrane. Calculations used $^{176}Lu/^{177}Hf$ ratio of 0.015 (Scherer et al. 2001) and a ^{176}Lu decay constant of 1.865×10^{-11}. Most model ages fall between 3750 and 3500 Ma. Separation of model ages by supersuite (b–d) shows increasing gaps between zircon crystallization ages and model ages from the older to the younger supersuites, indicating ongoing crustal recycling of similar old crustal sources with time (From Hickman 2021; with Geological Survey of Western Australia permission)

supracrustal basaltic to andesitic material similar to that of the Coonterunah Subgroup and enriched in K, LILE, Th, and LREE compared to most Paleoarchean basalts. Hf model ages from the zircons of Coonterunah age range from 3640 to 3580 Ma. The major Hf isotopic differences between these zircons and the >3540 Ma zircons are here interpreted to reveal a key event in the evolutionary history of the Pilbara Craton. Felsic components of the crust older than 3540 Ma (represented by the >3540 Ma zircons) were, assuming a significantly depleted mantle, derived from >3800 Ma sources, whereas the Coonterunah Subgroup was derived from sources younger than 3650 Ma.

3. Hf data from zircons dated between 3514 and 3485 Ma have very variable ε_{Hf} values suggesting both juvenile and evolved sources. A significant number of zircons with strongly negative ε_{Hf} values is consistent with sources as old as those for the >3540 Ma zircons.
4. Well-defined isotopic trends between 3460 and 3200 Ma include a change in mean ε_{Hf} values from moderately positive to moderately negative. The data are consistent with increasing reworking of older crust as the EPT evolved. Previous investigations of EPT granitic rocks have described a secular change from TTG between 3485 Ma and 3420 Ma to granodiorite and monzogranite from 3325 Ma to 3223 Ma (Bickle et al. 1993; Smithies 2000; Van Kranendonk et al. 2002, 2007a, b; Smithies et al. 2003, 2007b, 2009; Champion and Smithies 2007; Champion 2013; Champion and Huston 2016; Gardiner et al. 2017, 2018). Sm–Nd isotopic data indicate increased crustal reworking during crustal evolution of the EPT, with diminishing addition of juvenile material. To determine if Lu–Hf isotopic data show the same trends, isotopic data from the cognate zircons (crystallized at the same time as the host rock) of tonalites, granodiorites, and monzogranites were statistically processed separately. Cognate zircons in the tonalites and diorites have a mean ε_{Hf} value of +0.35 ($n = 196$), compared to those in granodiorite with a mean ε_{Hf} value of −0.14 ($n = 122$) and those in monzogranite with a mean ε_{Hf} value of −0.55 ($n = 157$). The mean ages of the same zircon groups confirm that tonalite zircons (mean age 3339 Ma) are appreciably older than monzogranite zircons (mean age 3272 Ma), with granodiorite zircons being intermediate in age (mean age 3326 Ma). However, all three zircon groups have very similar mean Hf $T_{DM}{}^2$ model age (tonalites, 3629 Ma; granodiorites, 3620 Ma; monzogranites, 3600 Ma) consistent with derivation from late Eoarchean–early Paleoarchean sources.
5. An abrupt switch to strongly positive ε_{Hf} values is observed at c. 3200 Ma. This indicates a change from Paleoarchean reworking of old crustal material to an early Mesoarchean influx of juvenile magma. This timing coincides with breakup of the EPT (Hickman 2021), plate separation, and intrusion of the Mount Billroth Supersuite between 3199 and 3164 Ma. If two strongly positive ε_{Hf} values above 'MORB-DM' are omitted from the calculation, the average ε_{Hf} value of 18 analyses is +2.6.

The positive ε_{Hf} values in zircons of early Mesoarchean rocks in the East Pilbara are accompanied by strongly positive ε_{Hf} values in volcanic rocks of the same age in

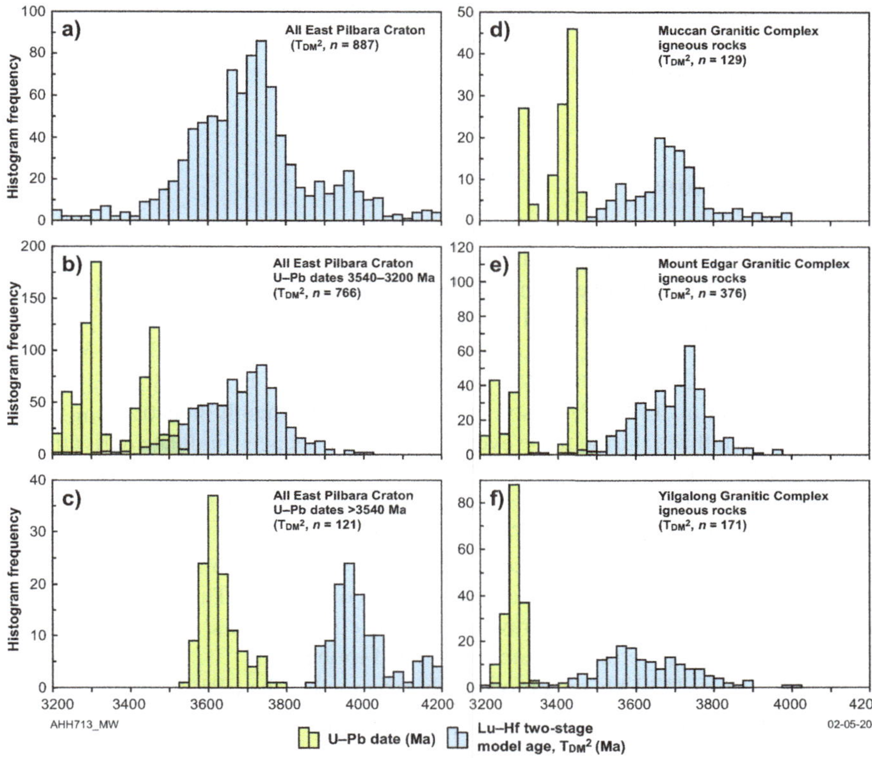

Fig. 2.11 Histograms of two-stage Hf model ages (T_{DM}^2) for igneous and detrital zircons from Paleoarchean and Mesoarchean rocks of the east part of the Northern Pilbara Craton. Calculations used $^{176}Lu/^{177}Hf$ ratio of 0.015 and a Scherer et al. (2001) ^{176} Lu decay constant (1.865×10^{-11}). (a) Model ages for all detrital and igneous zircons; (**b**) model ages for 3540–3200 igneous and detrital zircons; (**c**) model ages for >3540 Ma zircons (note the much older model ages, and the greater separation of crystallization ages and model ages, compared to <3540 Ma zircons); (**d**, **e**, **f**) model ages from three different granitic cores of East Pilbara domes, showing no major differences (From Hickman 2021; with Geological Survey of Western Australia permission)

the northwest Pilbara. Nebel et al. (2014) reported that ε_{Hf} values of five whole-rock komatiite samples from the 3200–3160 Ma Regal Formation, interpreted from other evidence to be oceanic-like crust (Ohta et al. 1996; Sun and Hickman 1998, 1999), ranged from +5.9 to −0.2, and averaged +3.5.

In summary, the zircon Hf isotope evidence suggests that 3530–3460 volcanics and granitic rocks were derived from more juvenile sources than applied to pre-3530 Ma felsic igneous rocks (Fig. 2.9). The apparently abrupt change in magma sources between 3550 and 3530 Ma indicates a major event in the crustal evolution of the Pilbara Craton, although the nature of this event (e.g. continental rifting, plate collision, or mantle melting due to asteroid impact) is unknown. The trend of increasing crustal reworking from 3530 Ma to 3220 Ma was terminated by breakup of the Pilbara Craton at c. 3220 Ma (Chap. 6). Juvenile magma sources contributed to volcanism and granitic intrusion after 3220 Ma, although some post-3220 Ma igneous rocks were partly derived from much older crust.

2.3 Conclusions

Dates on pre-3530 Ma xenocrystic zircons in felsic igneous rocks (Table 2.1) and detrital zircons in Paleoarchean and Mesoarchean metasedimentary rocks (Figs. 2.2 and 2.5) suggest that the pre-Pilbara Supergroup sialic crust evolved from about 3800 Ma. There is some suggestion of episodic magmatic events at 3760–3700 Ma (Petersson et al. 2019b), 3650 Ma (Kemp et al. 2015a, b), and 3590–3570 Ma (McNaughton et al. 1988; Petersson et al. 2019a). The mainly negative ε_{Hf} values in zircons crystallized during these events, and Hf model ages, indicate derivation of melts from considerably older sources (4000–3800 Ma). There are major differences of Hf model ages between Pilbara zircons older than 3530 Ma and zircons younger than 3530 Ma. It is evident that Eoarchean and early Paleoarchean granitic rocks, from which the pre-3530 Ma zircons were derived, came from entirely different sources than those of the <3530 Ma granitic rocks of the EPT. The conclusion is that the sialic crust pre-dating the Warrawoona Group comprised one or more Eoarchean to early Paleoarchean terranes entirely unrelated to the EPT.

Although no c. 3530 Ma unconformity is preserved in the EPT, it follows that the Warrawoona Group was deposited unconformably on c. 3800–3530 Ma sialic crust. This crust formed an extensive basement to the Pilbara greenstone succession in the Archean. The scarcity of remnants of the early crust outcropping in the Pilbara Craton today is mainly a consequence of successive episodes of Paleoarchean vertical crustal reworking and post-3530 Ma granitic intrusion.

References

Allwood AC, Walter MR, Kamber BS, Marshall CP, Burch IW (2006) Stromatolite reef from the Early Archaean era of Australia. Nature 441:714–717

Allwood AC, Burch I, Walter MR (2007a) Stratigraphy and facies of the 3.43 Ga Strelley Pool Chert in the southwestern North Pole Dome, Pilbara Craton, Western Australia. Record, 2007/11. Geological Survey of Western Australia

Allwood AC, Walter MR, Burch IW, Kamber BS (2007b) 3.43 billion-year-old stromatolite reef from the Pilbara Craton of Western Australia: ecosystem-scale insights to early life on earth. Precambrian Res 158:198–227

Amelin Y, Lee D-C, Halliday AN, Pidgeon RT (1999) Nature of the Earth's earliest crust from hafnium isotopes in single detrital zircons. Nature 399:252–255. https://doi.org/10.1038/20426

Amelin Y, Lee D-C, Halliday AN (2000) Early-middle Archaean crustal evolution deduced from Lu–Hf and U–Pb isotopic studies of single zircon grains. Geochim Cosmochim Acta 62:4205–4225

Anhaeusser CR (1971a) Cyclic volcanicity and sedimentation in the evolutionary development of Archaean greenstone belts of shield areas. In: Glover JE (ed) Symposium on Archaean Rocks, Perth, 23–26 May 1970. Geological Society of Australia, Perth, Western Australia, pp 57–70

Anhaeusser CR (1971b) The Barberton Mountain Land, South Africa–a guide to the understanding of the Archaean geology of Western Australia. In: Glover JE (ed) Symposium on Archaean rocks, Perth, 23–26 May 1970. Geological Society of Australia, Perth, Western Australia, pp 103–119

Anhaeusser CR, Mason R, Viljoen MJ, Viljoen RP (1969) A reappraisal of some aspects of Precambrian shield geology. Geol Soc Am Bull 80:2175–2200

Arndt NT, Goldstein SL (1987) Use and abuse of crust-formation ages. Geology 15:893–895

Arndt NT, Bruzak G, Reischmann T (2001) The oldest continental and oceanic plateaus: geochemistry of basalts and komatiites of the Pilbara Craton, Australia. In: Ernst RE, Buchan K (eds) Mantle plumes: their identification through time. Geological Society of America, pp 359–387

Bagas L, Farrell TR, Nelson DR (2004) The age and provenance of the Mosquito Creek Formation. In: Geological Survey of Western Australia Annual Review, vol 2003–04, pp 62–70

Bagas L, Bierlein FP, Bodorkos S, Nelson DR (2008) Tectonic setting, evolution, and orogenic gold potential of the late Mesoarchaean Mosquito Creek Basin, North Pilbara Craton, Western Australia. Precambrian Res 160:237–244

Beintema KA (2003) Geodynamic evolution of the West and Central Pilbara Craton: a mid-Archean active continental margin. Geologica Ultraiectina, Utrecht University, Utrecht, The Netherlands, PhD thesis (unpublished), 248p

Beintema KA, de Leeuw GAM, White SH, Hein KAA (2001) Tabba Tabba shear: a crustal-scale structure in the Pilbara Craton, WA. In: Cassidy KF, Dunphy JM, Van Kranendonk MJ (eds) Extended Abstracts–4th International Archaean Symposium, Perth, 24–28 September 2001. AGSO (Geoscience Australia), pp. 285–287

Beintema KA, Mason PRD, Nelson DR, White SH, Wijbrans JR (2003) New constraints on the timing of tectonic activity in the Archaean central Pilbara Craton, Western Australia. J Virtual Explor 13. https://doi.org/10.3809/jvirtex.vol.2003.013

Bennett VC, Nutman AP, McCulloch MT (1993) Nd isotopic evidence for transient, highly depleted mantle reservoirs in the early history of the Earth. Earth Planet Sci Lett 119:299–317. https://doi.org/10.1016/0012-821X(93)90140-5

Bickle MJ, Bettenay LF, Chapman HJ, Groves DI, McNaughton NJ, Campbell IH, de Laeter JR (1989) The age and origin of younger granitic plutons of the Shaw batholith in the Archean Pilbara Block, Western Australia. Contrib Mineral Petrol 101:361–376

Bickle MJ, Bettenay LF, Chapman HJ, Groves DI, McNaughton NJ, Campbell IH, de Laeter JR (1993) Origin of the 3500–3300 Ma calc-alkaline rocks in the Pilbara Archaean: isotopic and geochemical constraints from the Shaw Batholith. Precambrian Res 60:117–149

Brauhart CW, Huston D, Andrew AS (2000) Oxygen isotope mapping in the Panorama VMS district, Pilbara Craton, Western Australia: applications to estimating temperatures of alteration and to exploration. Mineralium Deposita 35:727–740

Buick R, Barnes KR (1984) Cherts in the Warrawoona Group: early Archaean silicified sediments deposited in shallow-water environments. In: Muhling JR, Groves DI, Blake TS (eds) Archaean and Proterozoic basins of the Pilbara, Western Australia: evolution and mineralization potential. The University of Western Australia, Geology Department and University Extension, pp. 37–53

Buick R, Thornett J, McNaughton N, Smith JB, Barley ME, Savage MD (1995) Record of emergent continental crust ~3.5 billion years ago in the Pilbara Craton of Australia. Nature 375:574–577

Buick R, Brauhart CW, Morant P, Thornett JR, Maniw JG, Archibald NJ, Doepel MG, Fletcher IR, Pickard AL, Smith JB, Barley ME, McNaughton NJ, Groves DI (2002) Geochronology and stratigraphic relationships of the Sulphur Springs Group and Strelley Granite: a temporally distinct igneous province in the Archean Pilbara Craton, Australia. Precambrian Res 114:87–120

Byerly GR, Lowe DR, Heubeck C (2019) Chapter 24–geologic evolution of the Barberton Greenstone Belt–a unique record of crustal development, surface processes, and early life

3.55–3.20Ga. In: Van Kranendonk MJ, Bennett VC, Hoffmann JE (eds) Earth's oldest rocks, 2nd edn. Elsevier B.V, pp 569–613

Byerly GR, Lowe DR, Wooden JL, Xie X (2002) An Archean impact layer from the Pilbara and Kaapvaal cratons. Science 297:1325–1327

Cawood PA, Hawkesworth CJ, Dhuime B (2012) Detrital zircon record and tectonic setting. Geology 40:875–878

Champion DC (2013) Neodymium depleted mantle model age map of Australia: explanatory notes and user guide. Record, 2013/44. Geoscience Australia

Champion DC, Huston DL (2016) Radiogenic isotopes, ore deposits and metallogenic terranes: novel approaches based on regional isotopic maps and the mineral systems concept. Ore Geol Rev 76:229–256. https://doi.org/10.1016/j.oregeorev.2015.09.025

Champion DC, Smithies RH (2007) Geochemistry of Paleoarchean granites of the East Pilbara Terrane, Pilbara Craton, Western Australia: implications for early Archean crustal growth. In: Van Kranendonk MJ, Bennett VC, Smithies RH (eds) Earth's oldest rocks. Elsevier B.V, Burlington, pp 369–410

Collerson KD, McCulloch MT (1983) Field and Sr–Nd isotopic constraints on Archaean crust and mantle evolution in the East Pilbara Block, Western Australia. In: Australian National University, Research School of Earth Sciences, Annual Report 1982. Canberra, ACT, pp 167–168

Collins WJ (1993) Melting of Archaean silicic crust under high aH_2O conditions: genesis of 3300 Ma Na-rich granitoids in the Mount Edgar Batholith, Pilbara Block, Western Australia. Precambrian Res 60:151–174

Compston W, Kröner A (1988) Multiple zircon growth within early Archaean tonalitic gneiss from the ancient gneiss complex, Swaziland. Earth Planet Sci Lett 87:13–28

Dawes PR, Smithies RH, Centofanti J, Podmore DC (1995a) Sunrise Hill unconformity: a newly discovered regional hiatus between Archaean granites and greenstones in the northeastern Pilbara Craton. Aust J Earth Sci 42:635–639

Dawes PR, Smithies RH, Centofanti J, Podmore DC (1995b) Unconformable contact relationships between the Muccan and Warrawagine batholiths and the Archaean Gorge Creek Group in the Yarrie mine area, northeast Pilbara. Record, 1994/3. Geological Survey of Western Australia

De Vries ST, Nijman W, Wijbrans JR, Nelson DR (2006) Stratigraphic continuity and early deformation of the central part of the Coppin gap Greenstone Belt, Pilbara, Western Australia. Precambrian Res 147:1–27

de Wit MJ, Ashwal LD (eds) (1997) Greenstone belts. Clarendon Press, Oxford

de Wit MJ (1983) Notes on a preliminary 1:25,000 geological map of the southern part of the Barberton Greenstone Belt. In: Anhaeusser CR (ed) Contributions to the geology of the Barberton Mountain land. Spec. Publ. Geol. Soc, South Africa, pp 185–187

de Wit MJ (1998) On Archean granites, greenstones, cratons and tectonics: does the evidence demand a verdict? Precambrian Res 91:181–226

de Wit MJ, Furnes H, Robins B (2011) Geology and tectonostratigraphy of the Onverwacht suite. Barberton Greenstone Belt, South Africa Precambr Res 186:1–27

de Wit MJ, Roering C, Hart RJ, Armstrong RA, de Ronde CEJ, Green RWE, Tredoux M, Peberdy E, Hart RA (1992) Formation of an Archaean continent. Nature 357:553–562

Délor C, Burg JP, Clarke G (1991) Relations diapirisme-metamorphisme dans la Province du Pilbara (Australie occidentale): implications pour les régimes thermiques et tectoniques à la Archéen. Compte Rendu Academie de Sciences Paris 312:257–263

DePaolo DJ (1988) Neodymium isotope geochemistry: an introduction. Minerals and rocks series, vol no. 20. Springer-Verlag, Berlin, Germany

DiMarco MJ, Lowe DR (1989a) Shallow-water volcaniclastic deposition in the early Archean Panorama Formation, Warrawoona Group, Eastern Pilbara Block, Western Australia. Sediment Geol 64:43–63

DiMarco MJ, Lowe DR (1989b) Stratigraphy and sedimentology of an Early Archean felsic volcanic sequence, Eastern Pilbara Block, Western Australia, with special reference to the Duffer Formation and implications for crustal evolution. Precambrian Res 44:147–169

Drabon N, Lowe DR, Byerly GR, Harrington JA (2017) Detrital zircon geochronology of sandstones of the 3.6–3.2 Ga Barberton greenstone belt: no evidence for older continental crust. Geology 45:803–806

Eriksson KA (1981) Archaean platform-to-trough sedimentation, East Pilbara Block, Australia. In: Glover JE, Groves DI (eds) Spec. Publ. Geol. Soc. Aust, vol 7, pp 235–244

Fisher CM, Vervoort JD (2018) Using the magmatic record to constrain the growth of continental crust–the Eoarchean zircon Hf record of Greenland. Earth Planet Sci Lett 488:79–91

François C, Philippot P, Rey PF, Rubatto D (2014) Burial and exhumation during Archean sagduction in the East Pilbara granite–greenstone terrane. Earth Planet Sci Lett 396:235–251

Furnes H, de Wit MJ, Robins B (2012) A review of new interpretations of the tectonostratigraphy, geochemistry and evolution of the Onverwacht suite, Barberton Greenstone Belt, South Africa. Gondwana Res 23:403–428

Gardiner NJ, Hickman AH, Kirkland CL, Lu Y, Johnson T, Zhao J-X (2017) Processes of crust formation in the early Earth imaged through Hf isotopes from the East Pilbara Terrane. Precambrian Res 297:56–76. https://doi.org/10.1016/j.precamres.2017.05.004

Gardiner NJ, Hickman AH, Kirkland CL, Lu Y, Johnson TE, Wingate MTD (2018) New Hf isotope insights into the Paleoarchean magmatic evolution of the Mount Edgar dome, Pilbara Craton: implications for early Earth and crust formation processes. Report 181. Geological Survey of Western Australia

Glikson AY (1972) Early Precambrian evidence of a primitive ocean crust and island nuclei of sodic granite. Geol Soc Am Bull 83:3323–3344

Glikson AY (1979) Early Precambrian tonalite-trondhjemite sialic nuclei. Earth Sci Rev 15:1–73

Glikson AY, Hickman AH (1981) Geochemistry of Archaean volcanic successions, Eastern Pilbara Block, Western Australia. Bur Min Res Geol Geophys 36

Glikson AY, Lambert IB (1976) Vertical zonation and petrogenesis of the early Precambrian crust in Western Australia. Tectonophysics 30:55–89

Green MG, Sylvester PJ, Buick R (2000) Growth and recycling of early Archaean continental crust: geochemical evidence from the Coonterunah and Warrawoona Groups, Pilbara Craton, Australia. Tectonophysics 322:69–88

Griffin WL, Pearson NJ, Belousova EA, Jackson SR, van Achterbergh E, O'Reilly SY, Shee SR (2000) The Hf isotope composition of cratonic mantle: LAM-MC-ICPMS analysis of zircon megacrysts in kimberlites. Geochim Cosmochim Acta 64:133–147

Gruau G, Jahn BM, Glikson AY, Davy R, Hickman AH, Chauvel C (1987) Age of the Archean Talga-Talga subgroup, Pilbara Block, Western Australia, and the early evolution of the mantle: new Sm–Nd isotopic evidence. Earth Planet Sci Lett 85:105–116

Guitreau M, Blichert-Toft J, Martin H, Mojzsis SJ, Albarède F (2012) Hafnium isotope evidence from Archean granitic rocks for deep-mantle origin of continental crust. Earth Planet Sci Lett 337:211–223

Hamilton PJ, Evensen NM, O'Nions RK, Glikson AY, Hickman AH (1980) Sm-Nd dating of the Talga Talga subgroup, Warrawoona Group, Pilbara Block, Western Australia. In second International Archaean Symposium, Perth 1980, extended abstracts, JE Glover and DI Groves (eds). Geological Society of Australia and IGCP Archaean Geochemistry Project, Perth 1980, pp. 11–12

Hamilton PJ, Evensen NM, O'Nions RK, Glikson AY, Hickman AH (1981) Sm–Nd dating of the north star basalt, Warrawoona Group, Pilbara Block, Western Australia. In: Glover JE, Groves DI (eds) Geol. Soc. Aust. Spec. Publ, vol 7, pp 187–192

Hasenstab E, Tusch J, Schnabel C, Marien CS, Van Kranendonk MJ, Smithies H, Howard H, Maier WD, Munker C (2021) Evolution of the early to late Archean mantle from Hf-Nd-Ce isotope systematics in basalts and komatiites from the Pilbara Craton. Earth Planet Sci Lett 553(116627):1–13

Hickman AH (1975) Explanatory notes on the Nullagine 1:250 000 geological sheet. Record, 1975/5. Geological Survey of Western Australia

Hickman AH (1980a) Archaean stratigraphic successions in various parts of the Pilbara Block (correlation chart). Geology of the Pilbara Block and its environs. Geological Survey of Western Australia
Hickman AH (1980b) Excursion guide–Archaean geology of the Pilbara Block: second International Archaean Symposium, Perth 1980. Geological Society of Australia, W.A. Division, Perth, Western Australia
Hickman AH (1981) Crustal evolution of the Pilbara Block. In: Glover JE, Groves DI (eds) Archaean Geology: second International Symposium, Perth 1980. Geological Society of Australia, pp. 57–69
Hickman AH (1983) Geology of the Pilbara Block and its environs. In: Bulletin Geological Survey of Western Australia, vol 127
Hickman AH (1984) Archaean diapirism in the Pilbara Block, Western Australia. In: Kröner A, Greiling R (eds) Precambrian tectonics illustrated. Schweizerbart Science Publishers, Stuttgart, Germany, pp 113–127
Hickman AH (2008) Regional review of the 3426–3350 Ma Strelley Pool Formation, Pilbara Craton, Western Australia. Record, 2008/15. Geological Survey of Western Australia
Hickman AH (2012) Review of the Pilbara Craton and Fortescue Basin, Western Australia: crustal evolution providing environments for early life. Island Arc 21:1–31
Hickman AH (2016) Northwest Pilbara Craton: a record of 450 million years in the growth of Archean continental crust. Report, 160. Geological Survey of Western Australia
Hickman AH (2021) East Pilbara Craton: a record of one billion years in the growth of Archean continental crust. Report 143, Geological Survey of Western Australia
Hickman AH, Lipple SL (1975) Explanatory notes on the Marble Bar 1:250 000 geological sheet, Western Australia. Record, 1974/20. Geological Survey of Western Australia
Hickman AH, Lipple SL (1978) Marble Bar, WA Sheet SF50–8, 2nd edn. Geological Survey of Western Australia
Hickman AH, Van Kranendonk MJ (2004) Diapiric processes in the formation of Archaean continental crust, East Pilbara granite–greenstone terrane, Australia. In: Eriksson PG, Altermann W, Nelson DR, Mueller WU, Catuneanu O (eds) The Precambrian earth: tempos and events. Elsevier, Amsterdam, The Netherlands, pp 54–75
Hoffmann JE, Kroner A (2019) Chapter 23, Early Archean Crustal evolution in Southern Africa–an updated record of the ancient gneiss complex of Swaziland. In: Van Kranendonk MJ, Bennett VC, Hoffmann JE (eds) Earth's oldest rocks, 2nd edn. Elsevier B.V, pp 437–462. https://doi.org/10.1016/B978-0-444-63901-1.00023-X
Hofmann AW (1997) Mantle geochemistry: the message from oceanic volcanism. Nature 385:219–229
Jahn B, Glikson AY, Peucat JJ, Hickman AH (1981) REE geochemistry and isotopic data of Archean silicic volcanics and granitoids from the Pilbara Block, Western Australia: implications for the early crustal evolution. Geochim Cosmochim Acta 45:1633–1652
Johnson SP, Sheppard S, Rasmussen B, Wingate MTD, Kirkland CL, Muhling JR, Fletcher IR, Belousova EA (2011) Two collisions, two sutures: punctuated pre−1950 Ma assembly of the West Australian Craton during the Ophthalmian and Glenburgh Orogenies. Precambrian Res 189:239–262. https://doi.org/10.1016/j.precamres.2011.07.011
Johnson SP, Thorne AM, Tyler IM, Korsch RJ, Kennett BLN, Cutten HN, Goodwin J, Blay OA, Blewett RS, Joly A, Dentith MC, Aitken ARA, Holzschuh J, Salmon M, Reading A, Heinson G, Boren G, Ross J, Costelloe RD, Fomin T (2013) Crustal architecture of the Capricorn Orogen, Western Australia and associated metallogeny. Aust J Earth Sci 60:681–705. https://doi.org/10.1080/08120099.2013.826735
Kabashima T, Isozaki Y, Hirata T, Maruyama S, Kiminami K, Nishimura Y (2003) U–Pb zircon ages by laser ablation ICP-MS techniques from the Archean felsic intrusive and volcanic rocks in the North Pole Area, Western Australia. J Geogr 112:1–17
Kamo SL, Davis DW (1994) Reassessment of Archean crustal development in the Barberton Mountain land, South Africa, based on U–Pb dating. Tectonics 13:167–192

Kemp AIS, Hickman AH, Kirkland CL (2015a) Early evolution of the Pilbara Craton, Western Australia, from hafnium isotopes in detrital and inherited zircons. Report 151. Geological Survey of Western Australia

Kemp AIS, Hickman AH, Kirkland CL, Vervoort JD (2015b) Hf isotopes in detrital and inherited zircons of the Pilbara Craton provide no evidence for Hadean continents. Precambrian Res 261: 112–126

Kemp AIS, Vervoort JD, Bjorkman KE, Iaccheri LM (2017) Hafnium isotope characteristics of Palaeoarchaean zircon OG1/OGC from the Owens Gully Diorite, Pilbara Craton, Western Australia. Geostand Geoanal Res 41:659–673. https://doi.org/10.1111/ggr.12182

Kemp AIS, Vervoort JD, Hickman AH, Smithies RH, Wyche S, Wingate MTD, Kirkland CL (2014) Growing ancient Australia: hafnium and neodymium isotope constraints from the Yilgarn and Pilbara cratons. In: Dynamic planet–Precambrian geochronology (Abstracts), pp. 205–206

Kemp AIS, Vervoort JD, Smithies RH, Hickman AH, Van Kranendonk MJ (2011) Recoupling the Nd–Hf isotope record of the early Earth? Evidence from the Pilbara Craton, Western Australia (Abstract V44B-04). In: Abstracts

Kisters AFM, Belcher RW, Poujol M, Dziggel A (2010) Continental growth and convergence-related arc plutonism in the Mesoarchean: evidence from the Barberton granitoid-greenstone terrain, South Africa. Precambrian Res 178:15–26

Kitajima, K, Hirata, T, Maruyama, S, Yamanashi, T., Sano, Y. and Liou, JG (2008) U–Pb zircon geochronology using LA-ICP-MS in the North Pole Dome, Pilbara Craton, Western Australia: a new tectonic growth model for the Archean chert/greenstone succession, v. 80, p. 1–14

Komiya T, Maruyama S (2007) A very hydrous mantle under the western Pacific region: implications for formation of marginal basins and style of Archean plate tectonics. Gondwana Res 11: 132–147

Krapež B (1984) Sedimentation in a small, fault-bounded basin: the Lalla Rookh sandstone, East Pilbara Block. In: Muhling JR, Groves DI, Blake TS (eds) Archaean and Proterozoic basins of the Pilbara, Western Australia: evolution and mineralization potential. The University of Western Australia, Geology Department and University Extension, pp. 89–110

Krapež B, Barley ME (1987) Archaean strike-slip faulting and related ensialic basins: evidence from the Pilbara Block. Australia Geol Mag 124:555–567. https://doi.org/10.1017/S0016756800017386

Kröner A, Anhaeusser CR, Hoffmann JE, Wong J, Geng H, Hegner E, Xie H, Yang J, Liu D (2016) Chronology of the oldest supracrustal sequences in the Palaeoarchaean Barberton Greenstone Belt, South Africa and Swaziland. Precambrian Res 279:123–143

Kröner A, Compston W, Williams IS (1989) Growth of early Archaean crust in the Ancient Gneiss complex of Swaziland as revealed by single zircon dating. Tectonophysics 161:271–298

Kröner A, Hegner E, Wendt JI, Byerly GR (1996) The oldest part of the Barberton granitoid-greenstone terrain, South Africa: evidence for crust formation between 3.5 and 3.7 Ga. Precambrian Res 78:105–124

Kröner A, Hoffmann JE, Xie H, Munker C, Hegner E, Wan Y, Hofmann AJ, Liu D, Yang J (2014) Generation of early Archaean grey gneisses through melting of older crust in the Eastern Kaapvaal Craton, southern Africa. Precambrian Res 255:823–846

McCulloch MT (1987) Sm–Nd isotopic constraints on the evolution of Precambrian crust in the Australian continent. In: Kröner A (ed) Proterozoic lithosphere evolution. American Geophysical Union, pp 111–120

McCulloch MT, Wasserburg GJ (1978) Sm–Nd and Rb–Sr chronology of continental crust formation. Science 200:1003–1011

McCulloch MT, Bennett VC (1993) Evolution of the early earth: constraints from ^{143}Nd–^{142}Nd isotope systematics. Lithos 30:237–255

McNaughton NJ, Green MD, Compston W, Williams IS (1988) Are anorthositic rocks basement to the Pilbara Craton? Geol Soc Aust Abstr 21:272–273

Moyen J-F, Stevens G, Kisters AFM (2006) Record of mid-Archaean subduction and metamorphism in the Barberton terrain, South Africa. Nature 442:559–562

Nebel O, Campbell IH, Sossi PA, Van Kranendonk MJ (2014) Hafnium and iron isotopes in early Archean komatiites record a plume-driven convection cycle in the hadean earth. Earth Planet Sci Lett 397:111–120

Nebel-Jacobsen Y, Munker C, Nebel O, Gerdes A, Mezger K, Nelson DR (2010) Reworking of Earth's first crust: constraints from Hf isotopes in Archean zircons from Mt. Narryer. Australia Precambr Res 182:175–186

Nelson DR (1996) Compilation of SHRIMP U-Pb zircon geochronology data, 1995. Record, 1996/5. Geological Survey of Western Australia

Nelson DR (1998) Compilation of SHRIMP U-Pb zircon geochronology data, 1997. Record, 1998/2. Geological Survey of Western Australia

Nelson DR (1999) Compilation of geochronology data, 1998. Record, 1999/2. Geological Survey of Western Australia

Nelson DR (2000) Compilation of geochronology data, 1999. Geological Survey of Western Australia, Record 2000(2):251p

Nelson DR (2001) Compilation of geochronology data, 2000. Record, 2001/2. Geological Survey of Western Australia

Nelson DR (2002) Compilation of geochronology data, 2001. Record, 2002/2. Geological Survey of Western Australia

Nelson DR (2005) 178031: porphyritic biotite granodiorite. Mulgandoona Hill, Record 556. Geological Survey of Western Australia

Nelson DR (2006) Compilation of geochronology data 2006, June 2006 update. Geological Survey of Western Australia

Ohta H, Maruyama S, Takahashi E, Watanabe Y, Kato Y (1996) Field occurrence, geochemistry and petrogenesis of the Archaean mid-oceanic ridge basalts (AMORBs) of the Cleaverville area, Pilbara Craton, Western Australia. Lithos 37:199–221

Olivier N, Dromart G, Coltice N, Flament N, Rey P, Sauvestre R (2012) A deep subaqueous fan depositional model for the Palaeoarchean (3.46 Ga) marble Bar Cherts, Warrawoona Group. Western Australia Geol Mag 149:743–749

Pawley MJ, Van Kranendonk MJ, Collins WJ (2004) Interplay between deformation and magmatism during doming of the Archaean Shaw granitoid complex, Pilbara Craton, Western Australia. Precambrian Res 131:213–230

Petersson A, Kemp AIS, Hickman AH, Van Kranendonk MJ, Whitehouse MJ, Martin L, Gray CM (2019a) A new 3.59 Ga magmatic suite and a chondritic source to the East Pilbara Craton. Chem Geol 511:51–70. https://doi.org/10.1016/j.chemgeo.2019.01.021

Petersson A, Kemp AIS, Whitehouse MJ (2019b) A Yilgarn seed to the Pilbara Craton (Australia)? Evidence from inherited zircons. Geology 47:1098–1102. https://doi.org/10.1130/G46696.1

Petersson A, Kemp AIS, Gray CM, Whitehouse MJ (2020) Formation of early Archean granite–greenstone terranes from a globally chondritic mantle: insights from igneous rocks of the Pilbara Craton. Western Australia Chemical Geology 551. https://doi.org/10.1016/j.chemgeo.2020.119757

Pidgeon RT (1978) 3450-m.y.-old volcanics in the Archaean layered greenstone succession of the Pilbara Block, Western Australia. Earth Planet Sci Lett 37:421–428. https://doi.org/10.1016/0012-821X(78)90057-2

Roberts NMW, Spencer CJ (2015) The zircon archive of continent formation through time. In: Continent formation through time, NMW Roberts, MJ Van Kranendonk S Parman, S Shirley, PD Clift (eds). Geological Society, London, Special Publications 389:197–225

Scherer E, Münker C, Mezger K (2001) Calibration of the Lutetium-Hafnium Clock. Science 293 (5530):683–687. https://doi.org/10.1126/science.1061372

Schoene B, Bowring SA (2010) Rates and mechanisms of Mesoarchean magmatic arc construction, Eastern Kaapvaal Craton, Swaziland. GSA Bull 122:408–429. https://doi.org/10.1130/B26501.1

Schoene B, de Wit MJ, Bowring SA (2008) Mesoarchean assembly and stabilization of the Eastern Kaapvaal Craton: a structural–thermochronological perspective. Tectonics 27:1–27. https://doi.org/10.1029/2008TC002267

Sheppard S, Occhipinti SA, Nelson DR (2005) Intracontinental reworking in the Capricorn Orogen, Western Australia: the 1680–1620 Ma Mangaroon Orogeny. Aust J Earth Sci 52:443–460

Smirnov AV, Evans ADA, Ernst RE, Söderlund U, Li ZX (2013) Trading partners: tectonic ancestry of southern Africa and western Australia, in Archean supercratons Vaalbara and Zimgarn. Precambrian Res 224:11–12

Smith JB, Barley ME, Groves DI, Krapež B, McNaughton NJ, Bickle MJ, Chapman HJ (1998) The Sholl Shear Zone, West Pilbara; evidence for a domain boundary structure from integrated tectonostratigraphic analysis, SHRIMP U–Pb dating and isotopic and geochemical data of granitoids. Precambrian Res 88:143–172

Smithies RH (2000) The Archaean tonalite–trondhjemite–granodiorite (TTG) series is not an analogue of Cenozoic adakite. Earth Planet Sci Lett 182:115–125

Smithies RH, Champion DC, Cassidy KF (2003) Formation of Earth's early Archaean continental crust. Precambrian Res 127:89–101

Smithies RH, Champion DC, Sun S-S (2004) Evidence for early LREE-enriched mantle source regions: diverse magmas from the c. 3.0 Ga Mallina Basin, Pilbara Craton, NW Australia. J Petrol 45:1515–1537

Smithies RH, Champion DC, Van Kranendonk MJ (2007a) The oldest well-preserved volcanic rocks on earth: geochemical clues to the early evolution of the Pilbara Supergroup and implications for the growth of a Paleoarchean continent. In: Van Kranendonk MJ, Bennett VC, Smithies RH (eds) Earth's oldest rocks. Elsevier B.V, Burlington, Massachusetts, USA, pp 339–367

Smithies RH, Champion DC, Van Kranendonk MJ (2009) Formation of Paleoarchean continental crust through infracrustal melting of enriched basalt. Earth Planet Sci Lett 281:298–306

Smithies RH, Champion DC, Van Kranendonk MJ, Hickman AH (2007b) Geochemistry of volcanic rocks of the northern Pilbara Craton, Western Australia. Report, vol 104. Geological Survey of Western Australia

Smithies RH, Van Kranendonk MJ, Champion DC (2007c) The Mesoarchean emergence of modern-style subduction. In: Maruyama S, Santosh M (eds) Island arcs: past and present, pp 50–68

Spaggiari CV, Wartho J-A, Wilde SA (2008) Proterozoic deformation in the northwest of the Archean Yilgarn Craton, Western Australia. Precambrian Res 162:354–384

Suhr N, Hoffmann JE, Kröner A, Schröder S (2015) Archaean granulite-facies paragneisses from Central Swaziland: inferences on Palaeoarchaean crustal reworking and a complex metamorphic history. J Geol Soc 172:139–152. https://doi.org/10.1144/jgs2014-007

Sun S-S, Hickman AH (1998) New Nd-isotopic and geochemical data from the west Pilbara–implications for Archaean crustal accretion and shear zone development. Australian Geological Survey Organisation, Research Newsletter:25–29

Sun S-S, Hickman AH (1999) Geochemical characteristics of ca 3.0-Ga Cleaverville greenstones and later mafic dykes, west Pilbara: implication for Archaean crustal accretion. Australian Geological Survey Organisation, Research Newsletter:23–29

Taylor J, Stevens G, Buick IS, Lana C (2012) Successive mid-crustal, high-grade metamorphic events provide insight into Mid-Archean mountain building along the SE margin of the proto–Kaapvaal Craton. Geological Society America Bulletin 124:1191–1211. https://doi.org/10.1130/B30543.1

Tessalina SG, Bourdon B, Van Kranendonk M, Birck JL, Philippot P (2010) Influence of Hadean crust evident in basalts and cherts from the Pilbara Craton. Nat Geosci 3:214–217

Thorpe R, Hickman AH, Davis D, Morton JGG, Trendall AF (1992) U–Pb zircon geochronology of Archaean felsic units in the Marble Bar region, Pilbara Craton, Western Australia. Precambrian Res 56:169–189

Tyler IM, Fletcher IR, de Laeter JR, Williams IR, Libby WG (1992) Isotope and rare earth element evidence for a late Archaean terrane boundary in the southeastern Pilbara Craton, Western Australia. Precambrian Res 54:211–229

Van Kranendonk MJ (2000) Geology of the North Shaw 1:100 000 sheet. 1:100 000 Geological Series Explanatory Notes. Geological Survey of Western Australia

Van Kranendonk MJ (2011) Cool greenstone drips and the role of partial convective overturn in Barberton greenstone belt evolution. J Afr Earth Sci 60:346–352. https://doi.org/10.1016/j.jafrearsci.2011.03.012

Van Kranendonk MJ, Collins WJ, Hickman AH, Pawley MJ (2004a) Critical tests of vertical vs horizontal tectonic models for the Archean East Pilbara granite–greenstone Terrane, Pilbara Craton, Western Australia. Precambrian Res 131:173–211

Van Kranendonk MJ, Hickman AH, Smithies RH, Champion DC (2007a) Paleoarchean development of a continental nucleus: the East Pilbara Terrane of the Pilbara Craton, Western Australia. In: Van Kranendonk MJ, Bennett VC, Smithies RH (eds) Earth's oldest rocks. Elsevier B.V, Burlington, Massachusetts, USA, pp 307–337

Van Kranendonk MJ, Hickman AH, Smithies RH, Nelson DN, Pike G (2002) Geology and tectonic evolution of the Archaean North Pilbara terrain, Pilbara Craton, Western Australia. Econ Geol 97:695–732. https://doi.org/10.2113/gsecongeo.97.4.695

Van Kranendonk MJ, Hickman AH, Smithies RH, Williams IR, Bagas L, Farrell TR (2006) Revised lithostratigraphy of Archean supracrustal and intrusive rocks in the northern Pilbara Craton, Western Australia. Record, 2006/15. Geological Survey of Western Australia

Van Kranendonk MJ, Kirkland CL, Cliff J (2015b) Oxygen isotopes in Pilbara Craton zircons support a global increase in crustal recycling at 3.2 Ga. Lithos 228-229:90–98

Van Kranendonk MJ, Kröner A, Hoffman JE, Nagel T, Anhaeusser CR (2014) Just another drip: re-analysis of a proposed Mesoarchean suture from the Barberton Mountain Land, South Africa. Precambrian Res 254:19–35. https://doi.org/10.1016/j.precamres.2014.07.022

Van Kranendonk MJ, Smithies RH, Griffin WL, Huston DL, Hickman AH, Champion DC, Anhaeusser CR, Pirajno F (2015a) Making it thick: a volcanic plateau origin of Paleoarchean continental lithosphere of the Pilbara and Kaapvaal cratons. Geol Soc Lond, Spec Publ 389:89–111. https://doi.org/10.1144/SP389.12

Van Kranendonk MJ, Smithies RH, Hickman AH, Bagas L, Williams IR, Farrell TR (2004b) Event stratigraphy applied to 700 million years of Archean crustal evolution, Pilbara Craton, Western Australia. Geological Survey of Western Australia Annual Review 2003–04:49–61

Van Kranendonk MJ, Smithies RH, Hickman AH, Champion DC (2007b) Review: secular tectonic evolution of Archean continental crust: interplay between horizontal and vertical processes in the formation of the Pilbara Craton, Australia. Terra Nova 19:1–38

Vervoort JD (2011) Evolution of the depleted mantle–revisited. In: Abstracts, abstract ID V43E-06

Vervoort JD, Blichert-Toft J (1999) Evolution of the depleted mantle: Hf isotope evidence from juvenile rocks through time. Geochim Cosmochim Acta 63:533–556

Vervoort JD, Kemp AIS (2016) Clarifying the zircon Hf isotope record of crust–mantle evolution. Chem Geol 425:65–75. https://doi.org/10.1016/j.chemgeo.2016.01.023

Viljoen MJ, Viljoen, RP (1971) The geological and geochemical evolution of the Onverwacht volcanic group of the Barberton Mountain Land, South Africa. In Symposium on Archaean rocks, Glover JE. (ed). Geol. Soc. Australia Special Pub. No. 3:133–149

Wacey D, McLoughlin N, Stoakes CA, Kilburn MR, Green OR, Brasier MD (2010) The 3426–3350 Ma Strelley Pool Formation in the East Strelley greenstone belt–a field and petrographic guide. Record, 2010/10. Geological Survey of Western Australia

White WM (2013) Geochemistry. Wiley-Blackwell, Chichester, UK

Wiemer D, Schrank CE, Murphy DT, Hickman AH (2016) Lithostratigraphy and structure of the early Archaean Doolena Gap greenstone belt, East Pilbara Terrane, Western Australia. Precambrian Res 282:121–138

Wiemer D, Schrank CE, Murphy DT, Wenham L, Allen CM (2018) Earth's oldest stable crust in the Pilbara Craton formed by cyclic gravitational overturns. Nat Geosci 11:357–361. https://doi.org/10.1038/s41561-018-0105-9

Wilhelmij HR, Dunlop JSR (1984) A genetic stratigraphic investigation of the Gorge Creek Group in the Pilgangoora syncline. In: Muhling JR, Groves DI, Blake TS (eds) Archaean and Proterozoic basins of the Pilbara, Western Australia: evolution and mineralization potential. The University of Western Australia, Geology Department and University Extension, pp. 68–88

Windley BF (1976) New tectonic models for the evolution of Archaean continents and oceans. In: Windley BF (ed) The early history of the Earth. NATO Advanced Study Institute, University of Leicester, pp 105–111

Wingate MTD, Bodorkos S, Van Kranendonk MJ (2009a) 180070: volcaniclastic sandstone, drillcore PDP2c, Dresser Mine

Wingate MTD, Bodorkos S, Van Kranendonk MJ (2009b) 180057: tonalitic orthogneiss, Big Junction Well

Wingate MTD, Kirkland CL, Bodorkos S, Hickman AH (2010) 160258: felsic metavolcanic rock. Orchard Well

Zeh A, Gerdes A, Millonig L (2011) Hafnium isotope record of the ancient gneiss complex, Swaziland, southern Africa: evidence for Archaean crust–mantle formation and crust reworking between 3.66 and 2.73 Ga. J Geol Soc Lond 168:953–963

Chapter 3
Warrawoona Large Igneous Province, 3530–3427 Ma

Abstract The 3530–3427 Ma Warrawoona Group is the oldest of the three groups that make up the Pilbara Supergroup. This 3530–3235 Ma supergroup comprises the Paleoarchean greenstone succession of the East Pilbara Terrane and is preserved in 20 greenstone belts. The group was erupted across the entire area of the 3800–3530 Ma Pilbara crust (Chap. 2), and its thickness varies between 10 and 15 km. Variations are partly due to local erosional unconformities that were formed from 3460 Ma onwards when granite–greenstone domes of the terrane began to rise at different rates. Apart from thin sedimentary units deposited between 3490 and 3474 Ma and between 3459 and 3450 Ma, the Warrawoona Group is volcanic. The succession is composed of successive ultramafic–mafic–felsic volcanic cycles in which felsic volcanism was contemporaneous with intrusion of tonalite–trondhjemite–granodiorite (TTG). Field exposures reveal that many of the TTG intrusions were subvolcanic to the felsic volcanic formations.

The East Pilbara Terrane is now exposed across 40,000 km^2 of the northeast section of the Pilbara Craton, with concealed parts estimated to occupy an additional 60,000 km^2. With an interpreted total volume of volcanic rocks exceeding 1,000,000 km^3, the Warrawoona Group easily meets the volume requirement for a large igneous province (LIP). The 3530–3300 Ma stratigraphic successions of the Pilbara and Kaapvaal Cratons are remarkably similar, even including a common volcanic hiatus between about 3426 and 3350 Ma. In the Pilbara, two lines of evidence indicate a magmatic event commencing abruptly at 3530 Ma: firstly, a major peak in the frequency of zircon aged between 3530 and 3490 Ma, preceded by an almost total lack of zircons dated between 3550 and 3530 Ma; and secondly, Lu–Hf isotope evidence for a surge of mantle-derived juvenile magmas between 3530 and 3490 Ma. This sudden magmatic activity is interpreted to coincide with the arrival of the first of a series of mantle plumes that had major impacts on the Paleoarchean crustal evolution of the craton.

Keywords Paleoarchean · Large igneous province · Pilbara Supergroup · Volcanic cycles · Zircon geochronology · Lu–Hf isotope data

A. H. Hickman, *Archean Evolution of the Pilbara Craton and Fortescue Basin*, Modern Approaches in Solid Earth Sciences 24,
https://doi.org/10.1007/978-3-031-18007-1_3

3.1 Introduction

Evidence in Chap. 2 establishes that the Paleoarchean greenstone succession of the Pilbara Craton, the Pilbara Supergroup (Fig. 3.1), was deposited on the older sialic crust. This crust was formed by at least three magmatic events between 3800 Ma and 3530 Ma. U–Pb zircon ages prove the existence of Eoarchean crust as old as 3760 Ma and Sm–Nd and Lu–Hf isotope data indicate that parts of this early crust were derived from sources extracted from the mantle at or before 3800 Ma. The minimum age of 3800 Ma is calculated irrespective of applying either depleted mantle or chondritic mantle models to the Lu–Hf data. The evidence for the 3800–3530 Ma crust has come from many different parts of the Pilbara Craton so it was evidently widespread at 3530 Ma. There is no evidence for a single sialic nucleus suggested in some interpretations.

Recognition of the existence of widespread 3800–3530 Ma crust provides important constraints for interpretation of the craton's crustal evolution. For example, along with stratigraphic, geochemical, and isotopic evidence from within the Warrawoona Group, and discussed in this chapter, the existence of extensive pre-3530 Ma sialic crust precludes interpretations that the Warrawoona Group evolved within an oceanic plate. Several oceanic plate interpretations based on Phanerozoic-style plate-tectonic models have been applied to some sections of the East Pilbara Terrane (EPT). However, apart from the existence of the early continental crust, other evidence from the PCMP (Chap. 1) is inconsistent with plate tectonic interpretations (Van Kranendonk et al. 2002, 2004; 2006a, b, 2007a, b, 2015, 2019; Hickman 2004, 2012, 2021; Hickman and Van Kranendonk 2004, 2012; Champion and Smithies 2007, 2019; Smithies et al. 2005a, b, 2019). Other research has also indicated the existence of early continental crust (Thébaud and Rey 2013; Wiemer et al. 2016, 2018; Gardiner et al. 2017, 2018; Johnson et al. 2017).

The Pilbara Supergroup comprises three groups: in ascending stratigraphic order, the 3530–3427 Ma Warrawoona Group, 3350–3315 Ma Kelly Group, and the 3280–3235 Ma Sulphur Springs Group (Fig. 3.1). Whereas the Kelly and Sulphur Springs Group are each composed of a single ultramafic–mafic–felsic volcanic cycle, the Warrawoona Group contains evidence of six cycles (Hickman 2011, 2012; Hickman and Van Kranendonk 2012). The six cycles are identified in four subgroups, in ascending stratigraphic order: the 3530–3490 Ma Talga Talga and Coonterunah Subgroups (correlated units in different greenstone belts), 3474–3459 Ma Coongan Subgroup, and the 3450–3427 Ma Salgash Subgroup. Due to post-3530 Ma granitic intrusion and vertical deformation, the original stratigraphic contact of the Warrawoona Group with the pre-3530 Ma crust has not been preserved. However, the consistent stratigraphy of the Talga Talga and Coonterunah Subgroups across most greenstone belts indicates that, prior to 3460–3426 Ma uplift and erosion, these groups formed an essentially 'layer-cake' succession across large areas of the 3800–3530 Ma crust (Hickman 1981, 1983, 2021).

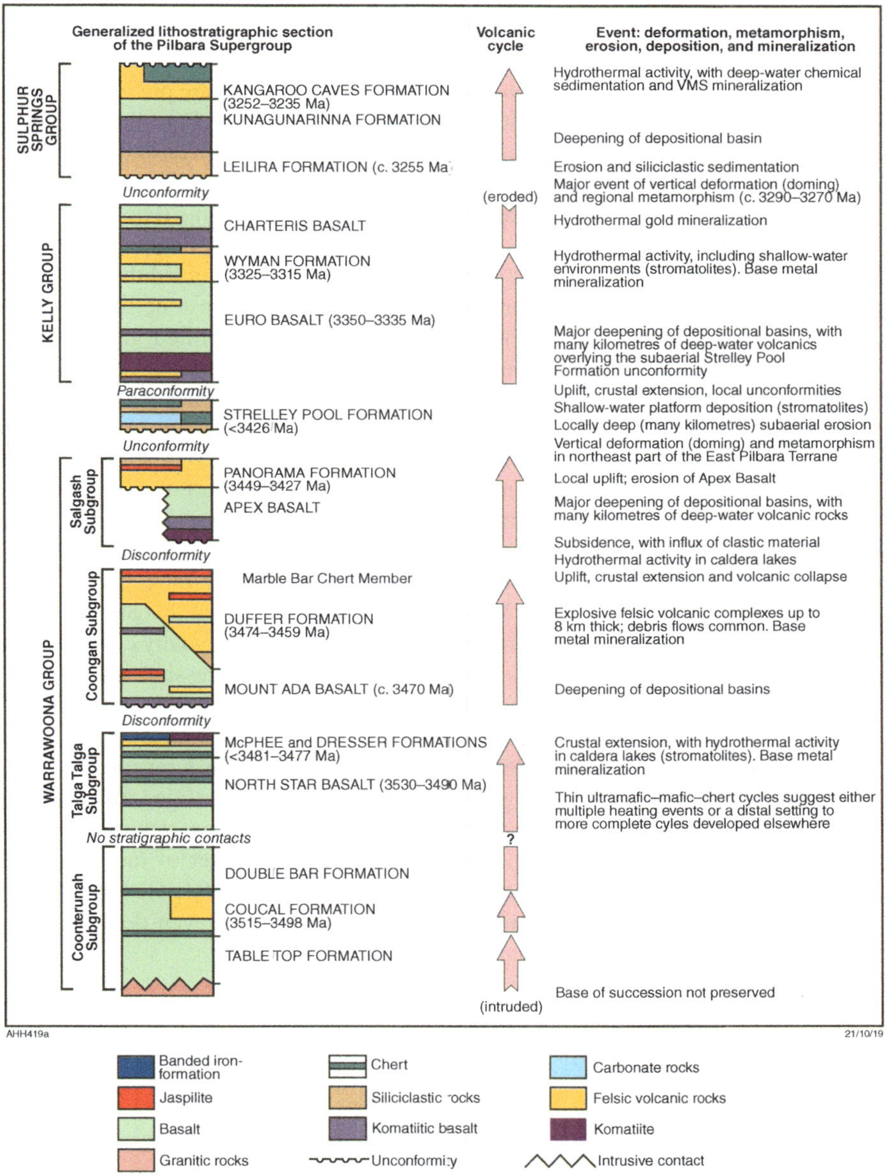

Fig. 3.1 Generalized lithostratigraphy of the Pilbara Supergroup. The succession is composed of multiple volcanic cycles, mostly separated by unconformities. Geochronology indicates that the Coonterunah and Talga Talga Subgroups are the same age, and each contains three volcanic cycles. Events of deformation, metamorphism, erosion, and mineralization are summarized. (Modified from Hickman 2011; with Geological Survey of Western Australia permission)

3.1.1 *Preservation of the Warrawoona Group*

Paleoarchean greenstone successions of the Pilbara Supergroup are preserved within the EPT of the East Pilbara and within the Karratha Terrane (KT) of the northwest Pilbara (Fig. 1.3). The exposed part of the EPT contains 20 greenstone belts that separate domal granitic complexes, and together the greenstone belts and granitic complexes form eleven domes separated by major boundary faults (Hickman 2001, 2012, 2021; Hickman and Van Kranendonk 2004) (Fig. 1.7). Following nomenclature used in other Archean terranes worldwide, it was first interpreted that the linear and arcuate outcrops of volcanic and sedimentary rocks ('greenstone belts') in the EPT are contained within synclinal structures between the uplifted granitic complexes (Hickman 1981, 1983). Elsewhere in the world, such 'synclinal greenstone belts' had commonly been interpreted to correspond to separate depositional basins (Anhaeusser et al. 1968, 1969; Glikson 1970, 1971, 1972; Goodwin and Ridler 1970; Stowe 1971; Windley and Bridgwater 1971; Windley 1973). The first detailed stratigraphic mapping of the EPT in the 1970s indicated that the Pilbara greenstone belts originated by local down-folding (later referred to as 'sagduction') of a laterally continuous greenstone succession between areas of gravity-driven diapiric doming (Hickman 1975, 1981, 1983, 1984; Hickman and Van Kranendonk 2004). More detailed mapping during the 1994–2005 PCMP (Chap. 1) revealed that almost all the 'greenstone belts' of the EPT are not simply synclines but are actually composed of structurally juxtaposed sections of the regional greenstone succession separated by faults with major vertical displacements (Van Kranendonk 1998; Hickman 2001). Therefore, the previously named greenstone belts (Hickman 1980b) were separated into a greater number of newly defined greenstone belts (Van Kranendonk et al. 2002) (Fig. 1.7). Each of these greenstone belts has a succession that closely correlates with the granitic stratigraphy of the immediately adjacent granitic complex (Hickman 2021). Consequently, the EPT is essentially a cluster of fault-bounded granite–greenstone domes (Hickman and Van Kranendonk 2004; Hickman 2011, 2021), although due to limited uplift and erosion two of the eleven domes (North Pole and McPhee) do not expose large granitic cores.

3.2 Stratigraphy

The Warrawoona Group comprises four subgroups that collectively contain ten formations (Table 3.1). The Coonterunah Subgroup is restricted to greenstone belts in the northwest part of the EPT where the Coongan and Salgash Subgroups are almost entirely absent due to 3460–3420 Ma erosion. The type area of the Warrawoona Group is in the Marble Bar greenstone belt, where all units except the Coonterunah Subgroup are well exposed and form a right-way-up succession 10–15 km thick.

Table 3.1 Generalized lithostratigraphy of the East Pilbara Craton (Modified from Hickman 2021; with Geological Survey of Western Australia permission)

Supergroup	Group (Basin/LIP)	Subgroup	Formation	Age (Ma)*
	Nullagine Group		Mosquito Creek Formation	2980–2930
	(Mosquito Creek Basin)		*PARACONFORMITY*	
			Coondamar Formation	3220–3175
	No contact			
De Grey Supergroup	Croydon Group		Lalla Rookh Sandstone	3015–2940
	(Mallina Basin)		Cattle Well Formation	<2988
	Disconformity			
			Coonieena Basalt	3015–2950
	Erosional unconformity			
	Gorge Creek Group		Cundaline Formation	c. 3015
	(Gorge Creek Basin)		Cleaverville Formation	3022–3015
			Farrel Quartzite	3067–3022
	Unconformity			
	(Budjan Creek Basin)		Budjan Creek Formation	3223–3165
	No contact			
			Empress Formation	3176–3165
			Hong Kong Chert	3176–3165
	Soanesville Group		Pyramid Hill Formation	3176–3165
	(Soanesville Basin)		Honeyeater Basalt	3185–3176
			Paddy Market Formation	3190–3185
			Corboy Formation	c. 3190
			Cardinal Formation	3223–3190
	Unconformity			
Pilbara Supergroup	Sulphur Springs Group		Kangaroo Caves Formation	3252–3235
	(Sulphur Springs Basin)		Kunagunarrina Formation	3275–3253
			Leilira Formation	3290–3755

(continued)

Table 3.1 (continued)

Supergroup	Group (Basin/LIP)	Subgroup	Formation	Age (Ma)*
	Erosional unconformity			
	Kelly Group		Charteris Basalt	c. 3315
			Wyman Formation	3325–3315
			LOCAL UNCONFORMITY	
	Kelly LIP		Euro Basalt	3350–3335
	Disconformity			
			Strelley Pool Formation	3426–3350
	Erosional unconformity—Paraconformity			
	Warrawoona Group	Salgash Subgroup	Panorama Formation	3449–3427
	(Warrawoona LIP)		*LOCAL UNCONFORMITY*	
			Apex Basalt	3455–3441
		Disconformity		
		Coongan Subgroup	Duffer Formation	3474–3459
			Mount Ada Basalt	3474–3469
		Disconformity		
		Talga Talga Subgroup	Dresser/McPhee Formation	3481–3477
			North Star Basalt	3530–3490
		No contact		
		Coonterunah Subgroup	Double Bar Formation	3498–3490
			Coucal Formation	3515–3498
			Table Top Formation	3530–3515
	Inferred unconformity (not exposed)			
Early crust				3800–3530

* Age ranges are inferred from isotopic data and stratigraphic relations

The Warrawoona LIP includes granitic rocks related to the same mantle plume events as the volcanic succession, and in some instances visibly crystallized in magma chambers that fed overlying felsic volcanics. Geochemical and isotopic data from the granites is therefore relevant to the origin of the volcanics, and is noted during the description of the Warrawoona Group. Detailed accounts of the geochemistry of the granitic rocks in the LIP are available elsewhere (Champion and Smithies 2007, 2019; Smithies et al. 2003), but a general review of the granites is included at the end of this chapter.

3.2.1 Coonterunah Subgroup

The Coonterunah and Talga Talga Subgroups are composed of ultramafic–mafic volcanic cycles in which basal komatiites pass upwards into komatiitic basalts and tholeiites. Relatively minor thicknesses of felsic volcanic rocks occur in the central part of the Coonterunah Subgroup (Smithies et al. 2007, 2009) where they have been dated between 3515 Ma (Buick et al. 1995) and 3498 Ma (Nelson 2002). Granitic intrusions contemporaneous with these felsic volcanic units have been dated between 3523 and 3496 Ma (Wiemer et al. 2018; Petersson et al. 2020).

In the East Strelley and western Warralong greenstone belts on the east and south sides of the Carlindi Dome (Fig. 1.7), the Coonterunah Subgroup is a 6-km-thick volcanic succession composed of tholeiitic basalt with lesser amounts of komatiite, komatiitic basalt, and felsic volcanic rocks. The subgroup is correlated with the Talga Subgroup in other greenstone belts of the EPT, and might also be present in the Goldsworthy and Pilbara Well greenstone belts. The subgroup comprises three conformable formations: in ascending stratigraphic order, the Table Top, Coucal, and Double Bar Formations. The succession was folded, metamorphosed to amphibolite facies, intruded by granitic rocks of the Mulgandoona Supersuite (3530–3490 Ma) and Callina Supersuite (3484–3462 Ma), and deeply eroded prior to deposition of the immediately overlying 3426–3350 Ma Strelley Pool Formation (Chap. 4). These geological features establish that the tectonic setting was not oceanic at c. 3426 Ma. In the East Strelley greenstone belt, uplift and erosion between 3450 and 3426 Ma are interpreted to have removed the Coongan and Salgash Subgroups (upper Warrawoona Group) that overlay the Coonterunah Subgroup. Parts of that younger succession are preserved in the upper part of the Goldsworthy greenstone belt on the northern side of the Carlindi Dome (Fig. 1.7).

Enrichment in K, LILE, Th, and LREE in the basaltic to andesitic rocks of the Coucal Formation might not be the result of crustal assimilation (Smithies et al. 2009). These enrichments might alternatively have been inherited from a mantle source that was enriched in recycled crustal components. Smithies et al. (2007) classified the komatiites at the base of the subgroup as Al-undepleted, with flat normalized trace element patterns. They reported Al_2O_3/TiO_2 ratios between 20 and 24 and noted that Gd/Yb ratios are chondritic (1.1 to 1.3). Arndt and Lesher (2004) recorded that Al-undepleted komatiites, which they referred to as 'Munro-type', typically have Al_2O_3/TiO_2 ratios close to 20 and near-chondritic Gd/Yb ratios, whereas Al-depleted komatiites ('Barberton-type') have ratios close to 10, with higher Gd/Yb ratios approximately 1.3–1.5.

Sm–Nd isotope data suggest that magmas for both the Coonterunah and the Talga Talga Subgroups were derived from mixed sources including the 3530–3490 Ma mantle and the pre-3530 Ma crust. Samples of komatiitic basalt, basalt, andesite, dacite, and rhyolite ($n = 20$, data reviewed in Hickman 2021) have provided Nd $T_{DM}2$ model ages ranging from 3670 to 3550 Ma and ε_{Nd} values varying between +1.71 and + 0.27 (Table 2.2). Nd model ages suggest the times at which the sources of rock-forming magmas were extracted from the mantle (McCulloch and

Wasserburg 1978) but crystallizing magmas in continental settings were rarely derived from a single mantle source. Most were derived from multiple sources that had been extracted from the mantle at different times. In this situation, individual Nd model ages indicate only the average times from which materials in particular samples had been resident in the continental crust (Arndt and Goldstein 1987). All Nd model ages from the Warrawoona Group and contemporaneous granitic rocks are less than 3700 Ma ($n = 67$, reviewed in Hickman 2021). These Nd model ages from the EPT are in marked contrast to Hf model ages from zircons that crystallized before 3530 Ma indicate sources substantially older than 3800 Ma (Chap. 2) and suggest an influx of juvenile magma at about 3530 Ma (Hickman 2021).

Four samples of komatiite from the Coonterunah Subgroup (Nebel et al. 2014) and one komatiite sample from the Talga Talga Subgroup (Hamilton et al. 1981) provided ε_{Nd} values averaging +3.5. These strongly positive ε_{Nd} values suggest a mantle source. However, the komatiite data are probably not representative of the entire volcanic succession because Nebel et al. (2014) reported similar strongly positive ε_{Hf} values from a much wider age range of Pilbara komatiites ($n = 21$) and explained this as evidence that all the komatiite lavas were derived from the lower mantle. Blichert-Toft and Arndt (1999) also obtained consistently high positive ε_{Hf} and ε_{Nd} values from komatiites of the Barberton Greenstone Belt in South Africa, and from various komatiites of other Archean greenstone belts. These results suggest that Sm–Nd and Lu–Hf isotope data from komatiites should be considered separately from Sm–Nd and Lu–Hf data obtained from mafic and felsic rocks.

3.2.1.1 Table Top Formation

The oldest formation of the Coonterunah Subgroup, the Table Top Formation, is up to 3.45 km thick (Van Kranendonk 2000) and comprises a volcanic succession of massive and pillowed tholeiitic basalt with komatiite, komatiitic basalt, high-Al basalt, andesite, chert, jaspilitic, and minor BIF (Glikson et al. 1986; Van Kranendonk 2000; Smithies et al. 2007). Sills of dolerite and gabbro are common throughout the Table Top Formation and much of the succession is metamorphosed to amphibolite facies. The formation outcrops on the southeast side of the Carlindi Dome (Fig. 1.7) where it occupies parts of the East Strelley and Warralong greenstone belts. It might also form part of the Pilbara Well greenstone belt on the western side of the Yule Dome (Hickman 2021). In the East Strelley greenstone belt, the formation is extensively intruded by granitic rocks of the 3484–3462 Ma Callina Supersuite, confirming its age as >3484 Ma. U–Pb zircon dating on the conformably overlying Coucal Formation indicates that the depositional age of the Table Top Formation is >3515 Ma (Buick et al. 1995). Correlation of the Coonterunah and Talga Talga Subgroups (Hickman 2021) implies that the Table Top Formation originated as part of an extensive basaltic succession deposited on pre-3530 Ma sialic crust. Green et al. (2000) reported the presence of undated granitic boulders in the Coonterunah Subgroup.

3.2.1.2 Coucal Formation

Conformably overlying the Table Top Formation, the Coucal Formation is up to 1.34 km thick (Van Kranendonk 2000) and comprises metamorphosed tholeiitic basalt, andesitic to dacitic volcanic and volcaniclastic units, massive porphyritic dacite, vitric tuff, and sandstone. Metamorphosed chert units, varying in thickness up to 10 m, are locally associated with BIF, silicified pelitic and layered carbonate rocks, and outcrop over strike lengths up to 5 km. Buick et al. (1995) dated a sample of brecciated hyaloclastic rhyolite at 3515 ± 3 Ma, whereas a quartz-phyric rhyolite was dated at 3498 ± 2 Ma (Nelson 2002).

The Coucal Formation is geochemically distinct from the Table Top Formation in that the tholeiitic basalts have higher concentrations of incompatible trace elements (Th, U, Nb, Zr) and LREE. Smithies et al. (2007) recognized two geochemical series of volcanic rocks: F1 ranging from andesite to dacite and dominating the lower half of the sequence, and F2 ranging from basalt to andesite in the upper half. F1 rocks have SiO_2 contents between 55 and 65 wt.% whereas SiO_2 in F2 is between 48 and 58 wt.%. Both series have relatively low K_2O (<1.0 wt. %) and are Fe-rich. Compared to the F2 series, Coonterunah F1 rocks have slightly higher La/Sm, La/Nb, and La/Yb ratios, possibly due to greater magma interaction with the felsic crust. Smithies et al. (2007) concluded that geochemical data indicate that the felsic volcanic rocks of the Coucal Formation are andesites and dacites derived from fractionation of tholeiitic parental magmas, variably contaminated by crustal material. Smithies et al. (2009) used the unusual chemical compositions of the Coucal Formation basalts to explain Paleoarchean TTG generation in the absence of subduction. In their model, Paleoarchean TTG magmas were derived through infracrustal partial melting of basaltic crust that was similar in composition to the Coonterunah F2 series (enriched K, LILE, Th, and LREE). They proposed that a thick mafic protocrust, dominated by F2-type basalts, formed in the Pilbara Craton between 3730 and 3530 Ma. This early crust was possibly by melting of primitive asthenospheric mantle enriched in recycled felsic crust. The protocrust probably included more typical Archean basalts, fractionated basaltic andesite, and local intrusions of low-Al TTG (Smithies et al. 2009). Nd model ages and ε_{Nd} values are very similar to those of the Table Top Formation, and to those of basalts in the Talga Talga Subgroup.

3.2.1.3 Double Bar Formation

The Double Bar Formation is up to 1.87 km thick (Van Kranendonk 2000) and comprises a succession of tholeiitic pillow basalt, basaltic volcaniclastic rocks, massive basalt, thin chert units, and dolerite sills. Folding prior to erosion and deposition of the Strelley Pool Formation has caused the formation to be absent between Sulphur Springs Creek (Zone 50, MGA 727170E 7,665,340 N) and Strelley Pool (Zone 50, MGA 722270E 7,664,080 N). Sericitization and silicification of the basalt for depths up to 20 m underlying the unconformity has been interpreted as representing a Paleoarchean paleosol (Altinok 2006). Lindsay et al. (2005) interpreted hydrothermal alteration at this level, but the 100 million year interval between Double Bar volcanism and deposition of the Strelley Pool Formation, and the angular nature of the unconformity, indicate that any hydrothermal alteration must have been no older than the Strelley Pool Formation.

3.2.2 *Talga Talga Subgroup*

In the type area of the Talga Talga Subgroup, around McPhee Reward mine 20 km north of Marble Bar (Marble Bar greenstone belt), the subgroup comprises the 3530–3490 Ma North Star Basalt and the c. 3477 Ma McPhee Formation. The same succession is exposed in the Coongan, North Shaw, Doolena Gap, and Panorama greenstone belts, although in the latter the McPhee Formation is exposed as the lithologically similar 3481 Ma Dresser Formation. Two $^{40}Ar/^{39}Ar$ dates of 3520 Ma for hornblende from amphibolite-facies metabasalts of the North Star Basalt in the Coongan and North Shaw greenstone belts (Davids et al. 1997; Zegers et al. 1999) suggest that the North Star Basalt and Coonterunah Subgroup are similar in depositional age.

The McPhee Formation comprises grey and white banded chert (interpreted as a silicified fine-grained volcaniclastic or carbonate rock), iron formation, metapelite, basalt, dolerite, and carbonated ultramafic rocks. The ultramafic rocks are interpreted to be komatiite sills that were intruded during eruption of the first lavas of the overlying Mount Ada Basalt (Coongan Subgroup). Silicified felsic tuff in the McPhee Formation of the Marble Bar greenstone belt was dated at 3477 ± 2 Ma (Nelson 2000). The Dresser Formation, which is confined to the Panorama greenstone belt, is composed of grey and white banded chert, basalt, and minor felsic volcaniclastic and carbonate rocks. Dolerite sills are likely to be subvolcanic intrusions of the overlying Mount Ada Basalt. A felsic volcaniclastic unit was dated at 3481 ± 3 Ma (Wingate et al. 2009b), although this date is based on only two analyses of a single zircon.

3.2.2.1 North Star Basalt

In well-exposed sections between the Talga Talga and Coongan Rivers 20 km north of Marble Bar, the North Star Basalt is over 2 km thick, and varies in metamorphic grade from amphibolite facies in its lower part to greenschist facies at higher stratigraphic levels. In these sections, the formation is composed of three volcanic cycles (Van Kranendonk et al. 2006a). The lower cycle of the North Star Basalt is a succession of metamorphosed pillow basalts and massive basalt with thin units of altered komatiite adjacent to intrusive granitic rocks of the 3484–3462 Ma Callina Supersuite. The stratigraphic thickness of the formation varies between 1 and 2 km thick in different greenstone belts. Metamorphic grade increases to amphibolite facies close to basal contacts with the Callina Supersuite, and the lowermost section of the formation north of Marble Bar is a complex assemblage of sheared amphibolite intruded by numerous veins of granite and pegmatite. The lower cycle is incomplete and its thickness is uncertain due to deformation and granitic intrusion.

The middle cycle is about 500 m thick and consists of a basal unit of carbonated and serpentinized olivine-cumulate peridotite overlain by metapyroxenite and metadolerite. Above the metadolerite, the remainder of the cycle is composed of less metamorphosed, massive, and pillowed tholeiitic basalt overlain by an upper unit of grey and white banded chert and BIF about 10 m thick. This chert unit, which is inferred to be older than 3490 Ma, might be similar in age to chert and BIF in the Coucal Formation, and therefore one of the oldest sedimentary units in the Warrawoona Group. Unsuccessful attempts have been made to find evidence of early life in the Coucal Formation, but there have been no similar studies on cherts of the North Star Basalt. Thick sills and veins of porphyritic dacite and rhyolite which intrude the basalts of the middle cycle have not been dated but are interpreted to belong to the c. 3465 Ma Homeward Bound Granite.

Above the chert and BIF at the top of the second cycle, the upper cycle is 400 m thick and mainly composed of komatiitic basalt and tholeiitic basalt intruded by sills of dolerite and gabbro. The stratigraphic base of this cycle is composed of komatiite and the top of the cycle consists of extremely silicified basalt (hydrothermal alteration) and local lenses of dacite beneath the thick basal chert unit of the overlying McPhee Formation.

In the southern Coongan greenstone belt (Fig. 1.7), Bagas et al. (2004) described the North Star Basalt as composed of metamorphosed tholeiitic basalt with several units of chlorite–tremolite schist and talc–carbonate rock. These Mg-rich units, which are 20–50 m thick, most likely represent komatiitic basalt and komatiite, although amphibolite facies metamorphism has destroyed primary textures through much of the succession. The upper part of the formation locally includes a 100–200-m-thick unit of metamorphosed komatiite and komatiitic basalt overlain by banded chert, BIF, and units of quartz-rich sandstone (Bagas et al. 2004). This sandstone may be of similar depositional age to sandstone and pelite in the Coucal Formation. The ultramafic unit is lenticular, probably due to tectonic attenuation, and was mapped for only 8 km along strike but the overlying chert extends for 13 km. It is overlain by komatiitic basalt and tholeiitic basalt, which might be equivalent to the upper volcanic cycle of the North Star Basalt in the Marble Bar greenstone belt.

In the Panorama greenstone belt, the North Star Basalt occupies the core of the North Pole Dome where it is intruded by the 3440-Ma North Pole Monzogranite. In this area, the formation is at least 2 km thick and composed of weakly metamorphosed massive and pillowed basalt, komatiitic basalt, basaltic hyaloclastite, gabbro, and dolerite. The formation is disconformably overlain by the Dresser Formation dated at about 3481 Ma but its stratigraphic base is not exposed. As in the North Star Basalt of the Marble Bar greenstone belt and the Double Bar Formation of the Coonterunah Subgroup of the East Strelley greenstone belt, komatiitic basalt at the top of the formation is extensively altered, and in this locality the alteration is more conclusively hydrothermal (Kitajima et al. 2001; Terabayashi et al. 2003; Van Kranendonk and Pirajno 2004; Brown et al. 2005, 2011; Van Kranendonk 2006; Van Kranendonk et al. 2006b, 2008). The metamorphic grade of the North Star Basalt is mainly lower greenschist facies except close to the contact with the North Pole Monzogranite where a lower amphibolite-facies metamorphic aureole is developed (Van Kranendonk et al. 2006b).

Excluding data from one sample of komatiite, whole-rock Nd ($T_{DM}{}^2$) model ages on the North Star Basalt (Smithies et al. 2007) indicate average crustal source ages of 3670–3550 Ma. The komatiite has a strongly positive ε_{Nd} value of +2.31 and a juvenile Nd model age of 3510 Ma suggesting that, unlike the basalts, its source was the lower mantle (Nebel et al. 2014). Basalts of the North Star Basalt are high-Ti and include a minority of rocks geochemically similar to the Coonterunah F2 series (enriched K, LILE, Th, and LREE). Geochemical traverses in the Marble Bar greenstone belt (Glikson and Hickman 1981a, b) revealed upward geochemical trends in the subgroup (Figs. 3.2 and 3.3) and local andesitic and dacitic rocks at present near the top of the North Star Basalt.

3.2.2.2 Dresser Formation

The Dresser Formation is a very important stratigraphic unit of the Warrawoona Group for two reasons: firstly, it contains the best evidence for 3480 Ma of life on Earth; and secondly, it is the oldest formation to reveal that deposition of the Warrawoona Group included shallow-water and subaerial environments. In terms of its lithological composition, most of the <1000-m thickness of the formation is composed of pillowed komatiitic basalt. However, the Dresser Formation also contains up to three thick chert units, the lowest of which is fossiliferous and locally contains large deposits of barite. The lower chert also contains thin clastic and chemical sedimentary rocks and includes evaporites.

Fossils of the Dresser Formation include stromatolites (Walter et al. 1980; Buick et al. 1981; Groves et al. 1981; Walter 1983; Schopf and Walter 1983; Van Kranendonk 2000, 2006, 2007, 2010a; Schopf 2006; Van Kranendonk et al. 2008; Djokic et al. 2017), microfossils (Dunlop et al. 1978; Schopf and Walter 1983; Ueno et al. 2001a, b, 2004, 2006; Schopf 2006; Philippot et al. 2007; Glikson et al. 2008, 2012; Djokic et al. 2017), and microbially induced sedimentary structures (Noffke et al. 2013). Stromatolites are locally exposed on bedding planes and in cross-sections (Fig. 3.4).

The formation provides valuable insights into the range of depositional environments in the Warrawoona Group at 3480 Ma (Dunlop 1978; Barley et al. 1979; Walter et al. 1980; Dunlop and Buick 1981; Buick et al. 1984; Buick and Dunlop 1990; Nijman et al. 1998, 1999; Van Kranendonk 2000, 2006, 2007; Van Kranendonk and Pirajno 2004; Van Kranendonk et al. 2007a, b; Djokic et al. 2017). Reviews of the fossil and geochemical evidence for ancient life in the Dresser Formation have been provided by Schopf (2006), Van Kranendonk (2007, 2010a, b), Ueno (2007), and Van Kranendonk et al. (2008).

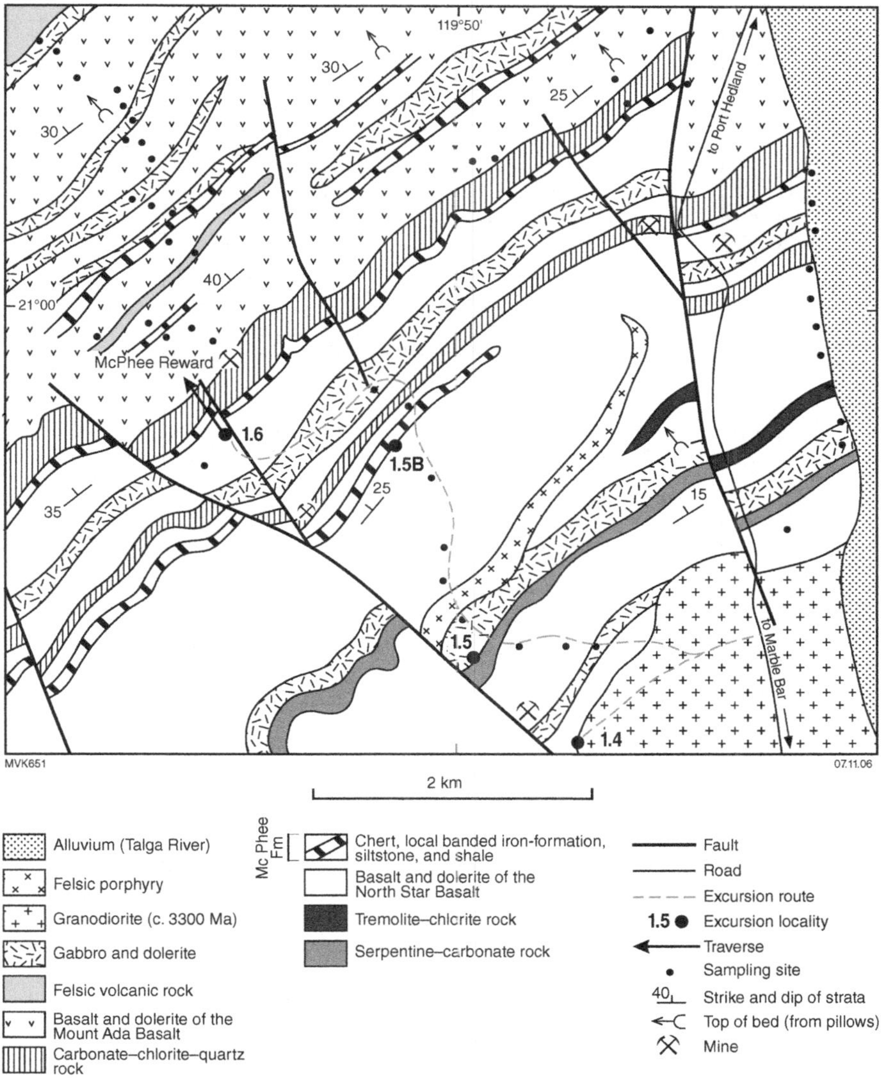

Fig. 3.2 Geological map of the Talga Talga Subgroup in the McPhee Reward area showing geochemical sample sites and localities visited on a field excursion (From Hickman 1980a; with Geological Survey of Western Australia permission)

The lower chert unit, also informally referred to as the 'chert–barite unit', varies between 50 and 8 m thick (Van Kranendonk et al. 2008). This lithologically diverse member of the Dresser Formation comprises secondary chert (silicified sedimentary carbonate rocks), sandstone (including silicified carbonate grainstone), conglomerate, detrital carbonate rocks, hydrothermal chert (stratiform and in veins), veins and

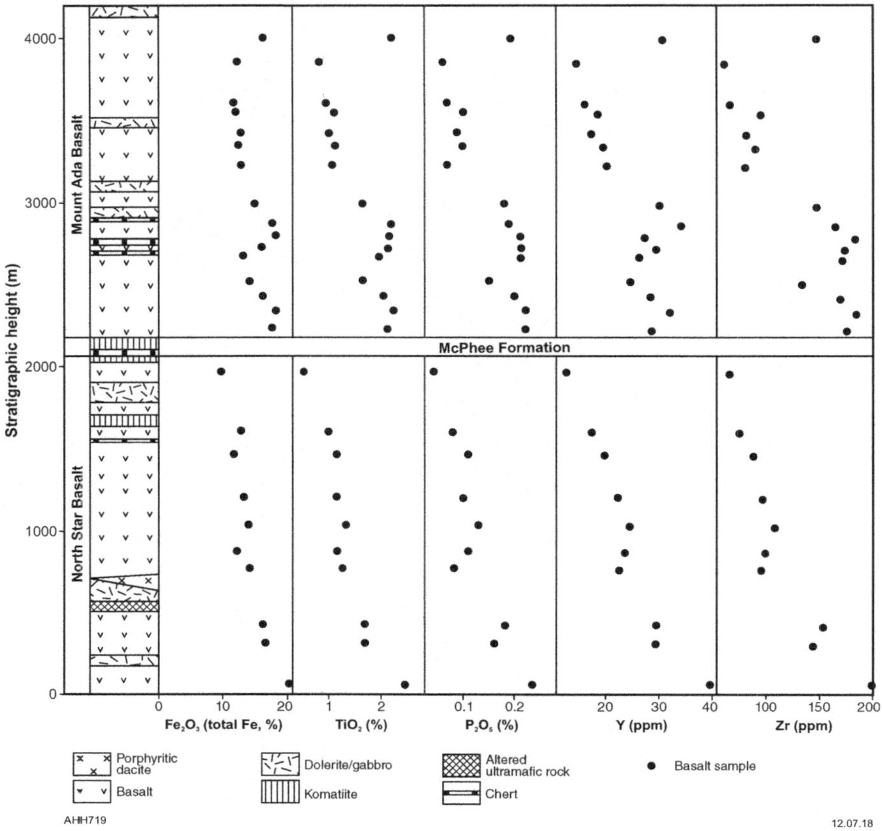

Fig. 3.3 Vertical geochemical trends in basaltic rocks of the North Star and Mount Ada Basalts, Marble Bar greenstone belt. The data indicate two similar volcanic cycles separated by the McPhee Formation. Geochronology establishes a continuous stratigraphic succession without structural repetition (Modified from Hickman 1980a; with Geological Survey of Western Australia permission)

layers of barite, and thin felsic volcaniclastic units. Very minor sulphide mineralization (pyrite, sphalerite, and galena) is present within the chert–barite sections of the formation, with pyrite replacing carbonate. Investigations of the formation were initially related to its mineral potential following the discovery of exceptionally large deposits of barite (Hickman 1973, 1983; Abeysingh and Fetherston 1997). In the early 1970s, it was thought possible that, in addition to barite, the formation might contain significant volcanogenic massive sulphide (VMS) mineralization; however, subsequent exploration did not identify any potentially economic VMS deposits.

The origin of the barite has been controversial, with interpretations varying between primary sedimentary deposition (Hickman 1973; Perry et al. 1975), replacement of gypsum in evaporites (Dunlop 1978; Lambert et al. 1976; Barley et al. 1979;

Fig. 3.4 Stromatolites and microbial mats in the c. 3481 Ma Dresser Formation. (**a**) Outcrop section through domical stromatolite overlying a bed composed of silicified weakly laminated microbial mats; (**b**) Dresser domical stromatolite enlargement (centre right on **a**); (**c**) Bedding plane view of a domical stromatolite (From Van Kranendonk et al. 2008; with Geological Survey of Western Australia permission)

Dunlop et al. 1978; Buick and Barnes 1984; Dunlop and Buick 1981; Buick and Dunlop 1990), and syn-depositional hydrothermal mineralization (Nijman et al. 1998, 1999; Van Kranendonk 2000, 2006; Nijman et al. 2001; Runnegar et al. 2001; Van Kranendonk and Pirajno 2004; Brown et al. 2006, 2011; Van Kranendonk et al. 2008; Harris et al. 2009; Djokic et al. 2017). As interpretations of the origin of the barite changed so did interpretations of the depositional environment of the Dresser Formation. The interpretation of primary gypsum implied shallow-water, intertidal to supratidal sedimentary deposition in a series of shallow-water evaporative basins or lagoons marginal to a larger ocean, whereas the hydrothermal interpretation suggested deposition around hydrothermal vents and hot springs, possibly within one or more volcanic calderas.

Opinions still differ concerning the relative importance of sedimentary and hydrothermal environments for the Dresser Formation, with Noffke et al. (2013) arguing for shallow-water sedimentary deposition whereas Djokic et al. (2017) present evidence for an on-land hydrothermal hot spring environment. In fact, the sedimentary and hydrothermal models are not mutually exclusive because large barite deposits are restricted to the Dresser Formation on the northeast side of the

North Pole Dome, and in that area they are concentrated in three or four main clusters. Thus, the more recent emphasis on hydrothermal deposition is arguably biased from studies that have been focused on the most mineralized sections of the formation. The predominant depositional environment of the Dresser Formation across most of the Panorama greenstone belt might have been similar to that of the Strelley Pool Formation which, as described by Hickman (2010), consisted of shallow-water sandstone and platform carbonate deposition during a lengthy break in volcanic activity. Intense hydrothermal activity in the Dresser Formation was apparently restricted to zones of extensive syn-depositional faulting (Nijman et al. 1998, 1999).

Correlation of the Dresser Formation with the c. 3477 Ma McPhee Formation favours fault control of the barite because the McPhee Formation, deposited over a much wider area of the EPT, does not contain growth faults and does not contain barite. Based on regional stratigraphy and available geochronology, the Dresser and McPhee Formations together represent a significant break in volcanism between the Talga Talga and Coongan volcanic cycles (Fig. 3.1). The protoliths of the individual secondary chert members of the Dresser Formation were most likely deposited over millions of years, as interpreted for the Marble Bar Chert Member of the Duffer Formation (Glikson et al. 2016).

The stratigraphy of the Dresser Formation around the North Pole Dome is complicated by extensive faulting. Some of the faults are growth faults (Nijman et al. 1998) but others cross-cut the Dresser Formation. Hickman (1973) explained these faults, which statistical analysis indicated to be approximately radial around the dome, as extensional structures related to doming. Alternatively, a crustal extension might have occurred above the head of the mantle plume interpreted to be responsible for eruption of the overlying Coongan Subgroup. Extensional fracture systems were also developed immediately beneath the 3474–3459 Ma volcanic successions of the Salgash Subgroup and the 3350–3315 Ma Kelly Group (Kloppenburg 2003; Van Kranendonk et al. 2006a; Hickman and Van Kranendonk 2008a, b; Hickman 2012), both of which are attributed to mantle plumes.

As noted above, the first detailed studies of the sedimentology, mineralogy, and paleontology of the formation indicated tidal to subaerial deposition, possibly along the shore of an ocean or in shallow-water evaporative basins or lagoons (Barley 1978; Dunlop 1978; Lambert et al. 1976; Barley et al. 1979; Dunlop et al. 1978; Dunlop and Buick 1981; Groves et al. 1981; Buick and Barnes 1984; Buick and Dunlop 1990). These workers interpreted barite interbedded with chert in the Dresser Formation to have originated by diagenetic replacement of gypsum in evaporites. However, Runnegar et al. (2001) used X-ray computerized tomography to image the barite crystals and concluded that the primary mineralogy was barite, not gypsum. Other investigations of the Dresser Formation interpreted widespread hydrothermal deposits within and adjacent to the extensional faults (Nijman et al. 1998; Van Kranendonk 2000, 2006, 2010a, b; Ueno et al. 2001a, b; Van Kranendonk et al. 2002, 2008; Van Kranendonk and Pirajno 2004; Pirajno and Van Kranendonk 2005; Harris et al. 2009). A conspicuous feature of the barite, both in discordant veins along extensional faults and in bedding parallel deposits within the formation, is that it forms layers of crystals, generally 10–20 cm thick, separated either by mm thick stylolite seams or chert layers of similar thickness to the barite layers (Hickman 1973). Harris et al. (2009) interpreted geochronological evidence for pulses of hydrothermal activity during and after deposition of the Dresser Formation, with activity continuing until c. 3420 Ma.

The shallow-water depositional environment of the Dresser formation combined with it being immediately underlain and overlain by pillow basalt formations several km thick indicates alternating crustal uplift and subsidence during deposition of the Warrawoona Group. Syn-depositional extensional faulting (Nijman et al. 1998) and the occurrence of locally derived conglomerate in the formation (Van Kranendonk et al. 2008) testify to crustal instability and periodic high-energy depositional environments. The vast majority of the zircons in a sandstone of the formation were dated at c. 3525 Ma (Wingate et al. 2009b) but there are no exposed felsic rocks of this age in the North Pole Dome. A detrital interpretation for the zircons would therefore imply derivation from older greenstones or granitic rocks from outside the North Pole Dome area, most likely the Carlindi Dome to the northwest (Fig. 1.7).

3.2.2.3 McPhee Formation

The McPhee Formation is exposed in the Marble Bar, Coongan, North Shaw, and Warralong greenstone belts of the EPT (Fig. 1.7) (Van Kranendonk 2010a, b; Hickman 2011). The distinguishing feature of the McPhee Formation which, like the Dresser Formation in the Panorama greenstone belt, separates thick basaltic successions of the North Star and Mount Ada Basalts, is its thin and lithologically mixed succession of grey- and white-banded chert, metapelite, BIF, ultramafic rocks, basalt, and dolerite. The ultramafic and mafic rocks are interpreted to include sills related to the overlying Mount Ada Basalt. Because the stratigraphic contact at the top of the McPhee Formation is commonly sheared, some of the altered komatiitic units might be basal lavas of the Mount Ada Basalt. Evidence of felsic volcanic activity during deposition of the McPhee Formation is provided by silicified felsic tuff dated at 3477 ± 2 Ma (Fig. 3.5) (Nelson 2000, GSWA 148498). The lithological diversity of the McPhee Formation, with resulting variations in rock competence, led to the formation being extensively sheared, folded, and faulting during doming.

3.2.3 Coongan Subgroup

The Coongan Subgroup (Van Kranendonk et al. 2006a) represents a single volcanic cycle in which the Mount Ada Basalt, containing komatiite, komatiitic basalt, and tholeiite, passes upwards and laterally into andesite, dacite, and rhyolite of the Duffer Formation (Table 3.1). U–Pb zircon dating, summarized in Fig. 3.6, indicates that felsic volcanism of the Duffer Formation commenced at 3474 Ma (Nelson 2000)

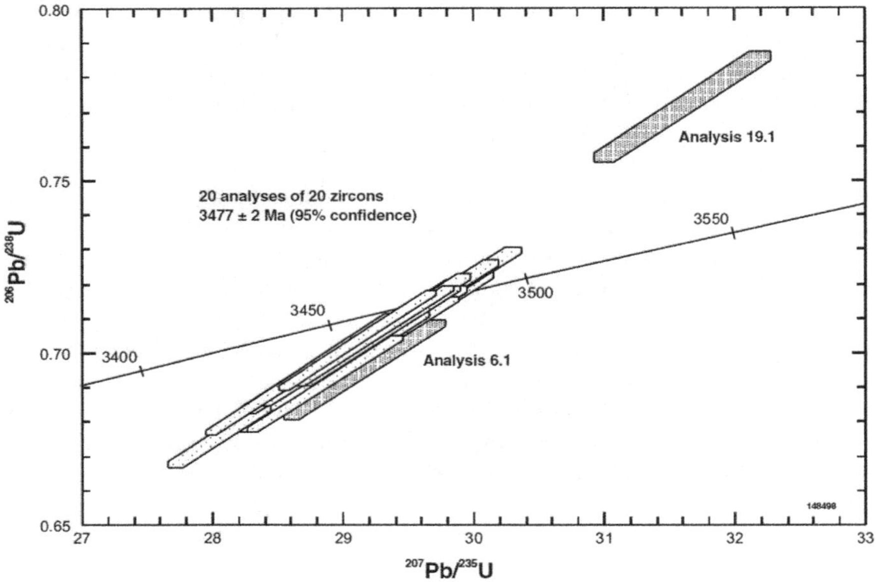

Fig. 3.5 Concordia plot of U–Pb zircon data from a sample of pale green chert (silicified felsic tuff, Geological Survey of Western Australia 148498) of the McPhee Formation, Eight Mile Bore south of McPhee Reward mine. (From Nelson 2000; with Geological Survey of Western Australia permission)

and continued until 3459 Ma (De Vries et al. 2006). Felsic units in the Mount Ada Basalt have been dated between c. 3470 and 3469 Ma, although some of the dated units might be intrusions related to the overlying Duffer Formation. Deposition of the Marble Bar Chert Member, at the top of the Duffer Formation, is inferred to have occurred between 3459 and 3450 Ma.

In the Marble Bar greenstone belt the maximum stratigraphic thickness of the subgroup is about 10 km, largely due to an exceptionally thick development (up to 8 km) of the Duffer Formation at Marble Bar (Fig. 3.7). In the northern Coongan greenstone belt, 30 km southwest from Marble Bar, the subgroup is about 5 km thick but elsewhere its thickness is generally less than 5 km. The Coongan Subgroup forms parts of the Marble Bar, Coongan, North Shaw, Panorama, Warralong, Doolena Gap, and East Strelley greenstone belts (Fig. 1.7). The extremely limited preservation of the subgroup in the East Strelley greenstone belt is attributed to uplift and deep erosion of the Carlindi Dome before 3426 Ma (Hickman 2010). In that dome, only remnants of the Duffer Formation (c.3467 Ma felsic volcaniclastic rocks and high-level porphyritic felsic intrusive rocks) are still preserved under the unconformity at the base of the Strelley Pool Formation (Van Kranendonk 2000). Exposed stratigraphic evidence indicates that the Coongan Subgroup was deposited across at least 30,000 km^2 of the EPT.

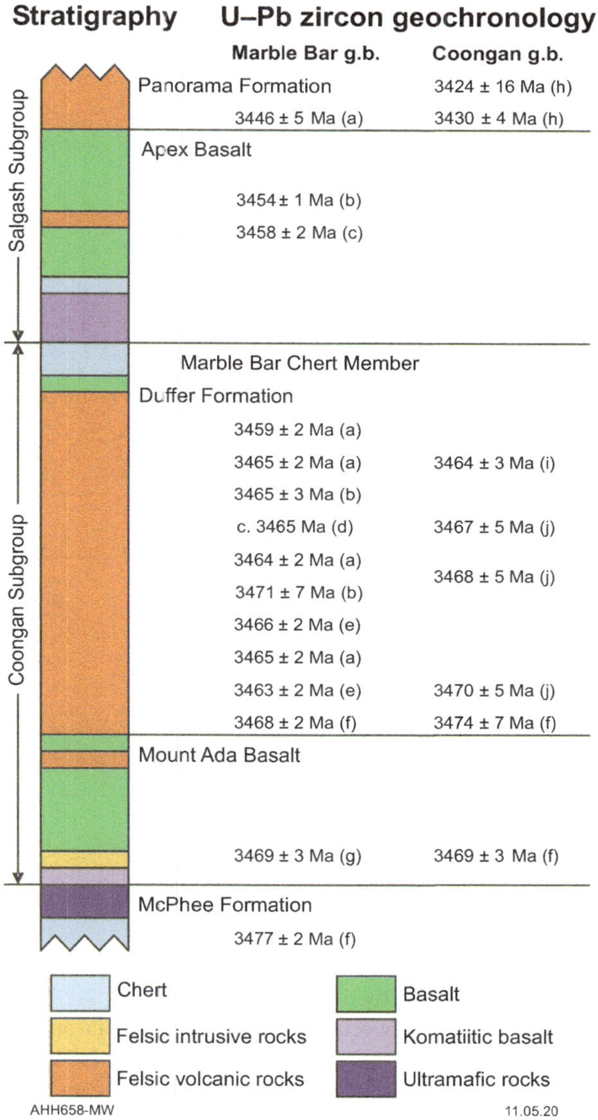

Fig. 3.6 U–Pb zircon geochronology of igneous rocks of the Coongan and Salgash Subgroups that constrain the depositional age of the Marble Bar Chert Member. Sources of data: (**a**) De Vries et al. 2006; (**b**) RI Thorpe, writ. Comm. 1991; (**c**) Thorpe et al. 1990; (**d**) Thorpe et al. 1992a, b; (**e**) McNaughton et al. 1993; (**f**) Nelson 2000; (**g**) Nelson 1999; (**h**) Nelson 2002; (**i**) Nelson 2004; (**j**) Nelson 2001 (Modified from Glikson et al. 2016; with Geological Survey of Western Australia permission)

Deposition of the Coongan Subgroup commenced with the eruption of komatiite and komatiitic basalt. The existence of komatiite at the base of volcanic cycles provides strong evidence for a mantle plume origin because the generation of komatiitic magmas indicates mantle melting at very high temperatures (Arndt et al. 1997, 2001; Condie 2001; Van Kranendonk et al. 2002, 2006a; Smithies et al. 2005b; Pirajno 2007a, b). The basal komatiites and komatiitic basalts of the

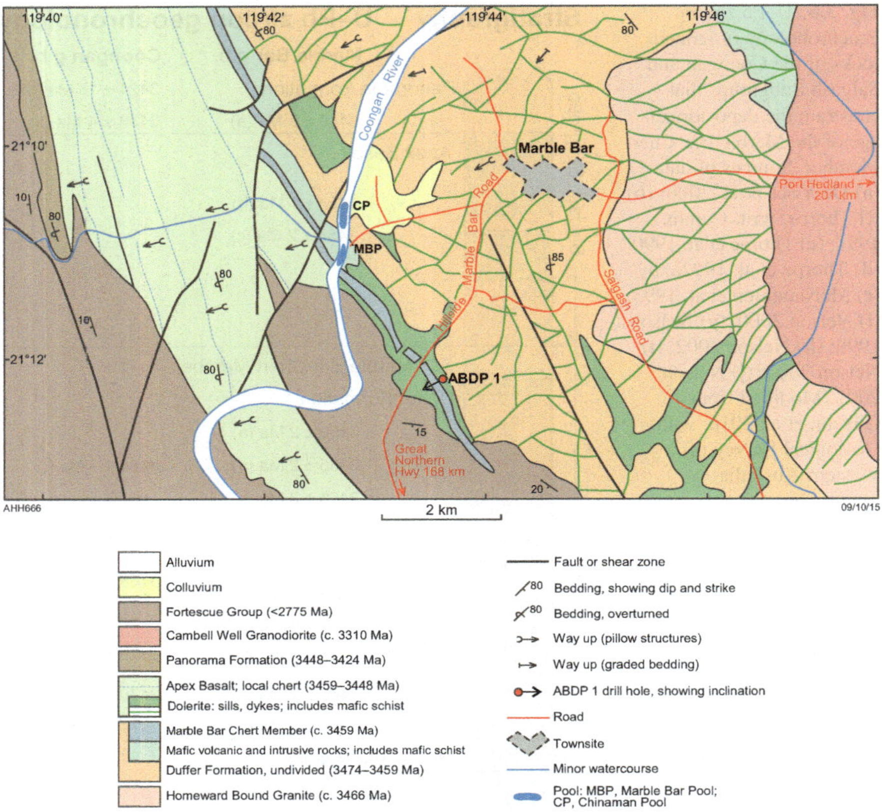

Fig. 3.7 Geological sketch map of the Marble Bar area showing the exceptional thickness of the Duffer Formation, and a swarm of dolerite dykes and sills that were feeders to the Apex Basalt (From Hickman 2021; with Geological Survey of Western Australia permission)

Mount Ada Basalt are invariably overlain by much thicker units of pillowed tholeiitic basalt intruded by dolerite sills.

The contact between the Mount Ada Basalt and the Duffer Formation is conformable in most greenstone belts, although it is locally difficult to define owing to lateral intercalation of basalt and felsic volcanic rocks in some sections (Van Kranendonk et al. 2006a). In the northwest Marble Bar greenstone belt, west and north of the Talga Talga mining area, the contact is marked by a thin unit of chert, BIF, volcaniclastic siltstone and sandstone, shale, and felsic tuff up to 10 m thick (Hickman 1983, p. 91; Van Kranendonk 2010a, b). In the Doolena Gap greenstone belt (Fig. 1.7), pillow basalts of the Mount Ada Basalt are separated from overlying felsic volcanic rocks of the Duffer Formation by a siltstone unit up to 200 m thick (Wiemer et al. 2016). At the base of this siltstone there is a 1-m-thick unit comprising repeated cm thick upward-fining graded layers of conglomerate and siltstone. The

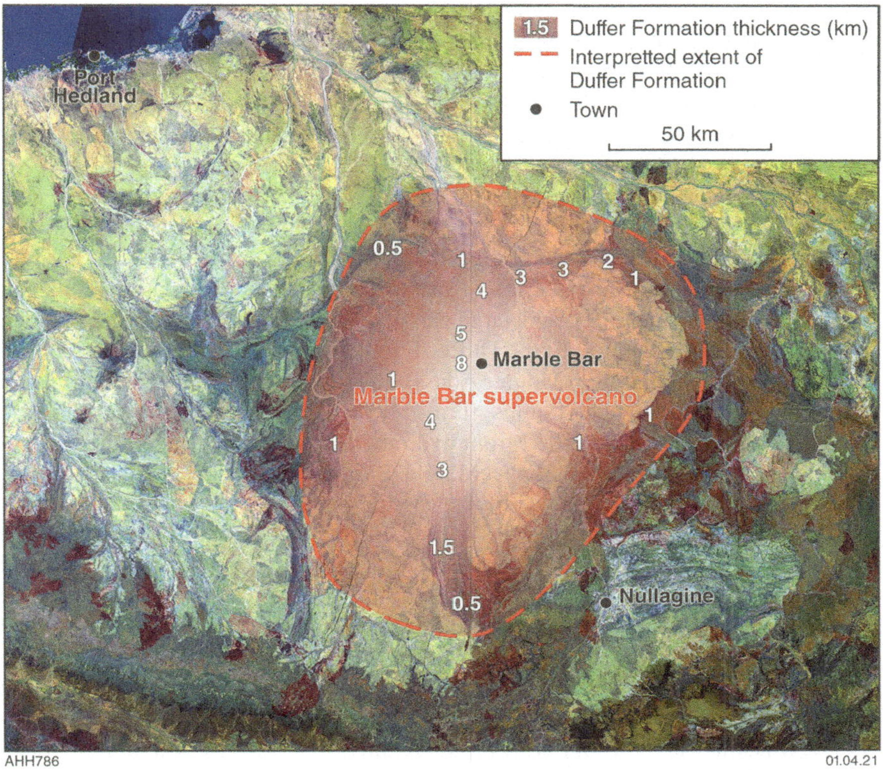

Fig. 3.8 Depositional extent of the Duffer Formation in the East Pilbara Terrane showing diminishing stratigraphic thicknesses away from an 8-km thickness at Marble Bar. Based on the depositional area and the total volume of felsic volcanics erupted, the Duffer Formation evolved in a c. 3465-Ma supervolcano (With Geological Survey of Western Australia permission)

existence of this sedimentary unit indicates a brief volcanic hiatus, at least in the Marble Bar–Doolena Gap area.

Regional thickness and facies variations in the Duffer Formation indicate that it was erupted in a 15,000 km^2 supervolcano centred on the area now situated between the Mount Edgar and Shaw Domes (Fig. 3.8). The maximum stratigraphic thicknesses of 8 km at Marble Bar and 5 km in the northern Coongan greenstone belt decrease to about 1 km on the eastern side of the Mount Edgar Dome, the southern Coongan greenstone belt, and the Panorama, North Shaw, and Doolena Gap greenstone belts (Fig. 1.7). Elsewhere, the Duffer Formation is less than 100 m thick or entirely absent from the Warrawoona Group succession. Since in most areas the Duffer Formation is overlain by the shallow-water Marble Bar Chert Member (MBCM), the Duffer Formation must, before deposition of the MBCM, have occupied a depositional basin that subsided by up to 8 km in the large area of greenstones west of Marble Bar. This major subsidence provides the first evidence of sagduction in the EPT, and was contemporaneous with 3474–3459 Ma diapiric uplift that occurred in surrounding areas such as the area now occupied by the Carlindi Dome (Hickman 1984; Van Kranendonk 2000; Van Kranendonk et al. 2006a; Hickman and Van Kranendonk 2008a), the Muccan Dome (Wiemer et al. 2016), the Shaw Dome (Van Kranendonk 2000), and the Mount Edgar Dome (Collins 1989; Hickman and Van Kranendonk 2004).

The felsic volcanic succession of the Duffer Formation is separated from the next ultramafic–mafic–felsic volcanic cycle (Salgash Subgroup) by the 100-m-thick Marble Bar Chert Member (Duffer Formation) (Van Kranendonk et al. 2006a). This sedimentary unit, which for many years was ranked as a formation (Hickman 1977, 1983, 2004), outcrops over a large part of the EPT (Fig. 3.9). Van Kranendonk et al. (2006a) linked the unit to the Duffer Formation on the grounds that volcanic cycles, such as that of the Mount Ada Basalt and the Duffer Formation, are commonly 'capped' by chert units formed by hydrothermal activity following volcanism (i.e. deposition of chert from hydrothermal vents). However, the original lithologies of the lower part of the Marble Bar Chert Member were sandstone, conglomerate, shale, carbonate rocks, and fine-grained volcaniclastic rocks (Orberger et al. 2006; Olivier et al. 2012; Glikson et al. 2016; Hickman 2021). Hydrothermal silification of these clastic sedimentary rocks occurred during late in the deposition of the unit, perhaps partly during eruption of the overlying Apex Basalt. Trace element abundances indicate that the hydrothermal fluids were derived from mafic sources.

Geochemical data from the Coongan Subgroup have been reported by Hickman and Lipple (1975), Glikson and Hickman (1981a, b), Jahn et al. (1981), Hickman (1983), Glikson et al. (1987), Barley (1993), Cullers et al. (1993), and Smithies et al. (2005a, b, 2007). Glikson and Hickman (1981a) described the Mount Ada Basalt in conjunction with komatiitic and basaltic rocks in the Talga Talga Subgroup. Glikson and Hickman (1981b) reported data from 50 samples of the Duffer Formation and 17 samples from the Panorama Formation. To examine vertical geochemical trends in the Duffer Formation, 41 samples were collected from a 4-km stratigraphic section north of Marble Bar. Vertical profiles of Zr/TiO_2, Th, La, Ce (increasing), and Ni and V (decreasing) reveal a progressive upward trend from andesite, through dacite, to rhyolite (Fig. 3.10). The compositional trend was also evident on a Zr/TiO_2: Nb/Y plot (Fig. 3.11).

Barley (1993) interpreted the Mount Ada Basalt and Duffer Formation to be petrogenetically related to the basalts derived from partial melting of a mantle source enriched in the more incompatible elements (K, Rb, Zr, LREE), followed by fractional crystallization of olivine and pyroxene. The felsic rocks of the Duffer Formation were then derived from the basaltic magmas by fractional crystallization of the observed phenocryst phases (pyroxene + plagioclase + iron-titanium oxide ± amphibole). Barley (1993) observed that simple incremental melting of mafic crust cannot explain the full range of compositions observed, and favoured a petrogenetic model involving both mantle-derived basaltic magmas and intermediate to silicic melts of mafic crust, with variable amounts of magma mixing and fractional crystallization, with or without crustal contamination.

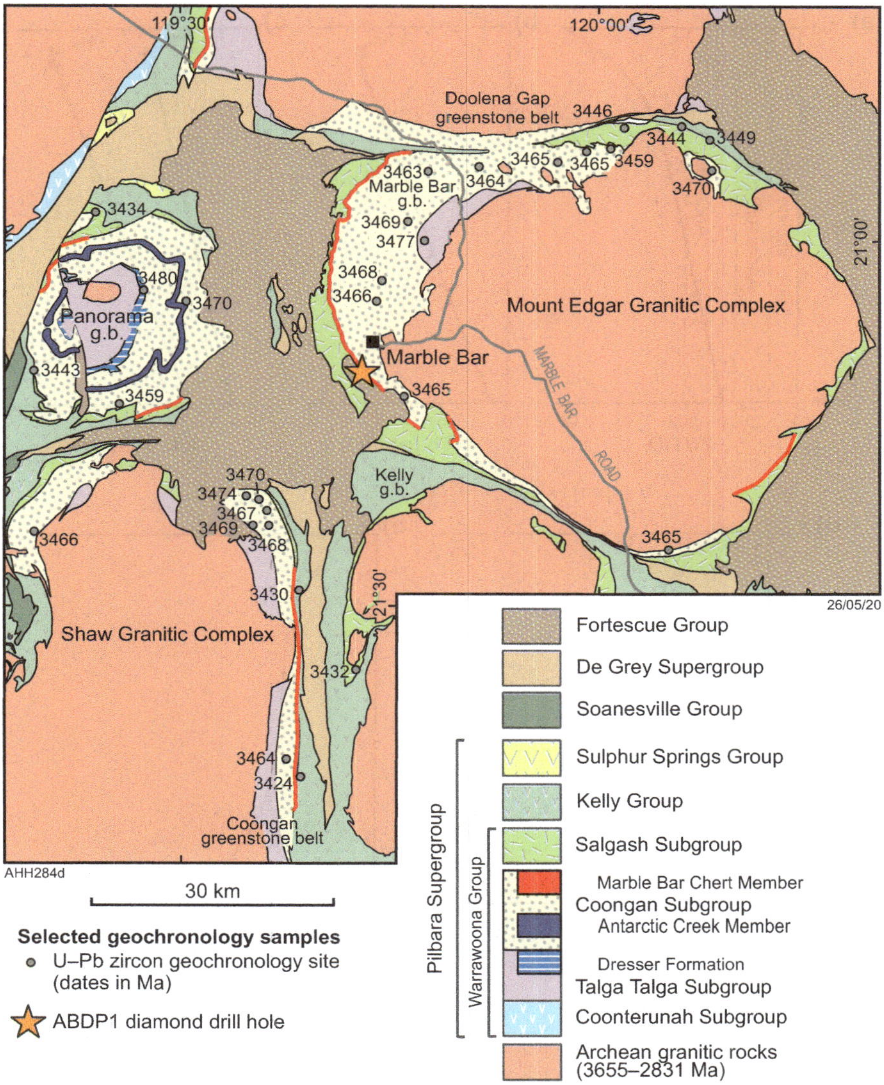

Fig. 3.9 Geological map of part of the East Pilbara Terrane showing outcrops of the Marble Bar Chert Member (red lines) within the stratigraphy of the Pilbara Supergroup. U–Pb zircon dates from the Pilbara Supergroup is summarized, and the location of stratigraphic drill hole ABDP 1 is shown (Modified from Glikson et al. 2016; with Geological Survey of Western Australia permission)

Smithies et al. (2007) reported geochemical data from the Mount Ada Basalt and Duffer Formation north of Marble Bar. They stated that the Mount Ada Basalt in this area is mainly composed of high-Ti tholeiites. Normalized trace element patterns for these rocks were found to be similar to those of the North Star Basalt. Low-Ti basalts

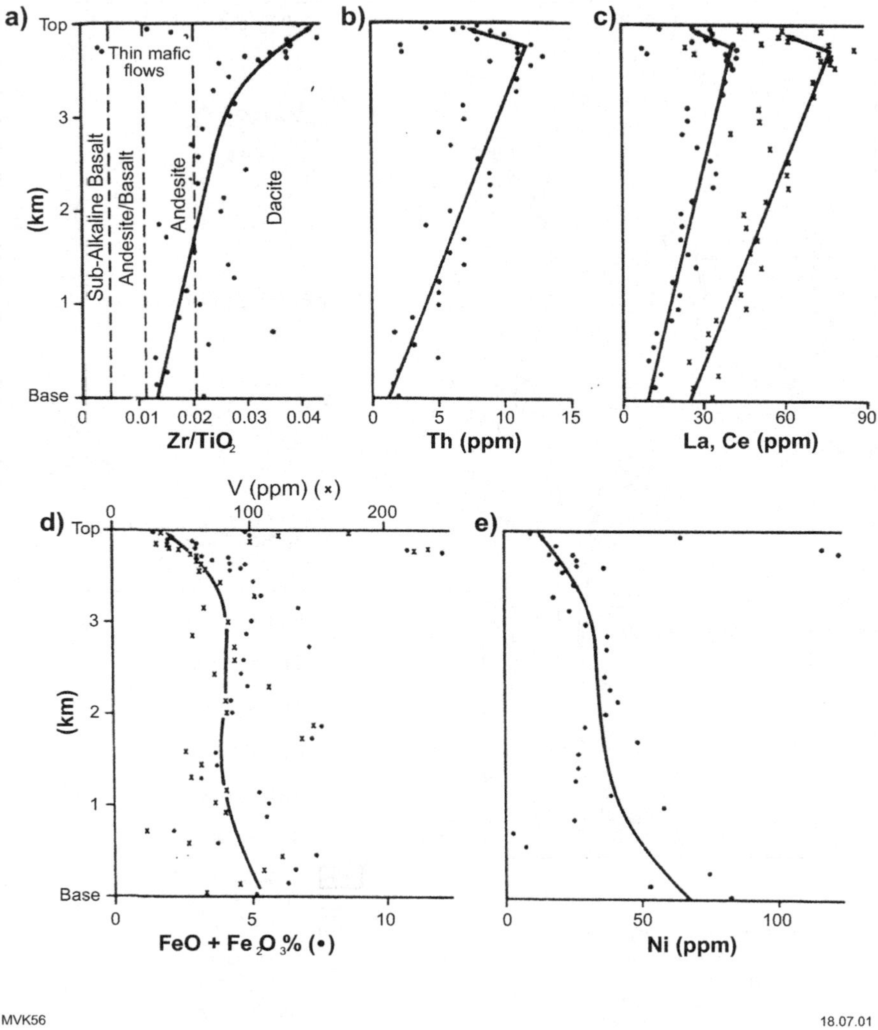

Fig. 3.10 Vertical differentiation trends in the Duffer Formation north of Marble Bar. Systematic geochemical sampling was conducted through the 4-km stratigraphic thickness of the formation from the Coongan River west to Bowls Gorge (From Hickman 1983; with Geological Survey of Western Australia permission)

have lower incompatible trace element concentrations than the high-Ti rocks and were interpreted to have been derived from a more depleted source. Tholeiitic basalts were reported to be interleaved with felsic volcanic rocks in the lower half of the Duffer Formation, and the high-Ti tholeiites in the Duffer Formation were noted to be identical in trace element concentration and normalized patterns to high-Ti

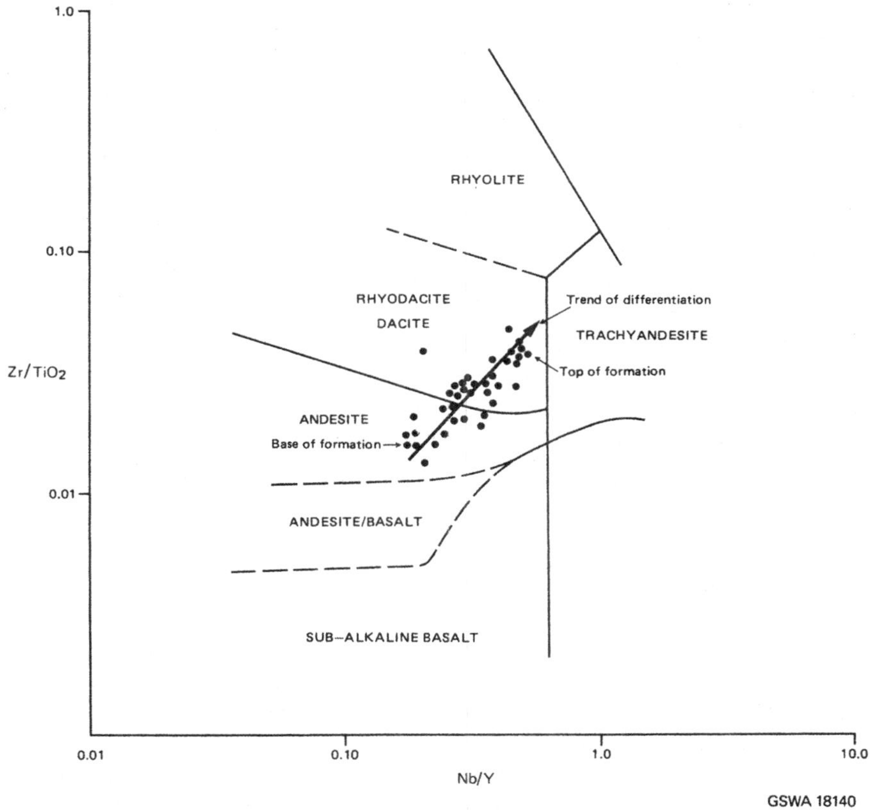

Fig. 3.11 Zr/TiO_2:Nb/Y diagram (after Winchester and Floyd 1977) for all samples of the Duffer Formation collected between the Coongan River and Bowls Gorge (From Hickman 1983; with Geological Survey of Western Australia permission)

tholeiites in the underlying Mount Ada Basalt. Smithies et al. (2007) reported that the main part of the Duffer Formation is composed of two sodic ($K_2O/Na_2O = <0.5$) basalt to dacite series, and that these are chemically similar to Archean TTGs, with features including high La/Nb, Th/Nb, and La/Yb ratios. In contrast, rhyolites in the upper Duffer Formation have far higher K_2O/Na_2O (1.5–2.6), and high Fe, HREE, Zr, and Nb, suggesting a strongly fractionated tholeiitic magma. The dacites of the Duffer Formation are chemically similar to Archean TTGs. However, a comparison between Sm–Nd data from c. 3465 Ma TTGs of the Shaw Dome (North Shaw Suite, Bickle et al. 1993) and from the c. 3465 Ma Duffer Formation of the Mount Edgar Dome indicated differences that led Smithies et al. (2007) to argue for different sources. Smithies et al. (2007) interpreted these differences to mean that the Duffer Formation magmas were generated through fractionation of tholeiitic parent magma contaminated by crustal TTG, whereas the contemporaneous TTG (Callina Supersuite) was derived through melting of basaltic crust and older TTG rocks.

Assuming a 3470-Ma depleted mantle, Nd $T_{DM}2$ model ages from basalts, andesites, and dacites of the subgroup (9 samples) indicate average source ages between 3620 and 3520 Ma. ε_{Nd} values of +1.73 to +0.45 suggest magma derivation both from the mantle and from crustal recycling. Similar isotopic data are recorded from the Callina Supersuite which comprises granitic intrusions contemporaneous with the Coongan Subgroup. Nd $T_{DM}2$ model ages from Callina Supersuite intrusions of the Shaw, Mount Edgar, Carlindi, and Muccan Domes range between 3770 and 3530 Ma, and ε_{Nd} values vary between +1.69 and − 1.49 (data in Hickman 2021). The mean Hf $T_{DM}2$ model age of zircons with Callina Supersuite crystallization ages (3484–3462 Ma) is 3739 Ma and the mean ε_{Hf} value is +0.85 ($n = 111$). Some very variable ε_{Hf} values (+3.8 to −5.9, data in Hickman 2021) suggest that granitic magmas were derived from a mix of juvenile and older crustal sources. As in the Coonterunah and Talga Talga Subgroups, komatiites have provided more positive whole rock ε_{Hf} values than other lithologies, two samples from the Mount Ada Basalt having ε_{Hf} values of +4.3 and + 3.2 (Gruau et al. 1987).

3.2.3.1 Mount Ada Basalt

The Mount Ada Basalt is a 2–3-km-thick formation forming the lower part of an ultramafic–mafic–felsic volcanic cycle up to 10 km thick. Available geochronology suggests that mafic section of the cycle was erupted rapidly at c. 3470 Ma (data in Hickman 2021). The formation is mainly composed of pillowed and massive basalt and dolerite sills, except that the basal 100–300 m of the formation is typically composed of komatiite and pyroxene spinifex-textured komatiitic basalt. Komatiite 12 km north-northeast of Marble Bar contains up to 31.16% MgO (recalculated volatile-free), up to 5650 ppm Cr and up to 2093 ppm Ni (samples 75040034c-f, Glikson and Hickman 1981a, b). Minor clastic sedimentary and chert units are present in the central and upper basaltic parts of the formation, which also include rare felsic and mafic volcaniclastic units.

In most greenstone belts, the Mount Ada Basalt disconformably overlies the McPhee Formation, although in the Panorama greenstone belt it disconformably overlies the lateral equivalent of that formation, the Dresser Formation. In the Marble Bar and Doolena Gap greenstone belts, the Mount Ada Basalt and Duffer Formation are separated by up to 200 m of basaltic to felsic volcanoclastic rocks, indicating a local break in volcanism. The metamorphic grade of the formation varies between lower greenstone facies and lower amphibolite facies.

3.2.3.1.1 Earth's Oldest Asteroid Impact Spherules

The Mount Ada Basalt contains evidence of Earth's oldest asteroid impact event. Lowe and Byerly (1986) discovered a distal ejecta layer containing impact spherules in the Antarctic Creek Member of the formation at North Pole (Fig 3.12a). Later descriptions were provided by Byerly et al. (2002), Glikson (2004), Glikson et al. (2004), and Glikson and Pirajno (2018); the main unit containing the spherules became known as the 'ACM-1 layer'. The spherules, mainly 0.10–0.75 mm in diameter (Fig. 3.12b), are contained within a unit of chert breccia and conglomerate interpreted to be a tsunami deposit (Fig. 3.12c). Spherules are present with two 15–50 cm-thick units of chert and sandstone separated by a 200-m-thick dolerite sill (Glikson et al. 2004). The Antarctic Creek Member is underlain and overlain by thick succession of basaltic volcanic rocks. Some doubt exists as to whether the separate spherule layers in the Antarctic Creek Member were deposited from different impact events or if the upper layer represents a sedimentary reworking of the lower deposit (ACM-1 layer). Microscopic examination shows the spherules to be microkrystites with quench textures including cloudy to opaque needle-shaped pseudomorphs, most likely replacing olivine or pyroxene, and lath-like pseudomorphs of quartz probably replacing plagioclase (Fig. 3.12d). Devitrification features include radiating fan-shaped aggregates of sericite.

Geochemical analysis of spherule-bearing chert from the ACM layer provided results consistent with an impact spherule layer, including Ir, 42 ppb; Pd, 9 ppb; Pt, 96 ppb; Os, 21 ppb; Rh, 16 ppb; Ru, 69 ppb; and a Ir/Pd ratio of 4.67 (unpublished data, AY Glikson, written communication 2019). The anomalously high platinum

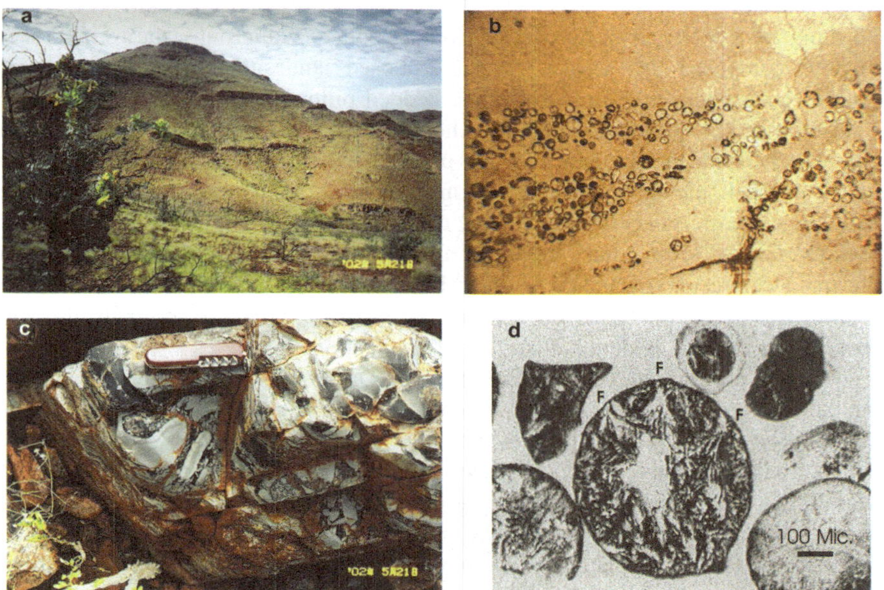

Fig. 3.12 The ACM-1 impact ejecta in the c. 3470-Ma Antarctic Creek Member of the Mount Ada Basalt at Miralga Creek, North Pole: (**a**) outcrops of the Antarctic Creek Member north of Miralga Creek; (**b**) lens of impact spherules within chert; (**c**) chert breccia; (**d**) microkrystite spherule showing quench textures (All photographs are kindly provided by AY Glikson)

element (PGE) concentrations in ACM-1 are accompanied by low-to-moderate Ni and Cr contents (Ni, 14 ppm; Cr, 262 ppm). Based on this single analysis of the ACM-1 layer, its Ir content (an indicator of an extraterrestrial component in impact spherules) is higher than in most recorded Archean and Proterozoic spherule layers (data reviewed by Glass and Simonson 2013). Byerly et al. (2002) dated the ACM-1 ejecta layer at 3470 ± 2 Ma.

In the Barberton Greenstone Belt of South Africa, a similar spherule layer dated at 3470 ± 2 Ma (Byerly et al. 2002) has been traced over a strike length of 13 km (Lowe and Byerly 1986; Lowe 1999; Walsh and Lowe 1999; Lowe et al. 2003; Glass and Simonson 2013). This is referred to as the 'S1 layer' and is preserved in thin sandstone and chert units within komatiite and basalt of the Hooggenoeg Formation (Lowe et al. 2003). In view of S1 and ACM layers being the same age, and having other features in common (Glass and Simonson 2013), several workers have suggested that the layers contain ejecta from the same major asteroid impact (Byerly et al. 2002; Simonson and Hassler 2002; Glikson and Vickers 2004; Glass and Simonson 2013).

3.2.3.2 Duffer Formation

The Duffer Formation varies in thickness between 8 km and 100 m, and is exposed in a cluster of greenstone belts of the EPT over an area of 15,000 km^2 (Fig. 3.9). Forming the upper part of a volcanic cycle (Coongan Subgroup), the formation is mainly composed of weakly metamorphosed andesitic to rhyolitic felsic volcaniclastic rocks and lava flows, interbedded pillow basalt units, volcaniclastic sandstone and conglomerate, and thin chert units. Subvolcanic felsic intrusions are common beneath the thickest eruptive parts of the formation. In the Marble Bar greenstone belt, the upper part of the formation includes thick units of conglomerate and volcaniclastic sandstone (Fig. 3.13) underlying the Marble Bar Chert Member. Metamorphosed sedimentary rocks are also present in distal sections remote from the main centre of volcanism around Marble Bar. The stratigraphic thickness at Marble Bar is 8 km, based on continuous exposure across a vertically dipping succession, with no evidence of structural duplication.

Volcanic and sedimentary facies within the Duffer Formation provide evidence on the depositional settings of individual sections. Coarse volcaniclastic facies including agglomerate are interpreted to be proximal to large volcanic vents, and this facies is restricted to the thickest sections of the formation within about 40 km of Marble Bar. More distal facies are represented by much thinner sections of ash-fall deposits, fine-grained volcaniclastic sedimentary rocks, quartz sandstone, and shale. In the Doolena Gap greenstone belt, which is interpreted to lie close to the northern limit of the formation, the Duffer Formation shows a lateral change of facies from felsic lavas and pyroclastic rocks in the east to siltstones and turbidites in the west (Wiemer et al. 2016). On the southeast side of the volcanic pile, in the Yandicoogina area of the Mount Edgar Dome, the Duffer Formation is sheared and metamorphosed (Williams and Bagas 2007), but its composition is consistent with relatively distal

Fig. 3.13 Stretching of pebbles in conglomerate of the upper Duffer Formation resulting from vertical deformation (sagduction) near Salgash on the northwest side of the Warrawoona Syncline (MGA Zone 50, 789930E 7645930 N) (Previously unpublished photograph; with Geological Survey of Western Australia permission)

facies including volcaniclastic sedimentary rocks, sandstone, and shale. In the North Shaw greenstone belt (Fig. 1.7), where the Duffer Formation is about 1 km thick, the formation comprises metamorphosed felsic volcaniclastic rocks, ignimbrites (Fig. 3.14a, b), sandstone, and conglomerate (Van Kranendonk 2000). These lithologies suggest that the western limit of the Duffer Formation was west of North Shaw.

3.2.3.2.1 Marble Bar Chert Member

The Marble Bar Chert Member (MBCM) is currently defined as the uppermost stratigraphic member of the Duffer Formation (Van Kranendonk et al. 2006a). Before 2006, this lithologically distinctive unit was referred to as the 'Towers Formation' (Hickman 1977), and was considered to be an important stratigraphic marker unit in the Warrawoona Group (Hickman 1983). Subsequently, more detailed investigations than were possible during the rapid 1:250,000 scale mapping in the 1970s indicated that some of the first regional correlations between major cherts of the EPT were incorrect (DiMarco and Lowe 1989a, b; Van Kranendonk and Morant 1999; Van Kranendonk et al. 2002). Even so, the PCMP confirmed that the 'Towers Formation' was a stratigraphic marker in four large greenstone belts of the

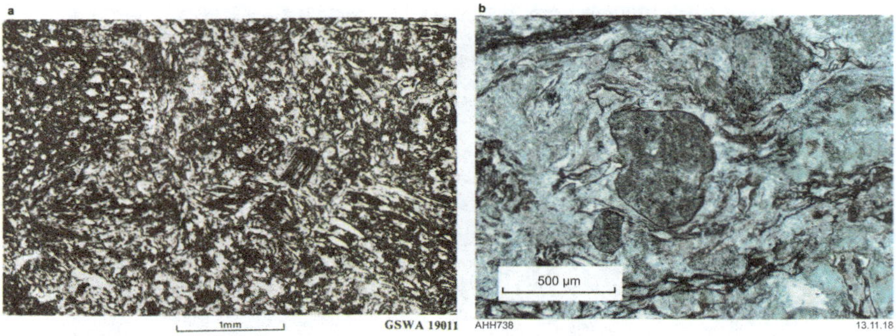

Fig. 3.14 Thin-section views (plane-polarized light) of ignimbrite in the upper part of the Duffer Formation north of Marble Bar: (**a**) flow lamination is deflected around a rounded fragment of porphyritic dacite; (**b**) devitrified glass shards with fragments of pumice and lava. (**a**, from Hickman 2021; **b**, from Hickman 1983; both with Geological Survey of Western Australia permission)

EPT (Marble Bar, Coongan, Panorama, and Warralong). Based on an interpretation that the chert unit is essentially a hydrothermal siliceous cap to the felsic volcanic Duffer Formation, Van Kranendonk et al. (2006a) changed the stratigraphic rank of the 'Towers Formation' by renaming and redefining it as the Marble Bar Chert Member of the Duffer Formation. With increased knowledge of the stratigraphy of the MBCM, the case for it being reinstated as a separate formation has been strengthened (Hickman 2021). The unit is now known to be a silicified clastic sedimentary unit rather than a hydrothermal chert, and marks a break in volcanism between the Coongan and Salgash volcanic cycles that lasted up to 10 Ma.

The type section of the MBCM is located at Marble Bar Pool on the Coongan River (Fig. 3.7). Here, extensive outcrops of white and grey banded chert, in addition to red jaspilitic chert (Fig. 3.15a, b), are cross-cut by irregular veins and layers of dark grey hydrothermal chert (Fig. 3.16). Despite some early questions over its regional extent, the member is now known to be exposed in the Marble Bar, Coongan, Panorama, and Warralong greenstone belts (Fig. 3.9). Its depositional extent was probably greater, but some greenstone belts (e.g. Kelly, McPhee, and Mount Elsie) do not expose stratigraphic levels beneath the Salgash Subgroup, whereas in others the Duffer Formation and most parts of the overlying Salgash Subgroup are absent due to Paleoarchean erosion. The maximum depositional age of the member is closely constrained by the youngest published date on volcanic rocks of the Duffer Formation, 3459 ± 2 Ma (De Vries et al. 2006), and by other dates that are only slightly older (Fig. 3.6).

At Marble Bar Pool, the MBCM is represented by a 60-m-thick stratigraphic section (Kato and Nakamura 2003). However, at this locality the base of the unit is faulted against pillowed basalt and dolerite, and part of it is missing. Evidence from diamond drilling 2.5 km to the southeast of the pool (drill core ABDP 1: Hickman 2005; Hoashi et al. 2009) indicates that the faulting at Marble Bar Pool removed the lower 30 m of the member (Glikson et al. 2016). In the more complete stratigraphic

Fig. 3.15 Outcrops of jaspilitic chert of the Marble Bar Chert Member at Marble Bar Pool: (**a**) view (looking south) of an exposure of the upper 10 m of the member on the east bank of the Coongan River, 70 m south of Marble Bar Pool. Grey rocks to the right of the chert, and in the bed of the river, are pillowed basalt flows at the base of the Apex Basalt; (**b**) close-up view showing alternating layers of red, white, and grey chert with a central bed of fragmented chert. Notably, the grey chert (hydrothermal), and some of the white chert (partly replacing red chert), locally cut across layers of red chert. The red chert contains fine-scale microbanding whereas the grey and white chert units are massive. Outcrop width (foreground) two metres (From Hickman 2021; with Geological Survey of Western Australia permission)

section provided by ABDP 1 (Fig. 3.17), this lower section of the stratigraphy is mainly composed of silicified clastic sedimentary rocks including diamictite and two thin layers of asteroid impact spherules (Glikson et al. 2016). This lower clastic section is now interpreted to be exposed at Chinaman Pool (Hickman 2021) (Fig. 3.7) where a 20-m-thick succession, previously named the Chinaman Pool Chert Member (Van Kranendonk et al. 2001a, b), overlies the felsic volcanic rocks and volcaniclastic sedimentary rocks of the Duffer Formation. The Chinaman Pool Chert Member (CPCM) includes polymictic conglomerate and pebbly sandstones deposited from debris flow during erosion of the Duffer Formation (Olivier et al. 2012). These sandstone and conglomerate units probably include the diamictites identified in ABDP 1, and might contain the same impact spherule layers identified in the drill core (Glikson et al. 2016). However, impact spherule layers are commonly lenticular due to erosion and reworking, and the $<$1-mm diameter spherules are difficult to observe in outcrop. The CPCM represents a transition sequence from high-energy, clastic deposition at the top of the felsic volcanic pile of the Duffer Formation to low-energy deposition of fine-grained clastic sediments, carbonates, and primary hydrothermal deposits exposed in the upper section of the member at Marble Bar Pool.

The upward-fining succession of the member, from the coarse clastic facies at Chinaman Pool to silicified muds and hydrothermal chemical precipitates at the top of the Marble Bar Pool section, has led to diverse interpretations of the composition and origin of the MBCM (DiMarco and Lowe 1989b; Sugitani 1992; Minami et al. 1995; Kojima et al. 1998; Van Kranendonk et al. 2001a, b; Kato and Nakamura 2003; Oberger et al. 2006; Suganama et al. 2006; Van Kranendonk 2006, 2010a, b; Van den Boorn et al. 2007, 2010; Hoashi et al. 2009; Olivier et al. 2012; Li et al.

Fig. 3.16 Hydrothermal intrusion and brecciation of the Marble Bar Chert Member and underlying altered volcanic rock of the Duffer Formation at Marble Bar Pool: (**a**) bedded jaspilitic chert partly replaced by veins and pods of massive or weakly layered white chert; (**b**) network of dark grey hydrothermal chert veins intruding and fragmenting bleached basaltic rocks of the Duffer Formation immediately east of the Marble Bar Chert Member; (**c**) dark grey hydrothermal chert breccia cutting through layered grey and white chert; (**d**) sills of grey and white chert emanating from a crosscutting feeder vein (Modified from Van Kranendonk 2010b; with Geological Survey of Western Australia permission)

2013; Rasmussen et al. 2014a, b; Glikson et al. 2016). In the Marble Bar area, the initial depositional environment of the member was similar to that interpreted by Olivier et al. (2012), marginal to an emerging landmass and including fluviatile channels, deltas, and submarine fans. Uplift and erosion of the Duffer Formation were most likely related to the rising mantle plume that subsequently evolved to cause volcanism of the Salgash Subgroup. However, water depths appear to have increased during deposition of the member, at least in the Marble Bar Pool area. This might have been due to gravitational collapse of the abnormally thick underlying volcanic pile of the Duffer Formation. Hoashi et al. (2009) related the upper jaspilitic section of the member to submarine hydrothermal vents. Analyses by Sugitani (1992) revealed that the jaspilitic chert contains up to 40 wt.% Fe_2O_3, although <10 wt. % Fe_2O_3 is normal.

Evidence of hydrothermal activity is present throughout the member at Marble Bar Pool and in the ABDP 1 core. Units of layered grey and white chert and red and white chert are veined and fragmented by dark grey chert containing numerous angular blocks of the banded chert (Fig. 3.16) (Oliver and Cawood 2001; Van Kranendonk et al. 2001b, 2006b; Van Kranendonk 2006, 2010a, b). The outcrops

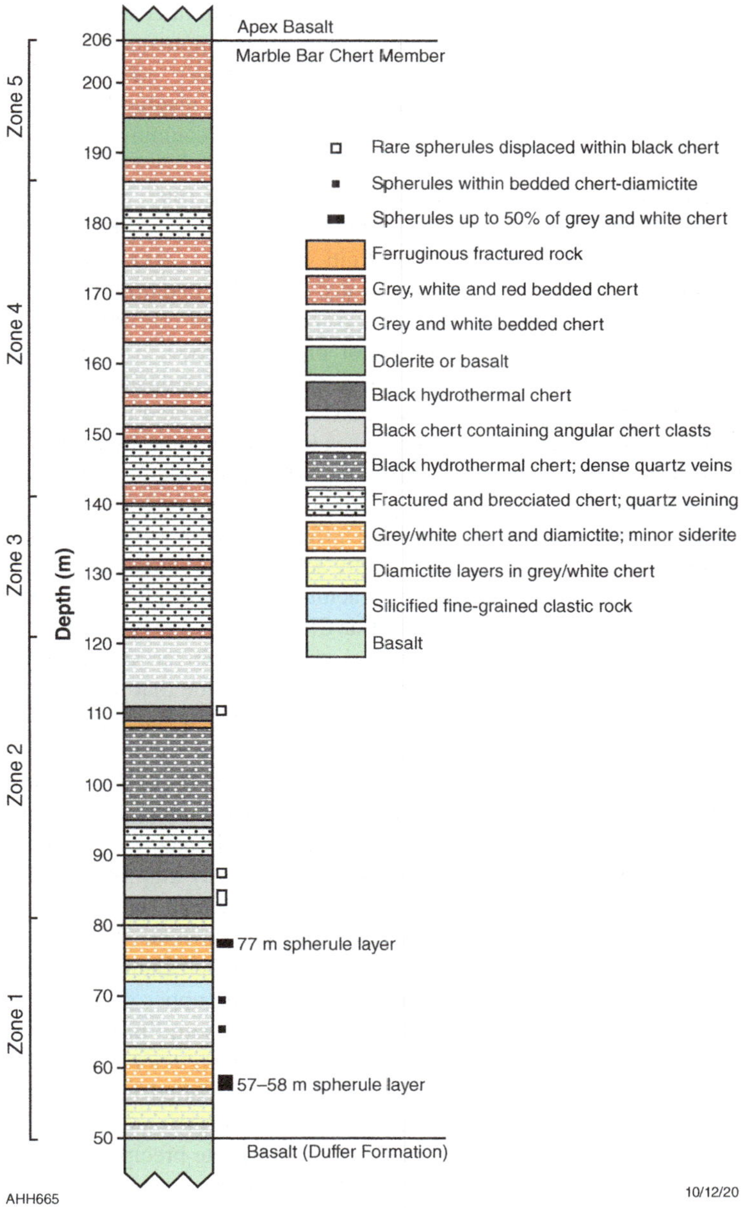

Fig. 3.17 Stratigraphy of the Marble Bar Chert Member intersected in drill hole ABDP 1. Drilling intersected bedding at an angle of about 35°, indicating a true stratigraphic thickness of about 110 m. Zones shown alongside the column are from the interpretation of Hoashi et al. (2009). Zones 2 and 3 are dominated by fractured chert veined by dark grey hydrothermal chert and are interpreted to have no stratigraphic significance. Notable stratigraphic features are the presence of siderite zones, diamictite, and impact spherule layers in Zone 1, and the almost complete restriction of jaspilitic chert to Zones 4 and 5. The presence of hematite at depths greater than 200 m has been interpreted as evidence of oxygen in the Paleoarchean depositional environment (Hoashi et al.,

show many examples of hydrothermal chert veins having injected and forced apart the layered chert to produce a mixture of angular chert fragments within a matrix of massive, cryptocrystalline, and dark grey chert. Oliver and Cawood (2001) documented rock pavements at Marble Bar Pool, which display blocks of banded chert up to 10 m across within a matrix of dark grey chert containing chert fragments down to <1 mm in diameter.

Geochemical and petrographic investigations (Sugitani 1992; Minami et al. 1995; Kojima et al. 1998; Kato and Nakamura 2003; Oberger et al. 2006; van den Boorn et al. 2007, 2010; Hoashi et al. 2009; Pinti et al. 2007, 2009a; Li et al. 2013; Rasmussen et al., 2014a, b) have been variously aimed at determining the protoliths of the member, its depositional setting, the oxygen content of Earth's Paleoarchean atmosphere, and searching for evidence of organic carbon. The majority of the studies concluded that a large part of the member was deposited as fine-grained mafic and felsic detritus, and that the succession was subjected to pervasive hydrothermal alteration. Trace element abundances indicate that the hydrothermal fluids were derived from mafic sources, which suggests an association with the dolerite dykes and sills that intruded the Duffer Formation and fed into the overlying Apex Basalt (see Apex Basalt). The underlying Duffer Formation of the Marble Bar area contains thick dolerite sills that preferentially accumulated immediately beneath the member, and this might have influenced fracturing and hydrothermal alteration.

A controversial feature of the Marble Bar Chert Member is the origin and age of the hematite in the red and white banded chert. Hoashi et al. (2009) provided transmission electron microscope (TEM) evidence to argue that the hematite in the chert was a primary precipitate, with no evidence that it had replaced previous minerals. Hoashi et al. (2009) used this conclusion to argue for the existence of oxygenated seawater at 3460 Ma. However, this conclusion conflicts with interpretations that the Earth's atmosphere and hydrosphere were essentially anoxic until the Great Oxidation Event, an event that was recently estimated to have occurred at c. 2310 Ma (Philippot et al. 2018). Based on exceptionally high δ^{56}Fe values and a low oceanic U content in the member, Li et al. (2013) argued for very low levels of oxidation during its deposition. Subsequently, Rasmussen et al. (2014b) argued that the hematite was not primary but had replaced magnetite, greenalite, and siderite. As noted by Hoashi et al. (2009), total iron contents in the upper part of the member range between 2 and 25 wt. % Fe_2O_3, which ranks some parts of the unit as low-grade banded iron formation (BIF). Recent interpretations of Archean and Proterozoic BIF elsewhere suggest that it was not deposited in its present form, but that the primary lithologies (protoliths) were iron–silicate mud, Fe-rich clay such as greenalite, ferric oxyhydroxides, and Fe-rich carbonate precipitates (Rasmussen et al. 2014a, 2016; Gauger et al. 2015; Smith 2015; Konhausser et al. 2017).

Fig. 3.17 (continued) 2009) (Modified from Glikson et al. 2016; with Geological Survey of Western Australia permission)

Rasmussen et al. (2014b) interpreted lateral colour changes between white and red chert within individual mm to cm thick layers to be the result of late-stage oxidation of greenalite and magnetite in the white layers to hematite, thus producing the red layers. Apparently set against this interpretation, Van Kranendonk (2010b) provided photographic evidence of red chert having been replaced by massive white hydrothermal chert (Fig. 3.16a). In other parts of the member, this hydrothermal white chert contains blocks of red and white banded chert (Fig. 3.16d) suggesting that hematite crystallization was older than hydrothermal veining. Much of the hydrothermal veining apparently occurred between 3460 and 3450 Ma because the hydrothermal veins occupy brittle fractures that penetrate the entire thickness of the member but do not continue into the overlying Apex Basalt (Van Kranendonk 2006). An explanation of these apparent inconsistencies is that two generations of white chert are present: an early, finely layered white chert containing magnetite and greenalite, and a younger hydrothermal white chert. However, intrusive relations show that hematite in the red chert is older than the hydrothermal chert.

The thickest individual chert units of the Pilbara Supergroup are interpreted to be silicified deposits deposited over millions of years. As such, these units provide good targets for the detection of major asteroid impacts because ejecta fallout from such events would have been global. On the likelihood that the Marble Bar Chert Member was deposited over millions of years, Glikson et al. (2016) examined the ABDP 1 core and identified two thin layers of silicified impact spherules. Both spherule layers are contained within diamictites that Glikson et al. (2016) interpreted to be tsunami deposits. Unlike the polymictic conglomerates in the debris flows of the Chinaman Pool facies, the ABDP 1 diamictite is entirely composed of angular fragments of banded chert and sandstone. The diamictite is also lithologically very different from the hydrothermal chert breccia in irregular dykes and veins at higher stratigraphic levels.

3.2.4 Salgash Subgroup

The youngest subgroup of the Warrawoona Group, the 3450–3427 Ma Salgash Subgroup comprises komatiite, komatiitic basalt, and tholeiite of the Apex Basalt overlain by dacite, rhyolite, and volcaniclastic rocks of the Panorama Formation. In the Marble Bar greenstone belt, the Apex Basalt is up to 3.25 km thick whereas the overlying Panorama Formation is less than 500 m thick. However, in other greenstone belts the Panorama Formation is up to 1.5 km thick. Regional variability in the thickness and composition of the Panorama Formation results from its eruption at separate felsic volcanic centres (Barley 1981; Barley et al. 1984; DiMarco 1986; DiMarco and Lowe 1989a, b; Van Kranendonk 1999, 2000; Smithies et al. 2007; Van Kranendonk et al. 2007a, b), and from locally deep erosion of the subgroup prior to deposition of the 3426–3350 Ma Strelley Pool Formation. Both formations of the subgroup are absent from greenstone belts in the Carlindi, Yule, and Muccan Domes, and incompletely preserved in the Panorama greenstone belt of the North

Pole Dome (Fig. 1.7). However, the Panorama Formation is likely to have been present in all these domes prior to c. 3426 Ma erosion because the granitic cores include large volumes of the 3451–3416 Ma Tambina Supersuite. Low metamorphic grades in the subgroup (lower greenschist to prehnite–pumpellyite facies) are attributed to its high stratigraphic position in the Warrawoona Group and to its remoteness from contacts with intrusive granites.

As for other subgroups of the Warrawoona Group, the Salgash Subgroup was erupted on a volcanic plateau underlain by pre-3530 Ma sialic crust. The ultramafic–mafic–felsic volcanic succession of the subgroup represents a 30-million-year volcanic cycle interpreted to be related to a mantle plume event. The subgroup was deposited during diapiric deformation of both the underlying Warrawoona Group and the pre-3530 Ma sialic crust. Evidence of 3450 Ma uplift and erosion of the pre-3530 Ma crust is provided by detrital zircon ages between 3650 and 3592 Ma in sandstone within the 3450 Ma Apex Basalt (Chap. 2).

U–Pb zircon geochronology on the Panorama Formation suggests that felsic volcanism might have occurred in two separate stages: 3448–3446 Ma and 3434–3427 Ma (Hickman 2021). However, there are two alternative possibilities: 1, the apparent break in felsic volcanism from 3446 to 3434 Ma might be due to c. 3440 Ma erosion of the lower Panorama Formation; or 2, the relatively few available dates interpreted to represent the first stage were obtained either from high-level intrusive rocks of the Tambina Supersuite or from volcaniclastic units containing mainly detrital zircons. One line of evidence indicating that there were no breaks in felsic magmatism between 3448 and 3427 Ma is provided by detrital zircon dates from east Pilbara Craton sedimentary rocks (Fig. 2.2). Additionally, indirect evidence for felsic volcanism between 3448 and 3434 Ma is provided by 14 published U–Pb zircon dates between 3451 and 3435 Ma for granitic rocks of the Tambina Supersuite. Some granitic intrusions of the Tambina Supersuite were subvolcanic, either intruding the base of the Panorama Formation or intruding the underlying Apex Basalt, and can reasonably be inferred to have been feeders to the volcanic rocks. For example, dykes and stocks of c. 3445 Ma monzogranite radiate outwards from the North Pole Monzogranite in the North Pole Dome and terminate in the Panorama Formation (Thorpe et al. 1992a, b; Amelin et al. 2000; Brown et al. 2006, 2011; Harris et al. 2009).

Tholeiitic basalts of the Apex Basalt are dominated by high-Ti compositions although minor low-Ti tholeiites are also present (Smithies et al. 2007). Normalized trace element patterns for the high-Ti basalts are very similar to those from basalts of the underlying subgroups and are up to 15 times primitive mantle values, with wider ranges in $[La/Yb]_{PM}$ (0.87–2.20) and $[La/Nb]_{PM}$ (1.00–1.62) (Smithies et al. 2007). Low-Ti basalts have lower incompatible trace element concentrations (mainly <4 times primitive mantle values) and show no negative Nb anomaly on primitive mantle-normalized plots. Smithies et al. (2005b, 2007) interpreted low $[La/Yb]_{PM}$ ratios (0.77–0.89) and very low $[La/Sm]_{PM}$ ratios (0.73–0.84) to reflect derivation from a more depleted source than that for the high-Ti basalts. Alternatively, such compositional differences between high-Ti and low-Ti basalts might be related to different degrees of partial melting of a plume or asthenospheric source (Arndt et al. 1993).

Cullers et al. (1993) compared the geochemistry of the Duffer and Panorama Formations, although the samples considered to be representative of these formations came from different parts of the EPT. Based partly on high La/Lu ratios and a lack of Eu anomalies, Cullers et al. (1993) interpreted felsic volcanic rocks of the Panorama Formation in the North Pole Dome to have been derived by melting of eclogite. These North Pole data were compared with previous analyses of felsic volcanic rocks (Barley et al. 1984) interpreted to be from the Duffer Formation in the McPhee and Kelly greenstone belts (Hickman 1980a; Barley 1993), and Cullers et al. (1993) interpreted the latter to have been derived by fractional crystallization of basaltic magmas. However, subsequent geochronology has established that the Duffer Formation is not exposed in the McPhee and Kelly greenstone belts, and the geochemical comparisons were actually made with other units of the Panorama Formation. Analytical data confirms that the geochemistry of the Panorama Formation is regionally variable (Smithies et al. 2005b, 2007), and that the Panorama Formation was erupted from compositionally distinct volcanic centres. Although there is some compositional overlap with the Duffer Formation, the Panorama Formation volcanic rocks typically have higher K_2O/Na_2O of 0.35–0.70 (Smithies et al. 2007). Smithies et al. (2007) reported that rocks of the Panorama Formation have higher normalized La/Yb and La/Nb ratios compared to felsic rocks of the Duffer and Coucal Formations and suggested this might reflect higher pressures of magma genesis or that the magmas had a greater contribution of a source component formed at high pressure (e.g. TTG).

Assuming a c. 3450-Ma depleted mantle, Nd $T_{DM}2$ model ages from the Apex Basalt and Panorama Formation indicate average crustal source ages between 3580 and 3530 Ma ($n = 4$, data in Smithies et al. 2007). On the same assumption, ε_{Nd} values between +1.21 and + 0.72 suggest a mix of juvenile magma from the mantle and from crustal recycling of older crust. The mean Hf T_{DM}^2 model age of zircons with Panorama Formation crystallization ages (3449–3427 Ma) is c. 3675 Ma, and the mean ε_{Hf} value is +0.82 ($n = 82$). Similar isotopic data are recorded from the Tambina Supersuite which comprises granitic intrusions contemporaneous with the Salgash Subgroup. Nd $T_{DM}2$ model ages from Tambina Supersuite intrusions of the Carlindi, Yule, and Mount Edgar Domes range between 3710 and 3540 Ma, and ε_{Nd} values vary between −1.34 and + 0.40 (data reviewed in Hickman 2021). The mean Hf T_{DM}^2 model age of zircons with Tambina Supersuite crystallization ages (3451–3416 Ma) is 3689 Ma and the mean ε_{Hf} value is +1.26 ($n = 100$). Assuming a c. 3450 Ma depleted mantle, the data suggest Eoarchean sources and lesser contributions of juvenile magma than in older units of the EPT.

3.2.4.1 Apex Basalt

The Apex Basalt is exposed in seven greenstone belts of the EPT (the Marble Bar, Kelly, McPhee, Western Shaw, Tambina, Emerald Mine, and Warralong greenstone belts) where it is up to 3.25 km thick and comprises successions of komatiite, komatiitic basalt, tholeiite, and thin units of chert, sandstone, black shale, and felsic volcaniclastic rocks. In the Marble Bar greenstone belt, on most sides of the Mount Edgar Dome, numerous dykes and sills of dolerite and gabbro of the Apex Basalt intrude the underlying Coongan Subgroup and extend upwards into the formation (Fig. 3.17) (Kloppenburg 2003; Van Kranendonk et al. 2006a; Hickman and Van Kranendonk 2008b; Hickman 2012, 2021; Glikson et al. 2016). About 45 km northeast of Marble Bar, large ultramafic–mafic layered intrusions include the Strutton, Gap, and Nobb Well Intrusions (Williams 1999) which are interpreted to be subvolcanic intrusions related to the Apex Basalt. Such dykes and layered intrusions were probably widespread in the crust underlying the Apex Basalt, although most are likely to have been intruded and effectively replaced by granitic intrusions of the Tambina, Emu Pool, and Cleland Supersuites.

Outside the Marble Bar greenstone belt, the formation has a maximum thickness of approximately 2 km. The Apex Basalt lies disconformably on the Duffer Formation, and in the Marble Bar greenstone belt it is conformably to unconformably overlain by the Panorama Formation. The absence of the Apex Basalt in the Doolena Gap greenstone belt (Muccan Dome) is due to a down-cutting erosional contact beneath the unconformably overlying Strelley Pool Formation (Van Kranendonk 2010a), and the formation is also absent beneath this 3426 Ma unconformity in the Shaw, North Pole, Carlindi, and Yule Domes. In most sections, the Apex Basalt disconformably overlies the Marble Bar Chert Member, although an erosional contact overlain by basal conglomerate and breccia is locally present 20 km north of Marble Bar (Van Kranendonk 2010a).

The Apex Basalt locally includes minor volcanic cycles consisting of ultramafic–mafic–chert sequences. In the Marble Bar greenstone belt, the first minor volcanic cycle of the Apex Basalt commenced with eruption of komatiite and komatiitic basalt onto the MBCM (Duffer Formation). West of Marble Bar Pool, in the Chinaman Creek area, the first minor volcanic cycle is mainly composed of pillowed komatiitic basalt overlain by a unit of metamorphosed sandstone, black shale, felsic volcaniclastic rocks, and hydrothermal silica. Silicification of the clastic rocks by fluids from large underlying hydrothermal dykes resulted in a stratiform unit of 'chert', now widely referred to as the 'Apex chert' (informal name). Komatiite and komatiitic basalt of the second minor cycle of the Salgash Subgroup disconformably overlies the 'Apex chert', although in the Marble Bar greenstone belt komatiite is mainly restricted to the Comet and Salgash areas, 10–15 km south of Marble Bar (Hickman and Van Kranendonk 2008b). The second minor cycle is up to 1 km thick and capped by grey and white banded chert overlying hydrothermally altered pillowed basalt. The altered basalt is intruded by hydrothermal chert dykes extending upwards to the base of the chert (Van Kranendonk and Pirajno 2004; Van Kranendonk 2006). Komatiitic basalt forms the greater part of the Apex Basalt in the Warralong greenstone belt but minor komatiite is also present in the local succession (Van Kranendonk 2010a). The serpentized komatiite has layers of well-developed olivine spinifex texture separated by cumulate textured peridotite (Van Kranendonk 2010a) whereas the komatiitic basalt is pillowed and dominated by pyroxene spinifex texture. The upper part of the Apex Basalt comprises minor cycles of komatiitic basalt and chert in both the Marble Bar and Warralong greenstone belts (Fig. 1.7). At least two additional minor volcanic cycles are interpreted to overlie the second minor cycle in the Chinaman Creek section, and the top of the Apex Basalt is overlain by the Panorama Formation.

The 'Apex chert' received international attention following the discovery of carbonaceous microstructures that were interpreted to be early Paleoarchean microfossils (Schopf and Packer 1987; Schopf 1992, 1993). The c. 3450 Ma depositional age of the 'Apex chert' meant that for almost 20 years these 'microfossils' were widely regarded as the oldest fossils on Earth. Subsequently, the biogenicity of the microstructures was questioned (Brasier et al. 2002, 2005, 2011, 2015; Garcia-Ruiz et al. 2002, 2003; Van Kranendonk 2006; Pinti et al. 2009b; Marshall et al. 2011, 2014) and their origin has remained controversial. Research on the 'Apex chert' has provided a large amount of evidence on its deposition and processes of hydrothermal alteration in thin sedimentary units of the Pilbara Supergroup.

During investigations of the 'Apex chert', Brasier et al. (2013) discovered different evidence for bacterial life within beds of pumice interpreted to have accumulated on a 3450-Ma shoreline. Apart from the paleontological significance of this discovery, the evidence for shallow-water deposition of the 'Apex chert' testifies to uplift and temporary emergence of the Apex Basalt following subaqueous deposition of up to 1 km of pillowed komatiitic basalts and tholeiites in the first minor volcanic cycle. The protoliths of the 'Apex chert' were deposited at the same stratigraphic position as black shale and sandstone in the Salgash–Warrawoona area, 15 km south of Marble Bar (Fig. 3.18). Here, stratiform Cu–Zn deposits hosted by the metamorphosed black shales provide additional evidence of hydrothermal activity at the level of the 'Apex chert'. Five samples of the black shale analyzed by Wille et al. (2013) contained up to 13.4 wt.% organic carbon (average 5.45 wt.%) up to 6770 ppm Zn (average 2804 ppm), up to 1088 ppm Cu (average 562 ppm), and up to 421 ppm Ni (average 296 ppm). The shale samples have extremely high Na_2O contents (up to 9 wt.%), which suggest albite alteration in a hydrothermal setting. The 'Apex chert' outcrops over a strike length of 50 km on the western side of the Mount Edgar Dome, from the Warrawoona Syncline, where it becomes sheared out southeastwards (Hickman and Van Kranendonk 2008b), to the Doolena Gap area 30 km north of Marble Bar. In combination with the lateral extent of the underlying MBCM, this testifies to the high degree of lateral continuity of depositional environments between volcanic cycles.

The minimum age of the Apex Basalt in the Marble Bar greenstone belt is indicated by geochronology from the overlying Panorama Formation west of Coppin Gap area (Fig. 3.18) where a rhyolite immediately above the Apex Basalt was dated at 3446 ± 5 Ma (De Vries 2004). However, a c. 3450-Ma depositional age for the formation is apparently questioned by detrital zircon dates from sandstone in the lower Apex Basalt (Fig. 2.6) suggesting a maximum depositional age of 3441 ± 12 Ma. A c. 3441-Ma maximum depositional age for the Apex Basalt would be significant for two reasons: 1, it would imply that the 100-m-thick MBCM, which separates the volcanic part of the Duffer Formation

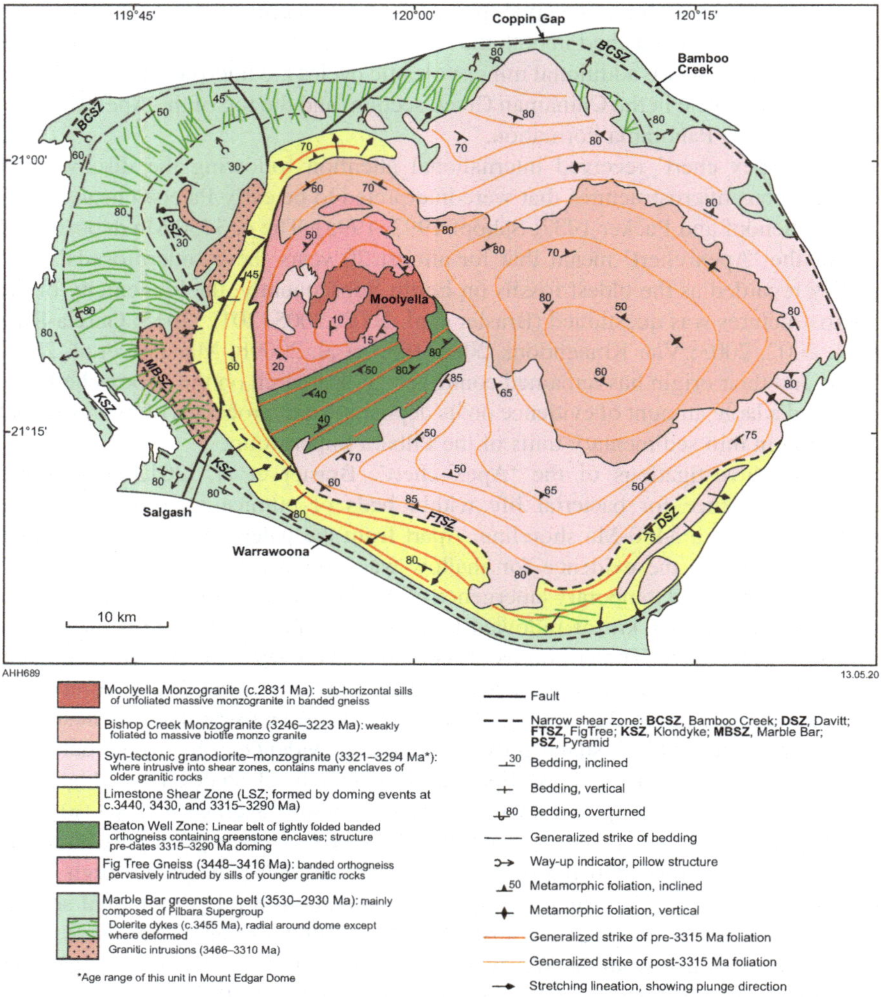

Fig. 3.18 Tectono-stratigraphic units of the Mount Edgar Dome, summarizing domal structure and geochronology. Structures characteristic of diapiric doming include radial outward-plunging stretching lineations within and adjacent to the 3440–3290 Ma Limestone Shear Zone (ring fault) and radial swarms of c. 3455 Ma dolerite dykes intrusive into extensional fractures formed by domal uplift. Place names (bold) are localities commonly cited in the text (From Gardiner et al. 2018; with Geological Survey of Western Australia permission)

(3474–3459 Ma) from the Apex Basalt, was deposited over almost 20 Ma; and 2, it would suggest that eruption of the Apex Basalt (komatiite and basalt) and the Panorama Formation (dacite and rhyolite) of the Salgash Subgroup overlapped by at least 10 Ma (3450–3440 Ma). Basaltic formations attributed to mantle plume events, as in the case of the Apex Basalt (Van Kranendonk et al. 2002, 2006, 2007;

Hickman 2011), are normally deposited quickly, typically over 1–5 Ma. A possible explanation for the c. 3441 Ma maximum depositional age based on dating of detrital zircons is all the zircons have suffered Pb loss.

3.2.4.2 Panorama Formation

The 3448–3427 Ma Panorama Formation is the youngest formation of the Warrawoona Group, and is exposed in the Marble Bar, Panorama, Coongan, Kelly, McPhee, Doolena Gap, and Warralong greenstone belts (Fig. 1.7). The formation reaches a maximum thickness of 1.5 km in the Kelly greenstone belt, and is over 1 km thick in the McPhee greenstone belt. Lithologies include rhyolitic and dacitic volcaniclastic rocks, felsic lava, porphyritic felsic intrusive rock, and clastic sedimentary rocks. Unlike the Duffer Formation, which was erupted in a single large felsic volcanic complex, the Panorama Formation was erupted during doming and, probably as a consequence of this, developed several separate and compositionally distinct eruptive centres. In most greenstone belts, the Panorama Formation conformably overlies the Apex Basalt, although there is geochronological and stratigraphic evidence to suggest that early 3448–3440 Ma felsic volcanics and sub-volcanic intrusions of the formation were contemporaneous with basaltic volcanism of the Apex Basalt. The Apex Basalt is locally absent from the Warrawoona Group due to 3450–3420 Ma erosion, and in these situations the Panorama Formation disconformably to unconformably overlies either the Duffer Formation or the Mount Ada Basalt. In different greenstone belts, the Panorama Formation is either disconformably or unconformably overlain by the Strelley Pool Formation. In the southeast part of the EPT, where there is no angular unconformity between the Panorama Formation and the Strelley Pool Formation, and where both formations contain sandstone and conglomerate, the contact between the formations is locally difficult to recognize.

The type area of the formation was originally designated as Panorama Ridge on the southern side of the North Pole Dome (Fig. 1.7) (Lipple 1975). However, the remoteness of Panorama Ridge has led to limited mapping and the local stratigraphy being incompletely understood. Geochronology on rhyolite on the northern slopes of the ridge has given conflicting dates of 3458 ± 2 Ma (Thorpe et al. 1992a, b) and 3451 ± 3 Ma (Petersson et al. 2019), and it is likely that at least one of the dated samples was from a felsic intrusion. A preferred type section is along Sandy Creek on the northeast side of the Corunna Downs Dome where the Panorama Formation has been mapped in more detail (Hickman 1980a), studied geochemically (Smithies et al. 2007), and dated between 3433 and 3427 Ma at four separate localities (Nelson 2000, 2001, 2005). Above an intrusive granitic contact, the lowest stratigraphic level of the formation that is preserved comprises interbedded dacitic and rhyolitic lavas and volcaniclastic rocks. Individual lava flows of plagioclase-phyric dacite and rhyolite up to 30 m thick are interbedded with lava breccias, clast-supported grain-flow breccias, graded pyroclastic flows, and hyaloclastic breccia (Smithies et al. 2007). The upper Panorama Formation at Sandy Creek is composed of felsic volcaniclastic units, epiclastic deposits, and minor rhyolitic lavas. The top of the formation is disconformably overlain by partly silicified stromatolitic carbonate rocks of the Strelley Pool Formation.

The first detailed descriptions of the Panorama Formation were provided by Barley (1981), Barley et al. (1984), and DiMarco (1986). Based on a regional stratigraphic interpretation by Hickman (1980b), Barley (1981) assigned thick successions of felsic volcanic rocks in the McPhee greenstone belt and the adjacent eastern Kelly greenstone belt (including Sandy Creek) to the Duffer Formation. Subsequent geochronology revealed that the depositional ages of the felsic successions in these areas are c. 3430 Ma and that they, therefore, belong to the Panorama Formation.

In the McPhee Dome, felsic volcanic rocks overlie a succession of plagioclase-porphyritic basalts and andesites (Barley 1981), and this has since been correlated with the Apex Basalt (Williams and Hickman 2000). From the description by Barley (1981), the Panorama Formation of the McPhee Dome is composed of dacitic lava overlain by felsic pyroclastic rocks that accumulated to form a volcanic centre. Debris flows and sedimentary rocks including turbidites were deposited on the flanks of this centre, and the formation is overlain by a thick chert unit (Strelley Pool Formation). Geochemical data from the section are available in Glikson and Hickman (1981a) and Smithies et al. (2007).

DiMarco (1986) investigated the stratigraphy and sedimentology of the Panorama Formation in parts of the Coongan, Kelly, Panorama, North Shaw, and East Strelley greenstone belts. He interpreted the Panorama and Duffer Formations to be members of a unit that he termed the 'Coongan Formation' (obsolete name) with the result that his descriptions of the 'Panorama Member' (obsolete name) in the Marble Bar and southern Panorama greenstone belts now apply to the Duffer Formation. DiMarco and Lowe (1989a, b) used the regional interpretation of DiMarco (1986) and, although reinstating the name Panorama Formation (from 'Panorama Member'), continued to place it directly above the Duffer Formation, thus mistakenly interpreting it to stratigraphically underlie the Marble Bar Chert Member (then referred to as the 'Towers Formation').

A detailed description of the Panorama Formation in the northwestern Panorama greenstone belt (Van Kranendonk 1999, 2000; Van Kranendonk and Hickman 2000) supported a previous conclusion that the formation was erupted from separate volcanic centres (Barley et al. 1984; DiMarco and Lowe 1989a, b). The succession assigned to the Panorama Formation in the northwest Panorama greenstone is unusual in several respects: 1, an absence of felsic lava; 2, a high proportion of sandstone units; and 3, reworked, as opposed to ash-fall, carbonate-bearing volcaniclastic units. A sample from the upper part of this succession, dated at 3434 ± 5 Ma (Nelson 2000), was interpreted to be a pervasively silicified felsic tuff, although with 85 vol.% quartz and 15 vol.% sericite, it is more likely to be a metamorphosed volcaniclastic sandstone. This succession of the Panorama Formation is also unusual in not being overlain by lithologies typical of the Strelley Pool Formation, raising the possibility that part of it might be laterally equivalent to that formation. Van Kranendonk (2000) remarked that carbonate-bearing felsic volcaniclastic rocks of the 'Panorama volcano' (northwest Panorama greenstone belt succession) are identical in appearance to sandstones and conglomerates in the basal Strelley Pool Formation. Retallack (2018) interpreted the succession of the Panorama Formation of this area to be a volcanic flank and floodplain facies in which layers containing barite nodules represent alluvial paleosols. The depositional environment described by Retallack (2018) is very similar to that interpreted for parts of the Strelley Pool Formation.

The sandstones and felsic tuffaceous units of the northwest Panorama greenstone belt (North Pole Dome) overlie a 40-m-thick jaspilitic iron formation, which was mapped as part of the Panorama Formation by Van Kranendonk (1999) because in 1999 the underlying basalt of the North Pole Dome was interpreted to be the Apex Basalt. With recognition that this basalt formation is actually the Mount Ada Basalt (Van Kranendonk et al. 2006b), the jaspilitic iron formation might be reasonably correlated with the Marble Bar Chert Member (Duffer Formation). Allwood et al. (2007) interpreted the iron formation to be conformable with the underlying Mount Ada Basalt which, in view of uplift and erosion of the Apex Basalt between 3450 and 3426 Ma, supports this correlation. Boulders and pebbles of the iron formation are included in the basal conglomerate of the overlying Strelley Pool Formation (Van Kranendonk 2000; Allwood et al. 2007).

The Panorama Formation was erupted from several volcanic centres (Barley 1981; Barley et al. 1984; DiMarco 1986; DiMarco and Lowe 1989a, b; Van Kranendonk 1999, 2000; Smithies et al. 2007; Van Kranendonk et al. 2007a, b). As with the Duffer Formation, lateral variations of volcanic and sedimentary facies provide the best evidence on the locations of these centres. Volcanic centres are interpreted for the McPhee Dome (Barley 1981) and North Pole Dome (DiMarco and Lowe 1989a, b; Van Kranendonk 2000). Thick successions of vent agglomerate in the northwest and east section of the Kelly greenstone belt suggest that a third centre was located in the Corunna Downs Dome. Conversely, sections of the Panorama Formation that are less than 500 m thick, and dominated by fine-grained felsic volcaniclastic rocks, sandstones, shales, or turbidites, represent more distal successions. Distal facies are exposed around the headwaters of Warralong Creek (northwest Marble Bar greenstone belt), south of Kittys Gap (north Marble Bar greenstone belt), south of Yandicoogina (southeast Marble Bar greenstone belt), and as thin units in the Goldsworthy, Warralong, Doolena Gap, North Shaw, and Coongan greenstone belts (Fig. 1.7). The conclusion that Panorama volcanic centres were located in areas now occupied by the granitic cores of the EPT domes (DiMarco and Lowe 1989a, b; Hickman and Van Kranendonk 2004; Williams and Bagas 2007) is consistent with the interpretation that from c. 3450 Ma onwards Paleoarchean granitic intrusion was concentrated into the centres of the evolving domes (Hickman and Van Kranendonk 2004).

Exposures of the Panorama Formation in the northwest Marble Bar greenstone belt are separated by faults and unconformably overlying formations of the Fortescue Group. In this area, around the headwaters of Warralong Creek, the lower part of the formation is composed of massive quartz-porphyritic rhyolite, fine-grained felsic volcaniclastic rocks, sandstone, and siltstone. The upper part of the formation comprises siltstone, shale, and iron formation overlying thin beds of pebbly sandstone and conglomerate (Van Kranendonk 2004). This upper unit is overlain by the Euro Basalt and might alternatively be interpreted as a facies of the Strelley Pool Formation. Near Kittys Gap, 35 km east of Warralong Creek, the Panorama Formation is mainly composed of rhyolitic agglomerate and flow-banded porphyritic rhyolite up to 800 m thick (Williams 1999). The rhyolitic succession is overlain by a silicified fine-grained volcaniclastic unit interpreted to be the Strelley Pool Formation. A porphyritic rhyolite near the base of the Panorama Formation was dated at 3446 ± 5 Ma (De Vries 2004). Farther east in the Marble Bar greenstone belt, southwest of Bamboo Creek mining area, lenticular units of the Panorama Formation separate the Apex and Euro Basalts along a disconformable contact (Williams 1999). Conventional U–Pb zircon geochronology was used to date this section of the formation at 3449 ± 3 Ma (Thorpe et al., 1990, 1992a, b). The geochemistry of the Panorama Formation in the Kittys Gap area is distinctly different from that of the Duffer Formation in the same greenstone belt (De Vries et al. 2006; Smithies et al. 2007). De Vries et al. (2006) reported that dacite of the Duffer Formation has substantially higher chondrite-normalized REE contents and lower normalized La/Yb than rhyodacite of the Panorama Formation, and in contrast to the Duffer Formation the Panorama Formation does not exhibit negative Eu-anomalies. Smithies et al. (2007) commented that normalized La/Yb and La/Nb were highest in the Panorama Formation, and their REE data indicate no negative Eu-anomalies.

Syn-depositional growth faulting has been interpreted in the Kittys Gap area of the northern Marble Bar greenstone belt (Nijman et al. 2001, 2017; Nijman and De Vries 2004; De Vries 2004; De Vries et al. 2006). The faults are visible on aerial photographs and Landsat imagery and, although younger faulting is also present in the area, lateral changes in the thicknesses of sedimentary units across the faults support some syndepositional movement. The faults are probably extensional structures related to uplift of the Mount Edgar Dome. Although Nijman et al. (2017) observed that the orientations of the syn-depositional faults are not consistent with the present geometry of the Mount Edgar Dome, this geometry was substantially formed by c. 3315 Ma doming (Collins et al. 1998). The absence of angular unconformities at the base of the Panorama Formation in the Mount Edgar, Corunna Downs, and McPhee Domes (Hickman 2010) suggests that these domes were not well-developed until after 3350 Ma.

3.3 Origin of the Warrawoona Group

The stratigraphy of the Warrawoona Group has been described and revised in numerous publications since its first recognition and definition (Hickman and Lipple 1975; Lipple 1975; Hickman 1980a, b, 1981, 1983). For example, the entire 3530–3235 Ma volcanic succession that was initially assigned to the Warrawoona Group (Lipple 1975) was later assigned to the Pilbara Supergroup (Van Kranendonk et al. 2006a), with the Warrawoona Group comprising only that part of the succession deposited before the 75-million-year break in volcanism (3426 and 3350 Ma). Its evolution has been linked to a succession of mantle plumes (Van Kranendonk et al. 2002, 2007a, b; Smithies et al. 2005a, b), and geochemical evidence has been used to interpret the sources of the magmas for its volcanic rocks and large volumes of contemporaneous granitic rocks (Champion and Smithies 1999, 2001, 2007, 2019; Bagas et al. 2003; Smithies et al. 2003, 2009). Recently, new isotopic evidence has provided significant insights into the transition of the craton from an Eoarchean–early Paleoarchean terrane of mafic crust and granitic gneiss to a Paleoarchean granite–greenstone terrane (Hickman 2021). This evidence indicates a major magmatic event that commenced abruptly at c. 3530 Ma, with widespread eruption of volcanic rocks on the scale of a large igneous province (LIP). Lu–Hf isotope data suggest that whereas the 3800–3530 Ma crust apparently evolved through reworking of older crust the Paleoarchean EPT evolved through the introduction of large volumes of juvenile, mantle-derived magmas.

Most workers interpret the Earth's mantle to have been significantly depleted by 3500 Ma (McCulloch and Wasserberg 1978; DePaolo 1980, 1988; Bennett et al. 1993; McCulloch and Bennett 1993; Hofmann 1997; Amelin et al. 1999, 2000; Griffin et al. 2000; Hasenstab et al. 2019, 2021). On this basis, Lu–Hf isotope data from all 126 Pilbara xenocrystic and detrital zircons dated between 3795 and 3540 Ma suggest mantle extraction ages greater than 3800 Ma (Fig. 2.9). These data provide no evidence for any significant additions of juvenile magma between 3800 and 3550 Ma. By contrast, most of the Pilbara igneous, xenocrystic, and detrital zircons dated between 3530 and 3400 Ma, and for which Lu–Hf data are available ($n = 309$, Hickman 2021), have ε_{Hf} values far more positive than CHUR (Fig. 2.9), suggesting an abrupt input of mantle-derived juvenile magmas commencing at 3530 Ma. Juvenile magmas continued to be added to the Pilbara crust until deposition of the Warrawoona Group ended with uplift, deformation, metamorphism, and erosion commencing at c. 3427 Ma. From 3530 Ma onwards, the addition of juvenile magma appears to have decreased while crustal reworking became increasingly significant (Fig. 2.9). The reason for the surge of volcanic and intrusive activity at 3530 Ma has not been established, although a likely explanation is that the pre-3530 Ma crust was uplifted and rifted above the first major Paleoarchean mantle plume to impact the Pilbara Craton. This plume was responsible for the eruption of the Talga Talga and Coonterunah Subgroups between 3530 and 3490 Ma, and for contemporaneous intrusion of the first granitic supersuite ('Mulgundoona supersuite', Petersson et al. 2020).

3.4 Evolution of the Warrawoona Group

Deposition of the Warrawoona Group occurred during a series of ultramafic–mafic–felsic volcanic cycles, each of about 15 million years duration (Hickman 2011, 2012; Hickman and Van Kranendonk 2012). The vertically repeated upward lithological changes from basal komatiite and komatiitic basalt to thicker units of tholeiitic basalt, and overlying units of andesite, dacite, and rhyolite, plus the depositional rates indicated by geochronology, are consistent with a mantle plume origin (Van Kranendonk et al. 2002, 2004, 2006b, 2007a, b; Smithies et al. 2005b, 2007). The tops of completely preserved volcanic cycles are overlain by thin units of metamorphosed volcaniclastic sandstone, shale, chert, and local jaspilitic chert or iron formation. Thin carbonate units (locally containing stromatolites and microbial mats) and evaporite deposits are locally present but are typically silicified.

The existence of the volcanic cycles in the Warrawoona Group (Anhaeusser 1971) led to an early definition of subgroups (Lipple 1975). The present subgroups, as defined by Van Kranendonk et al. (2006a) are the Coonterunah, Talga Talga, Coongan, and Salgash Subgroups. As illustrated in Fig. 3.1, the Coongan and Salgash Subgroups each represent a separate volcanic cycle whereas the Coonterunah and Talga Talga Subgroups include several incomplete cycles (Hickman 2011).

The Warrawoona Group displays an upward increase in the frequency and thickness of felsic volcanic units. The Coonterunah and Talga Talga Subgroups are mainly composed of ultramafic and mafic volcanics whereas the Coongan and Salgash Subgroups show more complete ultramafic–mafic–felsic cycles. Felsic volcanism increased with time (Smithies et al. 2007), although the eruption of komatiite and komatiitic basalt did not correspondingly decrease. Volumes of felsic intrusive rocks also appear to have increased with time, although this impression, gained from present exposures, is evidently partly a consequence of tectonic removal of large sections of the older granitic units by sagduction during gravity-driven diapiric doming. In those granite–greenstone domes that contain granites varying widely in age (e.g. the Mount Edgar, Shaw, Muccan, and Carlindi Domes), it is notable that the oldest granitic rocks are concentrated close to granite–greenstone contacts and, like the adjacent greenstones, are vertically sheared and attenuated (Hickman and Van Kranendonk 2004).

The depositional setting of the Warrawoona Group was a volcanic plateau underlain by the pre-3530 Ma sialic crust (Van Kranendonk et al. 2002; Hickman 2004). Thick successions of pillow basalt are evidence of subaqueous deposition, but not necessarily of deep-water deposition. Vertical deformation during deposition of the Warrawoona Group involved not only uplift of the domal granitic complexes but also increasing subsidence of the intervening depositional basins. Evidence of shallow-water deposition is provided by the sedimentology and palaeontology of the c. 3480 Ma Dresser Formation at the top of the Talga Talga Subgroup (Dunlop et al. 1978; Barley et al. 1979; Walter et al. 1980; Dunlop and Buick 1981; Buick and Barnes 1984; Buick and Dunlop 1990; Nijman et al. 1998, 1999; Van Kranendonk 2000; Noffke 2013), and by pumice beds in the 3450–3427 Ma Salgash Subgroup that Brasier et al. (2013) interpreted to be shoreline deposits. Episodes of subaerial erosion also occurred during deposition of the Mount Ada Basalt and the Duffer and Panorama Formations.

From c. 3470 Ma onwards, different areas of the volcanic plateau had distinctly different depositional environments. For example, the felsic volcanic complex of the

Duffer Formation (Coongan Subgroup), restricted to an area of c. 15,000 km^2 centred on what is now the Marble Bar area, formed a depositional environment quite different from the basaltic lava plains of the Mount Ada Basalt elsewhere on the volcanic plateau. Then, with the development of diapiric doming, various areas of the plateau were being uplifted, intruded by granites, and eroded while adjacent areas were subsiding and developing progressively deepening greenstone basins above zones of sagduction (Thébaud and Rey 2013). Once the locations of the first domes and basins had been established between 3450 and 3426 Ma, subsequent episodes of diapiric deformation accentuated the same structures (Hickman and Van Kranendonk 2004), effectively increasing lateral density contrasts across the EPT.

Between the volcanic cycles, erosion of uplifted areas led to deposition of clastic and chemical sedimentary successions, particularly along the margins of the basins. This type of sedimentary environment is interpreted to have applied to the Marble Bar Chert Member (at the top of the Coongan Subgroup), and possibly also to the Dresser Formation (at the top of the Talga Talga Subgroup), and perhaps to the 'Apex chert' (informal name) within the Salgash Subgroup. It also applied to deposition of the Strelley Pool Formation after eruption of the Salgash Subgroup. Thin chert units (silicified fine-grained clastic and volcaniclastic rocks) within the subgroups suggest 'second-order' volcanic cycles separated by periods of volcanic quiescence and hydrothermal activity.

A significant geochemical feature of both the Coonterunah and Talga Talga Subgroups, and in the Pilbara Supergroup generally, is a progressive upward decrease in high field strength elements (HFSE) and rare earth elements (REE) from the lowest exposed stratigraphic sections (Hickman 1980a; Glikson and Hickman 1981a, b; Smithies et al. 2005b, 2007). One possible explanation for this geochemical trend is decreasing crustal contamination of the basaltic magmas with time. Evidence for crustal contamination cited by Green et al. (2000) included upward geochemical trends such as relative enrichment in large ion lithophile elements (LILE), Th, U, and light rare earth elements (LREE). Green et al. (2000) observed that the same geochemical trend is present in the 3350–3335 Ma Euro Basalt that immediately overlies the Strelley Pool Formation. Smithies et al. (2007) also commented on strong geochemical similarities between the 3530–3515 Ma Table Top Formation and the Euro Basalt. The Euro Basalt was unconformably deposited on the granite–greenstone continental crust of the EPT that had formed from 3530 Ma onwards (Van Kranendonk et al., 2006a, 2007a, b; Hickman 2010, 2012), implying that the Table Top Formation was also deposited on continental crust (Table 1). The progressive upward decrease in high field strength elements (HFSE) and rare earth elements (REE) has also been explained as a consequence of increasingly depleted sources with time.

Relatively high levels of crustal contamination at the beginning of volcanic cycles might be explained by early wallrock interaction in conduits or by greater contamination along the Earth's surface because the first lava flows would have most likely interacted with any underlying sedimentary or felsic igneous rocks. Geochemical data from Glikson and Hickman (1981a, b) revealed very similar geochemical trends of upwardly decreasing HFSE and LILE for both the North Star and the Mount Ada

Basalts. Such repetition suggests that geochemical trends were confined to individual volcanic cycles rather than occurring over larger timeframes as represented by the entire Warrawoona Group, or even the Pilbara Supergroup. This conclusion is consistent with the observation by Smithies et al. (2005b) that the composition of high-Ti basalts did not change significantly during the 300-million-year eruption of the Pilbara Supergroup.

3.5 Large Igneous Province

The preserved thickness of the Warrawoona Group varies between 10 and 15 km and, allowing for concealment of the EPT by Neoarchean cover, its original depositional extent across the terrane is interpreted to have exceeded 100,000 km^2. Thus, with an estimated total volume of volcanic rocks exceeding 1,000,000 km^3, the Warrawoona Group easily meets the 100,000 km^3 volume requirement for a large igneous province (Bryan and Ernst 2008; Ernst 2014). Moreover, a close genetic relationship between felsic volcanic formations and contemporaneous granitic supersuites suggests that the considerable volumes of the latter (equal to or greater than the volcanic rocks) should be included in volume calculations for the Warrawoona Large Igneous Province. The group was erupted during several mantle plume events (Fig. 3.1) collectively spanning 100 million years. According to LIP criteria proposed by Bryan and Ernst (2008) this is too long for the formation of a single LIP. However, many continental LIPs are characterized by pulses of large-volume volcanic eruptions (Condie 2001; Ernst 2014). The South African equivalent of the Warrawoona Group, the Paleoarchean Onverwacht Group of the Kaapvaal Craton, has been described as a 'fragmented LIP' (Ernst 2007, 2014; Bryan and Ernst 2008). Available geochronology suggests that some basaltic formations of the Warrawoona Group were erupted in less than 5 Ma (e.g. the Mount Ada Basalt) whereas felsic volcanic formations, and contemporaneous granitic intrusions, spanned over 15 million years (e.g. the Duffer and Panorama Formations).

Because the subgroups of the Warrawoona Group were formed by separate mantle plume events, it might be argued that each subgroup constitutes a separate LIP. However, there is insufficient exposure to establish if all the individual subgroups would meet the volume criteria for LIPs. Moreover, adoption of this interpretation would have stratigraphic implications in suggesting that the present 'subgroups' should be redefined as 'groups', and such a significant stratigraphic revision is not proposed here. One consideration is that the subgroups are separated by short intervals of time (<ten million years) implying a close temporal relationship that is not typical of Proterozoic and Phanerozoic LIPs. It might be that the <50-million year criterion for a single LIP, proposed by Bryan and Ernst (2008) from Phanerozoic examples formed in plate tectonic settings, is not appropriate for the early Paleoarchean volcanic successions of the Pilbara Craton. A precedent for this argument is that the lower Onverwacht Group of the eastern Kaapvaal Craton, which is the same age as the Warrawoona Group and is composed of a very similar volcanic succession, has been described as a 'fragmented' LIP (Ernst 2007, 2014; Bryan and Ernst 2008). Thus, the Warrawoona Group is assigned to the Warrawoona LIP (Hickman 2012, 2021).

Mafic dyke swarms and large mafic intrusions commonly underlie eroded pre-Mesozoic continental LIPs (Ernst 2007), and examples of these features are well exposed beneath the Salgash Subgroup in the Marble Bar greenstone belt. Hundreds of closely spaced dolerite dykes intruded the >8-km-thick c. 3470 Ma felsic crust (Callina Supersuite and Duffer Formation) underlying the Apex Basalt (Fig. 3.7) (Kloppenburg 2003; Van Kranendonk et al. 2006b; Hickman and Van Kranendonk 2008a, b; Hickman 2012; Gardiner et al. 2017, 2018). The dykes cut through the Marble Bar Chert Member at the top of the Duffer Formation and feed into the lower lava flows of the Apex Basalt. The dykes radiate from the centre of the Mount Edgar Dome (Fig. 3.18) but due to younger granitic intrusion and shearing are only well preserved in relatively undeformed sections of the Marble Bar greenstone belt. This radial orientation suggests that the mafic magma intruded extensional fractures formed during uplift of the dome (Fig. 3.19).

3.6 Granitic Supersuites of the Warrawoona LIP

About 70 percent of the exposed area of the EPT is composed of granitic intrusions varying in age from 3530 to 3223 Ma. Geochronology and geochemistry have been used to assign over 70 individual intrusions to five supersuites (Van Kranendonk et al. 2004, 2006a; Petersson et al. 2020). U–Pb zircon dating and intrusive relationships have revealed that intrusions of all five supersuites are contemporaneous with felsic volcanic formations of the Pilbara Supergroup (Barley and Pickard 1999; Hickman 2004, 2012, 2021; Van Kranendonk et al. 2001a, 2002, 2004, 2006a; Hickman and Van Kranendonk 2004; Petersson et al. 2020). Granitic intrusions and contemporaneous felsic volcanic formations were products of the same mantle plume events (Van Kranendonk et al. 2002, 2006b, 2007a, b). In some instances, this is directly established from field evidence that felsic volcanic formations were fed from subvolcanic sills and granitic intrusions (Vearncombe and Kerrich 1999; Van Kranendonk 1999, 2000; Buick et al. 2002; Van Kranendonk et al. 2002, 2006a; Bagas et al. 2004; Van Kranendonk and Pirajno 2004; Brown et al. 2006, 2011; Hickman and Van Kranendonk 2008a; Hickman 2012, 2021).

Much felsic volcanism, particularly prior to c. 3420 Ma, occurred within plume-related ultramafic–mafic–felsic volcanic cycles in which magma was derived by melting of the mantle (Van Kranendonk et al. 2002, 2007a, b). Subvolcanic granitic rocks had the same origin, although geochemical data indicate that the majority of granitic magmas were derived by melting of older mafic and TTG crust (Collins 1993; Smithies et al. 2003, 2009; Champion and Smithies 2007, 2019). Crustally derived magmas were products of temperature increases beneath rapidly thickening volcanic successions and sagduction, both features directly related to mantle plumes. Accordingly, the granitic supersuites of the EPT are parts of the Warrawoona LIP.

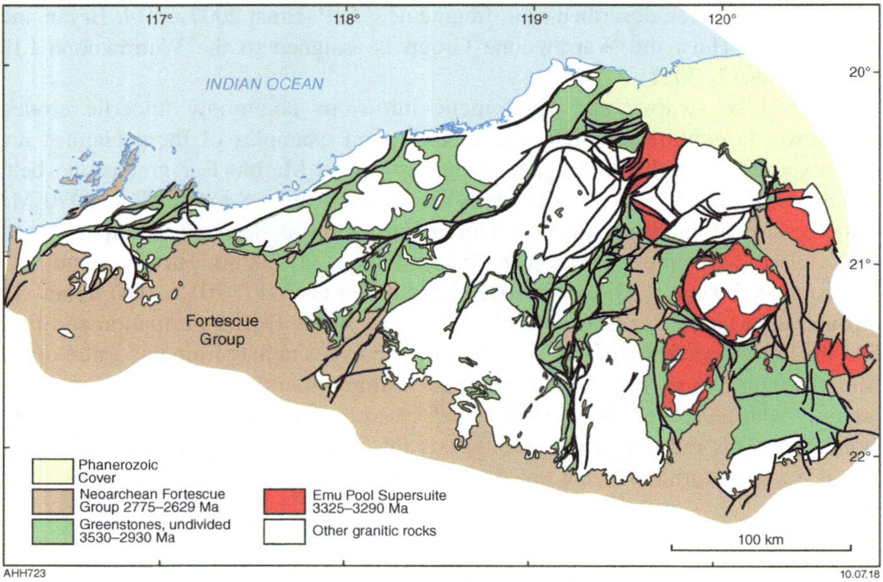

Fig. 3.19 Simplified geological map of the Northern Pilbara Craton showing the distribution of the Emu Pool Supersuite (Modified from Van Kranendonk et al. 2006a, b; with Geological Survey of Western Australia permission)

From oldest to youngest, the supersuites are: Mulgundoona Supersuite, emplaced during deposition of the Coucal Formation; Callina Supersuite, emplaced during volcanism of the Duffer Formation; Tambina Supersuite emplaced beneath the Panorama Formation; Emu Pool Supersuite emplaced during eruption of the Wyman Formation; and the Cleland Supersuite emplaced contemporaneously with volcanism of the Kangaroo Caves Formation. Each supersuite is composed of numerous individual intrusions. For example, the Emu Pool Supersuite comprises 24 named and unnamed intrusions ranging in age from c. 3324 to 3290 Ma, and varying in composition from tonalite to syenogranite (Table 3.2). Additionally, each of the supersuites is represented by intrusions in several EPT domes. The varying ages and compositions of the intrusions within individual supersuites, and their regional distributions, are important considerations before making any generalizations on subjects such as their magmatic sources. General geochemical and isotope trends are evident between the Callina and Cleland Supersuites but the degree of data scatter within these isotope trends (e.g. Fig. 2.9) might result from mixed juvenile and evolved sources for the different intrusions within any particular supersuite. In this scenario, mean isotope values such as ε_{Hf} are more likely to fall close to CHUR.

Table 3.2 Listing of all granitic intrusions of the East Pilbara Craton, showing age ranges from U–Pb zircon dating, and identifying the samples dated (From Hickman 2021; with Geological Survey of Western Australia permision)

Supersuite/ suite	Dome/terrane	Intrusion name/ location	Date (Ma)	Geochronology sample
Split Rock Supersuite			**2851–2831**	
	Mount Edgar Dome	Moolyella Monzogranite	c. 2831	169044
	McPhee Dome	Cookes Creek Monzogranite	c. 2837	178011
	Kurrana Terrane	Bonney Downs Monzogranite	c. 2838	178014
	Shaw Dome	Spear Hill Monzogranite	c. 2851	142879
	Shaw Dome	Cooglegong Monzogranite	n.d.	
	Yule Dome	Gillam Monzogranite	n.d.	
	Yule Dome	Kimmys Bore Monzogranite	n.d.	
	Yule Dome	Kadgewarrina Monzogranite	n.d.	
	Yule Dome	Minnamonica Monzogranite	n.d.	
	Corunna Downs Dome	Mondana Monzogranite	n.d.	
	Carlindi Dome	Numbana Monzogranite	n.d.	
	Carlindi Dome	Poocatche Monzogranite	n.d.	
	Tambourah Dome	Tambourah Monzogranite	n.d.	
Cutinduna Supersuite			**2897–2896**	
	Kurrana Terrane	Unnamed, Limestone Bore	c. 2896	178231
	Kurrana Terrane	Unnamed, Limestone Bore	c. 2897	178230
Sisters Supersuite			**2954*–2919**	
	Shaw Dome	Mulgandinnah Monzogranite	2934–2919	142882, 142883, 142965, 142967, T94/31(q)
	Carlindi Dome	Unnamed, Pilgangoora area	c. 2923	169021
	Shaw Dome	Bamboo Springs Monzogranite	2926–2923	178047, 178049
	Carlindi Dome	Woodstock Monzogranite	2933–2927	142884, 142885

(continued)

Table 3.2 (continued)

Supersuite/ suite	Dome/terrane	Intrusion name/ location	Date (Ma)	Geochronology sample
	Carlindi Dome	Unnamed granite	2944 -2931	KB351(b), KB746(b), KB770 (b)
	Yule Dome	Cheearra Monzogranite	2937–2933	142936, 160442
	Yule Dome	Powdar Monzogranite	c. 2935	142937
	Shaw Dome	Keep It Dark Monzogranite	c. 2936	Unpublished data
	Yule Dome	Mungarinya Monzogranite	c. 2938	142176
	Carlindi Dome	Petermarer Monzogranite	c. 2940	160745
	Yule Dome	Beabea Monzogranite	c. 2941	169018
	Yule Dome	Unnamed, Hong Kong Mine	c. 2942	180097
	Carlindi Dome	Unnamed aplite	c. 2944	KB351
	Yule Dome	Mungaroona Granodiorite	c. 2945	142938
	Carlindi Dome	Unnamed, Pilbara Well area	c. 2946	142941
	Yule Dome	Unnamed, Wodgina area	c. 2952	180038
	Yule Dome	Abydos Monzogranite	n.d.	
	Shaw Dome	Coondina Monzogranite	n.d.	
	Shaw Dome	Eley Monzogranite	n.d.	
	Yule Dome	Ellawarrina Monzogranite	n.d.	
	Yule Dome	Pilbara Creek Monzogranite	n.d.	
	Yule Dome	Pincunah Monzogranite	n.d.	
	Yule Dome	Yandeearra Granodiorite	n.d.	
Elizabeth Hill Supersuite			**3068–3066**	
	Yule Dome	Cockeraga Leucogranite	3068–3066	169016, 169014
Mount Billroth Supersuite			**3068–3066**	
	Yule Dome	Flat Rocks Tonalite	3166–3164	142946, 142948

(continued)

Table 3.2 (continued)

Supersuite/suite	Dome/terrane	Intrusion name/location	Date (Ma)	Geochronology sample
	Kurrana Terrane	Golden Eagle Orthogneiss	3199–3178	178012, 178013
Cleland Supersuite			**3270*–3223**	
	Mount Edgar Dome	Bishop Creek Monzogranite	3246–3223	142980, 142983, 169034, 178076, 178078
	Strelley Dome	Strelley Monzogranite	3239–3238	RM1(c), 203,366(c), 203,368 (c)
	Yule Dome	Kavir Granodiorite	3240–3240	Unpublished
	Warrawagine Dome	Unnamed, 6 Mile Well	c. 3244	142869
	Muccan Dome	Wolline Monzogranite	3252–3244	143805, 143810
	McPhee Dome	Unnamed, Spinaway Well	c. 3245	178188
	Muccan Dome	Wolline Monzogranite	3251–3250	178031, 178032
	Carlindi Dome	Unnamed, Camel Well	c. 3252	178040
	Carlindi Dome	Unnamed, Turner River	3257–3235	KB265(b), KB264(b), KB779 (b), KB810(b)
	Shaw Dome	Garden Creek Monzogranite	n.d.	
	Yule Dome	Siffleetes Granodiorite	n.d.	
	Mount Edgar Dome	Bullgarina Monzogranite	n.d.	
Unnamed suite			**3279–3274**	
	Yilgalong Dome	Unnamed, Yilgalong Creek area	3279–3274	169260, 178007, 178089
Emu Pool Supersuite			**3324–3290**	
	Yilgalong Dome	Elsie Creek Tonalite	3299–3290	169261, 169262, 169263, 178008, 144682
	Mount Edgar Dome	Zulu Granodiorite	3299–3294	169036, 178008
	Mount Edgar Dome	Mullugunya Granodiorite	3308–3297	178096, 178099
	Corunna Downs Dome	Nandingarra Granodiorite	3313–3300	160208, 160209, 160210, 160212, 168922
	Mount Edgar Dome	Johansen Monzogranite	3315–3303	178075, 178077

(continued)

Table 3.2 (continued)

Supersuite/ suite	Dome/terrane	Intrusion name/ location	Date (Ma)	Geochronology sample
	Muccan Dome	Unnamed, Five Mile Creek	c. 3315	178022
	Warrawagine Dome	Unnamed, Granite Well	c. 3303	142871
	Mount Edgar Dome	Joorina Granodiorite	3321–3304	142981, 142982, 142984, LTU5472(p)
	Mount Edgar Dome	Jenkin Granodiorite	3313–3307	169045, 169046, 178037
	Corunna Downs Dome	Boobina Porphyry	3324–3307	77713(k), 86473(a), 86483 (a), 103279(m)
	Mount Edgar Dome	Cotton Well Granodiorite	3318–3309	178034, CTW-1A(g)
	Mount Edgar Dome	Campbell Well Granodiorite	c. 3310	142974
	Mount Edgar Dome	Munganbrina Monzogranite	c. 3310	178095
	Corunna Downs Dome	Carbana Monzogranite	3317–3307	142978, 169027, 169028, 178001,
				178002, 178003, 178084, Mond(a), carb(a)
	Mount Edgar Dome	Wilina Granodiorite	3323–3310	169042, N/A(d)
	Mount Edgar Dome	Kennell Granodiorite	3315–3312	169038, 178038, ME-28(g)
	Muccan Dome	Unnamed, Don Well	c. 3313	143803
	McPhee Dome	Gobbos Granodiorite	c. 3313	180067, 86358(a)
	Mount Edgar Dome	Coppin Gap Granodiorite	3317–3314	LTU5577(p), 103279 (m)
	Mount Edgar Dome	Chessman Granodiorite	3318–3314	169040, 169041
	Mount Edgar Dome	Chimingadgi Trondhjemite	n.d.	
	Mount Edgar Dome	Davitt Syenogranite	n.d.	
	Corunna Downs Dome	Triberton Granodiorite	n.d.	
	Mount Edgar Dome	Walgunya Trondhjemite	n.d.	
Tambina Supersuite			**3451–3410**	
	Warrawagine Dome	Unnamed, 6 Mile Well	c. 3410	142870
	Mount Edgar Dome	Fig Tree Gneiss	3448–3416	169031, 169041, 178036, 178079, 178100,

(continued)

Table 3.2 (continued)

Supersuite/ suite	Dome/terrane	Intrusion name/ location	Date (Ma)	Geochronology sample
				LTU5416(p), LTU6417(p), LTU6419(p),
				P94-22(g),P95-gran(g)
	Shaw Dome	Tambina Supersuite	3445–3417	142878, 76340(j)
	Carlindi Dome	Wilson Well Gneiss	c. 3420	Unpublished
	Muccan Dome	Unnamed, Florries Well	c. 3420	178023
	Tambourah Dome	Petroglyph Gneiss	3428–3420	169019, unpublished
	Yule Dome	Yallingarrintha Tonalite	c. 3421	142170
	Warrawagine Dome	Unnamed, Big Junction Well	c. 3423	180057
	Shaw Dome	Abyssinia Well Orthogneiss	c. 3425	142968
	Corunna Downs Dome	Bookargemoona Tonalite	c. 3427	160211
	Shaw Dome	Unnamed, Unices Well	c. 3430	142966
	Shaw Dome	South Daltons Pluton	c. 3431	UWA-98053(i)
	Muccan Dome	Unnamed, Sunrise Hill	c. 3443	124755
	North Pole Dome	North pole Monzogranite	3446–3431	NP-1(f), NP-27(f), 109711 (m), unpublished
	Muccan Dome	Unnamed, Kennedy Gap	c. 3438	143807
	Mount Edgar Dome	Unnamed, Kitty's Gap	c. 3446	JW95-001(e)
	North Pole Dome	Unnamed, Miralga Creek	c. 3449	103283(p)
	Mount Edgar Dome	Unnamed, West Bamboo	c. 3449	94770
	Mount Edgar Dome	Unnamed, Pyramid Well	c. 3449	94750(p)
	Shaw Dome	Tambina Supersuite	c. 3451	T94/221(q)
	Mount Edgar Dome	Lady Adelaide Orthogneiss	n.d.	
Callina Supersuite			**3484–3462**	
	Shaw Dome	Unnamed, Fairwick Well	c. 3462	168923
	Mount Edgar Dome	Owens Gully Diorite	3465–3462	178035, OG1(n), CTW-36(g)

(continued)

Table 3.2 (continued)

Supersuite/ suite	Dome/terrane	Intrusion name/ location	Date (Ma)	Geochronology sample
	Shaw Dome	Unnamed, Coondina Pool	c. 3463	T94/222(q)
	Mount Edgar Dome	Homeward Bound Granite	c. 3466	142865
	Carlindi Dome	Unnamed, Strelley River	c. 3466	100698(l)
	Shaw Dome	Coolyia Creek Granodiorite	3469–3467	142962, T94/227(q), A224-1 (h)
	Shaw Dome	North Shaw Tonalite	3469–3468	178044, T94/193(q)
	Carlindi Dome	Unnamed, Coucal Ridge	3479–3468	94058, 95037(l)
	Carlindi Dome	Unnamed, Wilson Well	3484–3469	153188, 153190
	Shaw Dome	Unnamed, 26 Mile Well	c. 3469	142976
	Muccan Dome	Unnamed, Fred Well	c. 3470	142828
	Carlindi Dome	Motherin Monzogranite	c. 3475	Unpublished data
	Shaw Dome	Unnamed, Tambourah Creek	3485–3470	Shaw B gneiss (limited data, large age uncertainty)
	Mount Edgar Dome	Underwood Gneiss	n.d.	

* Maximum age of supersuite outside East Pilbara
n.d., not dated
Ranges (e.g. 2934–2919) represent the range of published crystallization ages for the unit
References for non-GSWA data
a, Barley and Pickard (1999)
b, Beintema (2003)
c, Buick et al. (2002)
d, Collins et al. (1998)
e, De Vries et al. (2006)
f, Harris et al. (2009)
g, Kloppenburg (2003)
h, McNaughton et al. (1988)
i, McNaughton et al. (1993)
j, Pidgeon (1978)
k, Pidgeon (1984)
l, R. Buick (writt. Comm. 2015)
m, R.I. Thorpe (writt. Comm. 1992)
n, Stern et al. (2009)
o, Thorpe et al. (1992a, b)
p, Williams and Collins (1990)
q, Zegers (1996)

3.6.1 *Mulgundoona Supersuite (3530–3490 Ma)*

Until 2016, the oldest date on any granitic intrusion of the EPT was 3484 ± 4 Ma (Nelson 1999). However, by analogy with four supersuites younger than 3490 Ma, it was inferred that felsic volcanism in the Coonterunah Subgroup had been accompanied by granitic intrusion between c. 3530 and 3490 Ma (Gardiner et al. 2017, 2018). Evidence supporting this concept was provided by an abundance of 3530–3490 Ma detrital zircons in Paleoarchean and Mesoarchean sedimentary rocks of the East Pilbara Craton. Because exposures of felsic volcanic rocks are relatively rare in the preserved outcrops of the Coonterunah and Talga Talga Subgroups, it is likely that most of the 3530–3490 Ma detrital zircons were derived by erosion of contemporaneous granitic rocks. In 2016, a diorite gneiss in this age range was identified in the Muccan Dome (Allen et al. 2016), and subsequent work identified other outcrops of granitic rocks older than 3490 Ma). Quartz diorite gneiss in the southern part of the Muccan Dome was dated at c. 3503 Ma (Wiemer et al. 2018). Petersson et al. (2020) reported other c. 3500 Ma intrusions in the Muccan and Carlindi Domes, and interpreted these to represent parts of a supersuite contemporaneous with felsic volcanic rocks of the Coucal Formation in the Coonterunah Subgroup. More detailed descriptions followed, including a granodiorite gneiss dated at 3523 ± 2 Ma, a tonalite gneiss at 3502 ± 3 Ma, and a quartz diorite gneiss dated at 3501 ± 1 Ma (Petersson et al. 2020). Additionally, in the northwestern Muccan Dome a c. 3312 Ma lithologically complex gneiss included eight zircons with large, rounded cores collectively dated at 3496 ± 7 Ma. Petersson et al. (2020) interpreted the >3490 Ma granitic intrusive rocks to belong to a 3530–3490 Ma supersuite and named this the Mulgundoona Supersuite. The present rarity of exposed 3530–3490 Ma granitic rocks is here attributed to the process of diapiric doming in which successive granitic supersuites were intruded into the central granitic cores of each dome (Hickman and Van Kranendonk 2004). As a generalization, each new intrusion displaced older intrusion outwards towards the margins of the granitic cores. Successive episodes of sagduction carried down the outer, older granitic components of each dome, along with adjacent basal greenstones, deep into the crust. Consequently, most remnants of pre-3490 Ma granitic rocks, and of pre-3530 Ma crust, have been discovered in sheared granite–greenstone contacts.

3.6.2 *Callina Supersuite (3484–3462 Ma)*

The Callina Supersuite is composed of sodic metagranitic rocks that outcrop in the Carlindi, Shaw, Mount Edgar, Yule, and Muccan Domes (Fig. 1.7). The dominant composition is mafic granodiorite but other compositions include hornblende diorite, biotite tonalite, and less commonly monzogranite and alkali granite. Alkali granite forms well-preserved subvolcanic intrusions in the Marble Bar area of the Mount Edgar Dome (Hickman and Lipple 1978; Van Kranendonk et al. 2006b; Hickman and Van Kranendonk 2008b). However, many other intrusions of the Callina Supersuite are migmatized and sheared in more strongly deformed zones near granite–greenstone contacts. Arguably the most studied components of the supersuite outcrop in the northern Shaw Dome, and include the North Shaw Suite (Bettenay et al. 1981; Bickle et al. 1983, 1985; Van Kranendonk 2000; Pawley et al. 2004), which comprises the North Shaw Tonalite (Hickman and Lipple 1978) and the Coolyia Creek Granodiorite (Van Kranendonk et al. 2001b).

Discussing the geochemistry of the Callina and Tambina Supersuites, Champion and Smithies (2007) recognized two geochemical groups: low-Al granites with low Al_2O_3, Na_2O, Sr, LREE/HREE and high FeO*, HREE, Y, and $(Gd/Yb)_N$; and high-Al granites with high Al_2O_3, Na_2O, Sr, LREE/HREE and low FeO*, HREE, Y, and $(Gd/Yb)_N$. They argued that the lack of significant Eu anomalies, especially in the high-Al subgroup, coupled with a lack of correlation between Eu/Eu* (measured europium divided by europium calculated from interpolation between samarium and gadolinium; Henderson 1984) with MgO or SiO_2, and moderate to high Mg#, was inconsistent with significant fractional crystallization. They concluded that the TTGs were derived by partial melting of mafic crust, although LILE- and Th-rich TTGs would have also required more enriched source components, such as intermediate and felsic rocks.

3.6.3 Tambina Supersuite (3451–3416 Ma)

The Tambina Supersuite is mainly composed of TTG gneiss and is well exposed along the west and east sides of the Shaw Dome and on the south and west sides of the Mount Edgar Dome (Fig. 1.7). Less well-exposed sections of the supersuite form parts of the Carlindi, Muccan, Warrawagine, North Pole, Corunna Downs, Tambourah, and Yule Domes. The North Pole Dome contains only monzogranite and small subvolcanic intrusions of porphyritic rhyolite and dacite. Monzogranite and leucogranite also form parts of the Shaw, Mount Edgar, and Yule Domes. In the centre of the Warrawagine Dome and on the western side of the Shaw Dome, the Tambina Supersuite contains xenoliths of granitic and mafic intrusive rocks that predate the EPT by between 135 and 48 Ma. Belts containing numerous greenstone xenoliths are also a feature of the Tambina Supersuite in the Mount Edgar, Muccan, Shaw, Yule, and Carlindi Domes. In some instances, these xenolith belts pass laterally into greenstone belts suggesting that they are the synformal root zones of eroded greenstone keels (Hickman 1975, 1983, 1984).

The Tambina Supersuite in the Shaw Dome is sodic, including tonalite and granodiorite, and does not display extensive migmatization or deformation. Swarms of leucogranite veins are interpreted as products of melting of the older Callina Supersuite (Van Kranendonk 2000; Pawley et al. 2004; Van Kranendonk et al. 2004). In the Mount Edgar Dome, the Fig Tree Gneiss is compositionally banded and strongly deformed, and is interpreted to have been emplaced at mid-crustal levels and subsequently buried to depths approaching 25 km with metamorphism at upper amphibolite facies. Most workers have interpreted the TTG magmas to have been derived either by partial melting of older felsic crust either, as in the case of leucogranite in the Shaw Dome (Pawley et al. 2004), or more generally by infracrustal melting of enriched basaltic crust (Champion and Smithies 2007, 2019; Smithies et al. 2009; Van Kranendonk et al. 2015).

3.6.4 Emu Pool Supersuite (3324–3290 Ma)

The Emu Pool Supersuite comprises intrusions of monzogranite, granodiorite, and rare tonalite within that part of the EPT lying east of the Yule, Tambourah, and Carlindi Domes (Fig. 1.7). No intrusions of this supersuite have been identified in the Shaw and North Pole Domes (Van Kranendonk et al. 2006a), and this is unlikely to be explained by insufficient dating because the genetically related Wyman Formation is almost entirely absent from the greenstone successions of these two domes. Since doming was accompanied by granitic intrusion (Collins et al. 1998; Hickman and Van Kranendonk 2004), these anomalies suggest that there was no significant c. 3315 Ma deformation in the Shaw and North Pole Domes.

The maximum age of the Emu Pool Supersuite is defined by a U–Pb zircon date of 3324 ± 6 Ma for the Wilina Granodiorite in the Mount Edgar Dome (Collins et al. 1998). This coincides with the three oldest U–Pb zircon dates of c. 3325 Ma obtained from the Wyman Formation (Thorpe et al. 1992a; McNaughton et al. 1993). All intrusions of the supersuite are older than 3290 Ma and most have provided U–Pb zircon dates between c. 3321 and 3300 Ma.

The full compositional range of the Emu Pool Supersuite is diorite to syenogranite. Several distinct compositional and textural phases and plutons have been identified in the Corunna Downs, Mount Edgar, Muccan, Warrawagine, and Yilgalong Domes (Fig. 1.7). In particular, many intrusions of granodiorite and monzogranite exhibit porphyritic texture in which abundant plagioclase or microcline phenocrysts are up to 5 cm long. In the Mount Edgar and Corunna Downs Domes, there are several examples of granodiorite intrusions immediately underlying subvolcanic porphyritic felsic intrusions and volcanic units. All intrusions of the supersuite are only moderately to weakly foliated because the EPT experienced relatively little deformation and metamorphism between 3290 Ma and the North Pilbara Orogeny commencing at c. 2955 Ma.

The Emu Pool Supersuite includes both high-Al and low-Al granitic rocks, similar to the pre-3420 Ma TTGs. High-Al granitic rocks are dominant in the Mount Edgar Dome (Collins 1993; Davy and Lewis 1986) whereas low-Al granites are more common in the Corunna Downs Dome (Champion and Smithies 2007). Average SiO_2 and K_2O contents are higher than in the pre-3420 Ma granitic rocks, and highest in the low-Al group. In this group, Na_2O is $<4.5\%$, and Na_2O/K_2O varies from c. 2.0 to 0.5. Unlike the high-Al group, the low-Al group has pronounced negative Eu anomalies (Eu/Eu* mostly 0.8 to <0.2). The high-Al granitic rocks exhibit a trend of increasingly negative ε_{Nd} values with higher K_2O, but this negative correlation is absent in the low-Al granitic rocks due to more compositional variations (Champion and Smithies 2007). Geochemical data suggest that the high-Al granites were most likely derived by partial melting of thickened mafic crust whereas derivation of the low-Al granites probably involved some fractional crystallization, especially for silica-rich granites. Nd and Hf isotope data indicate more recycling of older crust in the Emu Pool Supersuite than in the Callina Supersuite (Champion 2013, Champion and Huston 2016; Gardiner et al. 2017). ε_{Nd} values for the supersuite vary from +1.49 to −1.44 and Nd model ages range between 3630 and 3420 Ma (Smithies et al. 2007; data reviewed in Hickman 2021).

3.6.5 *Cleland Supersuite (3270–3223 Ma)*

The Cleland Supersuite comprises 3270–3223 Ma granitic intrusions in both the EPT of the northeast Pilbara and the Karratha Terrane of the northwest Pilbara. The presence of the supersuite in the Karratha Terrane suggests that its intrusion was related to the early stage of the 3280–3176 Ma East Pilbara Terrane Rifting Event. Geochronology in the EPT has revealed a concentration of granitic ages between 3257 and 3223 Ma. Most intrusions of the Cleland Supersuite are composed of monzogranite, although granodiorite and syenogranite are also locally present. One intrusion, the c. 3239 Ma Strelley Monzogranite, is a subvolcanic laccolith (Van Kranendonk 2000), and is atypical in that it exhibits rapakivi textures that are interpreted as evidence for mingling between basaltic and felsic magmas (Brauhart 1999). Heat from this intrusion drove hydrothermal circulation that precipitated volcanogenic massive sulphide (Cu–Zn) deposits (Morant 1998; Vearncombe et al. 1998; Huston et al. 2001; Van Kranendonk 2006).

Felsic magmas for the Cleland Supersuite were largely derived from melting of older granitic crust (Barley and Pickard 1999; Smithies et al. 2003; Champion and Smithies 2007, 2019; Van Kranendonk et al. 2007a, b; Champion and Huston 2016). High- and low-Al granites have similar characteristics to those of the Emu Pool Supersuite (Champion and Smithies 2007, 2019). ε_{Nd} values for the supersuite vary from +1.49 to −1.44, and Nd model ages range between 3560 and 3460 Ma (data reviewed in Hickman 2021).

3.7 Tectonic Setting of the Warrawoona LIP

Evidence in Chap. 2 establishes that the Warrawoona Group was deposited on 3800–3530 Ma continental crust that included both mafic and felsic components. Most of this early crust was destroyed by Paleoarchean tectonic processes and invasive granitic intrusion but small remnants remain in various parts of the EPT (McNaughton et al. 1988; Nelson 1999; Williams and Hickman 2000; Williams 2001; Wingate et al. 2009a; Kemp et al. 2015; Petersson et al. 2019, 2020).

Widespread pre-3530 Ma sialic crust provided a surface across which eruption of the Warrawoona Group formed a large volcanic plateau (Van Kranendonk et al. 2002, 2006a; Hickman 2004, 2012, 2021; Hickman and Van Kranendonk 2004). Importantly, the plateau was not only the site of volcanic deposition but also of extensive 3530–3416 Ma granitic intrusion. Available evidence suggests that prior to c. 3470 Ma, mafic and ultramafic volcanism was greater than felsic volcanism and granitic intrusion, but from 3470 Ma onwards volumes of new granitic and volcanic additions were similar. Together, these processes began to construct the 3530–3223 Ma granite–greenstone assemblage of the EPT. From 3470 Ma onwards, the plateau was deformed by gravity-driven doming and sagduction in response to increasing density inversion (a relatively dense ultramafic–mafic volcanic upper crust overlying a predominantly felsic lower crust) as first interpreted by the author (Hickman 1975, 1981, 1983, 1984). Deposition of the Warrawoona Group continued during the formation of dome–and–keel crustal architecture, resulting in the thickest greenstone successions accumulating between the rising domes.

3.7.1 Plate Tectonic Models

Prior to 2006, it had been argued that Phanerozoic plate tectonic models can explain geological observations in the EPT (Bickle et al. 1983; Barley 1993; Zegers et al. 1996; Kimura et al. 1991; van Haaften and White 1998; Barley and Pickard 1999; Kitajima et al. 2001; Komiya et al. 2002; Kato and Nakamura 2003; Kloppenburg 2003; Terabayashi et al. 2003). Between 1970 and 2000, plate tectonic models were widely applied to Archean granite–greenstone terranes worldwide, mostly based on the assumption that plate tectonic processes have operated throughout Earth's history. From about 2000 onwards, this belief became increasingly challenged, as reviewed by Cawood et al. (2018). In the case of the Pilbara Craton, plate tectonic interpretations in the Paleoarchean EPT were found to be inconsistent with many lines of evidence from the 1995–2005 PCMP (Van Kranendonk et al. 2002, 2004, 2006a, 2007a, b, 2015; Hickman 2004, 2012, 2021; Hickman and Van Kranendonk 2004, 2012; Champion and Smithies 2007, 2019; Smithies et al. 2007, 2009).

Early plate tectonic interpretations failed to consider the significance of a pre-Warrawoona Group sialic crust, and with two exceptions (Zegers 1996; Kloppenburg 2003) provided no explanations for the dome–and–keel crustal architecture of the EPT. Zegers (1996) and Kloppenburg (2003) argued that the dome–and–keel structures originated within metamorphic core complexes, but as explained by Hickman and Van Kranendonk (2004) and Van Kranendonk et al. (2004), there are many important differences between Phanerozoic metamorphic core complexes and the domes of the EPT. Kusky et al. (2021) argued that the dome–and–keel crustal architecture of the EPT originated as an intra-oceanic accretionary orogen that evolved into a continental margin arc. However, this conclusion was based on various incorrect interpretations made by previous workers who had applied plate tectonic models to small areas of the EPT, in particular the North Pole Dome

(Kitajima et al. 2001, 2008; Komiya et al. 2002; Terabayashi et al. 2003). These interpretations have already been critically assessed and corrected in the literature (Van Kranendonk and Collins 2001; Van Kranendonk et al. 2001a, 2002, 2004; Hickman and Van Kranendonk 2004; Hickman 2012), although Kusky et al. (2021) did not comment on these criticisms. Examples of key incorrect interpretations include: that the Warrawoona Group evolved as the oceanic crust (mid-ocean ridge and island arc); that the entire greenstone succession of the EPT (Pilbara Supergroup) was tectonically thickened by recumbent folding and thrusting; and that the EPT domes were produced by cross folding due to different phases of horizontal compression. All these possibilities were considered during the first detailed mapping of the EPT, and had to be rejected based on contrary evidence (Hickman 1983, 1984).

Active investigators of the EPT have reached the interpretation that the dome–and–keel crustal architecture was produced by gravity-driven vertical deformation, with contributing factors such as conductive incubation, thermal and mechanical crustal softening, and granitic intrusion (Hickman 1975, 1981, 1983, 1984, 2001; Collins 1989; Collins et al. 1998; Van Kranendonk 1998, 2000; Van Kranendonk and Collins 2001; Hickman and Van Kranendonk 2004; Pawley et al. 2004; Sandiford et al. 2004; Van Kranendonk et al. 2004; Bodorkos and Sandiford 2006; Thébaud and Rey 2013: François et al. 2014; Wiemer et al. 2016, 2018; Gardiner et al. 2017, 2018). This deformation occurred in stages during the magmatic evolution of a continental scale volcanic plateau from c. 3530 to 3223 Ma, and the preserved 100,000 km^2 EPT (including concealed areas) is merely a fragment of that Paleoarchean plateau.

3.7.2 *Oceanic Plateau?*

Some interpretations of the depositional setting of the Warrawoona Group have suggested that the volcanic plateau was oceanic similar to modern oceanic plateaus such as the Kerguelen Plateau (Arndt et al. 2001; Van Kranendonk and Pirajno 2004; Van Kranendonk et al. 2007a, b, 2015). The same interpretation was applied to greenstones in the Kaapvaal Craton of South Africa (De Wit et al. 1992). Although consideration of possible modern analogues has some merit, most geological features of the EPT are inconsistent with the oceanic plateau model. These features include: 1, the existence of pre-existing thick continental crust on which the Warrawoona Group was deposited (oceanic plateaus have little or no underlying continental crust); 2, the lithologically complex stratigraphy of the terrane which includes large volumes of granitic and felsic volcanic rocks contemporaneous with ultramafic–mafic volcanic rocks; 3, the dome-and-keel crustal architecture of the terrane, a feature absent from all modern oceanic plateaus; 4, evolution of the terrane over 300 million years, including eight mantle plume events, in contrast to most oceanic plateaus which were constructed over only a few million years during only one or two mantle plume events; 5, stratigraphic and geochronological evidence that during more than half of its 300-million year evolution the terrane was subject to subaerial erosion and shallow-water sedimentation (Dresser and McPhee Formations, 3490–3470 Ma; Strelley Pool Formation, 3426–3350 Ma; Leilira Formation, 3280–3255 Ma) (oceanic plateaus are mainly submerged); 6, major erosional unconformities (c. 3450, c. 3426, and c. 3290 Ma) following events of deformation, metamorphism, and granitic intrusion; 7, mafic dyke swarms beneath two of the thickest basaltic formations of the terrane (Apex Basalt and Euro Basalt), a feature characteristic of continental basalts but absent from oceanic plateaus (Kerr et al. 2000); 8, geochemical evidence of crustal contamination in the ultramafic–mafic volcanic rocks of the terrane (Glikson and Hickman 1981a; Green et al. 2000; Smithies et al. 2007).

From 3470 Ma onwards, the increasing volume of felsic igneous rocks in the EPT, combined with felsic components in the underlying 3800–3530 Ma continental crust, would have made it too buoyant for the deep oceanic submergence that is characteristic of oceanic plateaus. Locally thick pillow basalt formations in the Pilbara Supergroup (e.g. Apex Basalt and Euro Basalt) were mainly deposited in relatively small subsiding basins between the rising domes, and therefore do not provide evidence that the entire volcanic plateau was submerged.

References

Abeysinghe PB, Fetherston JM (1997) Barite and fluorite in Western Australia. Mineral Resources Bulletin, vol 17. Geological Survey of Western Australia

Allen CM, Wiemer D, Murphy DT (2016) Improving LA-ICPMS dating techniques: Experiments on zircon from a c. 3.51 Ga dioritic gneiss, East Pilbara Terrane, Western Australia. In: Session 35, T39, GSA Annual Meeting in Denver, Colorado, USA–2016

Allwood AC, Burch I, Walter MR (2007) Stratigraphy and facies of the 3.43 Ga Strelley Pool Chert in the southwestern North Pole Dome, Pilbara Craton, Western Australia. Record, 2007/11. Geological Survey of Western Australia

Altinok E, Ohmoto H (2006) Soil formation beneath the Earth's oldest known (3.46 Ga) unconformity? In: T105, Paleosols, proxies, and Paleoenvironments II, p 533

Amelin Y, Lee D-C, Halliday AN, Pidgeon RT (1999) Nature of the Earth's earliest crust from hafnium isotopes in single detrital zircons. Nature 399:252–255. https://doi.org/10.1038/20426

Amelin Y, Lee D-C, Halliday AN (2000) Early-middle Archaean crustal evolution deduced from Lu–Hf and U–Pb isotopic studies of single zircon grains. Geochim Cosmochim Acta 62:4205–4225

Anhaeusser CR (1971) Cyclic volcanicity and sedimentation in the evolutionary development of Archaean greenstone belts of shield areas. In: Glover JE (ed) Symposium on Archaean rocks, Perth, 23–26 May 1970. Geological Society of Australia, Perth, Western Australia, pp 57–70

Anhaeusser CR, Mason R, Viljoen MJ, Viljoen RP (1969) A reappraisal of some aspects of Precambrian shield geology. Geol Soc Am Bull 80:2175–2200

Anhaeusser CR, Roering C, Viljoen MJ, Viljoen RP (1968) The Barberton Mountain Land: a model of the elements and evolution of an Archaean fold belt. Geological Society of South Africa, Annexure 71:225–254

Arndt NT, Goldstein SL (1987) Use and abuse of crust-formation ages. Geology 15:893–895

Arndt NT, Czamanske GK, Wooden JL, Fedorenko VA (1993) Mantle and crustal contributions to continental flood volcanism. Tectonophysics 223:39–52

Arndt NT, Albarede F, Nisbet EG (1997) Mafic and ultramafic magmatism. In: de Wit MJ, Ashwal LD (eds) Greenstone belts. Oxford University Press, Oxford, pp 231–254

Arndt NT, Bruzak G, Reischmann T (2001) The oldest continental and oceanic plateaus: geochemistry of basalts and komatiites of the Pilbara Craton, Australia. In: Ernst RE, Buchan K (eds) Mantle plumes: their identification through time. Geological Society of America, pp 359–387

Arndt NT, Lesher CM (2004) Komatiite: encyclopedia of geology. Elsevier, pp 260–268

Bagas L, Smithies RH, Champion DC (2003) Geochemistry of the Corunna Downs Granitoid Complex, East Pilbara Granite-Greenstone Terrane. In: Geological Survey of Western Australia Annual Review 2001–02. Geological Survey of Western Australia, pp 61–69

Bagas L, Van Kranendonk MJ, Pawley MJ (2004) Geology of the Split Rock 1:100 000 sheet. 1: 100 000 Geological Series Explanatory Notes. Geological Survey of Western Australia

Barley ME (1978) Shallow-water sedimentation during deposition of the Archaean Warrawoona group, East Pilbara Block, Western Australia. In: Glover JE, Groves DI (eds) Archcan cherty metasediments: their sedimentology, micropalaeontology, biogeochemistry, and significance to mineralization. Univ. West, Australia Extension Service, pp 22–29

Barley ME (1981) Relations between volcanic rocks in the Warrawoona Group: continuous or cyclic evolution? Geological Society of Australia, Special Publication 7:263–273

Barley ME (1984) Volcanism and hydrothermal alteration, Warrawoona Group, East Pilbara. In: Muhling JR, Groves DI, Blake TS (eds) Archaean and Proterozoic basins of the Pilbara, Western Australia: evolution and mineralization potential. The University of Western Australia, Geology Department and University Extension, pp 23–36

Barley ME (1993) Volcanic, sedimentary and tectonostratigraphic environments of the ~3.46 Ga Warrawoona Megasequence: a review. Precambrian Res 60:47–67

Barley ME, Dunlop JSR, Glover JE, Groves D (1979) Sedimentary evidence for an Archean shallow-water volcanic-sedimentary facies, Eastern Pilbara Block, Western Australia. Earth Planet Sci Lett 43:74–84

Barley ME, Pickard AL (1999) An extensive, crustally-derived, 3325 to 3310 Ma silicic volcanoplutonic suite in the Eastern Pilbara Craton: evidence from the Kelly Belt, McPhee Dome and Corunna Downs Batholith. Precambrian Res 96:41–62

Beintema KA (2003) Geodynamic evolution of the West and Central Pilbara Craton: a mid-Archean active continental margin. Geologica Ultraiectina. Utrecht University, Utrecht

Bennett VC, Nutman AP, McCulloch MT (1993) Nd isotopic evidence for transient, highly depleted mantle reservoirs in the early history of the earth. Earth Planet Sci Lett 119:299–317. https://doi.org/10.1016/0012-821X(93)90140-5

Bettenay LF, Bickle MJ, Boulter CA, Groves DI, Morant P, Blake TS, James BA (1981) Evolution of the Shaw Batholith–An Archaean granitoid-gneiss dome in the Eastern Pilbara, Western Australia. In: Glover JE, Groves DI (eds) Archaean Geology: Second International Symposium, Perth 1980. Geological Society of Australia, pp 361–372

Bickle MJ, Bettenay LF, Barley ME, Chapman HJ, Groves DI, Campbell IH, de Laeter JR (1983) A 3500 Ma plutonic and volcanic calc-alkaline province in the Archean East Pilbara Block. Contrib Mineral Petrol 84:25–35

Bickle MJ, Morant P, Bettenay LF, Boulter CA, Blake TS, Groves DI (1985) Archaean tectonics of the Shaw Batholith, Pilbara Block, Western Australia: structural and metamorphic tests of the batholith concept. In: Ayers LD, Thurston PC, Card KD, Weber W (eds) Evolution of Archean supracrustal sequences. Geological Association of Canada, pp 325–341

Bickle MJ, Bettenay LF, Chapman HJ, Groves DI, McNaughton NJ, Campbell IH, de Laeter JR (1993) Origin of the 3500–3300 Ma calc-alkaline rocks in the Pilbara Archaean: isotopic and geochemical constraints from the Shaw Batholith. Precambrian Res 60:117–149

Blichert-Toft J, Arndt NT (1999) Hf isotope compositions of komatiites. Earth Planet Sci Lett 171: 439–451

Bodorkos S, Sandiford M (2006) Thermal and mechanical controls on the evolution of Archean crustal deformation: examples from Western Australia. In: Benn K, Mareschal J-C, Condie KC (eds) Archean geodynamics and environments. American Geophysical Union, monograph, vol 164, pp 131–147. https://doi.org/10.1029/164GM10

Brasier MD, Antcliffe J, Saunders M, Wacey D (2015) Changing the picture of Earth's earliest fossils (3.5–1.9 Ga) with new approaches and new discoveries. Proc Natl Acad Sci 112:4859–4864

Brasier MD, Green OR, Jephcoat AP, Kleppe AP, Van Kranendonk MJ, Lindsay JF, Steele A, Grassineau NV (2002) Questioning the evidence for Earth's oldest fossils. Nature 416:76–81

Brasier MD, Green OR, Lindsay JF, McLoughlin N, Steele A, Stokes C (2005) Critical testing of Earth's oldest putative fossil assemblage from the ~3.5 Ga apex chert, Chinaman Creek. Western Australia. Precambr Res 140:55–102

Brasier MD, Green OR, Lindsay JF, McLoughlin N, Stoakes CA, Brasier AT, Wacey D (2011) Geology and putative microfossil assemblage of the c. 3460 Ma 'Apex chert', Chinaman Creek, Western Australia–a field and petrographic guide. Record, 2011/7. Geological Survey of Western Australia

Brasier MD, Matthewman R, McMahon S, Kilburn MR, Wacey D (2013) Pumice from the ~3460 Ma Apex Basalt, Western Australia: a natural laboratory for the early biosphere. Precambrian Res 224:1–10

Brauhart C (1999) Regional alteration systems associated with Archean volcanogenic massive sulphide deposits at Panorama, Pilbara. PhD, The University of Western Australia, Western Australia

Brown A, Cudahy T, Walter MR, Hickman AH (2011) Hyperspectral alteration mapping of early Archean hydrothermal systems in the North Pole Dome, Pilbara Craton. Geological Survey of Western Australia, Record 2011/15

Brown AJ, Cudahy TJ, Walter MR (2006) Hydrothermal alteration at the Panorama Formation, North Pole Dome, Pilbara Craton, Western Australia. Precambrian Res 151:211–223

Brown AJ, Walter MR, Cudahy TJ (2005) Hyperspectral imaging spectroscopy of a Mars analogue environment at the North Pole Dome, Pilbara Craton, Western Australia. Aust J Earth Sci 52: 353–364

Bryan SE, Ernst RE (2008) Revised definition of large igneous provinces (LIP). Earth Sci Rev 86: 175–202

Buick R, Barnes KR (1984) Cherts in the Warrawoona Group: early Archaean silicified sediments deposited in shallow-water environments. In: Muhling JR, Groves DI, Blake TS (eds) Archaean and Proterozoic basins of the Pilbara, Western Australia: evolution and mineralization potential. The University of Western Australia, Geology Department and University Extension, pp 37–53

Buick R, Dunlop JSR (1990) Evaporitic sediments of early Archaean age from the Warrawoona Group, North Pole, Western Australia. Sedimentology 37:247–277. https://doi.org/10.1111/j.1365-3091.1990.tb00958.x

Buick R, Thornett J, McNaughton N, Smith JB, Barley ME, Savage MD (1995) Record of emergent continental crust ~3.5 billion years ago in the Pilbara Craton of Australia. Nature 375:574–577

Buick R, Dunlop JSR, Groves DI (1981) Stromatolite recognition in ancient rocks: an appraisal of irregular laminated structures in an early Archaean chert-barite unit form North Pole, Western Australia. Alcheringa 5:161–181

Buick R, Brauhart CW, Morant P, Thornett JR, Maniw JG, Archibald NJ, Doepel MG, Fletcher IR, Pickard AL, Smith JB, Barley ME, McNaughton NJ, Groves DI (2002) Geochronology and stratigraphic relationships of the Sulphur Springs Group and Strelley granite: a temporally distinct igneous province in the Archean Pilbara Craton, Australia. Precambrian Res 114:87–120

Byerly GR, Lowe DR, Wooden JL, Xie X (2002) An Archean impact layer from the Pilbara and Kaapvaal Cratons. Science 297:1325–1327

Cawood PA, Hawkesworth CJ, Pisarevsky SA, Dhuime B, Capitanio FA, Nebel O (2018) Geological archive of the onset of plate tectonics. Philos Trans A Math Phys Eng Sci 376. https://doi.org/10.1098/rsta.2017.0405

Champion DC (2013) Neodymium depleted mantle model age map of Australia: explanatory notes and user guide. Record, 2013/44. Geoscience Australia

Champion DC, Huston DL (2016) Radiogenic isotopes, ore deposits and metallogenic terranes: novel approaches based on regional isotopic maps and the mineral systems concept. Ore Geol Rev 76:229–256. https://doi.org/10.1016/j.oregeorev.2015.09.025

Champion DC, Smithies RH (1999) Archaean granites of the Yilgarn and Pilbara Cratons: secular changes. In: Barbarin B (ed) The origin of granites and related rocks, fourth Hutton symposium abstracts, 137. BRGM, Clermont-Ferrand

Champion DC, Smithies RH (2001) The geochemistry of the Yule Granitoid Complex, East Pilbara Granite-Greenstone Terrane: evidence for early felsic crust. In: Geological Survey of Western Australia annual review 1999-2000. Geological Survey of Western Australia, Perth, Western Australia, pp 42–48

Champion DC, Smithies RH (2007) Geochemistry of Paleoarchean granites of the East Pilbara Terrane, Pilbara Craton, Western Australia: implications for early Archean crustal growth. In: Van Kranendonk MJ, Bennett VC, Smithies RH (eds) Earth's oldest rocks. Elsevier B.V, Burlington, Massachusetts, USA, pp 369–410

Champion DC, Smithies RH (2019) Geochemistry of Paleoarchean granites of the East Pilbara Terrane, Pilbara Craton, Western Australia: implications for early Archean crustal growth. In: Van Kranendonk MJ, Bennett VC, Hoffmann JE (eds) Earth's oldest rocks, 2nd edn. Elsevier B. V, pp 487–518

Collins WJ, Van Kranendonk MJ, Teyssier C (1998) Partial convective overturn of Archaean crust in the East Pilbara Craton, Western Australia: driving mechanisms and tectonic implications. J Struct Geol 20:1405–1424. https://doi.org/10.1016/S0191-8141(98)00073-X

Collins WJ (1989) Polydiapirism of the Archean Mount Edgar Batholith, Pilbara Block, Western Australia. Precambrian Res 43:41–62

Collins WJ (1993) Melting of Archaean silicic crust under high aH_2O conditions: genesis of 3300 Ma Na-rich granitoids in the Mount Edgar Batholith, Pilbara Block, Western Australia. Precambrian Res 60:151–174

Condie KC (2001) Mantle plumes and their record in earth history. Cambridge University Press, Cambridge, UK

Cullers RL, DiMarco MJ, Lowe DR, Stone J (1993) Geochemistry of a silicified felsic volcanic suite from the early Archean Panorama Formation, Pilbara Block, Western Australia–an evaluation of depositional and post-depositional processes with special emphasis on the rare earth elements. Precambrian Res 60:99–116

Davids C, Wijbrans JR, White SH (1997) $^{40}Ar^{39}Ar$ laserprobe ages of metamorphic hornblendes from the Coongan Belt, Pilbara Western Australia. Precambrian Res 83:221–242

Davy R, Lewis JD (1986) The Mount Edgar Batholith, Pilbara area, Western Australia: geochemistry and petrography. Report, vol 17. Geological Survey of Western Australia

De Vries ST (2004) Early Archaean sedimentary basins: depositional environment and hydrothermal systems–examples from the Barbeton and Coppin Gap greenstone belts. Gologica Ultraiectina no. 244

De Vries ST, Nijman W, Wijbrans JR, Nelson DR (2006) Stratigraphic continuity and early deformation of the central part of the Coppin Gap Greenstone Belt, Pilbara, Western Australia. Precambrian Res 147:1–27

De Wit MJ, de Ronde CEJ, Tredoux M, Roering C, Hart RJ, Armstrong RA, Green RWE, Peberdy E, Hart RA (1992) Formation of an Archaean continent. Nature 357:553–562

DePaolo DJ (1980) Crustal growth and mantle evolution: inferences from models of element transport and Nd and Sr isotopes. Geochim Cosmochim Acta 44:1185–1196

DePaolo DJ (1988) Neodymium isotope geochemistry: an introduction. Minerals and rocks series no. 20. Springer-Verlag, Berlin, Germany

DiMarco MJ (1986) Stratigraphy, sedimentology, and sedimentary petrology of the early Archean Coongan Formation, Warrawoona Group, Eastern Pilbara Block. PhD, Louisiana State University, Western Australia

DiMarco MJ, Lowe DR (1989a) Shallow-water volcaniclastic deposition in the Early Archean Panorama Formation, Warrawoona Group, Eastern Pilbara Block, Western Australia. Sediment Geol 64:43–63

DiMarco MJ, Lowe DR (1989b) Stratigraphy and sedimentology of an early Archean felsic volcanic sequence, Eastern Pilbara Block, Western Australia, with special reference to the Duffer Formation and implications for crustal evolution. Precambrian Res 44:147–169

Djokic T, Van Kranendonk MJ, Campbell KA, Walter MR, Ward CR (2017) Earliest signs of life on land preserved in ca. 3.5 Ga hot spring deposits. Nat Commun:8. https://doi.org/10.1038/ncomms15263

Dunlop JSR (1978) Shallow-water sedimentation at North Pole, Pilbara, Western Australia. In: Glover JE, Groves DI (eds) Archaean cherty metasediments: their sedimentology, micropalaeontology, biogeochemistry and significance to mineralization. Extension Service, The University of Western Australia, Perth, pp. 30–38

Dunlop JSR, Buick R (1981) Archean epiclastic sediments derived from mafic volcanic rocks, North Pole, Pilbara Block, Western Australia. In: Glover JE, Groves DI (eds) Archaean geology: second international symposium, Perth 1980. Geological Society of Australia, pp 225–233

Ernst RE (2007) Mafic–ultramafic large igneous provinces (LIPs): importance of the pre-Mesozoic record. Episodes 30:108–114

Ernst RE (2014) Large igneous provinces. Cambridge University Press, Cambridge, UK

François C, Philippot P, Rey PF, Rubatto D (2014) Burial and exhumation during Archean sagduction in the East Pilbara Granite–Greenstone Terrane. Earth Planet Sci Lett 396:235–251

Garcia-Ruiz JM, Carnerup A, Christy AG, Welham NJ, Hyde ST (2002) Morphology: an ambiguous indicator of Biogenicity. Astrobiology 2:353–369

Garcia Ruiz JM, Hyde ST, Carnerup AM, Christy AG, Van Kranendonk MJ, Welham NJ (2003) Self-assembled silica-carbonate structures and detection of ancient microfossils. Science 302: 1194–1197

Gardiner NJ, Hickman AH, Kirkland CL, Lu Y, Johnson T, Zhao J-X (2017) Processes of crust formation in the early earth imaged through Hf isotopes from the East Pilbara Terrane. Precambrian Res 297:56–76. https://doi.org/10.1016/j.precamres.2017.05.004

Gardiner NJ, Hickman AH, Kirkland CL, Lu Y, Johnson TE, Wingate MTD (2018) New Hf isotope insights into the Paleoarchean magmatic evolution of the Mount Edgar Dome, Pilbara Craton: implications for early Earth and Crust Formation processes. Report, vol 181. Geological Survey of Western Australia

Gauger T, Konhauser K, Kappler A (2015) Protection of phototrophic iron(II)-oxidizing bacteria from UV irradiation by biogenic iron(III) minerals: implications for early Archean banded iron formation. Geology 43:1067–1170

Glass BP, Simonson BM (2013) Distal impact ejecta layers: a record of large impacts in sedimentary deposits. Impact Studies, Springer, Berlin, Germany

Glikson AY (1970) Geosynclinal evolution and geochemical affinities of early Precambrian systems. Tectonophysics 9:397–433

Glikson AY (1971) Primitive Archaean element distribution patterns: chemical evidence and geotectonic significance. Earth Planet Sci Lett 12:309–320

Glikson AY (1972) Early Precambrian evidence of a primitive ocean crust and island nuclei of sodic granite. Geol Soc Am Bull 83:3323–3344

Glikson AY (2004) Early Precambrian asteroid impact-triggered tsunami: excavated seabed debris flows exotic boulders and turbulence features associated with 3.47–2.47 Ga-old asteroid impact fallout units, Pilbara Craton. Western Australia Astrobiology 4:1–32

Glikson AY, Allen C, Vickers J (2004) Multiple 3.47-Ga-old asteroid impact fallout units, Pilbara Craton, Western Australia. Earth Planet Sci Lett 21:383–396

Glikson AY, Davy R, Hickman AH (1986) Geochemical data files of Archaean volcanic rocks, Pilbara Block, Western Australia. BMR (Geoscience Australia) Record, 1986/014

Glikson AY, Davy R, Hickman AH, Pride C, Jahn B (1987) Trace elements geochemistry and petrogenesis of Archaean felsic igneous units, Pilbara Block, Western Australia. Record, 1987/030. Bureau of Mineral Resources, Geology and Geophysics (Geoscience Australia)

Glikson AY, Hickman AH (1981a) Geochemistry of Archaean volcanic successions, Eastern Pilbara Block, Western Australia. BMR (Geoscience Australia) Record, 1981/036

Glikson AY, Hickman AH (1981b) Geochemical stratigraphy and petrogenesis of Archaean basic–ultrabasic volcanic units, Eastern Pilbara Block, Western Australia. In: Glover JE, Groves DI (eds) Archaean Geology: Second International Symposium, Perth 1980, vol 7. Geological Society of Australia, pp 287–300

Glikson AY, Hickman AH, Evans NJ, Kirkland CL, Park J-W, Rapp R, Romano SS (2016) A new ~3.46 Ga asteroid impact ejecta unit at Marble Bar, Pilbara Craton, Western Australia: a petrological, microprobe, and laser ablation ICPMS study. Precambrian Res 279:103–122

Glikson AY, Hickman AH, Vickers J (2008) Hickman crater, Ophthalmia range, Western Australia: evidence supporting a meteorite impact origin. Aust J Earth Sci 55:1107–1117

Glikson AY, Pirajno F (2018) Asteroids impacts, crustal evolution and related mineral systems with special reference to Australia. Springer International Publishing, Modern Approaches in Solid Earth Sciences, doi:https://doi.org/10.1007/978-3-319-74545-9

Glikson M, Hickman AH, Duck LJ, Golding SD, Webb RE (2012) Integration of observational and analytical methodologies to characterize organic matter in early Archaean rocks: distinguishing biological from abiotically synthesized carbonaceous matter structures. In: Golding SD, Glikson M (eds) Earliest life on earth: habitats, environments and methods of detection. Springer Science and Business Media, Amsterdam, The Netherlands, pp 209–238

Goodwin AM, Ridler RH (1970) The Abitibi orogenic belt. Canada Geological Survey Paper 70-40:1–24

Green MG, Sylvester PJ, Buick R (2000) Growth and recycling of early Archaean continental crust: geochemical evidence from the Coonterunah and Warrawoona Groups, Pilbara Craton, Australia. Tectonophysics 322:69–88

Griffin WL, Pearson NJ, Belousova EA, Jackson SR, van Achterbergh E, O'Reilly SY, Shee SR (2000) The Hf isotope composition of cratonic mantle: LAM-MC-ICPMS analysis of zircon megacrysts in kimberlites. Geochim Cosmochim Acta 64:133–147

Groves DI, Dunlop JSR, Buick R (1981) An early habitat of life. Sci Am 245:64–73

Gruau G, Jahn BM, Glikson AY, Davy R, Hickman AH, Chauvel C (1987) Age of the Archean Talga-Talga subgroup, Pilbara Block, Western Australia, and the early evolution of the mantle: new Sm–Nd isotopic evidence. Earth Planet Sci Lett 85:105–116

Hamilton PJ, Evensen NM, O'Nions RK, Glikson AY, Hickman AH (1981) Sm–Nd dating of the North Star Basalt, Warrawoona group, Pilbara Block, Western Australia. In: Glover JE, Groves DI (eds) Archaean geology: second International Symposium, Perth 1980. Geological Society of Australia, pp 187–192

Harris AC, White NC, McPhie J, Bull SW, Line MA, Skrzeczynski R, Mernagh TP, Tosdal RM (2009) Early Archean hot springs above epithermal veins, North Pole, Western Australia: new insights from fluid inclusions microanalysis. Econ Geol 104:793–814. https://doi.org/10.2113/gsecongeo.104.6.793

Hasenstab E, Tusch J, Achnabel C, Schmitt V, Marien CS, Van Kranendonk MJ, Munker C (2019) The evolution of the Archean mantle from combined isotope systematics in Pilbara Basalts and komatiites. Goldschmidt 2019, Abstract

Hasenstab E, Tusch J, Schnabel C, Marien CS, Van Kranendonk MJ, Smithies H, Howard H, Maier WD, Munker C (2021) Evolution of the early to late Archean mantle from Hf-Nd-Ce isotope systematics in Basalts and komatiites from the Pilbara Craton. Earth Planet Sci Lett 553(116627):1–13

Henderson P (1984) Rare earth elements geochemistry. Elsevier, Amsterdam, The Netherlands

Hickman AH (1973) The North Pole barite deposits, Pilbara Goldfield. In: Annual report for the year 1972. Geological Survey of Western Australia, Perth, Western Australia, pp 57–60

Hickman AH (1975) Precambrian structural geology of part of the Pilbara Region. In: Annual report for the year 1974. Geological Survey of Western Australia, pp 68–73

Hickman AH (1977) New and revised definitions of rock units in the Warrawoona Group, Pilbara Block. In: Annual report for the year 1976. Geological Survey of Western Australia, p 53

Hickman AH (1980a) Excursion guide–Archaean geology of the Pilbara Block: second International Archaean Symposium, Perth 1980. Geological Society of Australia, W.A. Division, Perth, Western Australia

Hickman AH (1980b) Lithological map and stratigraphic interpretation of the Pilbara Block. Geology if the Pilbara Block and its environs. Geological Survey of Western Australia

Hickman AH (1981) Crustal evolution of the Pilbara Block. In: Glover JE, Groves DI (eds) Archaean Geology: second International Symposium, Perth 1980. Geological Society of Australia, pp 57–69

Hickman AH (1983) Geology of the Pilbara Block and its environs. Bulletin, vol 127. Geological Survey of Western Australia

Hickman AH (1984) Archaean diapirism in the Pilbara Block, Western Australia. In: Kröner A, Greiling R (eds) Precambrian tectonics illustrated. Schweizerbart Science Publishers, Stuttgart, Germany, pp 113–127

Hickman AH (2001) East Pilbara diapirism: New evidence from mapping. In: Geological Survey of Western Australia (ed) GSWA 2001 extended abstracts: new geological data for WA explorers. Geological Survey of Western Australia, pp 23–25

Hickman AH (2004) Two contrasting granite–greenstones terranes in the Pilbara Craton, Australia: evidence for vertical and horizontal tectonic regimes prior to 2900 Ma. Precambrian Res 131: 153–172

Hickman AH (2005) Evidence of early life from international collaborative drilling in the Pilbara Craton. In: Geological Survey of Western Australia Annual Review 2004-05

Hickman AH (2010) Marble Bar, WA Sheet SF50-8, 3rd Geological Survey of Western Australia

Hickman AH (2011) Pilbara Supergroup of the East Pilbara Terrane, Pilbara Craton: updated lithostratigraphy and comments on the influence of vertical tectonics. In: Geological Survey of Western Australia Annual Review 2009–10. Geological Survey of Western Australia, pp. 50–59

Hickman AH (2012) Review of the Pilbara Craton and Fortescue Basin, Western Australia: crustal evolution providing environments for early life. Island Arc 21:1–31

Hickman AH (2021) East Pilbara Craton: a record of one billion years in the growth of Archean continental crust. Report 143, Geological Survey of Western Australia

Hickman AH, Lipple SL (1975) Explanatory notes on the Marble Bar 1:250 000 geological sheet, Western Australia. Record, 1974/20. Geological Survey of Western Australia

Hickman AH, Lipple SL (1978) Marble Bar, WA Sheet SF50-8, 2nd Geological Survey of Western Australia

Hickman AH, Van Kranendonk MJ (2004) Diapiric processes in the formation of Archaean continental crust, East Pilbara Granite–Greenstone Terrane, Australia. In: Eriksson PG, Altermann W, Nelson DR, Mueller WU, Catuneanu O (eds) The Precambrian earth: tempos and events. Elsevier, Amsterdam, The Netherlands, pp 54–75

Hickman AH, Van Kranendonk MJ (2008a) Archean crustal evolution and mineralization of the Northern Pilbara Craton–a field guide. Record, 2008/13. Geological Survey of Western Australia

Hickman AH, Van Kranendonk MJ (2008b) Marble Bar, WA Sheet 2855. Geological Survey of Western Australia

Hickman AH, Van Kranendonk MJ (2012) Early earth evolution: evidence from the 3.5–1.8 Ga geological history of the Pilbara region of Western Australia. Episodes 35:283–297. https://doi.org/10.18814/epiiugs/2012/v35i1/028

Hoashi M, Bevacqua DC, Otake T, Watanabe Y, Hickman AH, Utsunomiya S, Ohmoto H (2009) Primary hematite formation in an oxygenated sea 3.46 billion years ago. Nat Geosci 2:301–306

Hofmann AW (1997) Mantle geochemistry: the message from oceanic volcanism. Nature 385:219–229

Huston DL, Brauhart CW, Dreiberg SL, Davidson GJ, Groves DI (2001) Metal leaching and inorganic sulphate reduction in volcanic-hosted massive sulphide mineral systems: evidence from the paleo-Archean Panorama district, Western Australia. Geology 29:687–690

Jahn B, Glikson AY, Peucat JJ, Hickman AH (1981) REE geochemistry and isotopic data of Archean silicic volcanics and granitoids from the Pilbara Block, Western Australia: implications for the early crustal evolution. Geochim Cosmochim Acta 45:1633–1652

Johnson TE, Brown M, Gardiner NJ, Kirkland CL, Smithies RH (2017) Earth's first stable continents did not form by subduction. Nature 543:239–242

Kato Y, Nakamura K (2003) Origin and global tectonic significance of early Archean cherts from the Marble Bar greenstone belt, Pilbara Craton, Western Australia. Precambrian Res 125:191–243

Kemp AIS, Hickman AH, Kirkland CL (2015) Early evolution of the Pilbara Craton, Western Australia, from hafnium isotopes in detrital and inherited zircons. Report, vol 151. Geological Survey of Western Australia

Kerr AC, White RV, Saunders AD (2000) LIP Reading; recognizing oceanic Plateaux in the geological record. J Petrol 41:1041–1056

Kimura G, Maruyama S, Isozaki Y (1991) Early Archean accretionary complex in the Eastern Pilbara Craton, Western Australia–detailed field occurrence in the Marble Bar area. Eos 72:542

Kitajima K, Maruyama S, Utsonomita S, Liou JG (2001) Seafloor hydrothermal alteration at an Archaean mid-ocean ridge. J Metamorph Geol 19:581–597

Kitajima K, Hirata T, Maruyama S, Yamanashi T, Sano Y, Liou JG (2008) U–Pb zircon geochronology using LA-ICP-MS in the North Pole Dome, Pilbara Craton, Western Australia: a new tectonic growth model for the Archean chert/greenstone succession. Int Geol Rev 80:1–14

Kloppenburg A (2003) Structural evolution of the Marble Bar Domain, Pilbara Granite-Greenstone Terrain, Australia: the role of Archaean mid-crustal detachments. PhD, Utrecht University

Kojima S, Hanamuro T, Hayashi K, Haruna M, Ohmoto H (1998) Sulphide minerals in early Archean chemical sedimentary rocks of the Eastern Pilbara district, Western Australia. Mineral Petrol 64:219–235

Komiya T, Maruyama S, Hirata T, Yurimoto H (2002) Petrology and geochemistry of MORB and OIB in the mid-Archean North Pole region, Pilbara Craton, Western Australia: implications for the composition and temperature of the upper mantle at 3.5 Ga. Int Geol Rev 44:988–1016

Konhauser KO, Planavsky NJ, Hardisty DS, Robbins LJ, Warchola TJ, Haugaard R, Lalonde SV, Partin CA, Oonk P, Tsikos H, Lyons TW, Bekker A, Johnson CM (2017) Iron formations: a global record of Neoarchaean to Palaeoproterozoic environmental history. Earth Sci Rev 172: 140–177. https://doi.org/10.1016/j.earscirev.2017.06.012

Kusky T, Windley BF, Polat A, Wand L, Ning W, Zhong Y (2021) Archean dome-and-basin style structures form during growth and death of intra-oceanic and continental margin arcs in accretionary orogens. Earth-Sci Rev 220(103725):1–50. https://doi.org/10.1016/j.earscirev.2021.103725

Lambert R, St J, Chamberlain VE, Holland JG (1976) The geochemistry of Archaean rocks. In: Windley BF (ed) The early history of the earth: NATO advanced study institute. University of Leicester, pp 377–387

Li W, Czaja AD, Van Kranendonk MJ, Beard BL, Roden EE, Johnson CM (2013) An anoxic, Fe (II)-rich, U-poor ocean 3.46 billion years ago. Geochim Cosmochim Acta 120:65–79

Lindsay JF, Brasier MD, McLoughlin N, Green OR, Fogel M, Steele A, Mertzman SA (2005) The problem of deep carbon–an Archaean paradox. Precambrian Res 143:1–22

Lipple SL (1975) Definitions of new and revised stratigraphic units of the Eastern Pilbara Region. In: Annual report for the year 1974. Geological Survey of Western Australia, pp. 58–63

Lowe DR (1999) Petrology and sedimentology of cherts and related silicified sedimentary rocks in the Swaziland Supergroup. In: Lowe DR, Byerly GR (eds) Geologic evolution of the Barberton Greenstone Belt. Geological Society of America, South Africa, pp 83–114

Lowe DR, Byerly GR (1986) Early Archean silicate spherules of probable impact origin. Geology 14:83–86

Lowe DR, Byerly GR, Kyte FT, Shulkolyukov A, Asaro F, Krull A (2003) Characteristics, origin and implications of Archaean impact-produced spherule beds, 3.47–3.22 Ga, in the Barberton Greenstone Belt, South Africa: keys to the role of large impacts on the evolution of the early Earth. Astrobiology 3:7–48

Marshall AO, Jehlicka J, Rouzaud J-N, Marshall CP (2014) Multiple generations of carbonaceous material deposited in apex chert by basin-scale pervasive hydrothermal fluid flow. Gondwana Res 25:284–289

Marshall CP, Emry JR, Marshall AO (2011) Haematite pseudomicrofossils present in 3.5-billion-year-old apex chert. Nat Geosci 4:240–243

McCulloch MT, Bennett VC (1993) Evolution of the early earth: constraints from ^{143}Nd–^{142}Nd isotope systematics. Lithos 30:327–355

McCulloch MT, Wasserburg GJ (1978) Sm–Nd and Rb–Sr chronology of continental crust formation. Science 200:1003–1011

McNaughton NJ, Compston W, Barley ME (1993) Constraints on the age of the Warrawoona Group, Eastern Pilbara Block, Western Australia. Precambrian Res 60:69–98

McNaughton NJ, Green MD, Compston W, Williams IS (1988) Are anorthositic rocks basement to the Pilbara Craton? Geol Soc Aust Abstr 21:272–273

Minami M, Shimizu H, Masuda A, Adachi M (1995) Two Archean Sm-Nd ages of 3.2 and 2.5 Ga for the Marble Bar Chert, Warrawoona Group, Pilbara Block. Western Australia Geochemical Journal 29:347–362

Morant P (1998) Panorama zinc–copper deposits. In: Berkman DA, Mackenzie DH (eds) Geology of Australian and Papua new Guinean mineral deposits, The Australasian Institute of Mining and Metallurgy, vol 22. Melbourne, Australia, pp 287–292

Nebel O, Campbell IH, Sossi PA, Van Kranendonk MJ (2014) Hafnium and iron isotopes in early Archean komatiites record a plume-driven convection cycle in the Hadean Earth. Earth Planet Sci Lett 397:111–120

Nelson DR (1999) Compilation of geochronology data, 1998. Record, 1999/2. Geological Survey of Western Australia

Nelson DR (2000) Compilation of geochronology data, 1999. Record, 2000/2. Geological Survey of Western Australia

Nelson DR (2001) Compilation of geochronology data, 2000. Record, 2001/2. Geological Survey of Western Australia

Nelson DR (2002) Compilation of geochronology data, 2001. Record, 2002/2. Geological Survey of Western Australia

Nelson DR (2004) 169009: lithic tuff, Coongan Belt Well. Geochronology dataset 60. Compilation of geochronology data, June 2006 update. Geological Survey of Western Australia

Nelson DR (2005) 178083: metarhyolite, Gallop Well. Geochronology dataset 576. Compilation of geochronology data, June 2006 update. Geological Survey of Western Australia

Nijman W, De Vries ST (2004) Early Archaean crustal collapse structures and sedimentary basin dynamics. In: Eriksson PG, Altermann W, Nelson DR, Mueller WU, Catuneanu O (eds) The Precambrian earth: tempos and events. Elsevier, Amsterdam, The Netherlands, pp 139–154

Nijman W, de Bruijne KCH, Valkering ME (1998) Growth fault control of early Archaean cherts, barite mounds and chert–barite veins, North Pole Dome, Eastern Pilbara, Western Australia. Precambrian Res 88:25–52

Nijman W, de Bruijne KCH, Valkering ME (1999) Growth fault control of early Archaean cherts, barite mounds and chert-barite veins, North Pole Dome, Eastern Pilbara, Western Australia (Erratrum to Precambr Res, v. 88, p. 25–52). Precambrian Res 95:247–274

Nijman, W, de Vries ST, Houtzager O (2001) Earth's earliest sedimentary basins: the Lower Archaean of the Pilbara and Kaapvaal compared. In. Cassidy KF, Dunphy J, Van Kranendonk MJ (eds). AGSO (Geoscience Australia) Record 2001/37, Extended Abstracts–4th International Archaean Symposium, Perth, 24–28 September 2001, pp 520–522

Nijman W, Kloppenburg A, de Vries ST (2017) Archaean basin margin geology and crustal evolution: an East Pilbara traverse. J Geol Soc 174:1090–1112
Noffke N, Christiansen D, Wacey D, Hazen RM (2013) Microbially induced sedimentary structures recording an ancient ecosystem in the ca. 3.48 billion-year-old Dresser Formation, Pilbara. Western Australia. Astrobiology 13:1103–1124. https://doi.org/10.1089/ast.2013.1030
Orberger B, Rouchon V, Westall F, De Vries ST, Pinti DL, Wagner C, Wirth R, Hashizume K (2006) Microfacies and origin of some Archean cherts (Pilbara, Australia). In: Reimold WU, Gibson RL (eds) Processes on the early earth. Geological Society of America, pp 133–156
Oliver NHS, Cawood PA (2001) Early tectonic dewatering and brecciation on the overturned sequence at Marble Bar, Pilbara Craton, Western Australia: dome related or not? Precambrian Res 105:1–15
Olivier N, Dromart G, Coltice N, Flament N, Rey P, Sauvestre R (2012) A deep subaqueous fan depositional model for the Palaeoarchean (3.46 Ga) Marble Bar Cherts, Warrawoona group. Western Australia Geol Mag 149:743–749
Pawley MJ, Van Kranendonk MJ, Collins WJ (2004) Interplay between deformation and magmatism during doming of the Archaean Shaw granitoid complex, Pilbara Craton, Western Australia. Precambrian Res 131:213–230
Perry EC, Hickman AH, Barnes IL (1975) Archaean sedimentary barite, Pilbara Goldfield, Western Australia. In: program with abstracts. Geological Society of America, p 1226
Petersson A, Kemp AIS, Gray CM, Whitehouse MJ (2020) Formation of early Archean Granite–Greenstone Terranes from a globally chondritic mantle: insights from igneous rocks of the Pilbara Craton. Western Australia. Chemical Geology 551. https://doi.org/10.1016/j.chemgeo.2020.119757
Petersson A, Kemp AIS, Whitehouse MJ (2019) A Yilgarn seed to the Pilbara Craton (Australia)? Evidence from inherited zircons. Geology 47:1098–1102. https://doi.org/10.1130/G46696.1
Philippot P, Van Zuilen M, Lepot K, Thomazo C, Farquhar J, Van Kranendonk MJ (2007) Early Archaean microorganisms preferred elemental sulfur, not sulphate. Science 317:1534–1537
Philippot P, Avila JN, Killingsworth BA, Tessalina S, Baton F, Caquineau T, Muller E, Pecoits E, Cartigny P, Lalonde SV, Ireland TR, Thomazo C, Van Kranendonk MJ, Busigny V (2018) Globally asynchronous Sulphur isotope signals require re-definition of the great oxidation event. Nat Commun 9-2245. https://doi.org/10.1038/s41467-018-04621-x
Pidgeon RT (1978) Geochronological investigation of granite batholiths of the Archaean Granite–Greenstone Terrain of the Pilbara Block, Western Australia. In: Smith IEM, Williams JG (eds) Proceedings of the 1978 Archaean geochemistry conference. University of Toronto, Ontario, pp 360–362
Pidgeon RT (1984) Geochronological constraints on early volcanic evolution of the Pilbara Block, Western Australia. Aust J Earth Sci 31:237–242
Pinti DL, Hashizume K, Orberger B, Gallien J-P, Cloquet C, Massault M (2007) Biogenic nitrogen and carbon in Fe-Mn-oxyhydroxides from an Archean chert, Marble Bar. Western Australia Geochemistry Geophysics Geosystems 8. https://doi.org/10.1029/2006GC001394
Pinti DL, Hashizume K, Sugihara A, Massault M, Philippot P (2009a) Isotopic fractionation of nitrogen and carbon in Paleoarchean cherts from Pilbara Craton, Western Australia: origin of ^{15}N-depleted nitrogen. Geochim Cosmochim Acta 73:3818–3848
Pinti DL, Mineau R, Clement V (2009b) Hydrothermal alteration and microfossil artefacts of the 3,465-million-year-old apex chert. Nat Geosci 2:640–643
Pirajno F (2007a) Ancient to modern earth: the role of mantle plumes in the making of continental crust. In: Van Kranendonk MJ, Bennett VC, Smithies RH (eds) Earth's oldest rocks. Elsevier B. V, Burlington, Massachusetts, USA, pp 1037–1064
Pirajno F (2007b) Mantle plumes, associated intraplate tectono-magmatic processes and ore systems. Episodes 30:6–19
Pirajno F, Van Kranendonk MJ (2005) Review of hydrothermal processes and systems on earth and implications for Martian analogues. Aust J Earth Sci 52:329–351. https://doi.org/10.1080/08120090500134571

Rasmussen B, Krapež B, Meier DB (2014a) Replacement origin for hematite in 2.5 Ga banded iron formation: evidence for post-depositional oxidation of iron-bearing minerals. Geol Soc Am Bull 126:438–446

Rasmussen B, Krapež B, Muhling JR (2014b) Hematite replacement of iron-bearing precursor sediments in the 3.46-b.y.-old Marble Bar Chert, Pilbara Craton. Australia Geological Society of America Bulletin 126:1245–1258

Rasmussen B, Muhling JR, Suvorova A, Krapež B (2016) Dust to dust: evidence for the formation of "primary" hematite dust in banded iron formations via oxidation of iron silicate nanoparticles. Precambrian Res 284:49–63. https://doi.org/10.1016/j.precamres.2016.07.003

Retallack GJ (2018) The oldest known paleosol profiles on earth: 3.46 Ga Panorama Formation, Western Australia. Palaeogeogr Palaeoclimatol Palaeoecol 489:230–248

Runnegar B, Dollase WA, Ketcham RA, Colbert M, Carlson WD (2001) Early Archean sulfates from Western Australia first formed as hydrothermal barites not gypsum evaporites. GSA Annual Meeting November 5-8, 2001, paper no. 166-0

Sandiford M, Van Kranendonk MJ, Bodorkos S (2004) Conductive incubation and the origin of dome-and-keel structure in Archean Granite–Greenstone Terrains: a model based on the Eastern Pilbara Craton. Western Australia Tectonics 23. https://doi.org/10.1029/2002TC001452

Schopf JW, Packer BM (1987) Early Archaean (3.3-billion to 3.5-billion-year-old) microfossils from the Warrawoona Group. Australia Science 237:70–73

Schopf JW, Walter MR (1983) Archean microfossils: new evidence of ancient microbes. In: Schopf JW (ed) Earth's earliest biosphere: its origin and evolution. Princeton University Press, Princeton, NJ, pp 214–239

Schopf JW (1992) The oldest fossils and what they mean. In: Schopf JW (ed) Major events in the history of life. John and Bartlett, Boston, USA, pp 29–63

Schopf JW (1993) Microfossils of the early Archaean Apex Chert: new evidence of the antiquity of life. Science 260:640–646

Schopf JW (2006) Fossil evidence of Archaean life. Philosophical Transactions of the Royal Society B: Biological Sciences 361:869–885. https://doi.org/10.1098/rstb.2006.1834

Simonson BM, Hassler SW (2002) Revisiting an Archean impact layer – comment. Science 298:750

Smith AJB (2015) The life and times of banded iron formations. Geology 43:1111–1112. https://doi.org/10.1130/focus122015.1

Smithies RH, Champion DC, Cassidy KF (2003) Formation of Earth's early Archaean continental crust. Precambrian Res 127:89–101

Smithies RH, Champion DC, Van Kranendonk MJ (2009) Formation of Paleoarchean continental crust through infracrustal melting of enriched Basalt. Earth Planet Sci Lett 281:298–306

Smithies RH, Champion DC, Van Kranendonk MJ (2019) The oldest well-preserved felsic volcanic rocks on earth: geochemical clues to the early evolution of the Pilbara Supergroup and implications for the growth of a Paleoarchean protocontinent. In: Van Kranendonk MJ, Bennett VC, Hoffmann JE (eds) Earth's oldest rocks, 2nd edn. Elsevier B.V, pp 463–486

Smithies RH, Champion DC, Van Kranendonk MJ, Hickman AH (2007) Geochemistry of volcanic rocks of the Northern Pilbara Craton, Western Australia. Report, vol 104. Geological Survey of Western Australia

Smithies RH, Champion DC, Van Kranendonk MJ, Howard HM, Hickman AH (2005a) Modern-style subduction processes in the Mesoarchean: geochemical evidence from the 3.12 Ga Whundo intra-oceanic arc. Earth Planet Sci Lett 231:221–237

Smithies RH, Van Kranendonk MJ, Champion DC (2005b) It started with a plume–early Archaean basaltic proto-continental crust. Earth Planet Sci Lett 238:284–297

Stern RA, Bodorkos S, Kamo SL, Hickman AH, Corfu F (2009) Measurement of SIMS instrumental mass fractionation of Pb isotopes during zircon dating. Geostand Geoanal Res 33:145–168

Stowe CW (1971) Summary of the Tectonic Development of the Rhodesian Archaean Craton In: Glover JE (ed), Symposium on Archaean Rocks. Geological Society of Australia, Special Publication No. 3:377–383

Suganama Y, Hamano Y, Niitsuma S, Hoashi M, Hisamitsu T, Niitsuma N, Kodama K, Nedachi M (2006) Paleomagnetism of the Marble Bar Chert member, Western Australia: implications for apparent polar wander path for Pilbara Craton during Archean time. Earth Planet Sci Lett 252: 360–371

Sugitani K (1992) Geochemical characteristics of Archean cherts and other sedimentary rocks in the Pilbara Block, Western Australia: evidence for Archean seawater enriched in hydrothermally derived iron and silica. Precambrian Res 57:21–47

Terabayashi M, Masuda Y, Ozawa H (2003) Archean Ocean floor metamorphism in the North Pole area, Pilbara Craton, Western Australia. Precambrian Res 127:167–180

Thébaud N, Rey PF (2013) Archean gravity-driven tectonics on hot and flooded continents: controls on long-lived hydrothermal systems away from continental margins. Precambrian Res 229:93–104

Thorpe RI, Hickman AH, Davis DW, Mortensen JK, Trendall AF (1990) Application of recent zircon U–Pb geochronology in the Marble Bar region, Pilbara Craton, to modelling Archean lead evolution. In: Glover JE, Ho SE (eds), Extended abstracts volume, Third International Archaean Symposium, 17-21 September 1990. Geoconferences (W.A.) Inc. Perth, Western Australia, pp. 11–13

Thorpe R, Hickman AH, Davis D, Morton JGG, Trendall AF (1992a) U–Pb zircon geochronology of Archaean felsic units in the Marble Bar region, Pilbara Craton, Western Australia. Precambrian Res 56:169–189

Thorpe R, Hickman AH, Davis DW, Mortensen JK, Trendall AF (1992b) Constraints to models for Archaean lead evolution from precise U-Pb geochronology from the Marble Bar region, Pilbara Craton, Western Australia. In: Glover JE, Ho SE (eds) The Archaean: terrains, processes and metallogeny: proceedings for the third International Archaean Symposium, 17–21 September 1990. Geology Department and University Extension, The University of Western Australia, Perth, Western Australia, pp 395–408

Ueno Y (2007) Stable carbon and sulfur isotope geochemistry of the ca. 3490 Ma Dresser Formation hydrothermal deposit, Pilbara Craton, Western Australia. In: Van Kranendonk MJ, Bennett VC, Smithies RH (eds) Earth's oldest rocks. Elsevier B.V, Burlington, Massachusetts, USA, pp 203–236

Ueno Y, Isozaki Y, Yurimoto H, Maruyama S (2001a) Carbon isotopic signatures of individual Archaean microfossils(?) from Western Australia. Int Geol Rev 43:196–212

Ueno Y, Maruyama S, Isozaki Y, Yurimoto H (2001b) Early Archean (ca. 3.5 Ga) microfossils and ^{13}C-depleted carbonaceous matter in the North Pole area, Western Australia: field occurrence and geochemistry. In: Nakashima S, Maruyama S, Brack A, Windley BF (eds) Geochemistry and the origin of life. Universal Academic Press, Inc., Tokyo, Japan, pp 203–236

Ueno Y, Yoshioka H, Maruyama S, Isozaki Y (2004) Carbon isotopes and petrography of kerogens in ~3.5-Ga hydrothermal silica dikes in the North Pole area. Western Australia Geochimica et Cosmochimica Acta 68:573–589

Ueno Y, Isozaki Y, McNamara KJ (2006) Coccoid-like microstructures in a 3.0 Ga chert from Western Australia. Int Geol Rev 48:78–88

van den Boorn SHJM, van Bergen MJ, Nijman W, Vroon PZ (2007) Dual role of seawater and hydrothermal fluids in early Archean chert formation: evidence from silicon isotopes. Geology 35:939–942

van den Boorn SHJM, van Bergen MJ, Vroon PZ, De Vries ST, Nijman W (2010) Silicon isotope and trace element constraints on the origin of ~3.5 Ga cherts: implications for early Archean marine environments. Geochim Cosmochim Acta 74:1077–1103

van Haaften WM, White SH (1998) Evidence for multiphase deformation in the Archean basal Warrawoona Group in the Marble Bar area, East Pilbara, Western Australia. Precambrian Res 88:53–66

Van Kranendonk MJ (1998) Litho-tectonic and structural components of the North Shaw 1:100 000 sheet, Archaean Pilbara Craton. In: Geological Survey of Western Australia Annual Review 1997–98. Geological Survey of Western Australia, pp 63–70

Van Kranendonk MJ (1999) Two-stage degassing of the Archaean mantle: evidence from the 3.46 Ga Panorama volcano, Pilbara Craton, Western Australia. In: GSWA 99 extended abstracts: new geological data for WA explorers. Geological Survey of Western Australia, pp. 1–3

Van Kranendonk MJ (2000) Geology of the North Shaw 1:100 000 sheet. 1:100 000 Geological Series Explanatory Notes. Geological Survey of Western Australia

Van Kranendonk MJ (2004) Geology of the Carlindie 1:100 000 sheet. 1:100 000 geological series explanatory notes. Geological Survey of Western Australia

Van Kranendonk MJ (2006) Volcanic degassing, hydrothermal circulation and the flourishing of early life on earth: a review of the evidence from c. 3490–3240 Ma rocks of the Pilbara Supergroup, Pilbara Craton, Western Australia. Earth Sci Rev 74:197–240

Van Kranendonk MJ (2007) A review of the evidence for putative Paleoarchean life in the Pilbara Craton. In: Van Kranendonk MJ, Bennett VC, Smithies RH (eds) Earth's oldest rocks. Elsevier B.V, Burlington, Massachusetts, USA, pp 855–896

Van Kranendonk MJ (2010a) Geology of the Coongan 1:100 000 sheet. 1:100 000 Geological Series Explanatory Notes. Geological Survey of Western Australia

Van Kranendonk MJ (2010b) Three and a half billion years of life on earth: a transect back into deep time. Record, 2010/21. Geological Survey of Western Australia

Van Kranendonk MJ, Collins WJ (2001) A review of the evidence for vertical tectonics in the Archean Pilbara Craton, Western Australia. In: Cassidy KF, Dunphy JM, Van Kranendonk MJ (eds) Extended Abstracts-4th International Archaean Symposium, Perth, 24–28 September 2001. AGSO (Geoscience Australia), pp 365–367

Van Kranendonk MJ, Collins WJ, Hickman AH, Pawley MJ (2004) Critical tests of vertical vs horizontal tectonic models for the Archean East Pilbara Granite–Greenstone Terrane, Pilbara Craton, Western Australia. Precambrian Res 131:173–211

Van Kranendonk MJ, Hickman AH (2000) Archaean geology of the North Shaw region, East Pilbara Granite–Greenstone Terrane, Western Australia–a field guide. Record, 2000/5. Geological Survey of Western Australia

Van Kranendonk MJ, Hickman AH, Collins WJ (2001a) Comment on 'evidence for multiphase deformation in the Archean basal Warrawoona Group in the Marble Bar area, East Pilbara, Western Australia'. Precambrian Res 105:73–78

Van Kranendonk MJ, Hickman AH, Huston DL (2006b) Geology and mineralization of the East Pilbara–a field guide. Record, 2006/16. Geological Survey of Western Australia

Van Kranendonk MJ, Hickman AH, Smithies RH, Champion DC (2007a) Paleoarchean development of a continental nucleus: the East Pilbara Terrane of the Pilbara Craton, Western Australia. In: Van Kranendonk MJ, Bennett VC, Smithies RH (eds) Earth's oldest rocks. Elsevier B.V, Burlington, Massachusetts, USA, pp 307–337

Van Kranendonk MJ, Hickman AH, Smithies RH, Nelson DN, Pike G (2002) Geology and tectonic evolution of the Archaean North Pilbara Terrain, Pilbara Craton, Western Australia. Econ Geol 97:695–732. https://doi.org/10.2113/gsecongeo.97.4.695

Van Kranendonk MJ, Hickman AH, Smithies RH, Williams IR, Bagas L, Farrell TR (2006a) Revised lithostratigraphy of Archean supracrustal and intrusive rocks in the Northern Pilbara Craton, Western Australia. Record, 2006/15. Geological Survey of Western Australia

Van Kranendonk MJ, Hickman AH, Williams IR, Nijman W (2001b) Archaean geology of the East Pilbara Granite–Greenstone Terrane, Western Australia–a field guide. Record, 2001/9. Geological Survey of Western Australia

Van Kranendonk MJ, Morant P (1999) Revised Archaean stratigraphy of the North Shaw 1:100 000 sheet, Pilbara Craton. In: Geological Survey of Western Australia Annual Review 1997–98. Geological Survey of Western Australia, pp 55–62

Van Kranendonk MJ, Philippot P, Lepot K, Bodorkos S, Pirajno F (2008) Geological setting of Earth's oldest fossils in the c. 3.5 Ga Dresser Formation, Pilbara Craton. Western Australia Precambr Res 167:93–124

Van Kranendonk MJ, Pirajno F (2004) Geochemistry of metabasalts and hydrothermal alteration zones associated with c. 3.45 Ga chert and barite deposits: implications for the geological setting of the Warrawoona Group, Pilbara Craton, Australia. Geochemistry: exploration. Environment, Analysis 4:253–278

Van Kranendonk MJ, Smithies RH, Champion DC (2019) Chapter 19–Paleoarchean development of a continental nucleus: the East Pilbara Terrane of the Pilbara Craton, Western Australia. In: Van Kranendonk MJ, Bennett VC, Hoffmann JE (eds) Earth's oldest rocks, 2nd edn. Elsevier B. V, pp 437–462

Van Kranendonk MJ, Smithies RH, Griffin WL, Huston DL, Hickman AH, Champion DC, Anhaeusser CR, Pirajno F (2015) Making it thick: a volcanic plateau origin of Paleoarchean continental lithosphere of the Pilbara and Kaapvaal cratons. Geol Soc Lond, Spec Publ 389:89–111. https://doi.org/10.1144/SP389.12

Van Kranendonk MJ, Smithies RH, Hickman AH, Champion DC (2007b) Review: secular tectonic evolution of Archean continental crust: interplay between horizontal and vertical processes in the formation of the Pilbara Craton, Australia. Terra Nova 19:1–38

Vearncombe S, Kerrich R (1999) Geochemistry and geodynamic setting of volcanic and plutonic rocks associated with early Archaean volcanogenic massive sulphide mineralization, Pilbara Craton. Precambrian Res 98:243–270

Vearncombe S, Vearncombe JR, Barley ME (1998) Fault and stratigraphic controls on volcanogenic massive sulphide deposits in the Strelley belt, Pilbara Craton, Western Australia. Precambrian Res 88:67–82

Walsh MM, Lowe DR (1999) Modes of accumulation of carbonaceous matter in the early Archean: a petrographic and geochemical study of the carbonaceous cherts of the Swaziland Supergroup. In: Lowe DR, Byerly GR (eds), Geologic Evolution of the Barberton Greenstone Belt, South Africa. Geol Soc Am Spec Pap 329 pp. 115–132

Walter MR (1983) Archean Stromatolites: evidence of the Earth's earliest benthos. In: Schopf JW (ed) Earth's earliest biosphere: its origin and evolution. Princeton University Press, Princeton, pp 187–213

Walter MR, Buick R, Dunlop JSR (1980) Stromatolites 3,400–3,500 Myr old from the North Pole area, Western Australia. Nature 284:443–445

Wiemer D, Schrank CE, Murphy DT, Hickman AH (2016) Lithostratigraphy and structure of the early Archaean Doolena Gap greenstone belt, East Pilbara Terrane, Western Australia. Precambrian Res 282:121–138

Wiemer D, Schrank CE, Murphy DT, Wenham L, Allen CM (2018) Earth's oldest stable crust in the Pilbara Craton formed by cyclic gravitational overturns. Nat Geosci 11:357–361. https://doi.org/10.1038/s41561-018-0105-9

Wille M, Nebel O, Van Kranendonk MJ, Schoenber R, Kleinhanns IC, Ellwood MJ (2013) Mo–Cr isotope evidence for a reducing Archean atmosphere in 3.46–2.76 Ga black shales from the Pilbara. Western Australia Chemical Geology 340:68–76

Williams IS, Collins WJ (1990) Granite–Greenstone Terranes in the Pilbara Block, Australia, as coeval volcano-plutonic complexes: evidence from U–Pb zircon dating of the Mount Edgar batholith. Earth Planet Sci Lett 97:41–53. https://doi.org/10.1016/0012-821X(90)90097-H

Williams IR (1999) Geology of the Muccan 1:100 000 sheet. 1:100 000 Geological Series Explanatory Notes. Geological Survey of Western Australia

Williams IR (2001) Geology of the Warrawagine 1:100 000 sheet. 1:100 000 Geological Series Explanatory Notes. Geological Survey of Western Australia

Williams IR, Bagas L (2007) Geology of the Mount Edgar 1:100 000 sheet. 1:100 000 Geological Series Explanatory Notes. Geological Survey of Western Australia

Williams IR, Hickman AH (2000) Archaean geology of the Muccan region, East Pilbara Granite–Greenstone Terrane, Western Australia–a field guide. Record, 2000/4. Geological Survey of Western Australia

Winchester JA, Floyd PA (1977) Geochemical discrimination of different magma series and their differentiation products using immobile elements. Chem Geol 20:325–343

Windley BF (1973) A discussion on the evolution of the Precambrian crust–crustal development in the Precambrian. Philos Trans A Math Phys Eng Sci 273:321–341

Windley BF, Bridgwater D (1971) The evolution of Archaean low- and high-grade terrains. Geol Soc Aust Spec Publ 3:33–46

Wingate MTD, Bodorkos S, Van Kranendonk MJ (2009a) 180057: tonalitic orthogneiss, Big Junction Well

Wingate MTD, Bodorkos S, Van Kranendonk MJ (2009b) 180070: volcaniclastic sandstone, drillcore PDP2c, Dresser Mine

Zegers TE (1996) Structural, kinematic, and metallogenic evolution of selected domains of the Pilbara Granite–Greenstone Terrain. PhD, Geologica Ultraiectina, Utrecht University

Zegers TE, White SH, de Keijzer M, Dirks P (1996) Extensional structures during deposition of the 3460 Ma Warrawoona group in the Eastern Pilbara Craton, Western Australia. Precambrian Res 80:89–105

Zegers TE, Wijbrans JR, White SH (1999) $^{40}Ar/^{39}Ar$ age constraints on tectonothermal events in the Shaw area of the Eastern Pilbara Granite–Greenstone Terrain (Western Australia): 700 Ma of Archaean tectonic evolution. Tectonophysics 311:45–81

Chapter 4
Strelley Pool Formation: Continental Sedimentation Between Paleoarchean LIPs

Abstract The predominantly volcanic Warrawoona and Kelly Groups are separated by a regional unconformity overlain by the Strelley Pool Formation, a 20–1000 m thick, shallow-water succession of conglomerate, sandstone, silicified sedimentary carbonate rocks, evaporites, and volcaniclastic rocks. The formation was deposited during a 75-million-year break in volcanic activity between eruption of the Warrawoona and Kelly LIPs. A measure of the great lateral extent of the Strelley Pool Formation is that it is present in all greenstone belts of the Pilbara Craton that expose c. 3400 Ma stratigraphy. Prior to breakup of the Pilbara Craton, the formation might have been continuous with the Buck Reef Chert in the Onverwacht Group of South Africa. The formations are about the same age, and they separate very similar stratigraphic successions.

The Strelley Pool Formation was deposited in shallow-water marine, stromatolite reef, beach, sabkha, estuarine, fluviatile, and lacustrine environments. Extensive silicification of its carbonate and fine-grained clastic components has transformed much of the formation to 'chert', and the formation was once referred to as the 'Strelley Pool Chert'. The formation contains some of Earth's oldest and best-preserved fossil evidence of early life, including stromatolites, microbial mats, microfossils, and trace fossils.

Keywords Depositional environment · Early life · Stromatolite · Microbial mat · Microfossil

4.1 Introduction

The 3426–3350 Ma Strelley Pool Formation is a thin but regionally extensive sedimentary formation that separates the predominantly volcanic successions of the Warrawoona and Kelly Groups (Table 3.1). Following the final eruptions of the Panorama Formation at about 3427 Ma (Chap. 3), there was a 75-million-year period without volcanism. A similar break in volcanism is recorded in the Onverwacht Group of South Africa (Fig. 1.10). U–Pb zircon geochronology indicates that the last granitic intrusions of the Tambina Supersuite did not crystallize

A. H. Hickman, *Archean Evolution of the Pilbara Craton and Fortescue Basin*, Modern Approaches in Solid Earth Sciences 24,
https://doi.org/10.1007/978-3-031-18007-1_4

until about 3416 Ma. Although there were brief interludes of volcanic quiescence and erosion during deposition of the 3530–3427 Ma Warrawoona Group, the Strelley Pool Formation marks the first long-term period of crustal stability.

U–Pb zircon dates constraining the depositional age of the formation are provided by many dates between 3434 and 3427 Ma from the underlying Panorama Formation and dates between 3350 and 3335 Ma from the overlying Euro Basalt of the Kelly Group. Evidence of the absence of igneous activity during deposition of the Strelley Pool Formation is provided by detrital zircon ages from post-3350 Ma sedimentary formations in the northern Pilbara Craton (Fig. 2.2a). A few detrital zircon ages falling between 3420 and 3410 Ma are likely to be due to Pb loss, a common occurrence in clastic sedimentary rocks.

Although rarely more than 30 m thick, the formation is exposed in all those greenstone belts of the East Pilbara Terrane (EPT) that expose the upper part of the Warrawoona Group (Hickman 2008). Figure 4.1 shows outcrops of the Strelley Pool Formation within an 80 km radius of Marble Bar, and in other parts of the EPT the formation forms parts of the Tambina, Emerald Mine, Western Shaw, and Goldsworthy greenstone belts. Lateral facies changes associated with the shallow-water depositional environments of the formation are reflected in regional lithological variations, although in most areas basal sandstone and conglomerate are overlain by bedded carbonate rocks or secondary chert; mafic volcaniclastic rocks form the upper part of the formation. Less common lithologies include evaporites and jaspilite.

The Strelley Pool Formation unconformably overlies deformed and eroded formations of the Warrawoona Group (Buick et al. 1995), and unconformably to disconformably underlies the Kelly Group (Fig. 1.8). However, the nature of the basal unconformity varies across the EPT. In the northwest, the upper part of the Warrawoona Group was almost entirely removed by erosion prior to deposition of the Strelley Pool Formation. This differs from the stratigraphy in the central and southeast parts of the terrane where the Panorama Formation underlies the Strelley Pool Formation, and the contact is a paraconformity (Hickman 2008). The extent to which the Strelley Pool Formation was deposited across the granitic cores of the granite–greenstone domes of the EPT is unknown because its present outcrops almost invariably overlie greenstones. However, the Strelley Pool Formation of the Muccan, Carlindi, and Yule Domes contains >3530 Ma detrital zircons indicating erosion of not only the lower Warrawoona Group but also erosion of the 3800–3530 Ma crust that, based on present outcrops of the latter, was exposed in the granitic cores of the domes. It is likely that some areas of EPT granites and > 3530 Ma crust formed islands within the depositional basin of the formation.

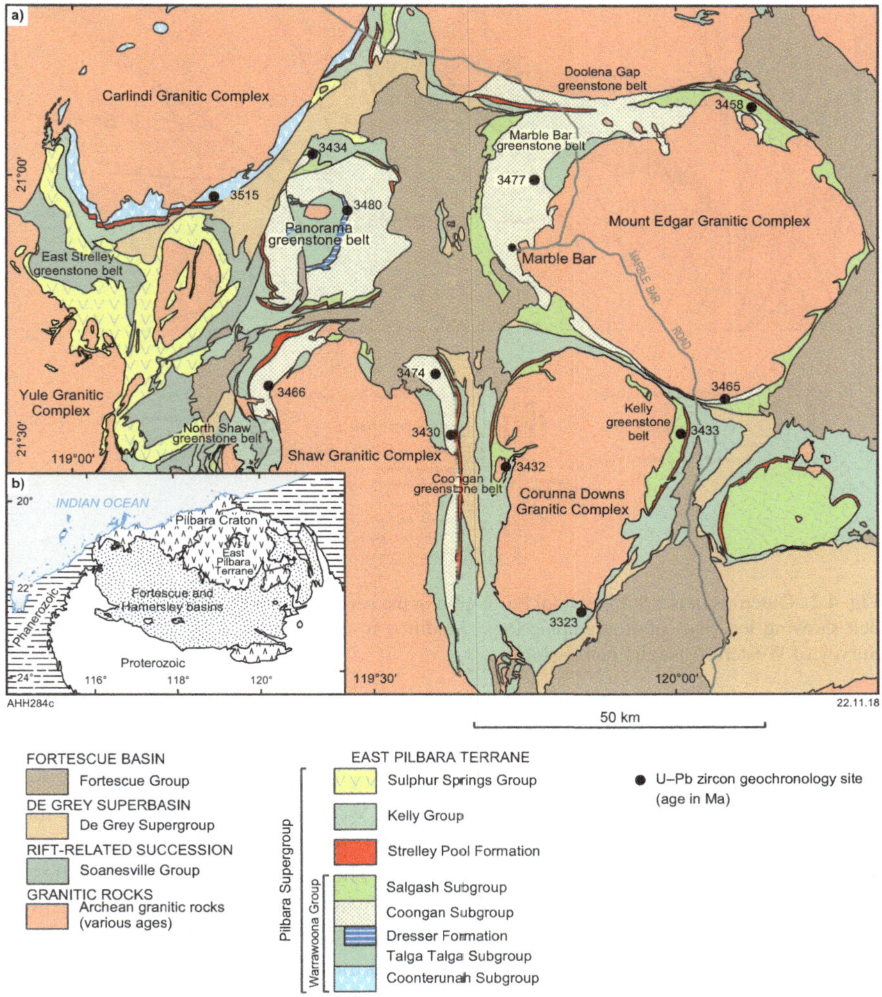

Fig. 4.1 Geological map of part of the East Pilbara Terrane showing outcrops of the Strelley Pool Formation within the stratigraphy of the Pilbara Supergroup. Although the formation is typically less than 100 m thick, it outcrops across most of the terrane, and marks a 75-Ma break between LIP-scale volcanism of the Warrawoona and Kelly Groups (From Hickman 2008; with Geological Survey of Western Australia permission)

4.2 Stratigraphy

The stratigraphy of the Strelley Pool Formation has been documented from several greenstone belts: the East Strelley greenstone belt (Lowe 1983; DiMarco and Lowe 1989; Van Kranendonk 2000, 2004; Wacey et al. 2010); the Panorama greenstone belt of the North Pole Dome (Lowe 1983; DiMarco and Lowe 1989; Hofmann et al.

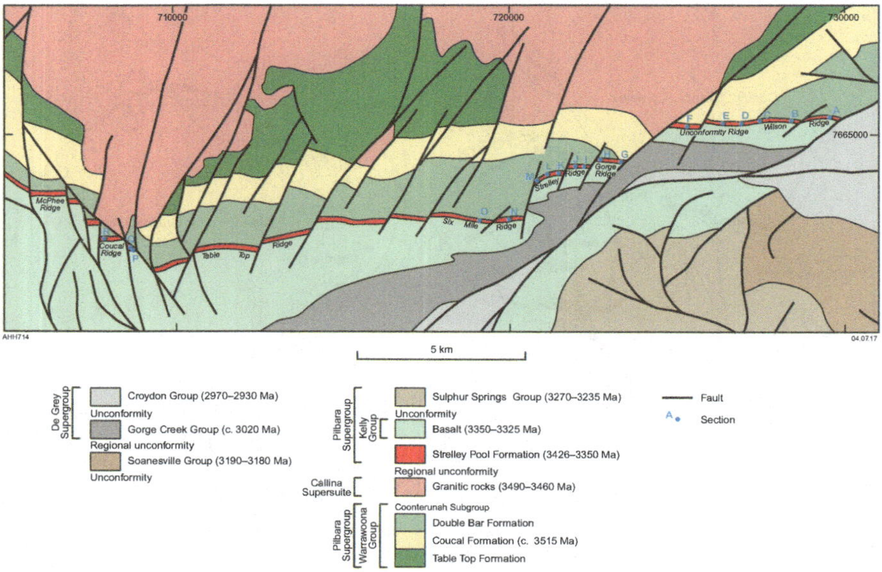

Fig. 4.2 Outcrops of the Strelley Pool Formation in the central part of the East Strelley greenstone belt showing locations of stratigraphic logs (Modified from Wacey et al. 2010; with Geological Survey of Western Australia permission)

1999; Van Kranendonk 2000; Allwood et al. 2004a, b, 2006, 2007a, b); the Marble Bar greenstone belt near Kittys Gap (Nijman et al. 2001; De Vries 2004; De Vries et al. 2010), the Kelly greenstone belt (Hickman 1980; Bagas 2003; Bagas et al. 2004); and the Doolena Gap greenstone belt (Wiemer et al. 2016). In the latter greenstone belt, the formation is exceptionally thick, consisting of 1000 m of sandstone and conglomerate capped by chert. The various sedimentary facies of the Strelley Pool Formation indicate shallow-water environments that included shallow-water marine, beach, estuarine, sabkha, stromatolite reef, fluviatile, and lacustrine (Allwood et al. 2006, 2007a, b; Hickman 2008). Where it overlies the Panorama Formation, the Strelley Pool Formation includes reworked volcaniclastic conglomerate and sandstone.

Investigations of the Strelley Pool Formation have revealed not only regional stratigraphic variations but also rapid lateral variations over distances of less than a few hundred metres (Allwood et al. 2007a, b; Hickman 2008; Wacey et al. 2010). Local variations were revealed by detailed stratigraphic studies of the formation in the western part of the Panorama greenstone belt (Allwood et al. 2007a) and in the East Strelley greenstone belt (Wacey et al. 2010). Figures 4.2 and 4.3 show the outcrop of the Strelley Pool Formation in the East Strelley greenstone belt and the positions of detailed stratigraphic logs made by Wacey et al. (2010). Lateral thickness and facies variations were illustrated by sections measured along a 25-km-long series of strike ridges east and west from Strelley Pool (Fig. 4.3). Variations in the lower part of the formation are likely to be consequences of migrating shallow-water

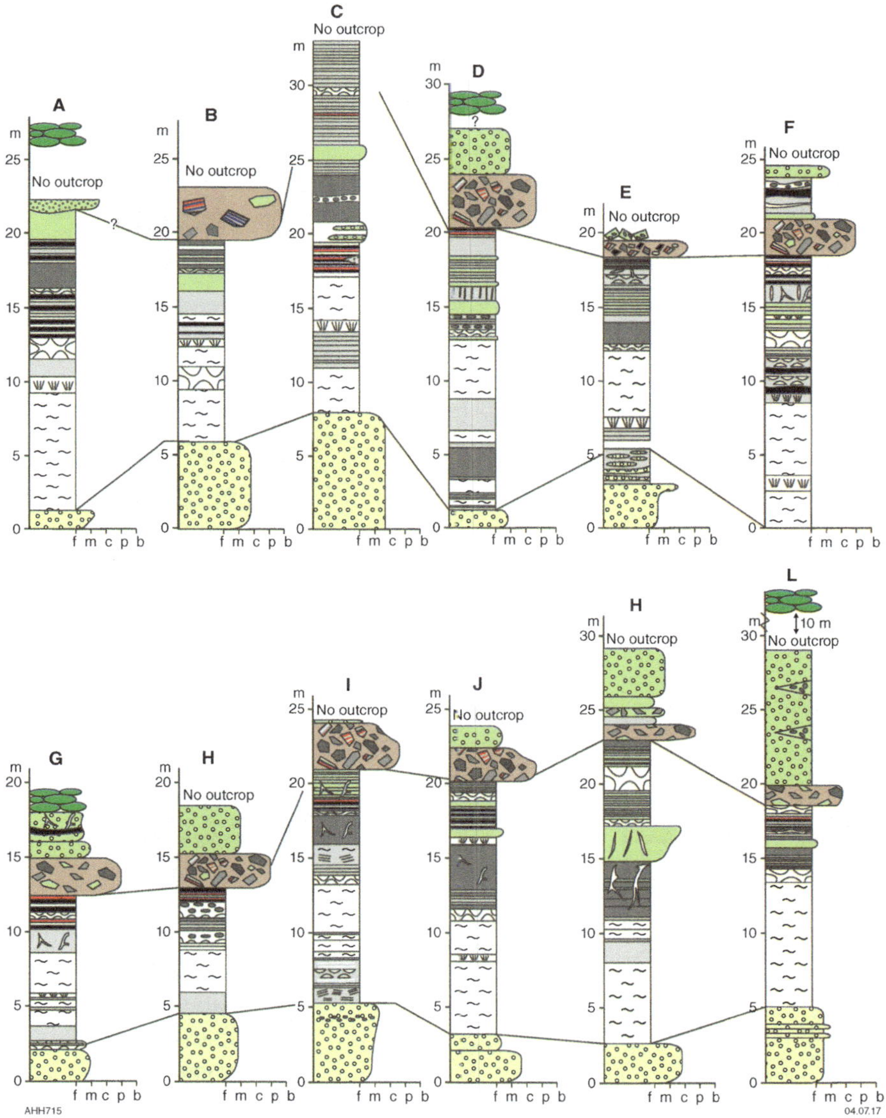

Fig. 4.3 Stratigraphic logs through the Strelley Pool Formation in the East Strelley greenstone belt. In detail, the logs show considerable lateral variations in thicknesses and sedimentary facies although the main features of the succession (basal sandstone, overlain by carbonate rocks, overlain by chert, overlain by conglomerate, overlain by basaltic volcaniclastic rocks) are laterally continuous except in the far western sections (logs P-R) where the basal sandstone is absent: (**a**) Logs A-L; (**b**) Logs M-R, Legend (Modified from Wacey et al. 2010; with Geological Survey of Western Australia permission)

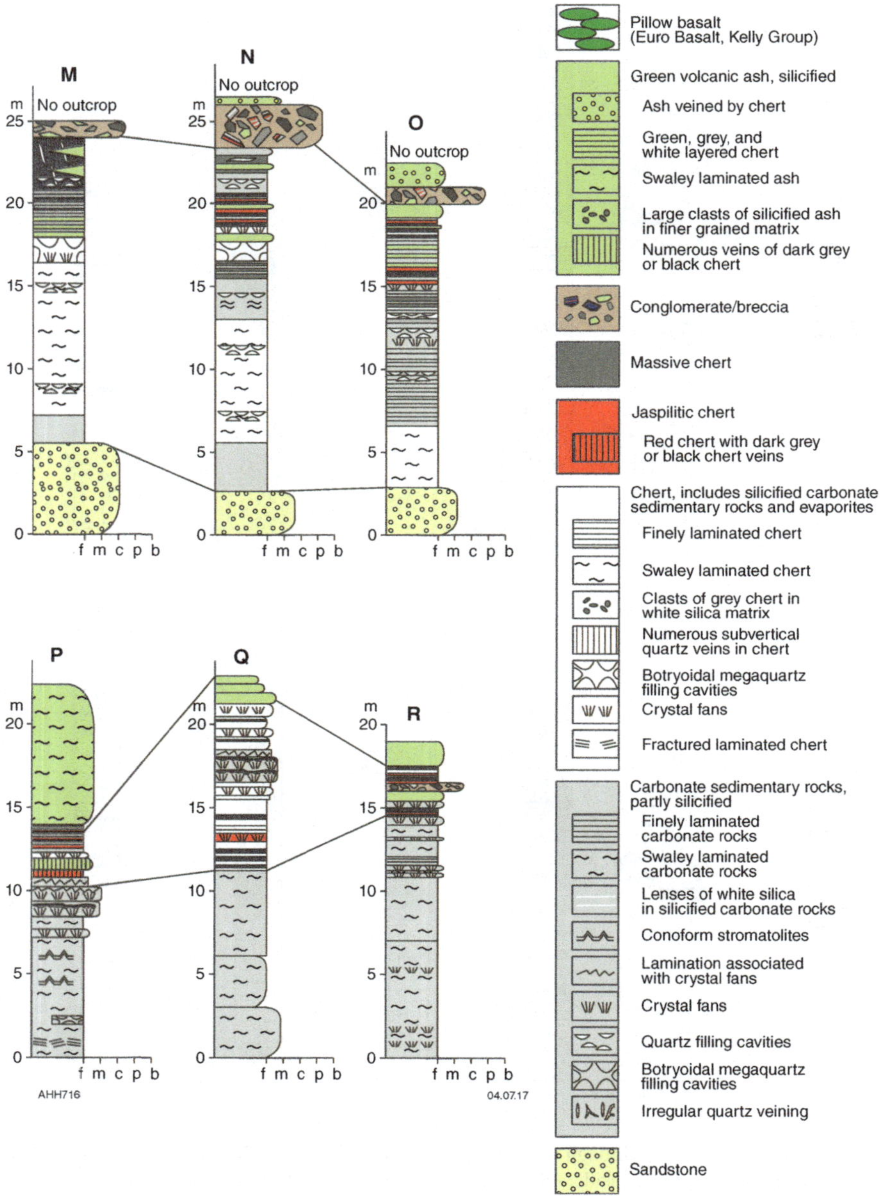

Fig. 4.3 (continued)

depositional systems (fluvial channels, sand bars etc.), and of controls by the topography of the underlying unconformity. An easterly transgression of the formation onto a previously exposed upland area of the Coonterunah Subgroup is

Table 4.1 Members (informal) and facies of the Strelley Pool Formation in the type section, Strelley Pool. From Lowe (1983)

Member	*Depositional environment*
Member 5	
Capping unit of volcaniclastic debris	Ashfall tuff and volcaniclastic, current-deposited beds (distal subaerial volcanic alluvial apron)
Member 4	
Unit of coarse, intraformational conglomerate and breccia	Debris flows and/or torrential flood deposits
Member 3	
Stratified unit of stromatolite, black chert, silicified evaporite, and activity evaporite-solution layers	Low-energy, clastic-starved environment subject to current and exposure. Modern analogues are intertidal flats and supratidal sabkhas
Member 2	
Massive-weathering cherty unit of laminated flat stromatolite, conical stromatolite, and silicified evaporite	Cyclic sedimentation in a shallow, hypersaline basin
Member 1	
Basal layer of quartzose sandstone	High-energy, shallow-water conditions

indicated by the absence of the basal sandstone unit east of Table Top Ridge (sections O–P, Fig. 4.3b). This eastern area also contains conoform stromatolites (Section P) suggesting especially shallow water, and evaporates have been recorded (Lowe 1980, 1983; Van Kranendonk 2000). Stromatolite occurrences in the Strelley Pool Formation of the Panorama greenstone belt also coincide with shallow-water environments (Hofmann et al. 1999; Van Kranendonk et al. 2003; Allwood et al. 2006, 2007a, b; Hickman et al. 2011).

Lowe (1983) divided the Strelley Pool Formation of the northwest East Pilbara Terrane into five informal members (Table 4.1) and subsequent workers used a similar subdivision (DiMarco and Lowe 1989; Van Kranendonk 2000; Allwood et al. 2007a; Wacey et al. 2010). However, the same subdivision has not been applied to the formation in the central and southeast East Pilbara Terrane where the succession is mainly composed of metasandstone and banded chert. The chert in these areas most likely represents silicified carbonate and fine-grained volcaniclastic rocks, although primary sedimentary precipitates are also possible. Wavy-laminated carbonate rocks are preserved in the Kelly and Coongan greenstone belts (Hickman 1980; Zegers et al. 1996; Grey et al. 2002, 2010, 2012; Bagas 2005; Williams and Bagas 2007). Lowe (1983) interpreted the northwest depositional environment as either a large hypersaline lake or a restricted shallow-water marine basin. He reported that basal sandstone fills channels cut into underlying parts of the Warrawoona Group. DiMarco and Lowe (1989) inferred that the Strelley Pool Formation overlies an erosional unconformity, although no exposures were cited.

4.2.1 Stratigraphic Rank: Formation or Group?

A detailed investigation into the regional stratigraphic variability of the Strelley Pool Formation would be likely to result in its redefinition as a group. Considerations suggesting eventual subdivision into several formations include:

- Deposition of the various sedimentary units now included within the Strelley Pool Formation spanned a time interval of approximately 75 Ma, an interval of time longer than for most individual periods of the Phanerozoic eon.
- Erosional breaks within the formation, and rapid lateral changes of sedimentary facies and succession thickness, are well documented. These features indicate local crustal instability during deposition, possibly related to uplift and extension caused by the mantle plume responsible for eruption of the Kelly Group.
- The Strelley Pool Formation is locally absent between the Warrawoona and Kelly Groups. Although this might be due to erosion prior to eruption of the Kelly Group, it is also likely that there were upland areas that were never covered by the formation.
- Certain distinctive components of the Strelley Pool Formation, such as the 1000-m-thick quartzite member in the Doolena Gap greenstone belt, and the laminated carbonate member of the East Strelley and Panorama greenstone belts, are mappable units that might be separated as formations.

4.2.2 Relations to the Panorama Formation

In the southeast part of the EPT, the Strelley Pool Formation is mainly composed of sandstone, grey and white banded chert, silicified fine-grained clastic rocks, and silicified carbonate rocks outcropping as laminated chert. The formation is underlain by relatively thick felsic volcaniclastic units of the Panorama Formation, the upper parts of which include volcaniclastic sandstone and chert. Until Bagas et al. (2004) reported local evidence of an erosional contact between the Panorama Formation and the 'Strelley Pool Chert' (obsolete name) in the Kelly greenstone belt, the units were interpreted to be conformable across the EPT (Van Kranendonk et al. 2002). Geochronology on the Panorama Formation in the eastern part of the EPT has provided dates consistently close to 3430 Ma and dating of the Strelley Pool Formation has given very similar maximum depositional ages (Hickman 2008). Additionally, similar sedimentary facies suggest that the lower parts of the Strelley Pool Formation and upper parts of the Panorama Formation were deposited in similar depositional environments. Contact relationships between the Panorama and Strelley Pool Formations in the Panorama greenstone belt have been documented by Van Kranendonk (2000) and Allwood et al. (2007a).

4.2.3 Unconformities

In the Panorama greenstone belt, the Strelley Pool Formation is underlain by two unconformities: the upper unconformity overlies a c. 3443-Ma thin, wedge-shaped unit of the Panorama Formation (Van Kranendonk 2011; Wingate et al. 2012), whereas the lower unconformity separates this unit from jaspilite that might be part of either the Panorama Formation (Van Kranendonk 2000) or the Marble Bar Chert Member (Hickman 2021). Consequently, in areas where the Panorama Formation is absent, it is likely that the unconformity at the base of the Strelley Pool Formation represents more than one erosional event, and there is evidence that deformation events, followed by erosion, occurred at 3490–3460 Ma (pre-3468 Ma deformation recorded by Buick et al. 1995, Baker and Collins, 2001, and Baker et al. 2002) and 3445–3420 Ma (Kloppenburg 2003; Pawley et al. 2004; Van Kranendonk et al. 2004).

In the East Strelley greenstone belt, Wacey et al. (2010) recorded thinning of the basal conglomerate–sandstone unit in a northwesterly direction. Farther to the west, the entire formation is missing between the Coonterunah Subgroup and the Euro Basalt (Blewett and Champion 2005), suggesting a northwest transgression of the Strelley Pool Formation onto a landmass composed of metamorphosed rocks of the Coonterunah Subgroup; this area could therefore be the northwestern margin of the depositional basin. Evidence of a third unconformity towards the top of the Strelley Pool Formation is provided by a laterally persistent unit of conglomerate in the Panorama and East Strelley greenstone belts (Lowe 1983; Allwood et al. 2007a; Hickman 2008; Wacey et al. 2010). This conglomerate contains boulders and pebbles eroded from the underlying parts of the formation and is overlain by mafic volcanic ash. The conglomerate and volcanic ash are conformable or paraconformable with the Euro Basalt, and probably represent a transition sequence into the basalt. The presence of the boulder conglomerate containing eroded blocks of the Strelley Pool Formation in the upper unit (Allwood et al. 2007a; Wacey et al. 2010) suggests a time break in which the main part of the formation was uplifted and eroded. The earliest effect of the mantle plume interpreted to have been responsible for eruption of the Euro Basalt (Van Kranendonk et al. 2004, 2006) was probably crustal uplift with extensional faulting. In both the East Strelley and Panorama greenstone belts the Strelley Pool Formation is fragmented by numerous small-scale extensional faults. Zegers et al. (1996) identified pre-Euro Basalt extensional faults in the Coongan greenstone belt, and Nijman et al. (2001) and De Vries (2004) described similar structures at the same stratigraphic level in the northern part of the Marble Bar greenstone belt. Extensional faulting of the Strelley Pool Formation would have resulted in erosion of the uplifted blocks and lenses of conglomerate and sandstone at the base of the Euro Basalt. In this scenario, the upper clastic unit of the Strelley Pool Formation might be a basal member of the Euro Basalt.

4.3 Geochronology

U–Pb zircon dating of the Strelley Pool Formation in the Gorge Creek area of the Warralong greenstone belt provided seven analyses with a weighted mean $^{207}Pb/^{206}Pb$ ratio corresponding to a date of 3426 ± 10 Ma (chi-squared = 0.84) (Nelson 1998, GSWA 142836), which was interpreted to indicate the maximum depositional age of the formation. The seven zircon grains were individually dated between 3444 and 3407 Ma. U–Pb zircon dating of the formation in the East Strelley greenstone belt led to an interpreted younger maximum depositional age of 3414 ± 34 Ma (Gardiner et al. 2019), the four zircon analyses used to calculate this age having a large range of individual ages to c. 3470 Ma. Although it is likely that deposition of the Strelley Pool Formation commenced at slightly different times in different areas of the EPT, the maximum depositional age in all areas except the northwest is constrained by U–Pb zircon dating of the underlying Panorama Formation at c. 3430 Ma.

Detrital zircon ages from samples of the Strelley Pool Formation collected in different areas of the EPT are locally older than c. 3550 Ma (Wingate et al. 2009, GSWA 180095; Nelson 1998, GSWA 142836; Wiemer et al. 2018; Gardiner et al. 2019). This might be evidence that the entire 10–15 km thickness of the Warrawoona Group had been locally removed by uplift and erosion prior to 3426 Ma. However, it is also possible that the >3550-Ma zircons were derived by recycling detrital and xenocrystic zircons from older rocks of the EPT.

4.4 World's Oldest Paleosols

Where the unconformity at the base of the Strelley Pool Formation is angular, as in the northwest part of the EPT, there are stratigraphic breaks of 90–20 million years: for example, between this formation and the c. 3515-Ma Coonterunah Subgroup. Although the Archean atmosphere is widely thought to have been anoxic, exposure to the Paleoarchean land surfaces for such long periods of time is likely to have resulted in significant chemical alteration. Weathering might have resulted from acid rain or from interaction with acidic surface waters. Alternatively, chemical weathering might have occurred under an atmosphere containing high levels of H_2 (Hao et al. 2017) related to volcanic activity (Walker 1978). In some parts of the northwest EPT, rocks immediately beneath the unconformity are altered to depths approaching 100 m (Buick et al. 1995; Van Kranendonk and Pirajno 2004; Allwood et al. 2006, 2007a, b; Altinok and Ohmoto 2006; Ohmoto et al. 2006; Johnson et al. 2008, 2009). Previously, this alteration has been attributed to hydrothermal activity (Brown et al. 2006, 2011; Van Kranendonk 2014) or to oxidation by an atmosphere containing O_2 levels similar to those of today (Ohmoto et al. 2006; Johnson et al. 2008, 2009; Hoashi et al. 2009). Hydrothermal activity immediately prior to deposition of the Strelley Pool Formation occurred during deposition of the Panorama

Formation in the Panorama greenstone belt (Brown et al. 2006), but the Panorama Formation is absent in the East Strelley greenstone belt where there is also well-developed alteration beneath the unconformity (Buick et al. 1995; Altinok and Ohmoto 2006; Wacey et al. 2010). Surface chemical weathering in the absence of atmospheric oxygen might explain much of the alteration beneath the unconformity. Retallack (2018) interpreted the existence of paleosol horizons in the Panorama Formation of the northwest Panorama greenstone belt, and the succession studied might have included the Strelley Pool Formation.

4.5 Suggestion of Hydrothermal Deposition

In view of the detailed sedimentological studies by Lowe (1980, 1983) and DiMarco and Lowe (1989), very few workers have suggested that the Strelley Pool Formation is a hydrothermal deposit. A hydrothermal model was first proposed by Van Kranendonk (2001) based on observations of dykes of massive chert extending from silicified clastic beds at the base of the formation into underlying volcanic rocks. This model was somewhat similar to that proposed to explain dykes and sills of chert and barite in the c. 3480-Ma Dresser Formation at North Pole (Nijman et al. 1998). The typical rock of the chert dykes contrasts with that of the bedded silicified rocks ('chert') of the formation in that it is massive, generally black or dark grey, and commonly contains breccia. Where such dykes meet the bedded rocks of the Strelley Pool Formation, the chert occupies transgressive sills, irregular vein systems, or takes the form of diffuse bodies of chert breccia. Van Kranendonk (2001) advocated a hydrothermal model to explain the source of chert and carbonate laminites in the Strelley Pool Formation of the Panorama greenstone belt. He interpreted the formation to have been deposited in several caldera lakes developed over the cooling but still hydrothermally active felsic volcanic pile of the Panorama Formation. Although there is no underlying Panorama Formation in the East Strelley greenstone belt, Lindsay et al. (2003, 2005) also interpreted hydrothermal deposition. They presented field and geochemical data to argue that the dykes represent the remnants of conduits of hydrothermal fluids from underlying volcanic and granitic rocks. However, this is impossible in the East Strelley greenstone belt where the Strelley Pool Formation is at least 80 million years younger than the granitic and volcanic rocks interpreted by Lindsay et al. (2003, 2005) to be the hydrothermal sources. The underlying granitic and volcanic rocks in the East Strelley greenstone belt were deformed, metamorphosed at amphibolite facies, and eroded prior to deposition of the Strelley Pool Formation. A hydrothermal origin for the stratigraphic succession of the Strelley Pool Formation is precluded by the fact that all its members were deposited as conglomerate, sandstone, carbonate rocks, evaporites, or volcaniclastic rocks. The rocks outwardly resemble chert only because of extensive secondary silicification. Wacey et al. (2010) noted that the black chert dykes in the East Strelley greenstone belt are too thin and limited in extent to be hydrothermal feeders for chert of the

Strelley Pool Formation, and this is also the situation in all other greenstone belts containing the formation.

The hydrothermal model was abandoned by Van Kranendonk et al. (2003) following the acquisition of geochemical data indicating marine deposition of sedimentary carbonate rocks. Chondrite normalized LREE depletion, elevated Y/Ho and positive La, Gd, and Er anomalies, all supported precipitation in the absence of hydrothermal fluids. The same conclusion was reached by Allwood et al. (2009) who reported on REE + Y patterns in both the carbonate and chert units of the Strelley Pool Formation and in chert of formations underlying the unconformity. They found that the 'stromatolite reef' carbonate facies in the lower part of the formation has REE + Y patterns similar to other Archean marine precipitates. However, they reported that chert in the volcaniclastic facies near the top of the formation shows patterns indicative of precipitation from hydrothermal and mixed marine hydrothermal fluids. This feature is consistent with the upper part of the Strelley Pool Formation, above the central unconformity in the Panorama greenstone belt, being deposited in a late hydrothermal environment. This suggests that this upper part of the formation is related to the overlying Euro Basalt.

4.6 Correlation with the Buck Reef Chert

As noted by Van Kranendonk et al. (2015), the same c. 75 million-year break in volcanic activity occurred in the Onverwacht Group of the eastern Kaapvaal Craton where the Buck Reef Chert separates the Hooggenoeg Formation (same age as the Mount Ada Basalt, Duffer Formation, and Panorama Formations) from komatiite and basalt of the Kromberg Formation (same age as the Euro Basalt). Comparison of the stratigraphic successions of the Pilbara Supergroup and Onverwacht Group (Fig. 1.10) illustrates these relationships. The Buck Reef Chert is lithologically and sedimentologically similar to the Strelley Pool Formation (De Vries 2004; Lowe and Byerly 2007; De Vries et al. 2010) and contains similar fossil evidence of early life (e.g. microbial mats, Tice and Lowe 2004; Duda et al. 2016; microfossils, Oehler et al. 2017). Unless these similarities are attributed to remarkable degrees of parallel crustal and biological evolution on separate Paleoarchean continents, they suggest that the Pilbara and Kaapvaal Cratons were parts of the same Paleoarchean–Paleoproterozoic continent, referred to by some workers as Vaalbara.

4.7 Fossil Record

The Strelley Pool Formation contains abundant evidence of Paleoarchean life, including stromatolites, microfossils, microbial mats, trace fossils, and kerogen. Stromatolite-like structures were first recorded at different localities by Lowe (1980) and Hickman (1980). In 1984, AF Trendall, then the Director of the

Fig. 4.4 Outcrop of chert of the Strelley Pool Formation on the east bank of the Shaw River, showing a bedding plane of conical 'egg carton' stromatolites exposed by the author and colleagues K Grey and HJ Hofmann in 1997. The largest cones are about 10 cm in diameter (From Geological Survey of Western Australia 1999; with Geological Survey of Western Australia permission)

Geological Survey of Western Australia, discovered stromatolite-like structures at another locality on the east bank of the Shaw River (Panorama greenstone belt), at a site now referred to as the 'Trendall Locality'. In 1990, Trendall, RI Thorpe, and Hickman revisited the site and interpreted some of the structures to be stromatolites. Thorpe sent photographs to an authority on stromatolites, HJ Hofmann, who in 1997 visited the site in company with GSWA paleontologist K Grey and Hickman. They identified numerous examples of stromatolite-like structures in partly silicified bedded carbonates, and also uncovered a 1-m-long bedding plane containing what has been become referred to as 'egg-carton' stromatolites (Figs. 4.4 and 4.5). Another outcrop of the formation on the ridge overlooking the discovery site exposes less regular forms of conical stromatolites (Fig. 4.6). Evidence for the biogenicity of the stromatolites was documented by Hofmann et al. (1999).

During a research project unrelated to fossils, 'stromatoloid structures' were discovered at another locality in the Coongan greenstone belt (Zegers et al. 1996). Geological mapping of the northern Pilbara Craton between 1994 and 2005 revealed stromatolite sites in five greenstone belts (Van Kranendonk 2000, 2001, 2007; Grey et al. 2002; Bagas et al. 2004); Smithies 2004; Williams and Bagas 2007).

Fig. 4.5 Close-up view of the stromatolites in Fig. 4.4 (Previously unpublished Geological Survey of Western Australia photograph; with Geological Survey of Western Australia permission)

Fig. 4.6 Natural exposure of conical stromatolites on a bedding plane in a cliff overlooking the outcrop in Fig. 4.4. The largest cones are about 10 cm in diameter (Previously unpublished photograph; with Geological Survey of Western Australia permission)

4.7.1 Stromatolites

Three of the best stromatolite sites became protected by State Geoheritage Reserves (Grey et al. 2010). Of these, the most documented site is the Trendall Locality

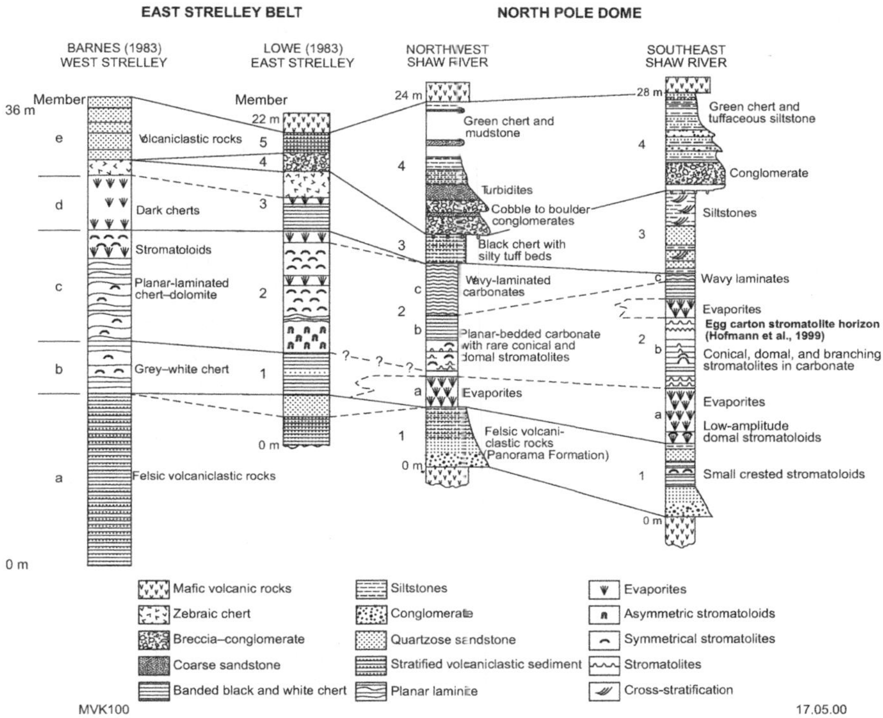

Fig. 4.7 Stratigraphic sections through the Strelley Pool Formation in the East Strelley and Panorama greenstone belts showing the positions of stromatolitic horizons (From Van Kranendonk 2000; with Geological Survey of Western Australia permission)

(Hofmann et al. 1999; Van Kranendonk 2000, 2007, 2011; Van Kranendonk et al. 2001, 2003; Allwood et al. 2004a, b, 2006, 2007a, b; Wacey et al. 2010; Hickman et al. 2011). Stratigraphic sections through the Strelley Pool Formation in the East Strelley greenstone belts and at the Trendall Locality in the North Pole Dome of the Panorama greenstone belt (Fig. 4.7) were provided by Van Kranendonk (2000).

A study of a 10-km strike length of the formation, north and south of the Trendall Locality (Fig. 4.8), provided 27 vertical stratigraphic sections that revealed a wide range of stromatolite morphologies and relationships to depositional environments (Allwood et al. 2007a). Stromatolite diversity and abundance were greatest in sedimentary facies lacking any evidence of hydrothermal, volcaniclastic, or terrigenous input.

Allwood et al. (2007a) subdivided the Strelley Pool Formation of the Shaw River area into four informal members, from base to top: Member 1, a basal unit of chert boulder conglomerate with carbonate matrix; Member 2, a laminated chert/carbonate (dolomite) unit with stromatolitic structures and beds of pseudomorphed evaporite crystals; Member 3, a unit of banded black and white chert with laminated iron-rich

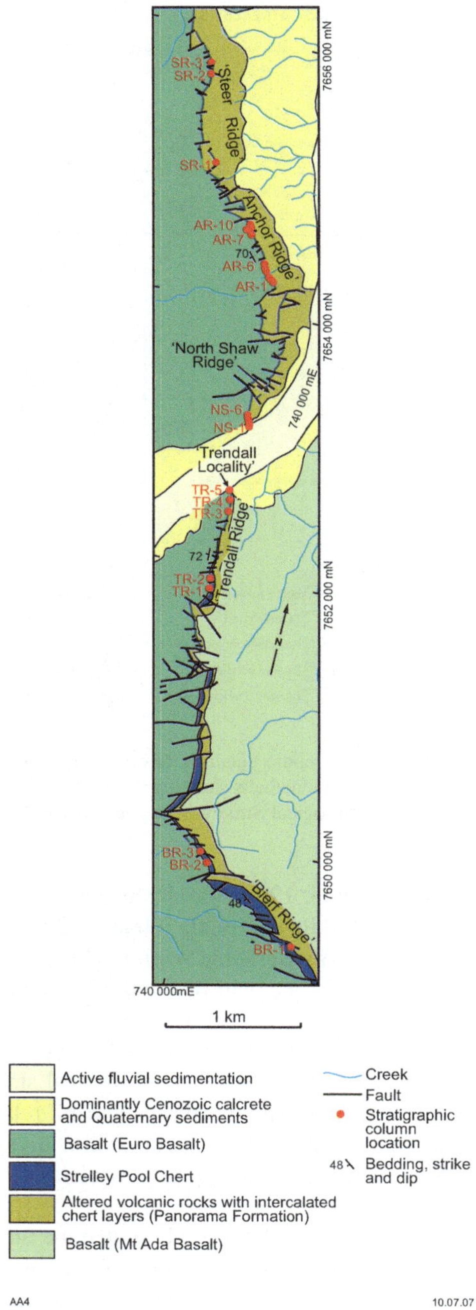

Fig. 4.8 Outcrops of the Strelley Pool Formation along ridges north and south of the Trendall Locality showing the locations of stratigraphic sections studied by A Allwood (From Allwood et al. 2007a; with Geological Survey of Western Australia permission)

stromatolitic structures; and Member 4, a clastic to volcaniclastic unit which fines upwards.

Member 1 is a laterally extensive but discontinuous chert boulder conglomerate interpreted as a high-energy, transgressive, and rocky coastline deposit (Allwood et al. 2007a). The conglomerate fills relict topography at the lower contact, including cliffs, fissures, and cavities. Bed thickness is highly variable and commonly reflects the size of the largest boulder in the bed. Bed thickness is also controlled by the topographic relief of the underlying Warrawoona Group.

Member 2 consists of stromatolitic carbonate and secondary chert interpreted as a peritidal carbonate platform succession. This member contains abundant and diverse stromatolitic facies, including large complex cones (Fig. 4.9), 'egg carton' stromatolites (Figs. 4.4 and 4.5), and a wavy laminite stromatolite facies (Allwood et al. 2007a).

Member 3 consists of banded black and white chert with up to 15% bundled iron-rich laminae and rare silicified chert-pebble conglomerates and crystal pseudomorph beds. The iron-rich laminae commonly form cm-scale wrinkles, bumps, domes, and pseudocolumns, with rare decimetre-scale cuspate domes.

Member 4 is a silicified, fining-upward, siliciclastic, and volcaniclastic succession that varies greatly in thickness across the study area. Stromatolites in this member include laminar mats and as small domical forms. These latter forms contain convincing morphological evidence for a biogenic origin, comparable to modern stromatolites.

4.7.1.1 Stromatolite Morphology

The main stromatolite morphology is stacked, generally slightly elliptical cones. The thickness of the laminae varies over the cones and in the troughs between the cones (Hofmann et al. 1999; Van Kranendonk 2007; Allwood et al. 2009; Hickman et al. 2011). The stromatolites initiate from points on a flat bedding plane, and develop structures that extend upwards for distances between a few centimetres and a metre. The conical stromatolites are up to 20 cm in diameter, and have steep lateral slopes that dip at angles between 70° and 80°, well in excess of the angle of repose for sedimentary grains. Individual bedding planes may contain many cones of similar dimension, or of varied sizes. Conical stromatolites at the Trendall Reserve are characteristically slightly elliptical in plan, probably due to a preferred current direction during their growth.

Most conical stromatolites throughout the fossil record contain an 'axial zone', as do the conical stromatolites of the Strelley Pool Formation (Hickman et al. 2011). Because axial zones were formed through biological activity (Walter et al. 1976), and not by physical and chemical processes, they further support the biological origin of the conical stromatolites. Some conical stromatolites show upwards branching and may form structures up to 35 cm in height. Others contain columnar branches off one side of the cone shape.

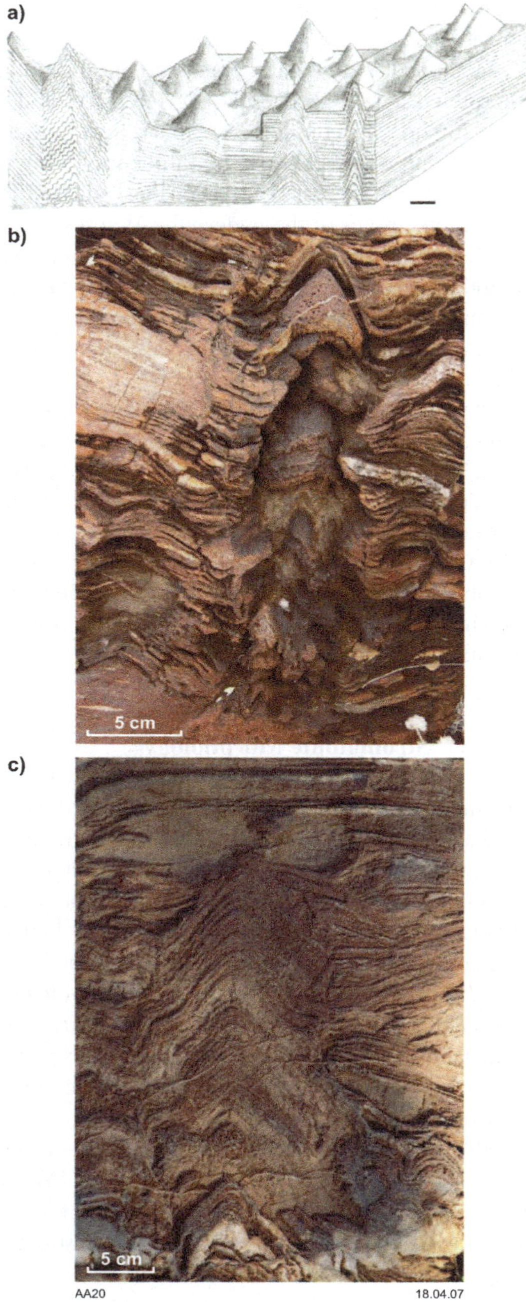

Fig. 4.9 Large conical stromatolites in the Strelley Pool Formation on the west side of the Panorama greenstone belt: (**a**) a 3D reconstruction of stromatolite morphology by AC Allwood (the scale bar is approximately 5 cm); (**b**) a conical stromatolite in outcrop at Trendall Ridge; (**c**) a conical stromatolite at North Shaw Ridge (From Allwood et al. 2007a; with Geological Survey of Western Australia permission)

4.7.2 Microfossils

The Strelley Pool Formation contains remarkably well-preserved microfossils near the Trendall Locality and in the Goldsworthy and Warralong greenstone belts (Sugitani et al. 2007, 2009, 2010, 2013, 2015; Van Kranendonk 2007; Alleon et al. 2018; Sugitani 2019). Morphologies include spheres and flattened spheres (Fig. 4.10), commonly linked into chains, and tubular structures, spindles, and film-like and thread-like structures. The sizes of the microfossils vary between 50 and 100 μm, although the spindles and spheres consistently have diameters of about 50 μm.

Alleon et al. (2018) applied Raman micro-spectroscopy, focused ion beam, scanning and transmission electron microscopy, and scanning transmission X-ray microscopy coupled with X-ray absorption near edge structure spectroscopy to study individual microfossils from the Strelley Pool Formation near the Trendall Locality. They confirmed the biogenicity of the structures and commented that they are even better preserved than the well-known c. 1900 Ma microfossils in the Gunflint Iron Formation of Canada. Spheroidal to ellipsoidal cell-like bodies and tubular sheath-like microstructures have been described on the carbonaceous margins of 10 μm diameter quartz grains, and in spaces between grains, in sandstone of the East Strelley greenstone belt (Wacey et al. 2011, 2012). Rather than being coatings around clastic quartz grains, the carbonaceous material is interpreted to represent the cell walls of microfossils, with the central areas of quartz having replaced the cell interiors.

4.7.3 Microbial Mats

Wavy laminated carbonate (or chert, where totally silicified) is a relatively common feature of the Strelley Pool Formation and might represent bacterial mats. Van Kranendonk and Nijman (2001) described ‘stromatolite mats’ in the upper clastic member of the Strelley Pool Formation. These mats are 2–5 mm thick layers of blue-black chert, in which a wavy lamination is defined by carbonaceous material and sand grains. Allwood et al. (2009) reported microscopic evidence suggesting microbial mat surfaces played a key role in the formation of stromatolites. In domical stromatolites, discrete layers of organic material at regular intervals through the structures. An absence of thickening of these layers in topographic lows was interpreted as evidence of adhesion of the organic material to the steep sides of the domes. Variable stromatolite morphology was interpreted to partly reflect differences in the proportions of microbial mat formation relative to other processes of deposition, and morphological diversity was therefore not necessarily evidence of biodiversity.

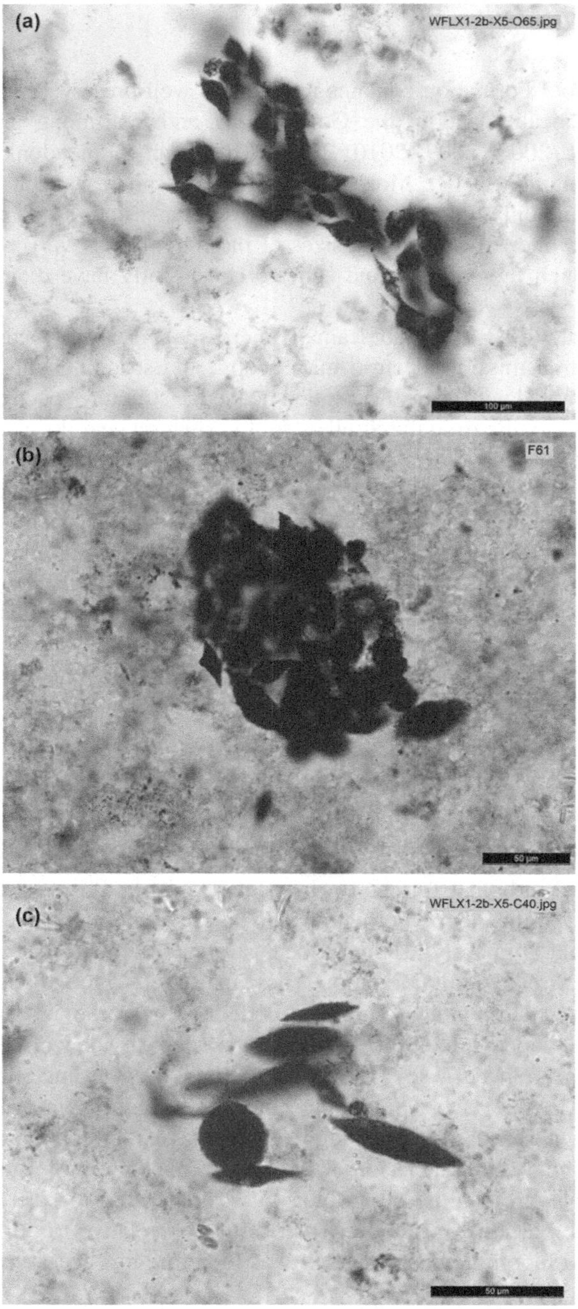

Fig. 4.10 Photomicrographs of carbonaceous microfossils in black chert of the Strelley Pool Formation, Goldsworthy greenstone belt: (**a**) a cluster of flanged lenticular structures, possibly originally connected in one or more chains; (**b**) similar to 'a' but forming a tighter cluster; c, polar and equatorial views of the microfossils (Previously unpublished photomicrography kindly provided by the discoverer of the microfossils, K Sugitani)

4.8 Significance to Crustal Evolution

The sedimentary facies of the Strelley Pool Formation, and its regional stratigraphic relations across 30,000 km^2 of the East Pilbara Terrane, establish that it was deposited above continental crust (Warrawoona Group, granitic rocks of the Callina and Tambina Supersuites, and older crust). This precludes a number of previous interpretations that the overlying Euro Basalt was deposited as deep-water oceanic crust (Isozaki et al. 1997; Kitajima et al. 2001; Furnes et al. 2007). Additionally, since the chemical composition of the Euro Basalt is very similar to that of basaltic formations of the underlying Warrawoona Group (Green et al. 2000; Smithies et al. 2007), various geochemical features previously used to interpret oceanic tectonic settings are evidently not applicable the Warrawoona Group. This conclusion is supported by Nd isotopic data from the Euro Basalt in which epsilon Nd (ε_{Nd}) values are approximately chondritic and Nd $T_{DM}{}^2$ model ages range between 3510 and 3470 Ma (Chap. 5). Since the theoretical depleted mantle ε_{Nd} at 3350–3315 Ma is approximately +2.8, the isotope data indicate melting of older crust and subcontinental lithospheric mantle.

4.9 Conclusions

1. Deposition of the Strelley Pool Formation between 3426 and 3350 Ma coincides with a c. 75-million-year break in volcanic activity in the Pilbara Supergroup (Van Kranendonk et al. 2007; Hickman 2008, 2012). A similar break in all magmatic activity across the EPT is indicated by U–Pb zircon dating of the Tambina Supersuite which is the intrusive equivalent of the Panorama Formation. Histogram plots of detrital zircon ages in Paleoarchean and Mesoarchean sedimentary rocks of the Pilbara Craton show major ‘peaks’ corresponding to the ages of the Warrawoona and Kelly Groups with a well-defined ‘trough’ between c. 3400 and c. 3350 Ma. Because zircon age spectra in sedimentary basins can provide a representative sampling of provenance regions (Taylor and McLennan 1995; Cawood et al. 2013), it can be inferred that the EPT does not contain 3400–3350 Ma igneous rocks. A contemporaneous break in volcanism and granitic intrusion occurred during deposition of the Onverwacht Group in the eastern Kaapvaal Craton (Van Kranendonk et al. 2015). In the Onverwacht Group, the Buck Reef Chert separates two formations (Hooggenoeg and Kromberg) with strong similarities to, respectively, the Panorama Formation and Euro Basalt. This suggests that the Strelley Pool—Buck Reef hiatus was significant on a continental scale, and supports correlations made in support of Vaalbara.
2. In the northwest EPT (Carlindi, Muccan, and North Pole Domes), the Strelley Pool Formation overlies a c. 3426-Ma angular uniformity formed by erosion of previously deformed and metamorphosed formations of the Warrawoona Group.

This establishes that by c. 3426 Ma the northwest EPT had a history of deformation, metamorphism, and deep erosion. In the central and southeast EPT, the unconformity is either a disconformity or a paraconformity. In most areas, it is not visibly angular and the Strelley Pool Formation always overlies the upper part of the Warrawoona Group, generally the Panorama Formation. Additionally, the Warrawoona Group is more metamorphosed in the northwest than in the southeast EPT. These regional variations indicate different structural and metamorphic histories across the terrane prior to 3426 Ma.

3. In some greenstone belts (most notably the East Strelley and Panorama greenstone belts), the Strelley Pool Formation contains a central unit of conglomerate (Lowe 1983; Van Kranendonk 2000; Allwood et al. 2006, 2007a, b; Wacey et al. 2010) providing evidence of an erosional unconformity within the formation (Hickman 2008). Conglomerate immediately overlying this unconformity includes clasts of consolidated sedimentary rock derived from erosion of the underlying formation. Above this conglomerate are volcaniclastic units that are mafic in composition suggesting initial stages in the eruption of the overlying Euro Basalt. Although this erosional unconformity may not be regional it reveals that deposition of the Strelley Pool Formation was protracted and included one or more periods of emergence and erosion.
4. Detrital zircon aged >3530 Ma from the Strelley Pool Formation suggest syn-depositional erosion of pre-Pilbara Supergroup felsic crust. This would imply depths of erosion exceeding 10 km (minimum stratigraphic thickness of the Warrawoona Group) although some of this erosion was probably related to uplift and erosion prior to 3450 Ma. The total depth of erosion, and exposure of pre-3530 Ma underlying felsic crust, supports a continental setting.
5. Sedimentological, paleontological, and geochemical features in many areas indicate shallow-water marine to subaerial depositional environments. Paleocurrent data are not recorded, although by analogy with the underlying Panorama Formation some of these environments were most likely marginal to partly eroded felsic volcanic piles near the centres of the EPT domes (DiMarco and Lowe 1989; Van Kranendonk 2000). Apparently, unfossiliferous banded cherts in some greenstone belts probably represent deposition in deeper water between the domes.
6. In the northwest EPT, the Strelley Pool Formation overlies a c. 3430-Ma paleoalterite (Allwood et al. 2007a) and it has been suggested that this unit, which is up to 100 m thick, is evidence of Archean weathering (Buick et al. 1995; Van Kranendonk and Pirajno 2004; Allwood et al. 2006, 2007a, b; Altinok and Ohmoto 2006; Ohmoto et al. 2006; Johnson et al. 2008, 2009). Geochemical data have been interpreted to support the existence of an oxygenated atmosphere one billion years before the Great Oxidation Event (Ohmoto et al. 2006; Johnson et al. 2008, 2009). Other workers have attributed the alteration to hydrothermal activity prior to deposition of the Strelley Pool Formation (Brown et al. 2006, 2011; Van Kranendonk 2014).
7. The formation contains some of Earth's most abundant and best-preserved evidence of early life, including stromatolites (Lowe 1980, 1983; Hofmann

et al. 1999; Van Kranendonk 2000, 2007; Grey et al. 2002, 2010, 2012; Allwood et al. 2004a, b, 2006, 2007a, b; Hickman et al. 2011; Bontognali et al. 2012), microbial mats (Allwood et al. 2009; Duda et al. 2016), microfossils (Sugitani et al. 2010, 2013, 2015; Alleon et al. 2018; Sugitani 2019), trace fossils (Wacey et al. 2011, 2012), and kerogen (Marshall et al. 2007). The abundance and diversity of the fossil record in the Strelley Pool Formation are attributed to widespread favourable depositional environments provided by the 75-million-year break in volcanic activity between eruption of the Warrawoona and Kelly Groups (Hickman 2012).

References

Alleon J, Bernard S, Le Guillou C, Beyssac O, Sugitani K, Robert F (2018) Chemical nature of the 3.4 Ga Strelley Pool microfossils. Geochemical Perspectives Letters 7:37–42. https://doi.org/10.7185/geochemlet.1817

Allwood AC, Burch I, Walter MR (2007a) Stratigraphy and facies of the 3.43 Ga Strelley Pool Chert in the southwestern North Pole Dome, Pilbara Craton, Western Australia. Record, 2007/11. Geological Survey of Western Australia

Allwood AC, Grotzinger JP, Knoll AH, Burch IW, Anderson MS, Coleman ML, Kanik I (2009) Controls on development and diversity of early Archean stromatolites. Proc Natl Acad Sci U S A 106:9548–9555

Allwood AC, Walter MR, Burch IW, Kamber BS (2007b) 3.43 billion-year-old stromatolite reef from the Pilbara Craton of Western Australia: ecosystem-scale insights to early life on earth. Precambrian Res 158:198–227

Allwood AC, Walter MR, Kamber BS, Marshall CP, Burch IW (2006) Stromatolite reef from the early Archaean era of Australia. Nature 441:714–717

Allwood AC, Walter MR, Marshall C, van Kranendonk M (2004a) Habit and habitat of earliest life on earth. Int J Astrobiol 3(S1):104

Allwood AC, Walter MR, Marshall C, Van Kranendonk MJ (2004b) Life at 3.4 Ga: paleobiology and paleoenvironment of the stromatolitic Strelley Pool Chert, Pilbara Craton, Western Australia. In: Session 197, T4, Precambrian geology, p 458

Altinok E, Ohmoto H (2006) Soil formation beneath the Earth's oldest known (3.46 Ga) unconformity? In: T105, Paleosols, Proxies, and Paleoenvironments II, p 533

Bagas L (2003) Stratigraphic revision of the Warrawoona and Gorge Creek Groups in the Kelly greenstone belt, Pilbara Craton, Western Australia. In: Geological Survey of Western Australia Annual Review 2001–02. Geological Survey of Western Australia, pp 53–60

Bagas L (2005) Geology of the Nullagine 1:100 000 sheet. 1:100 000 Geological series explanatory notes. Geological Survey of Western Australia

Bagas L, Van Kranendonk MJ, Pawley MJ (2004) Geology of the Split Rock 1:100 000 sheet. 1: 100 000 Geological Series Explanatory Notes. Geological Survey of Western Australia

Baker DEL, Collins WJ (2001) Structural history of the Northern East Strelley Belt, Pilbara Craton, Western Australia. In: Cassidy KF, Dunphy JM, Van Kranendonk MJ (eds) Extended Abstracts—4th International Archaean Symposium, Perth, 24–28 September 2001. AGSO (Geoscience Australia), pp 282–284

Baker DEL, Seccombe PK, Collins WJ (2002) Structural history and timing of gold mineralization in the Northern East Strelley Belt, Pilbara Craton, Western Australia. Econ Geol 97:775–785

Blewett RS, Champion DC (2005) Geology of the Wodgina 1:100 000 sheet. 1:100 000 Geological Series Explanatory Notes. Geological Survey of Western Australia

Bontognali TRR, Sessions AL, Allwood AC, Fischer WW, Grotzinger JP, Summons RE, Eiler JM (2012) Sulfur isotopes of organic matter preserved in 3.45-billion-year-old stromatolites reveal microbial metabolism. Proc Natl Acad Sci U S A 109:15146–15151

Brown AJ, Cudahy TJ, Walter MR (2006) Hydrothermal alteration at the Panorama Formation, North Pole Dome, Pilbara Craton, Western Australia. Precambrian Res 151:211–223

Brown A, Cudahy T, Walter MR, Hickman AH (2011) Hyperspectral alteration mapping of early Archean hydrothermal systems in the North Pole Dome, Pilbara Craton. Record, 2011/15. Geological Survey of Western Australia

Buick R, Thornett J, McNaughton N, Smith JB, Barley ME, Savage MD (1995) Record of emergent continental crust ~3.5 billion years ago in the Pilbara Craton of Australia. Nature 375:574–577

Cawood PA, Hawkesworth CJ, Dhuime B (2013) The continental record and the generation of continental crust. J Geol Soc 125:14–32. https://doi.org/10.1130/B30722.1

De Vries ST (2004) Early Archaean sedimentary basins: depositional environment and hydrothermal systems–examples from the Barbeton and Coppin Gap greenstone belts. Gologica Ultraiectina no. 244

De Vries ST, Nijman W, De Boer PL (2010) Sedimentary geology of the Palaeoarchaean Buck Ridge (South Africa) and Kittys Gap (Western Australia) volcano-sedimentary complexes. Precambrian Res 183:749–769

DiMarco MJ, Lowe DR (1989) Stratigraphy and sedimentology of an early Archean felsic volcanic sequence, Eastern Pilbara Block, Western Australia, with special reference to the Duffer Formation and implications for crustal evolution. Precambrian Res 44:147–169

Duda J-P, Van Kranendonk MJ, Thiel V, Ionescu D, Strauss H, Schafer N, Reitner N (2016) A rare glimpse of Paleoarchean life: geobiology of an exceptionally preserved microbial mat facies from the 3.4 Ga Strelley Pool Formation, Western Australia. PLoS One:11. https://doi.org/10.1371/journal.pone.0147629

Furnes H, Banerjee NR, Staudigel H, Muehlenbachs K, McLoughlin N, de Wit M, Van Kranendonk MJ (2007) Comparing petrographic signatures of bioalteration in recent to Mesoarchean pillow lavas: tracing subsurface life in oceanic igneous rocks. Precambrian Res 158:156–176

Gardiner NJ, Wacey D, Kirkland CL, Johnson TE, Jeon H (2019) Zircon U–Pb, Lu–Hf and O isotopes from the 3414 Ma Strelley Pool Formation, East Pilbara Terrane, and the Palaeoarchaean emergence of a cryptic cratonic core. Precambrian Res 321:64–84. https://doi.org/10.1016/j.precamres.2018.11.023

Geological Survey of Western Australia (1999) Geological Survey of Western Australia annual review 1998–99. Geological Survey of Western Australia, Frontispiece photograph, p ii

Green MG, Sylvester PJ, Buick R (2000) Growth and recycling of early Archaean continental crust: geochemical evidence from the Coonterunah and Warrawoona Groups, Pilbara Craton, Australia. Tectonophysics 322:69–88

Grey K, Clarke JDA, Hickman AH (2012) The proposed Dawn of Life geotourism trail, Marble Bar, Pilbara Craton, Western Australia – geology and evidence for early life. Record 2012/9. Geological Survey of Western Australia

Grey K, Hickman AH, Van Kranendonk MJ, Freeman MJ (2002) 3.45 billion year-old stromatolites in the Pilbara region of Western Australia: Proposals for site protection and public access. Record, 2002/17. Geological Survey of Western Australia

Grey K, Roberts FI, Freeman MJ, Hickman AH, Van Kranendonk MJ, Bevan AWR (2010) Management plan for state geoheritage reserves. Record, 2010/13. Geological Survey of Western Australia

Hao J, Sveriensky DA, Hazen RM (2017) A model for late Archean chemical weathering and world average river water. Earth Planet Sci Lett 457:191–203

Hickman AH (1980) Excursion guide–Archaean geology of the Pilbara Block: second International Archaean Symposium, Perth 1980. Geological Society of Australia, W.A. Division, Perth, Western Australia

Hickman AH (2008) Regional review of the 3426–3350 Ma Strelley Pool Formation, Pilbara Craton, Western Australia. Record, 2008/15. Geological Survey of Western Australia

Hickman AH (2012) Review of the Pilbara Craton and Fortescue Basin, Western Australia: crustal evolution providing environments for early life. Island Arc 21:1–31

Hickman AH (2021) East Pilbara Craton: a record of one billion years in the growth of Archean continental crust. Report 143, Geological Survey of Western Australia

Hickman AH, Van Kranendonk MJ, Grey K (2011) State Geoheritage Reserve R50149 (Trendall Reserve), North Pole, Pilbara Craton, Western Australia – geology and evidence for early Archean life. Geological Survey of Western Australia Record 2011/10

Hoashi M, Bevacqua DC, Otake T, Watanabe Y, Hickman AH, Utsunomiya S, Ohmoto H (2009) Primary hematite formation in an oxygenated sea 3.46 billion years ago. Nat Geosci 2:301–306

Hofmann HJ, Grey K, Hickman AH, Thorpe R (1999) Origin of 3.45 Ga coniform stromatolites in the Warrawoona Group, Western Australia. Geol Soc Am Bull 111:1256–1262

Isozaki Y, Kabashima T, Kitajima K, Maruyama S, Kato Y, Terabayashi M (1997) Early Archean mid-ocean ridge rocks and early life in the Pilbara Craton. Western Australia Eos 78:399

Johnson IM, Watanabe Y, Stewart B, Ohmoto H (2009) Earth's oldest (~3.4 Ga) lateritic paleosol in the Pilbara Craton, Western Australia. In: Challenges to Our Volatile Planet, A601

Johnson IM, Watanabe Y, Yamaguchi K, Hamasaki H, Ohmoto H (2008) Discovery of the oldest (~3.4 Ga) lateritic paleosols in the Pilbara Craton, Western Australia. In: Abstracts and Programs, p143

Kitajima K, Maruyama S, Utsunomiya S, Liou JG (2001) Seafloor hydrothermal alteration at an Archean mid-ocean ridge. J Metamorph Geol 19:583–599

Kloppenburg A (2003) Structural evolution of the Marble Bar Domain, Pilbara Granite-Greenstone Terrain, Australia: the role of Archaean mid-crustal detachments. PhD, Utrecht University

Lindsay JF, Brasier MD, McLoughlin N, Green OR, Fogel M, McNamara KM, Steele A, Mertzman SA (2003) Abiotic Earth–establishing a baseline for earliest Life, data from the Archean of Western Australia. In: Program and technical sessions 2003. Lunar and Planetary Institute, Abstract 1137

Lindsay JF, Brasier MD, McLoughlin N, Green OR, Fogel M, Steele A, Mertzman SA (2005) The problem of deep carbon—an Archaean paradox. Precambrian Res 143:1–22

Lowe DR (1980) Stromatolites 3,400-Myr old from the Archean of Western Australia. Nature 284: 441–443

Lowe DR (1983) Restricted shallow-water sedimentation of early Archean stromatolitic and evaporitic strata of the Strelley Pool Chert, Pilbara Block, Western Australia. Precambrian Res 19:239–283

Lowe DR, Byerly GR (2007) An overview of the geology of the Barberton Greenstone Belt and vicinity: implications for early crustal development. In: Van Kranendonk MJ, Bennett VC, Smithies RH (eds) Earth's oldest rocks. Elsevier B.V, Burlington, Massachusetts, USA, pp 481–526

Marshall CP, Love GD, Snape CE, Hill AC, Allwood AC, Walter MR, Van Kranendonk MJ, Bowden SA, Sylva SP, Summons RE (2007) Structural characterization of kerogen in 3.4 Ga Archaean cherts from the Pilbara Craton Western Australia. Precambrian Res 155:1–23

Nelson DR (1998) Compilation of SHRIMP U-Pb zircon geochronology data, 1997. Record 1998/ 2. Geological Survey of Western Australia

Nijman W, de Bruijne KCH, Valkering ME (1998) Growth fault control of early Archaean cherts, barite mounds and chert–barite veins, North Pole Dome, Eastern Pilbara, Western Australia. Precambrian Res 88:25–52

Nijman W, De Vries ST, Vos I, Louzada KL (2001) Synsedimentary collapse structures in the Warrawoona Group, Coppin Gap greenstone belt. In: Van Kranendonk MJ, Hickman AH, Williams IR, Nijman W (eds), Archean geology of the East Pilbara Granite–Greenstone Terrane, Western Australia–a field guide. Geological Survey of Western Australia, Record 2001/9

Oehler DZ, Walsh MM, Sugitani K, Liu M-C, House CH (2017) Large and robust lenticular microorganisms on the young earth. Precambrian Res 296:112–119

Ohmoto H, Watanabe Y, Allwood AC, Birch IW, Knauth P, Yamaguchi KE, Johnson I (2006) Discovery of probable >3.43 paleolaterites in the Pilbara Craton, Western Australia. In: Abstracts and Sessions, Abstract V24B-07

Pawley MJ, Van Kranendonk MJ, Collins WJ (2004) Interplay between deformation and magmatism during doming of the Archaean Shaw granitoid complex, Pilbara Craton, Western Australia. Precambrian Res 131:213–230

Retallack GJ (2018) The oldest known paleosol profiles on earth: 3.46 Ga Panorama Formation, Western Australia. Palaeogeogr Palaeoclimatol Palaeoecol 489:230–248

Smithies RH (2004) Geology of the De Grey and Pardoo 1:100 000 sheets. 1:100 000 Geological Series Explanatory Notes. Geological Survey of Western Australia

Smithies RH, Champion DC, Van Kranendonk MJ, Hickman AH (2007) Geochemistry of volcanic rocks of the Northern Pilbara Craton, Western Australia. Report, vol 104. Geological Survey of Western Australia

Sugitani K (2019) Early Archean (Pre-3.0 Ga) Cellularly preserved microfossils and microfossil-like structures from the Pilbara Craton, Western Australia – a review. In: Van Kranendonk MJ, Bennett VC, Hoffmann JE (eds) Earth's oldest rocks, 2nd edn. Elsevier B.V, pp 1007–1028

Sugitani K, Grey K, Allwood A, Nagaoka T, Mimura K, Minami M, Marshall CP, Van Kranendonk MJ, Walter MR (2007) Diverse microstructures from Archaean chert from the mount Goldsworthy–Mount Grant area, Pilbara Craton, Western Australia: microfossils, dubiofossils, or pseudofossils? Precambrian Res 158:228–262

Sugitani K, Grey K, Nagaoka T, Mimura K (2009) Three-dimensional morphological and textural complexity of Archean putative microfossils form the northeastern Pilbara Craton: indications of biogenicity of large (>15 μm) spheroidal and spindle-like structures. Astrobiology 9:603–615

Sugitani K, Lepot K, Mimura K, Van Kranendonk MJ, Oehler D, Walter MR (2010) Biogenicity of morphologically diverse carbonaceous microstructures from the ca. 3400 Ma Strelley Pool Formation, in the Pilbara Craton. Western Australia Astrobiology 10:899–920

Sugitani K, Mimura K, Nagaoka T, Lepot K, Takeuchi M (2013) Microfossil assemblage from the 3400 Ma Strelley Pool Formation in the Pilbara Craton, Western Australia: results from a new locality. Precambrian Res 226:59–74

Sugitani K, Mimura K, Takeuchi M, Lepot K, Ito S, Javaux EJ (2015) Early evolution of large micro-organisms with cytological complexity revealed by microanalyses of 3.4 Ga walled microfossils. Geobiology 13:507–521

Taylor SR, McLennan SM (1995) The geochemical evolution of the continental crust. Rev Geophys 33:241–265. https://doi.org/10.1029/95RG00262

Tice MM, Lowe DR (2004) Photosynthetic microbial mats in the 3,416-Myr-Old Ocean. Nature 43: 549–552

Van Kranendonk MJ (2001) Volcanic degassing, hydrothermal circulation and the flourishing of early life on Earth: new evidence from the c. 3.46 Ga Warrawoona Group, Pilbara Craton, Western Australia. The West Australian Geologist:3

Van Kranendonk MJ (2004) Geology of the Carlindie 1:100 000 sheet. 1:100 000 Geological Series Explanatory Notes. Geological Survey of Western Australia

Van Kranendonk MJ (2007) A review of the evidence for putative Paleoarchean life in the Pilbara Craton. In: Van Kranendonk MJ, Bennett VC, Smithies RH (eds) Earth's oldest rocks. Elsevier B.V., Burlington, Massachusetts, USA, pp855–896

Van Kranendonk MJ (2011) Stromatolite morphology as an indicator of biogenicity for Earth's oldest fossils from the 3.5 - 3.4 Ga Pilbara Craton, Western Australia. In: Reitner J, Queric N-V, Arp G (eds) Advances in Stromatolite Geobiology. Springer, pp 517–534

Van Kranendonk MJ (2014) Earth's early atmosphere and surface environments: a review. In: Shaw GH (ed) Earth's early atmosphere and surface environment. Geological Society of America, special paper, vol 504, pp 105–130. https://doi.org/10.1130/2014.2504(12)

Van Kranendonk MJ (2000) Geology of the North Shaw 1:100 000 sheet. 1:100 000 Geological Series Explanatory Notes. Geological Survey of Western Australia

Van Kranendonk MJ, Hickman AH, Smithies RH, Nelson DN, Pike G (2002) Geology and tectonic evolution of the Archaean North Pilbara Terrain, Pilbara Craton, Western Australia. Econ Geol 97:695–732. https://doi.org/10.2113/gsecongeo.97.4.695

Van Kranendonk MJ, Hickman AH, Smithies RH, Williams IR, Bagas L, Farrell TR (2006) Revised lithostratigraphy of Archean supracrustal and intrusive rocks in the Northern Pilbara Craton, Western Australia. Record, 2006/15. Geological Survey of Western Australia

Van Kranendonk MJ, Hickman AH, Williams IR, and Nijman, W. 2001, Archaean geology Van Kranendonk MJ, Hickman AH, Williams IR, Nijman W (2001) Archaean geology of the East Pilbara Granite–Greenstone Terrane, Western Australia–a field guide. Record, 2001/9. Geological Survey of Western Australia

Van Kranendonk MJ, Nijman W (2001) Geology of the North Pole Dome. In: Van Kranendonk MJ, Hickman AH, Williams IR, Nijman W (eds), Archaean geology of the East Pilbara Granite–Greenstone Terrane, Western Australia–a field guide. Geological Survey of Western Australia, Record 2001/9, pp. 53–72

Van Kranendonk MJ, Pirajno F (2004) Geochemistry of metabasalts and hydrothermal alteration zones associated with c. 3.45 Ga chert and barite deposits: implications for the geological setting of the Warrawoona Group, Pilbara Craton, Australia. Geochemistry: exploration. Environment, Analysis 4:253–278

Van Kranendonk MJ, Smithies RH, Hickman AH, Bagas L, Williams IR, Farrell TR (2004) Event stratigraphy applied to 700 million years of Archean crustal evolution, Pilbara Craton, Western Australia. Geological Survey of Western Australia Annual Review 2003–04:49–61

Van Kranendonk MJ, Smithies RH, Hickman AH, Champion DC (2007) Review: secular tectonic evolution of Archean continental crust: interplay between horizontal and vertical processes in the formation of the Pilbara Craton, Australia. Terra Nova 19:1–38

Van Kranendonk MJ, Webb GE, Kamber BS (2003) Geological and trace element evidence for a marine sedimentary environment of deposition and biogenicity of 3.45 Ga stromatolitic carbonates in the Pilbara Craton, and support for a reducing Archean Ocean. Geobiology 1:91–108

Van Kranendonk MJ, Smithies RH, Griffin WL, Huston DL, Hickman AH, Champion DC, Anhaeusser CR, Pirajno F (2015) Making it thick: a volcanic plateau origin of Palaeoarchean continental lithosphere of the Pilbara and Kaapvaal cratons. Geol Soc Lond Spec Publ 389 (1):83–111. https://doi.org/10.1144/SP389.12

Wacey D, Kilburn MR, Saunders M, Cliff J, Brasier MD (2011) Microfossils of Sulphur-metabolizing cells in 3.4-billion-year-old rocks of Western Australia. Nat Geosci 4:698–702

Wacey D, McLoughlin N, Stoakes CA, Kilburn MR, Green OR, Brasier MD (2010) The 3426–3350 Ma Strelley Pool Formation in the East Strelley greenstone belt–a field and petrographic guide. Record, 2010/10. Geological Survey of Western Australia

Wacey D, Menon S, Green L, Gerstmann D, Kong C, McLoughlin N, Saunders M, Brasier M (2012) Taphonomy of very ancient microfossils from the ~3400 Ma Strelley Pool Formation and ~1900 Ma Gunflint Formation: new insights using a focussed ion beam. Precambrian Res 220–221:234–250

Walker JCG (1978) Early history of oxygen and ozone in the atmosphere. Pure Appl Geophys 117: 498–512

Walter MR, Bauld J, Brock TD (1976) Microbiology and morphogenesis of columnar stromatolites (*Conophyton*, *Vacerrilla*) from hot springs in Yellowstone National Park. In: Walter MR (ed), Stromatolites, developments in sedimentology. 20 Elsevier, Amsterdam, The Netherlands, pp. 273–310

Wiemer D, Schrank CE, Murphy DT, Hickman AH (2016) Lithostratigraphy and structure of the early Archaean Doolena Gap greenstone belt, East Pilbara Terrane, Western Australia. Precambrian Res 282:121–138

Wiemer D, Schrank CE, Murphy DT, Wenham L, Allen CM (2018) Earth's oldest stable crust in the Pilbara Craton formed by cyclic gravitational overturns. Nat Geosci 11:357–361. https://doi.org/10.1038/s41561-018-0105-9

Williams IR, Bagas L (2007) Geology of the Mount Edgar 1:100 000 sheet. 1:100 000 Geological Series Explanatory Notes. Geological Survey of Western Australia
Wingate MTD, Bodorkos S, Van Kranendonk MJ (2009) 180095: felsic volcaniclastic metasandstone, Pioneer Mine
Wingate MTD, Kirkland CL, Bodorkos S, Van Kranendonk MJ, Hickman AH (2012) 160958: volcaniclastic metasandstone, Trendall Reserve
Zegers TE, White SH, de Keijzer M, Dirks P (1996) Extensional structures during deposition of the 3460 Ma Warrawoona Group in the Eastern Pilbara Craton, Western Australia. Precambrian Res 80:89–105

Chapter 5
Kelly Large Igneous Province, 3350–3315 Ma

Abstract Unconformably overlying thick continental crust, the Kelly Group comprises three formations, in ascending stratigraphic order: the 3350–3335 Ma Euro Basalt, up to 9 km thick, and composed of komatiite, basaltic komatiite, and tholeiite; the 3325–3315 Ma Wyman Formation, up to 2 km thick, and composed of rhyolite flows and subvolcanic rhyolite intrusions; and the undated Charteris Basalt, up to 2 km thick, and containing komatiite, basaltic komatiite, and tholeiite. With an average stratigraphic thickness of 4 km, and erupted across at least 100,000 km^2 of the Pilbara Craton, the Euro Basalt forms the main part of a Paleoarchean large igneous province, the Kelly LIP.

The plume-related ultramafic–mafic–felsic volcanic cycle that commenced with eruption of the Euro Basalt ended with eruption of the felsic volcanics of the Wyman Formation. However, unlike the Euro Basalt the Wyman Formation is restricted to the eastern half of the East Pilbara Terrane and was derived from partial melting of older felsic crust. Eruption of the Wyman Formation was accompanied by numerous granodiorite and monzogranite intrusions of the Emu Pool Supersuite. Geochronology indicates that in some of the East Pilbara domes there was a ten-million-year interval between eruption of the Euro Basalt and Wyman Formation, during which time some parts of the Euro Basalt were folded and eroded. The undated Charteris Basalt is lithologically similar to the Euro Basalt and might form part of a second volcanic cycle.

Keywords Euro Basalt · Komatiite · Mantle plume · Volcanic cycle · Continental crust

5.1 Introduction

The 3350–3315 Ma Kelly Group is a continental succession of komatiite, komatiitic basalt, tholeiite, and felsic volcanics up to 10 km thick. The group disconformably overlies the sedimentary succession of the 3426–3350 Ma Strelley Pool Formation (Chap. 4), which in turn unconformably overlies the 3530–3427 Ma Warrawoona Group and contemporaneous granitic intrusions (Chap. 3). The stratigraphy of the

A. H. Hickman, *Archean Evolution of the Pilbara Craton and Fortescue Basin*, Modern Approaches in Solid Earth Sciences 24,
https://doi.org/10.1007/978-3-031-18007-1_5

group is assigned to three formations, in ascending stratigraphic order: the 3350–3335 Ma Euro Basalt, up to 9 km thick, and composed of komatiite, basaltic komatiite, and tholeiite; the 3325–3315 Ma Wyman Formation, up to 2 km thick, and comprising rhyolite flows and subvolcanic rhyolite intrusions, rhyolitic volcaniclastic rocks, and minor sandstone, shale and chert; and the 3325–3315 Ma Charteris Basalt, up to 2 km thick, and composed of komatiite, basaltic komatiite, and tholeiite. The stratigraphic position of the Kelly Group within the Pilbara Supergroup is illustrated in Fig. 3.1 and Table 3.1.

The group was erupted during the 3350–3290 Ma evolution of a mantle plume in the East Pilbara Terrane (EPT). As in several previous mantle plume events to affect this terrane, magmatic activity commenced with the widespread eruption of komatiite, komatiitic basalt, and tholeiite. This ultramafic to mafic succession is assigned to the Euro Basalt which is exposed in 16 of the 20 greenstone belts in the EPT, and is interpreted to be concealed by younger formations in the remaining four. The exposed area of the EPT is only 40,000 km^2, but allowing for Neoarchean and Proterozoic cover, the total area of the terrane is about 100,000 km^2. At 3350 Ma, the terrane was very much larger than today because it included at least two sections of crust that were later separated as the Karratha Terrane (KT) and Kurrana Terrane (KUT) during the 3280–3165 Ma East Pilbara Terrane Rifting Event (Chap. 6). Still more parts of the EPT were removed as a result of the Neoarchean Pilbara Craton breakup between 2775 and 2501 Ma (Chap. 12).

5.1.1 Tectonic Setting

During eruption of the Kelly Group, the Pilbara Craton was composed of two crustal layers: lower crust, the 3800–3530 Ma terrane of mafic and felsic crust (Chap. 2); and upper crust, the 10–15 km thick, mafic–felsic volcanic Warrawoona Group and contemporaneous granitic intrusions (Chap. 3). The thickness of the 3800–3530 Ma lower crust has been estimated at between 30 and 45 km (Green et al. 2000; Van Kranendonk 2000a; Smithies et al. 2009; Thébaud and Rey 2013; François et al. 2014; Johnson et al. 2017; Wiemer et al. 2018). Based on numerical models for the evolution of diapirs (Mareschal and West 1980), and the dimensions of the dome-and-keel crustal architecture of the EPT, and metamorphic mineral assemblages that indicate pressures of about 6.5 kbar, Van Kranendonk (2000a) interpreted the total thickness of the Paleoarchean Pilbara crust (including the Warrawoona Group) to have been between 44 and 55 km. Later analysis of garnets in high-grade metamorphic rocks of the Mount Edgar Dome indicated Paleoarchean pressures between 6 and 11 kb and temperatures between 450 and 750 °C (François et al. 2014), consistent with a 30-km-thick crust. Both the Warrawoona Group and the underlying 3800–3530 Ma lower crust were intruded by voluminous Paleoarchean granitic supersuites, and both were deformed, metamorphosed, and eroded during several Paleoarchean tectonomagmatic events (Chap. 3). Therefore, there are many lines of evidence that the Kelly Group was deposited on thick (up to 45 km) 3800–3350 Ma continental crust.

By c. 3350 Ma, when eruption of the Kelly Group commenced, large areas of the pre-3426 Ma granite–greenstone domes of the EPT had been deeply eroded and overlain by the thin sedimentary succession of the Strelley Pool Formation (Chap. 4). The eroded surface of the terrane was relatively flat over the greenstone belts, and extended at least some way onto the granitic cores of the domes.

5.1.2 Kelly Large Igneous Province

The preserved thickness of the Kelly Group varies between 5 and 10 km, by far the thickest part of the succession being composed of the Euro Basalt. Allowing for concealment of the EPT by Neoarchean cover, the original depositional extent of the Euro Basalt is interpreted to have exceeded 100,000 km^2. With an estimated total volume of volcanic rocks exceeding 500,000 km^3, the Euro Basalt easily meets the 100,000 km^3 volume requirement for a large igneous province (LIP) (Bryan and Ernst 2008; Ernst 2014).

Unlike the Euro Basalt, the Wyman Formation is restricted to the eastern half of the EPT. Additionally, in most greenstone belts of this part of the EPT there is a compositional break between upper tholeiites of the Euro Basalt and the mainly rhyolitic succession of the Wyman Formation. On the interpretation that the Kelly Group was erupted during a single mantle plume event, parts of the succession linking the Euro Basalt to the Wyman Formation have not been identified, most likely due to erosion or incomplete exposure.

If the Pilbara and Kaapvaal Cratons evolved as adjacent parts of Vaalbara (Cheney et al. 1988; Cheney 1996; Zegers et al. 1998; Bleeker 2003; De Kock et al. 2009; Hickman 2012, 2016, 2021), there is a case to correlate the Euro Basalt with the Kromberg Formation of the Onverwacht Group. In this scenario, the original extent of the Euro Basalt/Kromberg Formation would have been at least 1,000,000 km^2. Eruption of the formation was accompanied by the intrusion of swarms of dolerite dykes into the underlying crust; such dykes intrude the Panorama Formation and Tambina Supersuite on the eastern, northern, and northwestern sides of the Corunna Downs Dome (Kloppenburg 2003; Hickman and Van Kranendonk 2008b; Hickman 2012, 2021) and ultramafic dykes intrude the Mount Ada Basalt and Duffer Formation beneath komatiite of the Euro Basalt in the southern Coongan greenstone belt between Withnell Creek and Triberton Creek (Bagas et al. 2004). Several large mafic intrusions into the Panorama Formation of the McPhee Dome are also now interpreted to have been feeders to the overlying Euro Basalt. Elsewhere in the EPT, the Euro Basalt commonly overlies thick mafic successions such as those of the Apex and Mount Ada Basalts in which younger dolerite dykes are difficult to identify.

5.2 Stratigraphy

The Kelly Group disconformably overlies the 3426–3350 Ma Strelley Pool Formation in most greenstone belts of the EPT (Fig. 4.1). Where the Strelley Pool Formation is absent, the Kelly Group overlies the Warrawoona Group across an erosional unconformity. The upper stratigraphic contact of the Kelly Group is an erosional unconformity, which in different areas underlies either the Sulphur Springs, Soanesville, or Gorge Creek Groups. These angular unconformities mostly resulted from deformation and metamorphism of the Kelly Group during the 3324–3290 Ma Emu Pool Event, followed by various periods of erosion.

5.2.1 Euro Basalt

The Euro Basalt is the lower formation of the Kelly Group (Van Kranendonk et al. 2006), and is exposed in 16 of the 20 greenstone belts in the EPT. In most areas, the formation is between 3 and 6 km thick but in the southwest Panorama greenstone belt (Fig. 1.7) it reaches a maximum thickness of 9.4 km (Van Kranendonk 2000b). The formation comprises successions dominated by spinifex-textured komatiitic basalt and pillowed tholeiite interlayered with thin units of grey and white banded chert. In the eastern part of the Kelly greenstone belt, basaltic fragmental pyroclastic rocks (Fig. 5.1) immediately overlie the Strelley Pool Formation. At other localities, fragmental basaltic rocks also form part of the upper Strelley Pool Formation (Chap. 4). However, a more typical feature of the basal part of the Euro Basalt is its content of komatiite (Figs. 5.2 and 5.3) and komatiitic basalt (Glikson and Hickman 1981a, b; Van Kranendonk et al. 2002, 2006, 2007a, b; Smithies et al. 2007). Five samples of a basal komatiite unit at Camel Creek (21 km south of Marble Bar) collected through a 400-m-thick vertical stratigraphic section, returned the following average composition: SiO_2, 42.36%; TiO_2, 0.18%; Al_2O_3, 4.10%, FeO (tot), 8.73%; MnO, 0.14%; MgO, 31.56%, CaO, 4.06%, Na_2O, 0.27%; K_2O, 0.02%; P_2O_5, 0.03%; $CO_2 + H_2O$, 8.55%, Cr, 1322 ppm; and Ni, 1435 ppm.

Komatiite lavas (>18% MgO) are interpreted to have been derived by high-temperature partial melting of the mantle (Campbell et al. 1989; Campbell and Griffiths 1990; Nebel et al. 2014; Sossi et al. 2016). Sm–Nd isotope data suggest that komatiite magmas were derived from the lower mantle (Nebel et al. 2014). Komatiitic basalt (>8% MgO), characterized by pyroxene spinifex texture (Fig. 5.4), is either pillowed or massive, and is composed of tremolite, albite, chlorite, epidote, and quartz, with clinopyroxene that is partly altered to chlorite and epidote. Ocelli texture (Fig. 5.5) is a feature of some komatiitic basalts and commonly rims pillow structures. Good examples of feathery spinifex texture (after pyroxene) and platy spinifex texture (after olivine) are exposed along Sandy Creek on the northeast side of the Corunna Downs Dome.

Fig. 5.1 Basaltic agglomerate and tuff at the stratigraphic base of the Euro Basalt in the Kelly greenstone belt: (**a**) blocks of basaltic breccia from a unit immediately overlying chert of the Strelley Pool Formation (scale shown by hammer); (**b**) close-up view of angular fragmental textures in one of the blocks (MGA Zone 51, 196600E 7622950N) (From Grey et al. 2012; with Geological Survey of Western Australia permission)

The upper part of the Euro Basalt is composed of tholeiitic basalt and thin units of chert, as is typical of other basaltic formations of the Pilbara Supergroup. As illustrated in Fig. 5.6, the primitive mantle-normalized trace element patterns for

Fig. 5.2 Olivine spinifex texture close to the top of a komatiite flow in the Euro Basalt at Coppin Gap. The rock is extensively carbonated. (MGA Zone 50, 200260E 7687830 N) (Modified from Van Kranendonk 2010b; with Geological Survey of Western Australia permission)

high-Ti tholeiites from the Euro Basalt are almost identical to those of high-Ti tholeiites in the Coonterunah Subgroup, Mount Ada Basalt, and Apex Basalt (Smithies et al. 2007). The similarity to the geochemistry of the Table Top Formation at the base of the Pilbara Supergroup is significant to the depositional environment of that formation because the Euro Basalt is known to have been deposited on continental crust up to 45 km thick.

Based on the presence of erosional surfaces and conglomerate within the upper part of the underlying Strelley Pool Formation (Chap. 4), much of the Euro Basalt was initially erupted close to sea level. However, the abnormal thickness of the formation combined with its content of pillow basalts (Fig. 5.7) establishes that during the 15-million-year period of eruption the basalts were deposited in one or more

Fig. 5.3 Partly silicified komatiite of the Euro Basalt, showing platy olivine spinifex texture, Mount Elsie greenstone belt (MGA Zone 51, 252360E 7610850N) (From Farrell 2006; with Geological Survey of Western Australia permission)

Fig. 5.4 Outcrop of coarse pyroxene spinifex texture in weakly metamorphosed komatiitic basalt in the Euro Basalt (MGA Zone 50, 798290E 7683455N) (Modified from Van Kranendonk 2010a; with Geological Survey of Western Australia permission)

rapidly subsiding depositional basins. Deep subsidence of ultramafic–mafic volcanic basins between rising granitic domes is a key feature in the diapiric model explaining the development of the dome–and–keel crustal architecture of the EPT (Hickman and Van Kranendonk 2004). Periodic shallowing of the basins between 3350 and

Fig. 5.5 Ocelli in komatiitic basalt of the Euro Basalt in the northwestern part of the McPhee greenstone belt (MGA Zone 51205250E 7612030N) (From Bagas 2005; with Geological Survey of Western Australia permission)

3335 Ma is suggested by the local presence of thin clastic sedimentary units. Mafic and ultramafic dykes through granites and felsic volcanics underlying the Euro Basalt suggest that subaerial eruption was included because the granites are likely to have formed upland areas. However, any Euro Basalt deposited on the granitic cores of domes would soon have been eroded during and following later doming. These considerations suggest that all of the presently exposed parts of the Euro Basalt were deposited in basins between the domes.

5.2.2 Wyman Formation

The 3325–3315 Ma Wyman Formation comprises weakly metamorphosed rhyolite flows, subvolcanic rhyolite intrusions, and felsic volcaniclastic rocks including agglomerate and tuff, and minor sandstone, shale, and chert. Clastic sedimentary rocks typically form units at the base of the formation, and in some areas include conglomerate. The formation is locally interlayered with pillowed komatiitic basalt similar to that in the overlying Charteris Basalt. The top of the formation includes stromatolitic carbonate rocks, black shale, and beds of barite up to 1 m thick (Williams and Bagas 2007).

An unusual feature of the Wyman Formation is the widespread columnar jointed porphyritic rhyolite (Fig. 5.8), either forming thick flows or high-level intrusions.

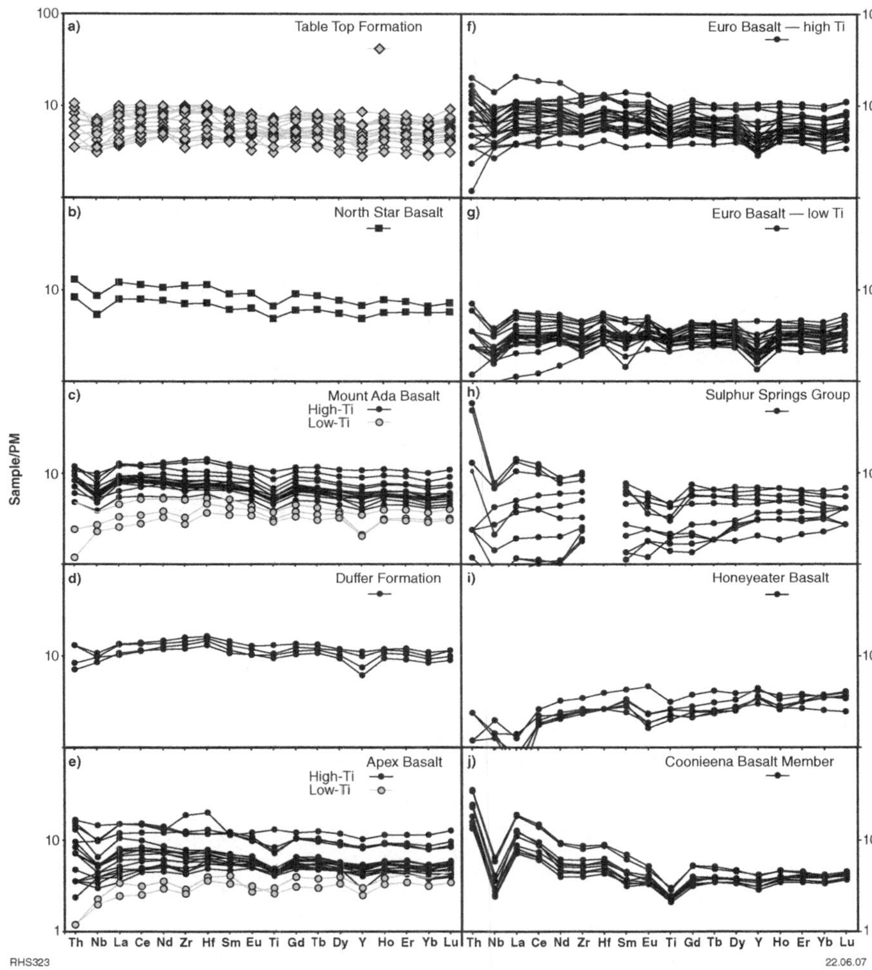

Fig. 5.6 Trace element plots normalized to primitive mantle for basalts of the Pilbara Supergroup, Honeyeater Basalt, and Coonieena Basalt (From Smithies et al. 2007; with Geological Survey of Western Australia permission)

Rhyolite and felsic volcaniclastic units commonly attain combined maximum thicknesses of between 1 and 2 km. In the Warralong greenstone belt, the formation is composed of clastic and basaltic rocks; correlation with the Wyman Formation is supported by a date of 3324 ± 4 Ma (GSWA 168992, Nelson 2004). This succession is located at the northwest limit of the formation's distribution in the EPT, and the clastic sedimentary rocks were probably deposited in outwash fans derived from the felsic volcanic complexes farther southeast.

The felsic volcanic units of the 3325–3315 Ma Wyman Formation are genetically related to monzogranite and granodiorite intrusions of the 3325–3290 Ma Emu Pool

Fig. 5.7 Pillow structures in the Euro Basalt in the northwestern part of the McPhee greenstone belt (MGA Zone 51202200E 7609100N) (From Bagas 2005; with Geological Survey of Western Australia permission)

Fig. 5.8 Columnar rhyolite in the Wyman Formation: (**a**) Camel Creek, Kelly greenstone belt (MGA 787454E, 7640098N); (**b**) northern McPhee greenstone belt (MGA 222200E, 7619200N) (From Hickman 2021; with Geological Survey of Western Australia permission)

Supersuite. Unlike older Paleoarchean granitic supersuites of the EPT, most intrusions of this supersuite were emplaced high into the Warrawoona Group and locally intrude the Kelly Group. Subvolcanic intrusions linking the granites to the volcanics are exposed in the Corunna Downs, Mount Edgar, and McPhee Domes.

5.2.3 Charteris Basalt

The Charteris Basalt is the uppermost formation of the Kelly Group, between 1 and 2 km thick, and predominantly composed of komatiitic basalt and tholeiite. The formation also includes fine-grained felsic volcaniclastic units, local komatiite, thin units of blue-black, grey-, and white-banded chert, and dolerite sills. The Charteris Basalt conformably overlies the Wyman Formation and is partly interlayered with it. However, in the type area at Charteris Creek, there is no evidence of lateral equivalence. Being at the top of the Kelly Group, the formation was extensively eroded during the 3325–3290 Ma Emu Pool Event. As a consequence, the Charteris Basalt, along with the other formations of the underlying Kelly Group, is unconformably overlain by different groups in different areas.

5.2.4 Unconformities within the Kelly Group

Early stratigraphic interpretations of the greenstone succession of the EPT described local unconformities between the Euro Basalt and the Wyman Formation (Lipple 1975; Hickman and Lipple 1978; Hickman 1981, 1983, 1990; Barley and Pickard 1999). Due to these unconformities, Hickman (1990) excluded the Wyman Formation from the Warrawoona Group, which at that time included the Euro Basalt. However, based on mapping of the southern Kelly greenstone belt, Bagas (2003) interpreted the Wyman Formation to be conformable on the Euro Basalt, and this relationship was adopted as part of a regional stratigraphic revision of the Northern Pilbara Craton (Van Kranendonk et al. 2004c, 2006). Even so, an angular unconformity separates the Euro Basalt from the Wyman Formation in the Warralong greenstone belt (Van Kranendonk 2004a), and unconformable relations exist in the McPhee greenstone belt (Barley and Pickard 1999) and in some sections of the Kelly greenstone belt (Hickman and Lipple 1978). Knowledge that the eleven granite–greenstone domes of the East Pilbara Terrane evolved at different rates (Hickman 1984, 2021; Van Kranendonk 1998; Hickman and Van Kranendonk 2004) explains different stratigraphic relations in the greenstone belts (Hickman 2021). Differing rates of dome uplift are illustrated by the fact that two of the terrane's eleven domes are greenstone domes with no exposure to a large central granitic complex. In the cases of these domes (North Pole and McPhee), there was an insufficient central granitic intrusion to cause the domes to rise as much as in adjacent domes.

5.3 Komatiite and Komatiitic Basalt in the Kelly Group

Volcanism in the Pilbara Supergroup is attributed to mantle plumes (Collins et al. 1998; Van Kranendonk et al. 2002, 2004a, b, 2006, 2007a, b; 2010b, 2015; Hickman 2004, 2011; Hickman 2021; Hickman and Van Kranendonk 2004; Smithies et al. 2005). The stratigraphic succession of the supergroup indicates up to eight ultramafic–mafic–felsic volcanic cycles (Hickman 2011, 2012), two of which might be represented in the Kelly Group (Fig. 3.1). The record is incomplete due to erosion of the upper parts of some cycles, as applies to the Charteris Basalt cycle. The lower sections of volcanic cycles typically include komatiite derived by high-temperature partial melting of the mantle (Campbell et al. 1989; Campbell and Griffiths 1990; Nebel et al. 2014; Sossi et al. 2016). Low viscosity, high-temperature komatiite lavas are thought to have been erupted above the axial regions of plumes. Models relating to the structure and evolution of mantle plumes suggest that an upward and lateral compositional changes from komatiites to basalts reflect melting under different temperatures and pressures in different sections of the plume (Campbell et al. 1989; Campbell and Griffiths 1990). Following the initial eruption of komatiites, mantle cooling led to the eruption of komatiitic basalts, followed by tholeiitic basalts.

Many workers have commented that komatiites are mainly confined to Archean successions, and attributed this to higher Archean mantle temperatures. In the Pilbara, komatiites and komatiitic basalts were erupted after periods of little or no volcanic activity, suggesting that in the absence of volcanism temperatures in the mantle might have risen due to radioactive decay. If so, it is likely that the especially long 75 million-year break in volcanism between 3426 and 3350 Ma (Chap. 4), and the existence of thick continental crust (Warrawoona Group, Paleoarchean granites, and pre-3530 Ma crust) acting as an insulating cap, contributed to the evolution of the Kelly plume and to the eruption of locally thick komatiite units.

Komatiites or komatiitic basalts are present in the lower stratigraphy of the Euro Basalt in the Kelly, Marble Bar, Coongan, Panorama, East Strelley, North Shaw, Western Shaw, Tambina, Emerald Mine, Doolena Gap, Warralong, McPhee, Mount Elsie, and Pilbara Well greenstone belts. Many komatiite units are extensively altered and now represented by talc–carbonate–chlorite–serpentine assemblages; in these cases it can be difficult to distinguish extrusive and intrusive lithologies. However, less altered komatiites, locally displaying bladed olivine spinifex textures (Fig. 5.4), are exposed at several localities, for example, (1) the northern Kelly greenstone belt at Camel Creek, 21 km south-southeast of Marble Bar (Glikson and Hickman 1981a, b; Smithies et al. 2007; Hickman and Van Kranendonk 2008b); (2) the northeast Marble Bar greenstone belt at Coppin Gap and Spinifex Ridge (Williams 1999; Van Kranendonk et al. 2006; Smithies et al. 2007) and in the Bamboo Creek mining area (Zegers et al. 2002; Williams 2003); (3) the northern Coongan greenstone belt east of Glen Herring Creek (Hickman and Van Kranendonk 2008b); (4) the southern Coongan greenstone belt 27 km east of Hillside Homestead (Bagas et al. 2004); (5) southeast of Mount Elsie in the Mount

Elsie greenstone belt (Farrell 2006); (6) the eastern Kelly greenstone belt at the Dawn of Life Trail near Gallop Well (Grey et al. 2012); and (7) in the Doolena Gap greenstone belt 700 m East of Pear Creek and 7 km South of the Marble Bar–Port Hedland Road (Van Kranendonk 2010a).

Based on near-chondritic Gd/Yb ratios and Al_2O_3/TiO_2 ratios between 20 and 24, Smithies et al. (2007) classified komatiites and komatiitic basalts in the northeast Marble Bar and Kelly greenstone belts as Al-undepleted. They noted that the komatiites of the Euro Basalt are geochemically similar to those of the Table Top Formation at the base of the Coonterunah Subgroup (Chap. 2).

Komatiitic basalt (8–18 wt. % MgO) overlies, and is locally intercalated with, komatiite flows in all the greenstone belts that contain komatiite. This type of high-Mg basalt, which is characterized by pyroxene spinifex texture, is widespread in all greenstone belts, including those where komatiite has not yet been identified. Geochemical data show a gradation of compositions between komatiites and komatiitic basalts indicating derivation from the same plume, consistent with an interpretation by Van Kranendonk et al. (2007b). Geochemical data from the komatiitic basalts have been reported from the northern and eastern Kelly greenstone belt at Camel Creek and Sandy Creek (Glikson and Hickman 1981a, b; Smithies et al. 2007). Additional analyses were reported from the southern Coongan greenstone belt 9 km south of Petermarina Pool and 4 km west of the Coongan River (Arndt et al. 2001). Nebel et al. (2014) obtained whole-rock Lu–Hf isotope data from seven samples of komatiitic basalt collected from Sandy Creek on the northeastern side of Corunna Downs Dome. Excluding one extremely altered sample, the ε_{Hf} values were strongly positive and plotted near or above the depleted mantle ε_{Hf} value at 3350 Ma. Nebel et al. (2014) interpreted this to indicate a component of old, melt-depleted reservoir in the mantle source.

5.4 Tholeiitic Basalt in the Kelly Group

Tholeiitic basalt forms the upper part of the Euro Basalt in most greenstone belts of the EPT, and has a much wider distribution than komatiite. As in basaltic formations of the Warrawoona Group, most of the basalt is pillowed and contains numerous thin intercalation of grey and white banded chert. Most dolerite sills are interpreted to be subvolcanic intrusions of the Euro Basalt. Pillows are commonly very well preserved, and include way-up indicators such as tails, convex tops, and lava drainage tubes filled with quartz and carbonate. Many pillows have vesicular or amygdaloidal margins, and show marginal alteration. Chloritic hyaloclastite separates most pillows. The chert units are generally less than 10 m thick, and most represent silicified fine-grained inter-flow volcaniclastic units or silicified shale.

Analysis of 61 samples of the Euro Basalt by Smithies et al. (2007) revealed that the formation includes both high- and low-Ti tholeiites (Fig. 5.6), and that the two types are commonly intercalated. High-Ti tholeiites ($TiO_2 > 0.8$ wt. %) have relatively high concentrations of HFSE and REE, whereas low-Ti tholeiites

(TiO_2 < 0.8 wt. %) have lower concentrations of these elements. Compared to the low-Ti tholeiites, the high-Ti tholeiites, which are most common in the Pilbara Supergroup, are generally more Fe-rich, have very low Al_2O_3/TiO_2 (18.7–8.9) and high Gd/Yb (1.12–2.23) ratios. Examination of the analytical data in Smithies et al. (2007) shows that the high-Ti tholeiites have considerably lower average contents of MgO, Cr, and Ni than the low-Ti tholeiites, a feature also noted by Glikson and Hickman (1981a). Smithies et al. (2005) observed that the composition of the high-Ti tholeiites did not change significantly during the 300-million-year evolution of the Pilbara Supergroup whereas low-Ti tholeiites show secular trends to lower concentrations of incompatible elements, and to lower ratios of La/Sm, La/Gd, La/Yb, and Gd/Yb.

Smithies et al. (2005) interpreted the high-Ti tholeiites to have been derived by plume-related partial melting of sources compositionally similar to primitive mantle, whereas low-Ti tholeiites were derived through melting a single source that became progressively more depleted during the evolution of the EPT. Alternatively, the geochemical differences between high-Ti and low-Ti tholeiites might relate to the structure and evolution of mantle plumes. Campbell and Griffiths (1990) proposed that high-Ti basalts were sourced from relatively cool, low-degree partial melting zones in plume heads whereas low-Ti basalts were derived from hotter axial regions of plumes with greater partial melting of less enriched lower mantle material. Progressively more low-Ti basalts might be erupted as the plume evolved, with flattening of the plume head and the introduction of more material from the lower mantle into the zone of melting (Campbell and Griffiths 1990). Smithies et al. (2007) reported that the concentration ranges of incompatible trace elements for high-Ti tholeiites from the Euro Basalt are almost identical to the high-Ti tholeiites of the Coonterunah Subgroup.

Whereas the Euro Basalt was derived by plume-related melting of the mantle (Van Kranendonk et al. 2002, 2004a, b, 2006, 2007a, b, 2010b, 2015; Hickman and Van Kranendonk 2004; Smithies et al. 2005), the geochemical compositions of the Wyman Formation, and the genetically-related Emu Pool Supersuite, indicate partial melting of felsic crust (Barley and Pickard 1999; Champion and Smithies 1999, 2007; Bagas et al. 2004; Van Kranendonk et al. 2007a, b, 2019), with fractionation of the resulting magma producing very potassic volcanic rocks (Champion and Smithies 2007). The present interpretation is that partial melting of the older crust to produce the Wyman Formation and the Emu Pool Supersuite was most likely due to heating by the same mantle plume that was responsible for earlier eruption of the Euro Basalt. In that case, the Euro Basalt and the Wyman Formation were products of the same mantle plume event, justifying their inclusion in the same group.

The Charteris Basalt is lithologically similar to the Euro Basalt, being almost entirely composed of komatiitic basalt and tholeiite. Interpreted komatiite (not analyzed) outcrops in the axial graben of the Warrawoona Syncline (Hickman and Van Kranendonk 2008b), but this might be a faulted slice of the Euro Basalt from the Marble Bar greenstone belt. From a limited number of analyses, the Charteris Basalt contains basalts lower TiO_2, P_2O_5, Ce, Y, Zr, and Sm, and higher K_2O, Ba, Rb, Sr, Cr, Ni, and Al_2O_3/TiO_2 than most basalts of the Euro Basalt (data in Glikson and

Hickman 1981b: Glikson et al. 1986). Samples analyzed by Smithies et al. (2007) confirmed low TiO_2 contents ($TiO_2 < 0.6$ wt. %). Low-Ti basalts of the Charteris Basalt are likely to have been derived from more depleted sources than the Euro Basalt. Crustal contamination of the low-Ti basaltic magmas, perhaps partly due to intrusion through the Wyman Formation and sills of the Emu Pool Supersuite, might account for relatively high levels of K_2O, Ba, and Rb in these basalts. In this scenario, the Charteris Basalt is likely to have been a late-stage product of the same mantle plume that produced the Euro Basalt. Alternatively, the Charteris Basalt might be the basal section of an overlying volcanic cycle (Hickman 2011, 2012).

5.5 Sm–Nd Isotope Data

Whole-rock Sm–Nd two-stage depleted mantle model ages (${T_{DM}}^2$) from the Euro Basalt range between c. 3510 and 3470 Ma, and ε_{Nd} values are approximately chondritic (Smithies et al. 2007). No Sm–Nd analyses are available for the Wyman Formation but granitic rocks of the contemporaneous and genetically related Emu Pool Supersuite have Nd model ages between c. 3630 and 3420 Ma, with ε_{Nd} values between +1.49 and − 1.44 (Smithies et al. 2007). These data suggest sources ranging in age from late Eoarchean to early Paleoarchean. Because the theoretical depleted mantle ε_{Nd} at 3350–3315 Ma is +2.8, the isotope data from the Kelly Group indicate mixed sources including melts from older crust.

5.6 Relevance to Continental Deposition of the Warrawoona Group

Despite evidence that the Euro Basalt was deposited on thick continental crust (e.g. Buick et al. 1995; Green et al. 2000), there have been contrary interpretations that Paleoarchean basaltic formations of the Pilbara Craton, including the Euro Basalt, originated as oceanic crust (Isozaki et al. 1997; Kitajima et al. 2001, 2008; Komiya et al. 2002; Kabashima et al. 2003; Terabayashi et al. 2003; Furness et al. 2007, 2015; Kusky et al. 2013). These interpretations were made using plate tectonic models in which the lithological association of pillow basalt and chert is related to mid-ocean ridge or island arc settings. Some workers have applied the concept that there is a characteristic 'oceanic plate stratigraphy (OPS)' (Isozaki et al. 1991; Matsuda and Isozaki 1991; Kitajima et al. 2001; Kusky et al. 2013) that can be used to identify oceanic crust in greenstones of any age. The 3530–3235 Ma Pilbara Supergroup is a succession of eight successive ultramafic–mafic–felsic volcanic cycles (Hickman 2011, 2012), has a total depositional thickness in excess of 15 km, and is intruded by five Paleoarchean granitic supersuites (3530–3223 Ma); as such, it bears no resemblance to OPS.

Green et al. (2000) reported geochemical evidence for crustal contamination in the Euro Basalt, including a trend of upward enrichment in LILE, Th, U, and LREE, and noted that the same trend was also present in the Table Top Formation at the base of the Warrawoona Group. Smithies et al. (2007) commented on strong geochemical similarities between the Table Top Formation and the Euro Basalt, and agreed that the latter was deposited on the older granite–greenstone crust of the EPT. The geochemical similarities between the Euro Basalt and the Table Top Formation add to the other evidence that the Warrawoona Group was deposited on thick continental crust.

5.7 Granitic Rocks of the Kelly LIP

About 70% of the exposed area of the EPT is composed of granitic intrusions varying in age from 3530 to 3223 Ma. Geochronology and geochemistry have been used to assign over 70 individual intrusions to five supersuites (Van Kranendonk et al. 2004b, c, 2006; Petersson et al. 2020). U–Pb zircon dating and intrusive relationships have revealed that intrusions of all five supersuites are contemporaneous with felsic volcanic formations of the Pilbara Supergroup (Barley and Pickard 1999; Hickman 2001, 2004, 2012, 2021; Van Kranendonk et al. 2001, 2002, 2004a, b, 2006; Hickman and Van Kranendonk 2004; Petersson et al. 2020). It is interpreted that episodes of granitic intrusion and felsic volcanism were genetically related to the same mantle plume events (Van Kranendonk et al. 2002, 2006, 2007a, b).

Field evidence confirms that granitic intrusions of various ages fed subvolcanic sills and overlying felsic volcanic formations (Vearncombe and Kerrich 1999; Van Kranendonk 2000b; Buick et al. 2002; Van Kranendonk et al. 2002, 2006; Bagas et al. 2004; Van Kranendonk and Pirajno 2004; Brown et al. 2006, 2011; Hickman and Van Kranendonk 2008a; Hickman 2012, 2021). Geological mapping has provided direct evidence that some granitic intrusions originated as magma chambers that fed subvolcanic laccoliths, and that these in turn released magma into overlying felsic volcanic centres. Examples of such relationships between intrusions of the Emu Pool Supersuite and the Wyman Formation include: (1) the 3317–3314 Ma Coppin Gap Granodiorite (Mount Edgar Dome) feeding into the ~3317 Ma Spinifex Ridge Cu–Mo porphyry and Wyman Formation at Coppin Gap (Hickman 1983; Van Kranendonk et al. 2006); (2) the 3313–3300 Ma Nandingarra Granodiorite (northwest Corunna Downs Dome) intruding the Kelly Group and feeding into the Wyman Formation in the Warrawoona Syncline; (3) the ~3314 Ma Carbana Monzogranite (east Corunna Downs Dome) being subvolcanic to the ~3315 Ma Boobina Porphyry and the Wyman Formation at Copper Hills (Barley and Pickard 1999; Bagas et al. 2004); and (4) the ~3313 Ma Gobbos Granodiorite (McPhee Dome) being subvolcanic to the Wyman Formation at Wallabirdee Ridge (Hickman 2021).

5.7.1 Emu Pool Supersuite (3324–3290 Ma)

The Emu Pool Supersuite was emplaced during eruption of the Wyman Formation and comprises 24 named and unnamed intrusions of monzogranite, granodiorite, and locally tonalite ranging in age from c. 3324 to 3290 Ma. Tonalite is largely confined to the Yilgalong Dome. No intrusions of this supersuite have been identified in the Shaw and North Pole Domes or any domes farther to the west. Except in the Yilgalong Dome (Fig. 1.7), all intrusions of the supersuite are older than c. 3294 Ma and most have U–Pb zircon crystallization ages between c. 3321 and 3300 Ma (Hickman 2021). The maximum age of the Emu Pool Supersuite is currently defined by a date of 3324 ± 6 Ma for the Wilina Granodiorite (unpublished data, Collins et al. 1998) in the Mount Edgar Dome. This date is questionable because Collins et al. (1998) described the intrusion as post-kinematic which is inconsistent with the main deformation in this dome occurring at c. 3315 Ma (Van Kranendonk et al. 2001, 2002, 2004a; Hickman and Van Kranendonk 2004). Other geochronology on the Wilina Granodiorite indicates that a date of 3310 ± 8 Ma (Nelson 2004) most likely represents its crystallization age.

The full compositional range of the Emu Pool Supersuite is diorite to syenogranite. Several distinct compositional and textural phases and plutons have been identified in the Corunna Downs, Mount Edgar, Muccan, Warrawagine, and Yilgalong Domes (Fig. 1.7). For example, many intrusions of granodiorite and monzogranite exhibit porphyritic texture in which abundant plagioclase or microcline phenocrysts are up to 5 cm long. Early mapping of the EPT revealed that feldspar phenocrysts were far more common within intrusions intermediate in age between the foliated TTGs and late-stage, post-tectonic granites associated with tin mineralization (Hickman 1983). Such porphyritic granitic rocks are now known to be mainly confined to the Emu Pool and Cleland Supersuites.

Intrusion of the Emu Pool Supersuite accompanied volcanic eruption of the Wyman Formation, and in the Mount Edgar and Corunna Downs Domes there are examples of granodiorite intrusions immediately underlying hypabyssal felsic intrusions and volcanic units. The supersuite consists of widespread and voluminous granitic rocks emplaced into the granitic cores of the domes during c. 3315 Ma doming (Hickman and Van Kranendonk 2004) of the Emu Pool Event. All intrusions of the supersuite are only moderately to weakly foliated because the EPT experienced relatively little deformation and metamorphism between the Emu Pool Event and the North Pilbara Orogeny at c. 2.95 Ga.

5.7.1.1 Geochemistry

The Emu Pool Supersuite is composed of both high-Al and low-Al granitic rocks, similar to pre-3420 Ma supersuites (Chap. 3). High-Al granitic rocks are dominant in the Mount Edgar Dome (Collins 1983; Davy and Lewis 1986) whereas low-Al granites are more common in the Corunna Downs Dome (Champion and Smithies

2007). Average SiO_2 and K_2O contents are higher than in the pre-3420 Ma granitic rocks, and highest in the low-Al group. In this group, Na_2O is <4.5%, and Na_2O/K_2O varies from c. 2.0 to 0.5. Unlike the high-Al group, the low-Al group has pronounced negative Eu anomalies (Eu/Eu* mostly 0.8 to <0.2). The high-Al granitic rocks exhibit a trend of increasingly negative ε_{Nd} values with higher K_2O, but this negative correlation is absent in the low-Al granitic rocks due to more compositional variations (Champion and Smithies 2007). The geochemical data suggest that the high-Al granites were most likely derived by partial melting of thickened crust whereas derivation of the low-Al granites probably involved some fractional crystallization, especially for silica-rich granites. Nd and Hf isotope data indicate more recycling of older crust in the Emu Pool Supersuite than in the Callina Supersuite (Champion 2013, Champion and Huston 2016; Gardiner et al. 2017). ε_{Nd} values for the supersuite vary from 1.49 to −1.44 and Nd model ages range between c. 3630 and 3420 Ma.

5.8 Emu Pool Event (3325–3290 Ma)

The Emu Pool Event (Hickman and Van Kranendonk 2008b) was an event of granitic intrusion, diapiric doming, and metamorphism in the eastern part of the EPT between c. 3325 and 3290 Ma. The diapiric uplift was considerable in some domes because it resulted in erosion of the Kelly Group, Strelley Pool Formation, Warrawoona Group, and pre-3530 Ma crust. An erosional unconformity is exposed between the Kelly Group and the Sulphur Springs Group in the Soanesville greenstone belt (Van Kranendonk 1999, 2000b; Van Kranendonk et al. 2006) and in the West Warralong greenstone belt (Fig. 5.9) where sandstone of the Leilira Formation unconformably overlies tightly folded units of the Wyman Formation (Van Kranendonk 2004b). Detrital zircon ages in the Leilira Formation (Sulphur Springs Group, Chap. 6) range up to c. 3589 Ma (Buick et al. 2002), suggesting exposure of the pre-3530 Ma crust of the Pilbara Craton during the Emu Pool Event.

The local angular unconformities between the Euro Basalt and the Wyman Formation testify to a deformation between ~3335 and 3325 Ma. More significant deformation followed at ~3315 Ma (Collins et al. 1998). On the interpretation that doming was accompanied by granitic intrusion (Collins et al. 1998; Van Kranendonk et al. 2002, 2007a, b; Hickman and Van Kranendonk 2004), the range of crystallization ages from the Emu Pool Supersuite provides evidence that different domes were uplifted at slightly different times, with the last doming occurring after 3300 Ma in the Yilgalong Dome. Several published estimates of the age of the doming at this time have been 3315–3310 Ma (Williams and Collins, 1990; Collins et al. 1998; Van Kranendonk et al. 2002; Kloppenburg, 2003; Francois et al. 2014), although this estimate was entirely derived from geochronology in the Mount Edgar Dome. The distribution of the Emu Pool Supersuite (Fig. 3.18) suggests that crustal thickening associated with the formation of the domes was restricted to the eastern half of the EPT. The western boundary of the supersuite

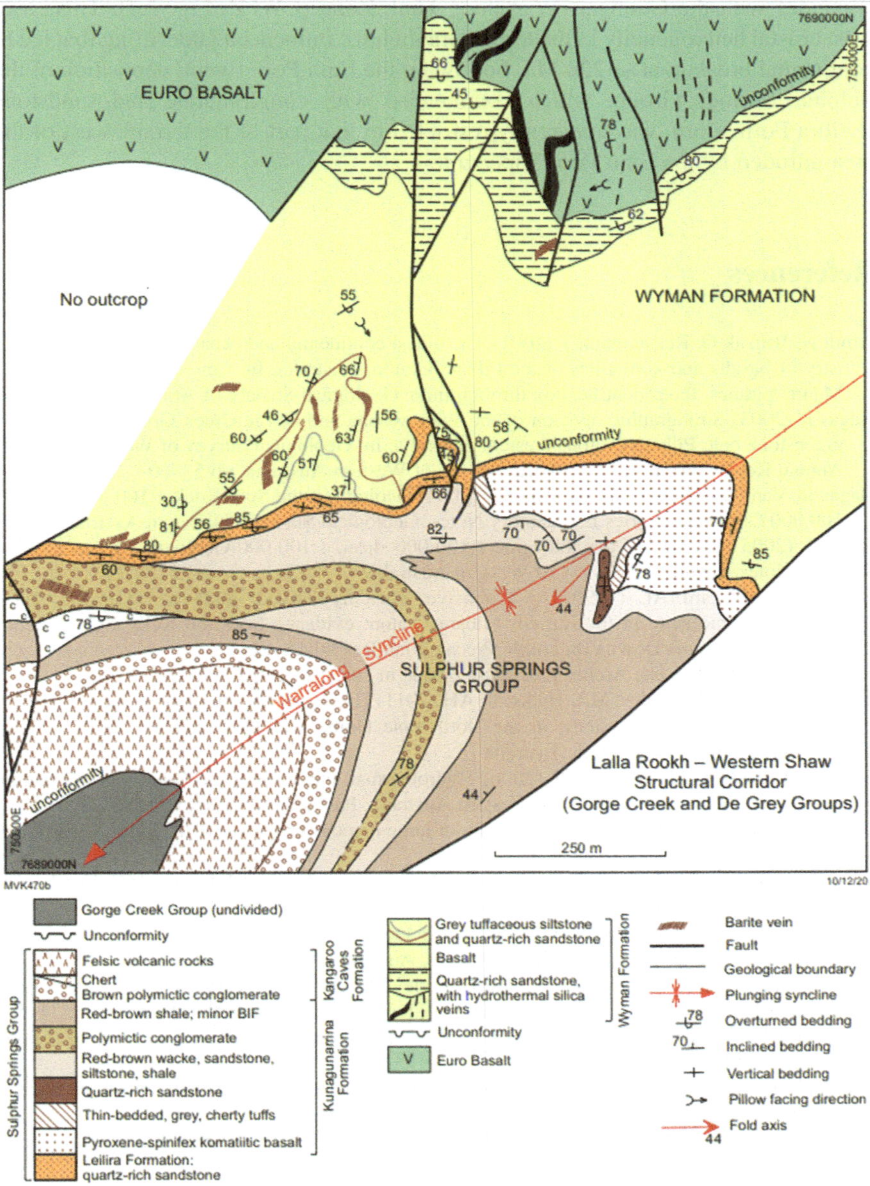

Fig. 5.9 Geological sketch map of the southern part of the Warralong greenstone belt showing angular unconformities between the Euro Basalt and the Wyman Formation, and between the Wyman Formation and the Sulphur Springs Group (locality centred at Zone 50, MGA 751600E, 7688350N) (From Van Kranendonk 2004a; with Geological Survey of Western Australia permission)

coincides with the Coongan–Warralong Fault Zone (CWFZ, Fig. 1.4). This east–west crustal heterogeneity in the terrane might have influenced later rifting that led to continental breakup at ~3220 Ma. Following the Emu Pool Event, deposition of the Sulphur Springs Group, which commenced with conglomerate and sandstone (Leilira Formation), was apparently restricted to that part of the terrane west of the area intruded by the Emu Pool Supersuite.

References

Arndt N, Bruzak G, Reischmann T (2001) The oldest continental and oceanic plateaus: geochemistry of basalts and komatiites of the Pilbara Craton, Australia. In: Ernst RE, Buchan K (eds) Mantle plumes: their identification through time. Geological Society of America, pp 359–387

Bagas L (2003) Stratigraphic revision of the Warrawoona and Gorge Creek Groups in the Kelly greenstone belt, Pilbara Craton, Western Australia. In: Geological Survey of Western Australia Annual Review 2001–02. Geological Survey of Western Australia, pp 53–60

Bagas L, Van Kranendonk MJ, Pawley MJ (2004) Geology of the Split Rock 1:100 000 sheet. 1: 100 000 Geological Series Explanatory Notes. Geological Survey of Western Australia

Bagas L (2005) Geology of the Nullagine 1:100 000 sheet. 1:100 000 Geological Series Explanatory Notes. Geological Survey of Western Australia

Barley ME, Pickard AL (1999) An extensive, crustally-derived, 3325 to 3310 Ma silicic volcanoplutonic suite in the Eastern Pilbara Craton: evidence from the Kelly Belt, McPhee Dome and Corunna Downs Batholith. Precambrian Res 96:41–62

Bleeker W (2003) The late Archean record: a puzzle in ca. 35 pieces. Lithos 71:99–134

Brown A, Cudahy T, Walter MR, Hickman AH (2011) Hyperspectral alteration mapping of early Archean hydrothermal systems in the North Pole Dome, Pilbara Craton. Record, 2011/15. Geological Survey of Western Australia

Brown AJ, Cudahy TJ, Walter MR (2006) Hydrothermal alteration at the Panorama Formation, North Pole Dome, Pilbara Craton, Western Australia. Precambrian Res 151:211–223

Bryan SE, Ernst RE (2008) Revised definition of large igneous provinces (LIP). Earth Sci Rev 86: 175–202

Buick R, Brauhart CW, Morant P, Thornett JR, Maniw JG, Archibald NJ, Doepel MG, Fletcher IR, Pickard AL, Smith JB, Barley ME, McNaughton NJ, Groves DI (2002) Geochronology and stratigraphic relationships of the Sulphur Springs group and Strelley granite: a temporally distinct igneous province in the Archean Pilbara Craton, Australia. Precambrian Res 114:87–120

Buick R, Thornett J, McNaughton N, Smith JB, Barley ME, Savage MD (1995) Record of emergent continental crust ~3.5 billion years ago in the Pilbara Craton of Australia. Nature 375:574–577

Campbell IH, Griffiths RW (1990) Implications of mantle plume structure for the evolution of flood basalts. Earth Planet Sci Lett 99:79–93

Campbell IH, Griffiths RW, Hill RI (1989) Melting in an Archaean mantle plume: heads it's basalts, tails it's komatiites. Nature 339:697–699

Champion DC (2013) Neodymium depleted mantle model age map of Australia: explanatory notes and user guide. Record, 2013/44. Geoscience Australia

Champion DC, Huston DL (2016) Radiogenic isotopes, ore deposits and metallogenic terranes: novel approaches based on regional isotopic maps and the mineral systems concept. Ore Geol Rev 76:229–256. https://doi.org/10.1016/j.oregeorev.2015.09.025

Champion DC, Smithies RH (1999) Archaean granites of the Yilgarn and Pilbara Cratons: secular changes. In: Barbarin B (ed) The origin of granites and related rocks, fourth Hutton symposium abstracts, 137. BRGM, Clermont-Ferrand

Champion DC, Smithies RH (2007) Geochemistry of Paleoarchean granites of the East Pilbara Terrane, Pilbara Craton, Western Australia: implications for early Archean crustal growth. In: Van Kranendonk MJ, Bennett VC, Smithies RH (eds) Earth's oldest rocks. Elsevier B.V, Burlington, Massachusetts, USA, pp 369–410

Cheney ES (1996) Sequence stratigraphy and plate tectonic significance of the Transvaal succession of southern Africa and its equivalent in Western Australia. Precambrian Res 79:3–24

Cheney ES, Roering C, Stettler E (1988) Valbara. In: Geocongress '88: extended abstracts. Geological Society of South Africa, pp 85–88

Collins WJ (1983) Geological evolution of an Archaean batholith. PhD, La Trobe University

Collins WJ, Van Kranendonk MJ, Teyssier C (1998) Partial convective overturn of Archaean crust in the East Pilbara Craton, Western Australia: driving mechanisms and tectonic implications. J Struct Geol 20:1405–1424. https://doi.org/10.1016/S0191-8141(98)00073-X

Davy R, Lewis JD (1986) The Mount Edgar Batholith, Pilbara area, Western Australia: Geochemistry and petrography. Report, vol 17. Geological Survey of Western Australia

De Kock MO, Evans DAD, Beukes NJ (2009) Validating the existence of Vaalbara in the Neoarchean. Precambrian Res 174:145–154

Ernst RE (2014) Large igneous provinces. Cambridge University Press, Cambridge, UK

Farrell TR (2006) Geology of the Eastern Creek 1:100 000 sheet. 1:100 000 Geological series explanatory notes. Geological Survey of Western Australia

François C, Philippot P, Rey PF, Rubatto D (2014) Burial and exhumation during Archean sagduction in the East Pilbara Granite–Greenstone Terrane. Earth Planet Sci Lett 396:235–251

Furnes H, Banerjee NR, Staudigel H, Muehlenbachs K, McLoughlin N, de Wit M, Van Kranendonk MJ (2007) Comparing petrographic signatures of bioalteration in recent to Mesoarchean pillow lavas: tracing subsurface life in oceanic igneous rocks. Precambrian Res 158:156–176

Furnes H, Dilek Y, de Wit M (2015) Precambrian greenstone sequences represent different ophiolite types. Gondwana Res 27:649–685. https://doi.org/10.1016/j.gr.2013.06.004

Gardiner NJ, Hickman AH, Kirkland CL, Lu Y, Johnson T, Zhao J-X (2017) Processes of crust formation in the early earth imaged through Hf isotopes from the East Pilbara Terrane. Precambrian Res 297:56–76. https://doi.org/10.1016/j.precamres.2017.05.004

Glikson AY, Davy R, Hickman AH (1986) Geochemical data files of Archaean volcanic rocks, Pilbara Block, Western Australia. BMR (Geoscience Australia) Record, 1986/014

Glikson AY, Hickman AH (1981a) Geochemistry of Archaean volcanic successions, Eastern Pilbara Block, Western Australia. BMR (Geoscience Australia) Record, 1981/036

Glikson AY, Hickman AH (1981b) Geochemical stratigraphy and petrogenesis of Archaean basic–ultrabasic volcanic units, Eastern Pilbara Block, Western Australia. In: Glover JE, Groves DI (eds) Archaean Geology: second International Symposium, Perth 1980, vol 7. Geological Society of Australia, pp 287–300

Green MG, Sylvester PJ, Buick R (2000) Growth and recycling of early Archaean continental crust: geochemical evidence from the Coonterunah and Warrawoona Groups, Pilbara Craton, Australia. Tectonophysics 322:69–88

Grey K, Clarke JDA, Hickman AH (2012) The proposed dawn of life geotourism trail, Marble Bar, Pilbara Craton, Western Australia–geology and evidence for early life. Record, 2012/9. Geological Survey of Western Australia

Hickman AH (1981) Crustal evolution of the Pilbara Block. In: Glover JE, Groves DI (eds) Archaean Geology: second International Symposium, Perth 1980. Geological Society of Australia, pp 57–69

Hickman AH (1983) Geology of the Pilbara Block and its environs. Bulletin, vol 127. Geological Survey of Western Australia

Hickman AH (1984) Archaean diapirism in the Pilbara Block, Western Australia. In: Kröner A, Greiling R (eds) Precambrian tectonics illustrated. Schweizerbart Science Publishers, Stuttgart, Germany, pp 113–127

Hickman AH (1990) Excursion no 5: Pilbara and Hamersley Basin. In: Ho SE, Glover JE, Myers JS, Muhling JR (eds) Excursion guidebook: third International Archaean Symposium, Perth

1990 (17–21 September). The University of Western Australia, Geology Department and University Extension, pp 1–60
Hickman AH (2001) East Pilbara diapirism: New evidence from mapping. In: Geological Survey of Western Australia (ed) GSWA 2001 extended abstracts: new geological data for WA explorers. Geological Survey of Western Australia, pp 23–25
Hickman AH (2004) Two contrasting Granite–Greenstones Terranes in the Pilbara Craton, Australia: evidence for vertical and horizontal tectonic regimes prior to 2900 Ma. Precambrian Res 131:153–172
Hickman AH (2011) Pilbara Supergroup of the East Pilbara Terrane, Pilbara Craton: Updated lithostratigraphy and comments on the influence of vertical tectonics. In: Geological Survey of Western Australia Annual Review 2009–10. Geological Survey of Western Australia, pp. 50–59
Hickman AH (2012) Review of the Pilbara Craton and Fortescue Basin, Western Australia: crustal evolution providing environments for early life. Island Arc 21:1–31
Hickman AH (2016) Northwest Pilbara Craton: a record of 450 million years in the growth of Archean continental crust. Report, 160. Geological Survey of Western Australia
Hickman AH (2021) East Pilbara Craton: a record of one billion years in the growth of Archean continental crust. Report 143, Geological Survey of Western Australia
Hickman AH, Lipple SL (1978) Marble Bar, WA sheet SF50–8, 2nd edn. Geological Survey of Western Australia
Hickman AH, Van Kranendonk MJ (2004) Diapiric processes in the formation of Archaean continental crust, East Pilbara Granite–Greenstone Terrane, Australia. In: Eriksson PG, Altermann W, Nelson DR, Mueller WU, Catuneanu O (eds) The Precambrian earth: tempos and events. Elsevier, Amsterdam, The Netherlands, pp 54–75
Hickman AH, Van Kranendonk MJ (2008a) Archean crustal evolution and mineralization of the Northern Pilbara Craton–a field guide. Record, 2008/13. Geological Survey of Western Australia
Hickman AH, Van Kranendonk MJ (2008b) Marble Bar, WA sheet 2855. Geological Survey of Western Australia
Isozaki Y, Kabashima T, Kitajima K, Maruyama S, Kato Y, Terabayashi M (1997) Early Archean Mid-Ocean ridge rocks and early life in the Pilbara Craton. Western Australia Eos 78:399
Isozaki Y, Maruyama S, Kimura G (1991) Middle Archean (3.3 Ga) Cleaverville accretionary complex in Northwestern Pilbara Block, Western Australia. Eos 72:542
Johnson TE, Brown M, Gardiner NJ, Kirkland CL, Smithies RH (2017) Earth's first stable continents did not form by subduction. Nature 543:239–242
Kabashima T, Isozaki Y, Hirata T, Maruyama S, Kiminami K, Nishimura Y (2003) U–Pb zircon ages by laser ablation ICP-MS techniques from the Archean felsic intrusive and volcanic rocks in the North Pole Area, Western Australia. J Geogr 112:1–17
Kitajima K, Hirata T, Maruyama S, Yamanashi T, Sano Y, Liou JG (2008) U–Pb zircon geochronology using LA-ICP-MS in the North Pole Dome, Pilbara Craton, Western Australia: a new tectonic growth model for the Archean chert/greenstone succession. Int Geol Rev 80:1–14
Kitajima K, Maruyama S, Utsunomiya S, Liou JG (2001) Seafloor hydrothermal alteration at an Archean Mid-Ocean ridge. J Metamorph Geol 19:583–599
Kloppenburg A (2003) Structural evolution of the Marble Bar Domain, Pilbara Granite-Greenstone Terrain, Australia: the role of Archaean mid-crustal detachments. PhD, Utrecht University
Komiya T, Maruyama S, Hirata T, Yurimoto H (2002) Petrology and geochemistry of MORB and OIB in the mid-Archean North Pole region, Pilbara Craton, Western Australia: implications for the composition and temperature of the upper mantle at 3.5 Ga. Int Geol Rev 44:988–1016
Kusky TM, Windley BF, Safonova I, Wakita K, Wakabayashi J, Polat A, Santosh M (2013) Recognition of ocean plate stratigraphy in accretionary orogens through earth history: a record of 3.8 billion years of sea floor spreading, subduction, and accretion. Gondwana Res 24:501–547
Lipple SL (1975) Definitions of new and revised stratigraphic units of the Eastern Pilbara region. In: Annual report for the year 1974. Geological Survey of Western Australia, pp. 58–63

Mareschal J-C, West GF (1980) A model for Archaean tectonism: numerical models of vertical tectonism in greenstone belts. Can J Earth Sci 17:60–71

Matsuda T, Isozaki Y (1991) Well-documented travel history of Mesozoic pelagic chert in Japan: from remote ocean to subduction zone. Tectonics 10:475–499

Nebel O, Campbell IH, Sossi PA, Van Kranendonk MJ (2014) Hafnium and iron isotopes in early Archean komatiites record a plume-driven convection cycle in the hadean earth. Earth Planet Sci Lett 397:111–120

Nelson DR (2004) Compilation of geochronology data, 2003. Record, 2004/2. Geological Survey of Western Australia

Petersson A, Kemp AIS, Gray CM, Whitehouse MJ (2020) Formation of early Archean Granite–Greenstone Terranes from a globally chondritic mantle: insights from igneous rocks of the Pilbara Craton. Western Australia Chemical Geology 551. https://doi.org/10.1016/j.chemgeo.2020.119757

Smithies RH, Champion DC, Van Kranendonk MJ (2009) Formation of Paleoarchean continental crust through infracrustal melting of enriched basalt. Earth Planet Sci Lett 281:298–306

Smithies RH, Champion DC, Van Kranendonk MJ, Hickman AH (2007) Geochemistry of volcanic rocks of the Northern Pilbara Craton, Western Australia. Report 104. Geological Survey of Western Australia

Smithies RH, Van Kranendonk MJ, Champion DC (2005) It started with a plume–early Archaean basaltic proto-continental crust. Earth Planet Sci Lett 238:284–297

Sossi PA, Eggins SM, Nesbitt RW, Nebel O, Hergt JM, Campbell IH, Neill HS, Van Kranendonk M, Davies DR (2016) Petrogenesis and geochemistry of Archean komatiites. J Petrol 57:147–184

Terabayashi M, Masuda Y, Ozawa H (2003) Archean Ocean floor metamorphism in the North Pole Area, Pilbara Craton, Western Australia. Precambrian Res 127:167–180

Thébaud N, Rey PF (2013) Archean gravity-driven tectonics on hot and flooded continents: controls on long-lived hydrothermal systems away from continental margins. Precambrian Res 229:93–104

Van Kranendonk MJ (1998) Litho-tectonic and structural components of the North Shaw 1:100 000 sheet, Archaean Pilbara Craton. In: Geological Survey of Western Australia Annual Review 1997–98. Geological Survey of Western Australia, pp 63–70

Van Kranendonk MJ (1999) North Shaw, WA Sheet 2755. Geological Survey of Western Australia

Van Kranendonk MJ (2000a) Evidence of thick Archaean crust in the East Pilbara. In: GSWA 2000 extended abstracts: geological data for WA explorers in the new millennium. Geological Survey of Western Australia, pp. 1–3

Van Kranendonk MJ (2000b) Geology of the North Shaw 1:100 000 sheet. 1:100 000 Geological Series Explanatory Notes. Geological Survey of Western Australia

Van Kranendonk MJ (2004a) Geology of the Carlindie 1:100 000 sheet. 1:100 000 Geological Series Explanatory Notes. Geological Survey of Western Australia

Van Kranendonk MJ (2004b) Coongan, WA Sheet 2856. Geological Survey of Western Australia

Van Kranendonk MJ (2010a) Geology of the Coongan 1:100 000 sheet. 1:100 000 Geological Series Explanatory Notes. Geological Survey of Western Australia

Van Kranendonk MJ (2010b) Three and a half billion years of life on earth: A transect back into deep time. Record, 2010/21. Geological Survey of Western Australia

Van Kranendonk MJ, Collins WJ, Hickman AH, Pawley MJ (2004a) Critical tests of vertical vs horizontal tectonic models for the Archean East Pilbara Granite–Greenstone Terrane, Pilbara Craton, Western Australia. Precambrian Res 131:173–211

Van Kranendonk MJ, Hickman AH, Smithies RH, Champion DC (2007a) Paleoarchean development of a continental nucleus: the East Pilbara Terrane of the Pilbara Craton, Western Australia. In: Van Kranendonk MJ, Bennett VC, Smithies RH (eds) Earth's oldest rocks. Elsevier B.V, Burlington, Massachusetts, USA, pp 307–337

Van Kranendonk MJ, Hickman AH, Smithies RH, Nelson DN, Pike G (2002) Geology and tectonic evolution of the Archaean North Pilbara Terrain, Pilbara Craton, Western Australia. Econ Geol 97:695–732. https://doi.org/10.2113/gsecongeo.97.4.695

Van Kranendonk MJ, Hickman AH, Smithies RH, Williams IR, Bagas L, Farrell TR (2006) Revised lithostratigraphy of Archean supracrustal and intrusive rocks in the Northern Pilbara Craton, Western Australia. Record, 2006/15. Geological Survey of Western Australia

Van Kranendonk MJ, Hickman AH, Williams IR, Nijman W (2001) Archaean geology of the East Pilbara Granite–Greenstone Terrane, Western Australia–a field guide. Record, 2001/9. Geological Survey of Western Australia

Van Kranendonk MJ, Pirajno F (2004) Geochemistry of metabasalts and hydrothermal alteration zones associated with c. 3.45 Ga chert and barite deposits: implications for the geological setting of the Warrawoona group, Pilbara Craton, Australia. Geochemistry: exploration. Environment, Analysis 4:253–278

Van Kranendonk MJ, Smithies RH, Champion DC (2019) Chapter 19–Paleoarchean development of a continental nucleus: the East Pilbara Terrane of the Pilbara Craton, Western Australia. In: Van Kranendonk MJ, Bennett VC, Hoffmann JE (eds) Earth's oldest rocks, 2nd edn. Elsevier B. V, pp 437–462

Van Kranendonk MJ, Smithies RH, Griffin WL, Huston DL, Hickman AH, Champion DC, Anhaeusser CR, Pirajno F (2015) Making it thick: a volcanic plateau origin of Paleoarchean continental lithosphere of the Pilbara and Kaapvaal Cratons. Geol Soc Lond, Spec Publ 389:89–111. https://doi.org/10.1144/SP389.12

Van Kranendonk MJ, Smithies RH, Hickman AH, Bagas L, Williams IR, Farrell TR (2004b) Event stratigraphy applied to 700 million years of Archean crustal evolution, Pilbara Craton, Western Australia. In: Geological Survey of Western Australia Annual Review 2003–04, pp. 49–61

Van Kranendonk MJ, Smithies RH, Hickman AH, Bagas L, Williams IR, Farrell TR (2004c) Archaean supergroups and supersuites in the Northern Pilbara Craton: the application of event stratigraphy to 1 Ga of crustal and metallogenic evolution. In: GSWA 2004 extended abstracts: promoting the prospectivity of Western Australia. Geological Survey of Western Australia, pp. 5–7

Van Kranendonk MJ, Smithies RH, Hickman AH, Champion DC (2007b) Review: secular tectonic evolution of Archean continental crust: interplay between horizontal and vertical processes in the formation of the Pilbara Craton, Australia. Terra Nova 19:1–38

Van Kranendonk MJ, Smithies RH, Hickman AH, Wingate MTD, Bodorkos S (2010) Evidence for Mesoarchean (~3.2 Ga) rifting of the Pilbara Craton: the missing link in an early Precambrian Wilson cycle. Precambrian Res 177:145–161

Vearncombe S, Kerrich R (1999) Geochemistry and geodynamic setting of volcanic and plutonic rocks associated with early Archaean volcanogenic massive sulphide mineralization, Pilbara Craton. Precambrian Res 98:243–270

Wiemer D, Schrank CE, Murphy DT, Wenham L, Allen CM (2018) Earth's oldest stable crust in the Pilbara Craton formed by cyclic gravitational overturns. Nat Geosci 11:357–361. https://doi.org/10.1038/s41561-018-0105-9

Williams IR (1999) Geology of the Muccan 1:100 000 sheet. 1:100 000 Geological Series Explanatory Notes. Geological Survey of Western Australia

Williams IR (2003) Yarrie, Western Australia, 3rd edn. 1:250 000 Geological Series Explanatory Notes. Geological Survey of Western Australia

Williams IR, Bagas L (2007) Geology of the Mount Edgar 1:100 000 sheet. 1:100 000 Geological Series Explanatory Notes. Geological Survey of Western Australia

Williams IS, Collins WJ (1990) Granite–Greenstone Terranes in the Pilbara Block, Australia, as coeval volcano-plutonic complexes: evidence from U–Pb zircon dating of the Mount Edgar batholith. Earth Planet Sci Lett 97:41–53. https://doi.org/10.1016/0012-821X(90)90097-H

Zegers TE, Barley ME, Groves DI, McNaughton NJ, White SH (2002) Oldest gold: deformation and hydrothermal alteration in the early Archean shear zone-hosted Bamboo Creek deposit, Pilbara, Western Australia. Econ Geol 97:757–773

Zegers TE, de Wit MJ, Dann J, White SH (1998) Vaalbara, Earth's oldest assembled continent? A combined structural, geochronological, and palaeomagnetic test. Terra Nova 10:250–259

Chapter 6
Paleoarchean Continental Breakup of the Pilbara Craton

Abstract Following deformation and magmatic activity of the 3325–3290 Ma Emu Pool Event (Chap. 5), deposition of the Sulphur Springs Group marked the beginning of crustal extension and rifting that led to the continental breakup of the Pilbara Craton. The extension and rifting are attributed to the arrival of the last major mantle plume to impact the Pilbara Craton. Melting of the mantle and crust resulted in an ultramafic–mafic–felsic volcanic cycle in the Sulphur Springs Group and the intrusion of granitic rocks of the 3274–3223 Ma Cleland Supersuite.

The Sulphur Springs Group and the Fig Tree Group of the eastern Kaapvaal Craton are transitional successions from Paleoarchean large igneous provinces to Mesoarchean sedimentary basins. Deposition of the Sulphur Springs Group ended with breakup of the Pilbara Craton at c. 3220 Ma. The breakup was followed by the separation of at least three plates of continental crust and the evolution of intervening basaltic basins. It marked the beginning of plate tectonic processes in which Paleoarchean vertical deformation and crustal recycling were replaced by Mesoarchean horizontal deformation and melts derived from plate separation, collision, and subduction.

Keywords Mantle plume · Crustal extension · Rifting · Plate separation · Plate tectonic processes

6.1 Introduction

Until the end of the 3325–3290 Ma Emu Pool Event (Chap. 5), the Paleoarchean volcanic plateau of the Pilbara Craton had evolved by successive episodes of volcanic eruption and granitic intrusion over a period of 240 million years (Chaps. 3–5). By 3290 Ma, a 15 km-thick volcanic succession (Warrawoona and Kelly Groups) had been deposited on the 3800–3530 Ma continental crust of the craton (Chap. 2). The Pilbara Craton had been intruded by four Paleoarchean granitic supersuites, so the total thickness of the Pilbara crust at 3290 Ma is likely to have been at least 40 km.

A. H. Hickman, *Archean Evolution of the Pilbara Craton and Fortescue Basin*, Modern Approaches in Solid Earth Sciences 24,
https://doi.org/10.1007/978-3-031-18007-1_6

Prior to 3325 Ma, the greenstone succession of the Pilbara Supergroup was essentially continuous across the craton due to the eruption of two large igneous provinces (Hickman 2012, 2021). However, eruption of the 3325–3315 Ma Wyman Formation and intrusion of the genetically related 3324–3290 Ma Emu Pool Supersuite were restricted to the eastern half of the East Pilbara Terrane (EPT) (Fig. 3.19). The western boundary of the 3325–3290 Ma felsic units coincides with the Coongan–Warralong Fault Zone (CWFZ, Fig. 1.9), although the sense of movement on this structure has not been established. Zegers, 1996; Zegers et al. (1999) referred to the southern section of the CWFZ as the 'Central Coongan Shear Zone', and Zegers et al. (1999) used $^{40}Ar/^{39}Ar$ dating to interpret activity between c. 3325 and 3197 Ma. Large ultramafic–mafic intrusions along the CWFZ have not been dated, but they intrude the Wyman Formation and are likely to have been emplaced during rifting.

The Emu Pool Event was followed by deposition of the Sulphur Springs Group and intrusion of the 3274–3223 Ma Cleland Supersuite. Although intrusions of the Cleland Supersuite were emplaced across most of the Northern Pilbara Craton, there is no evidence that the Sulphur Springs Group was deposited east of the CWFZ. Outcrops of the Sulphur Springs Group are confined to the Lalla Rookh–Western Shaw Structural Corridor (Fig. 1.9) and to areas farther west. All these western areas were subjected to crustal extension from c. 3290 Ma onwards.

The Sulphur Springs Group is correlated with the Roebourne Group of the Karratha Terrane (KT) in the Northwest Pilbara Craton (Sun and Hickman 1998; Hickman 2004, 2016). Felsic lava of the Sulphur Springs Group has been dated between 3253 ± 4 Ma (GSWA 160258, Wingate et al. 2010) and 3235 ± 3 Ma (sample 94002, Buick et al. 2002), and rhyolite in the Roebourne greenstone belt has been dated at 3251 ± 6 Ma (GSWA 118975, Nelson 1997). The current separation of these areas by more than 200 km is attributed to faulting (Sun and Hickman 1998), and the faults originated as part of a major rift basin (Hickman et al. 1998, 2001; Smithies et al. 1999, 2001a, b; Smithies and Champion 2002). Crustal extension between c. 3290 and 3165 Ma included continental breakup and plate separation during the East Pilbara Terrane Rifting Event (Hickman and Van Kranendonk 2008).

6.2 East Pilbara Terrane Rifting Event

The East Pilbara Terrane Rifting Event (EPTRE) resulted in a major change in the crustal evolution of the Pilbara Craton. During this event, mantle plume activity and vertical deformation were replaced by plate tectonic processes including subduction (Fig. 1.5). Paleoarchean melts derived from crustal recycling of older crust, still occurring during deposition of the Sulphur Springs Group and intrusion of the Cleland Supersuite, were succeeded by more juvenile Mesoarchean melts, initially derived from mantle-tapping rifts and later from subduction of the mafic crust within rift basins.

The EPTRE developed in three main stages:

1. 3280–3223 Ma deposition of the Sulphur Springs Group and intrusion of the Cleland Supersuite. This stage was marked by increasing crustal extension above the Sulphur Springs mantle plume. Following uplift of the EPT during the Emu Pool Event, rapid erosion resulted in clastic deposition of the Leilira Formation. At c. 3275 Ma, plume-related volcanism commenced with eruption of the ultramafic–mafic Kunagunarrina Formation followed by mafic–felsic volcanism of the Kangaroo Caves Formation. Granitic intrusion of the Strelley Monzogranite (Cleland Supersuite) accompanied felsic volcanism of the Kangaroo Caves Formation. Crustal extension increased with time, being particularly evident at c. 3235 Ma (Vearncombe et al. 1995, 1998; Buick et al. 2002) when deep rifting led to intrusion of ultramafic–mafic sills and dykes.
2. 3223–3200 Ma continental breakup of the Paleoarchean Pilbara Craton. At least three fragments of the Paleoarchean continental crust, which became the East Pilbara, Karratha, and Kurrana Terranes, began to separate (Hickman 2001, 2016; Van Kranendonk et al. 2002, 2007, 2010; Hickman et al. 2010). Other plates, possibly including crust of the South Pilbara Craton and the much larger plate of the East Kaapvaal Craton, might also have separated at this time. Rift basins of oceanic-like basaltic crust formed between the separating continental plates. The best-preserved evidence of this stage in the Pilbara Craton is provided by the juvenile basaltic crust of the 3200–3165 Ma Regal Formation (Sun and Hickman 1998, 1999). This formation was deposited between the EPT and KT (Van Kranendonk et al. 2006; Hickman 2016).
3. 3223–3165 Ma evolution of passive margin basins, typified by deposition of the Soanesville Group, along the margins of the separating plates (Chap. 7) (Van Kranendonk et al. 2006, 2007; Hickman et al. 2010; Hickman 2012, 2016, 2021). At c. 3185 Ma, clastic sedimentation of the lower Soanesville Group was followed by eruption of basaltic volcanic rocks (Honeyeater Basalt) and intrusion of ultramafic–mafic sills and dykes (Dalton Suite). In the Soanesville Basin, volcanism increased northwest towards the eastern margin of the Regal Basin. In the Mosquito Creek Basin, clastic sedimentation of the Budjan Creek Formation (Eriksson 1981; Bagas et al. 2004) was followed by mixed sedimentary and volcanic deposition of the Coondamar Formation (Bagas 2005; Farrell 2006).

6.3 Stratigraphy

Eight stratigraphic units were formed during the EPTRE: the Sulphur Springs Group and Cleland Supersuite during crustal extension and rifting prior to the breakup at c. 3220 Ma; the Regal Formation in the Regal Basin that was produced by the breakup; the Soanesville Group, Nickol River, Budjan Creek and Coondamar Formations that were deposited in passive margin basins along the sides of the three separating continental microplates (EPT, KT, and Kurrana Terrane, KUT); and, at the end of the EPTRE, the Mount Billroth Supersuite as the basaltic basins began to close. The Regal Basin and the passive margin basins are described in Chap. 7.

6.3.1 Sulphur Springs Group

The Sulphur Springs Group comprises three formations, in ascending stratigraphic order: the Leilira Formation, up to 1 km thick, and composed of wacke, quartz sandstone, pebbly sandstone, volcaniclastic sandstone, and shale; the Kunagunarrina Formation, up to 2.4 km thick, and composed of komatiite, komatiitic basalt, tholeiite, and chert; and the Kangaroo Caves Formation, up to 1.5 km thick, and composed of andesite and minor basalt, overlain by dacitic to rhyolitic volcaniclastic rocks and capped by chert. The stratigraphy of the group is extremely variable between different greenstone belts, with the result that the maximum thickness of the entire group in any single area is about 3 km.

Outcrops of the Sulphur Springs Group are restricted to the western half of the EPT, particularly within and adjacent to the Lalla Rookh–Western Shaw Structural Corridor. The combined stratigraphy of the Kunagunarrina and Kangaroo Caves Formations is that of an ultramafic–mafic–felsic volcanic cycle, like earlier plume-related volcanic cycles of the Warrawoona and Kelly Groups. The Sulphur Springs plume event was the last to affect the EPT and is interpreted to have triggered the crustal extension and rifting that led to the breakup of the volcanic plateau at c. 3220 Ma. Some basalts of the Sulphur Springs Group have fractionated LREE and prominent negative Eu anomalies. Other basalts have generally flat, but slightly depleted LREE, and some basalts contain a mixture of these two patterns. This characteristic suggests varying magmatic sources for group.

All three formations of the Sulphur Springs Group experienced syndepositional extensional faulting. Evidence of this was first recorded in the Kangaroo Caves Formation in the Soanesville greenstone belt (Vearncombe et al. 1995, 1998). Subsequently, detailed mapping of the western part of the Warralong greenstone belt by Van Kranendonk (2004a) revealed syndepositional extensional faults in the Leilira Formation. Crustal extension during deposition of the Sulphur Springs Group was a prelude to rifting and eventual continental breakup of the Paleoarchean Pilbara Craton at 3220 Ma (Van Kranendonk et al. 2010).

Deposition of the Leilira Formation is likely to have taken place in environments similar to those of other later clastic units in the East Pilbara such as the basal parts of the Soanesville and Gorge Creek Groups. Eriksson (1981, 1982) described the initial settings for these younger units as marine and lacustrine basins with a horst and graben topography. Komatiite in the Kunagunarrina Formation provides evidence of high-temperature melting of the mantle, and a mantle plume origin is interpreted for the formation (Van Kranendonk et al. 2002, 2006, 2007). Rising mantle plumes caused crustal uplift, extension, and thinning, which is consistent with stratigraphic evidence from the Sulphur Springs Group. Major lateral changes of thickness of the Kunagunarrina Formation between adjacent greenstone belts, and the presence of unconformities with the overlying Kangaroo Caves Formation indicate an unstable depositional environment that included deep erosion. Units of conglomerate along local unconformities at the basal contacts of the Kunagunarrina and Kangaroo Caves Formations (Van Kranendonk 2004a, b) support this interpretation.

Evidence of deposition above older continental crust includes Sm–Nd isotope data obtained by Brauhart (1999), which are like Sm–Nd data from the Warrawoona and Kelly Groups, and from older granitic rocks of the EPT. A sample of rhyolite from the Kangaroo Caves Formation gave a Sm–Nd two-stage depleted mantle model age ($T_{DM}2$) of c. 3500 Ma and a ε_{Nd} value of −0.48. Samples from the Strelley Monzogranite, comagmatic with the Kangaroo Caves Formation, have Nd $T_{DM}{}^{2}$ ages from c. 3540 to c. 3490 Ma and ε_{Nd} values between −0.41 and − 1.0. For reasons discussed by Champion (2013) and Champion and Huston (2016), these data are consistent with felsic magma derivation through partial melting of early Paleoarchean tonalite–trondhjemite–granodiorite (TTG) crust.

At the top of the group the stratigraphic contact between the Kangaroo Caves Formation and the Soanesville Group is an unconformity (Hickman and Lipple 1975; Hickman 1981, 1983; Wilhelmij and Dunlop 1984; Morant 1995, 1998; Glikson 2001; Buick et al. 2002; Glikson and Vickers 2006, 2010; Rasmussen et al. 2007). Available geochronology suggests a maximum depositional break of 45 million years (3235–3190 Ma), although the actual duration of the break was probably less.

6.3.1.1 Leilira Formation

The basal formation of the Sulphur Springs Group, the Leilira Formation is composed of wacke, conglomerate, sandstone, siltstone, shale, chert, and thin units of felsic volcaniclastic sandstone, and is preserved in five greenstone belts of the EPT (Soanesville, Pincunah, Panorama, East Strelley, and Warralong). Thick units of massive dacite and rhyolite in some greenstone belts have been interpreted as volcanic, although it is unlikely that felsic volcanism would have immediately preceded the ultramafic volcanism (Kunagunarrina Formation) at the beginning of an ultramafic–mafic–felsic volcanic cycle. In the absence of geochronology, an alternative interpretation is that these felsic units are sills related to the c. 3240 Ma Strelley Monzogranite which underlies and intrudes the Sulphur Springs Group. The Strelley Monzogranite was emplaced as a series of felsic sills after deposition of most of the Sulphur Springs Group (Brauhart 1999; Buick et al. 2002). The maximum thickness of the Leilira Formation is 3.9 km (Van Kranendonk and Morant 1998), although in most areas the formation is only 100–500 m thick.

Deposition of the Leilira Formation resulted from erosion of the Warrawoona and Kelly Groups and Paleoarchean granitic supersuites following the 3325–3290 Emu Pool Event (granitic intrusion, diapiric doming, and metamorphism). Although this predominantly sedimentary formation was included in the Sulphur Springs Group when the group was first defined (Van Kranendonk and Morant 1998), an alternative interpretation is that it is a separate formation. The relationship to the overlying ultramafic–mafic–felsic volcanic cycle of the group is dubious, and there is a reasonable argument that the Leilira Formation is analogous to the Strelley Pool Formation beneath the Kelly Group. Both formations are more closely related to erosion of underlying crust than to subsequent plume-related volcanism. All three formations of the Sulphur Springs Group experienced syndepositional extensional faulting. Evidence of this was first recorded in the Kangaroo Caves Formation in the Soanesville greenstone belt (Vearncombe et al. 1995, 1998). Subsequently, detailed mapping of the western part of the Warralong greenstone belt by Van Kranendonk (2004a, b) revealed syndepositional extensional faults in the Leilira Formation.

Apart from thin units of the formation in the far southwest of the Panorama greenstone belt, the formation is confined to the Lalla Rookh–Western Shaw Structural Corridor (Van Kranendonk 2006, 2008) and adjacent greenstone belts to the west. This suggests that, like overlying sedimentary formations of the Soanesville Group, the Leilira Formation was only deposited west of late Paleoarchean normal faults such as the Coongan–Warralong Fault Zone and the eastern boundary faults of the Lalla Rookh–Western Shaw Structural Corridor. Extension of the EPT crust west of the Coongan–Warralong Fault Zone is likely to have been accompanied by uplift of the eastern half of the EPT from c. 3325 Ma onwards.

6.3.1.2 Kunagunarrina Formation

The Kunagunarrina Formation is the central of three formations in the 3290–3235 Ma Sulphur Springs Group. Being mainly composed of komatiitic basalt (Fig. 6.1) with lesser amounts of komatiite and tholeiitic basalt, the formation is lithologically distinct from the successions of the underlying Leilira Formation (clastic sedimentary rocks) and the overlying Kangaroo Caves Formation (basalt–andesite–dacite–rhyolite). The Kunagunarrina Formation outcrops in four greenstone belts (Pincunah, East Strelley, Soanesville, and Warralong). Thin remnants of the formation are also mapped in the southwest of the Panorama greenstone belt, where the Sulphur Springs Group unconformably overlies the Euro Basalt near the Jameson's base metal prospect (Van Kranendonk 1997, 1999). With this exception, the formation has not been identified in the eastern half of the EPT.

A notable feature of the Kunagunarrina Formation is that although it is up to 2.4 km thick in the Pincunah and East Strelley greenstone belts, it is extremely thin in the type area of the Sulphur Springs Group in the Soanesville greenstone belt. This anomaly might be due to deep erosion of the Kunagunarrina Formation in the Soanesville greenstone belt prior to deposition of the Kangaroo Caves Formation. In the Warralong greenstone belt, the Kunagunarrina Formation was deeply eroded prior to deposition of felsic volcaniclastic sandstone of the Kangaroo Caves Formation. In the East Strelley greenstone belt, up to 2 km of pillowed and massive basalt with thin minor komatiitic basalt and thin beds of chert were mapped as part of the Kunagunarrina Formation, although these lithologies are the same as those in the immediately underlying Euro Basalt. This section of the East Strelley greenstone belt is strongly faulted and separation of the Kunagunarrina Formation from the Euro

Fig. 6.1 Komatiitic pillow basalt of the Kunagunarrina Formation (Zone 50, MGA 706659E 7649650N). (From Blewett and Champion 2005; with Geological Survey of Western Australia permission)

Basalt relies on the interpretation that an isolated unit of wacke is part of the Leilira Formation rather than a local clastic unit within the Euro Basalt.

The thickness and lithological composition of the Kunagunarrina Formation are laterally variable within and between different greenstone belts, although the formation is principally composed of komatiitic basalt and tholeiite. In the type section in the Pincunah greenstone belt, the formation overlies wacke, volcaniclastic, shale, and quartzite of the Leilira Formation, apparently along a tectonic contact. The basal unit of the Kunagunarrina Formation is locally composed of sheared carbonate-altered mafic volcanic rocks up to 500 m thick. This is overlain by white-, blue-black- and grey-layered chert. Some nearby sections east of the type section do not include the carbonate-altered rocks and in these locations the basal unit is komatiitic basalt. The chert is overlain by up to 1 km of komatiitic basalt, and this is overlain by a distinctive 10 m-thick, bright green, fuchsitic chert. Above the green chert is a dark brown weathered unit of komatiite and komatiitic basalt, approximately 850 m thick. In detail, this upper succession commences with a basal unit of olivine cumulate komatiite that is overlain by ocellar- and pillow-textured volcanic rocks with some coarse fragmental tuff. Above these komatiitic basalts are units with coarser spinifex textures, and komatiitic basalts with fragmented pillows.

In the southern part of the Soanesville greenstone belt, the Kunagunarrina Formation is composed of komatiite, komatiitic basalt, and tholeiitic basalt, which are metamorphosed in the greenschist facies. Van Kranendonk (2003) reported that on the southern limb of the Soanesville Syncline, the formation reaches a maximum thickness of 1.75 km. However, following additional mapping, Van Kranendonk (2006) reinterpreted a thick underlying unit on the western limb of the Soanesville

Syncline, previously mapped as unassigned peridotite (Van Kranendonk and Pawley 2002), as a 1 km-thick komatiite of the Kunagunarrina Formation. Above the komatiite is a unit of komatiitic basalt that is approximately 1 km thick and includes a central 50 m-thick unit of carbonate-altered mafic tuff and sandstone. This rock is composed of centimetre-sized mafic lapilli in a fine-grained, intermediate to mafic matrix, now composed of chlorite, carbonate, and quartz.

Relatively abrupt lateral changes of thickness and lithology in the Kunagunarrina Formation, and the presence of unconformities with the overlying Kangaroo Caves Formation, indicate an unstable depositional environment for the Sulphur Springs Group. All three formations of the Sulphur Springs Group experienced syndepositional extensional faulting. Tectonic activity resulted in units of conglomerate along local unconformities at the basal contacts of the Kunagunarrina and Kangaroo Caves Formations (Van Kranendonk 2004a, b).

Geochemical data from the Pincunah greenstone belt, the type area of the Kunagunarrina Formation, reveal that almost all of the basaltic rocks are low-Ti ($TiO_2 < 0.8$ wt.%) and that in these rocks the La/Sm, La/Gd, La/Yb, and Gd/Yb ratios are close to primitive mantle ratios (Smithies et al. 2007). This indicates a far more depleted source than for older basaltic rocks of the Pilbara Supergroup. About half the basalts (low- and high-Ti) show depleted light rare earth elements (LREE) patterns, similar to those characteristics of the rift-related Honeyeater Basalt in the overlying Soanesville Group.

The maximum age of the Kunagunarrina Formation is poorly constrained. The formation overlies the clastic sedimentary Leilira Formation, which was deposited after c. 3290 Ma, following the Emu Pool Event (Hickman and Van Kranendonk 2008). A magmatic event in the Yilgalong Dome of the EPT between 3277 and 3274 Ma (Hickman 2021) might mark the beginning of the Sulphur Springs volcanic cycle. Accordingly, the maximum age of the Kunagunarrina Formation is tentatively estimated at c. 3275 Ma. The minimum depositional age of the formation is loosely constrained by a date of 3253 ± 4 Ma (GSWA 160258, Wingate et al. 2010) in rhyolite of the Kangaroo Caves Formation within the Tabba Tabba Shear Zone.

6.3.1.3 Kangaroo Caves Formation

The Kangaroo Caves Formation is a c. 1.5 km-thick volcanic succession of basalt, andesite, dacite, and rhyolite forming the upper formation of the 3290–3235 Ma Sulphur Springs Group. In the Pincunah greenstone belt, the formation overlies the 2.4 km-thick Kunagunarrina Formation, but in several other areas, including the type area of the Sulphur Springs Group in the northern Soanesville greenstone belt, it overlies clastic sedimentary units of the Leilira Formation. The Kangaroo Caves Formation was erupted between c. 3253 and 3235 Ma during the emplacement of the c. 3240 Ma Strelley Monzogranite (Buick et al. 2002). Felsic volcanics at the top of the formation are capped by a laterally continuous unit of chert, informally referred to as the 'marker chert'. This grey-, blue-grey- and white-layered chert is locally up to 100 m thick and comprises silicified fine-grained volcaniclastic rocks and sandstone. Thin units of ferruginous chert and BIF are also locally present within the 'marker chert', along with late veins of black hydrothermal chert. Dykes and sills of rhyodacite and dacite from the Strelley Monzogranite intrude to the top of the formation and hydrothermal veining beneath the chert includes volcanogenic Cu–Zn mineralization.

The Strelley Monzogranite was initially intruded as a sill in the lower part of the Sulphur Springs Group, and it then expanded into a laccolith that domed the Kangaroo Caves Formation. In places, small intrusions of rhyodacite and dacite broke through the 'marker chert' forming mounds of brecciated felsic rock and small domes of porphyritic rhyodacite. Neptunian dykes of breccia penetrated the chert, and the mounds were silicified and eroded prior to unconformable deposition of the Soanesville Group.

The Sulphur Springs Group has been extensively studied and documented (Morant 1995, 1998; Vearncombe et al. 1995, 1998; Brauhart et al. 1998; Van Kranendonk and Morant 1998; Vearncombe and Kerrich 1999; Van Kranendonk 1999; Buick et al. 2002; van Ruitenbeek et al. 2012). The formation is laterally variable within and between different greenstone belts, but is dominated by andesitic to rhyolitic lithologies, including andesite and minor basalt, locally overlain by dacitic to rhyolitic volcaniclastic rocks, dacite and rhyodacite in flows and sills, rhyolite lava and pumice breccia, felsic volcanic rocks, and schist, felsic volcaniclastic breccia and volcaniclastic sandstone, andesitic volcaniclastic sandstone and shale and rhyolite. Plagioclase-porphyritic rhyodacite breccia at Kangaroo Caves was dated at 3235 $\pm$ 3 Ma (sample 94002, Buick et al. 2002).

In the eastern section of the Pincunah greenstone belt, the Kangaroo Caves Formation comprises andesitic tuff, quartz-porphyritic rhyolite, shale, sandstone, basalt, and chert (Van Kranendonk 1999). Farther west, the formation is mapped as wedging out across a system of faults (Blewett and Champion 2005). In the Warralong greenstone belt, the Kangaroo Caves Formation reaches a maximum thickness of 800 m and is mainly composed of massive felsic volcaniclastic sandstone underlain by polymictic conglomerate (Van Kranendonk 2004a, b). The conglomerate contains numerous shards of devitrified felsic volcanic glass in addition to clasts of komatiitic basalt, altered volcanic rocks, sandstone, felsite, and chert. The lower 200 m of the succession overlies the Kunagunarrina Formation across an erosional unconformity (Van Kranendonk 2004a, b).

In the Panorama greenstone belt, the Kangaroo Caves Formation is between 1 and 2 km thick, and mainly composed of rhyolite lava and pumice breccia. The rhyolite is undated and prior to the stratigraphic revision by Van Kranendonk and Morant (1998) was correlated with the Wyman Formation of the Kelly Group (Hickman 1980). Other lithologies in the Kangaroo Caves Formation of this area include dacitic to rhyolitic volcaniclastic rocks and andesitic volcaniclastic sandstone and shale, and the top of the formation is defined by white-, grey-, and black-layered chert correlated with the 'marker chert' of the Soanesville and Pincunah greenstone belts. Underlying the Kangaroo Caves Formation are clastic sedimentary rocks of the Leilira Formation, approximately 500 m thick, and dacite that is also interpreted to be part of the Leilira Formation.

The geochemistry of the Kangaroo Caves Formation was described by Brauhart (1999) who recognized two geochemical suites: one with a Zr/Th ratio of <23.5, and the other with a Zr/Th ratio of >23.5. Both suites were tholeiitic and interpreted to belong to a high-K magma series derived from a common partial melt source. The intermediate to felsic volcanic rocks of the Kangaroo Caves Formation have flat HREE profiles and more fractionated LREE profiles.

6.3.1.4 Analogies with the Fig Tree Group, Kaapvaal Craton

Although there are obvious and strong similarities between the stratigraphic successions of the Warrawoona and Kelly Groups in the Pilbara Craton and the Onverwacht Group in the Kaapvaal Craton (Fig. 1.10), this similarity has previously been thought to end at the Sulphur Springs Group. Whereas the Sulphur Springs Group has been portrayed as mainly volcanic, the contemporaneous Fig Tree Group in the Kaapvaal Craton is described as mainly sedimentary (Glikson and Vickers 2006; Glikson 2014; Van Kranendonk et al. 2015). Is this difference real, and if so, did the crustal evolution of the two cratons diverge at c. 3280 Ma?

The combined age range of the Sulphur Springs Group and the Cleland Supersite is c. 3280 to 3223 Ma, which is similar to the combined age range of c. 3260 to 3215 Ma for the Fig Tree Group and comagmatic granitic intrusions in the eastern Kaapvaal Craton (Poujol et al. 2003; Lowe and Byerly 2007; Kisters et al. 2010). Additionally, the diverse lithological composition of the Sulphur Springs Group (conglomerate, sandstone, wacke, shale, chert, BIF, volcaniclastic rocks, komatiitic volcanic rocks, and basalt–rhyolite) is similar to that of the Fig Tree Group (Lowe and Byerly 2007). Adjacent to the Barberton Greenstone Belt (BGB) in the eastern Kaapvaal Craton, large granitic intrusions (Kaap Valley, Stentor, Nelshoogte, Badplaas, Dalmein) that were probably comagmatic with felsic volcanic rocks of the Fig Tree Group have been dated between c. 3255 and 3215 Ma (Poujol et al. 2003; Van Kranendonk et al. 2014). Likewise, the Cleland Supersuite, which in the EPT has been dated at 3257–3223 Ma (27 dated samples, Hickman 2021), intrudes many of the domes (Van Kranendonk et al. 2006), and was comagmatic with felsic volcanic rocks of the Sulphur Springs Group (Buick et al. 2002).

Deposition of the Fig Tree Group marked a major change in the crustal evolution of the BGB (Lowe and Byerly 1999, 2007; Glikson 2001, 2005, 2008; Glikson and Vickers 2006). Mainly volcanic deposition of the Onverwacht Group was replaced by mainly sedimentary deposition of the Fig Tree Group, and subsequently by coarser clastic sedimentary rocks of the Moodies Group. This is analogous to a change recognized in the Pilbara Craton between the almost entirely volcanic successions of the Pilbara Supergroup and the mainly sedimentary formations of the Soanesville Group (Van Kranendonk et al. 2002, 2006, 2007, 2010; Hickman 2004, 2012). Most previous workers have interpreted the timing of these changes in the EPT and BGB to be different—c. 3225 in the EPT and c. 3260 in the BGB. However, the diverse lithological composition of the Sulphur Springs Group between EPT greenstone belts suggests that this difference may not be real. Recent detailed comparisons between the EPT and the BGB (Van Kranendonk et al. 2014, 2015) concluded that the terranes were formed by the same magmatic and tectonic processes operating over the same interval of time.

Regional investigations of the stratigraphy of the Sulphur Springs Group have revealed that it is not mainly volcanic. Likewise, in the BGB, komatiitic volcanic rocks assigned to the komatiitic Weltevreden Formation overlie sedimentary rocks in part of the Fig Tree Group (Lowe and Byerly 2007). Additionally, geochronological data reviewed by Lowe and Byerly (2007) suggest that the komatiitic Mendon Formation has an age range between c. 3298 and 3245 Ma. The sedimentary composition of the Sulphur Springs Group in some parts of the EPT is illustrated by its succession in the West Warralong greenstone belt (Fig. 5.9). Conglomerate, quartz sandstone, red-brown wacke and shale, BIF, chert and felsic volcanic rocks are predominant, with only minor komatiitic basalt (Van Kranendonk 2004b).

Because the Fig Tree Group is composed of interlayered volcaniclastic strata (marking the final stages of volcanism) and terrigenous clastic units eroded from uplifted portions of the Onverwacht Group, Lowe and Byerly (2007) interpreted it to be a 'transitional' unit between the Onverwacht Group (volcanic) and Moodies Group (clastic sedimentary). This reasoning can also be applied to the Sulphur Springs Group, with clastic material being derived from the uplifted Warrawoona and Kelly Groups, and from erosion of pre-3290 Ma granitic supersuites. The Sulphur Springs and Fig Tree Groups were deposited in similar tectonic settings of crustal extension (Van Kranendonk et al. 2015). Granitic intrusion between c. 3255 and 3220 Ma resulted from crustal melting associated with doming. The conclusion that both cratons were undergoing crustal extension between c. 3280 and 3220 Ma suggests that continental breakup occurred across the putative Vaalbara supercontinent at c. 3220 Ma.

6.3.2 Roebourne Group

The Roebourne Group is correlated with the Sulphur Springs Group because the successions are the same age, include the same lithologies, and prior to 3220 Ma breakup of the Paleoarchean Pilbara Craton are interpreted to have been contiguous units on the same volcanic plateau (Sun and Hickman 1998; Hickman 2001, 2004, 2016; Hickman et al. 2006, 2010). However, they became separated during the c. 3220 Ma breakup of the Paleoarchean volcanic plateau, after which the Regal Basin (a basaltic rift basin) opened between the East Pilbara and Karratha Terranes. Remnants of the Sulphur Springs Group and the Cleland Supersuite are present in the Tabba Tabba Shear Zone along the southeast margin of the Regal Basin, whereas the Roebourne Group and Karratha Granodiorite (Cleland Supersuite) are in contact with the northwest margin of the Sholl Shear Zone along the northwest margin of the Regal Basin.

In the Northwest Pilbara Craton, the Karratha Terrane is partly composed of mafic and ultramafic rocks of the Roebourne Group and partly of TTG of the Karratha Granodiorite. The group comprises two conformable formations, the Ruth Well Formation overlain by the Weerianna Basalt. The Ruth Well Formation is a mixed assemblage of komatiite, komatiitic basalt, tholeiite, chert, and sills of peridotite and dolerite, and the Weerianna Basalt is composed of tholeiitic and komatiitic basalt. Most of the group has been metamorphosed to lower amphibolite facies. Although an absence of felsic volcanic rocks has precluded dating by the SHRIMP U–Pb zircon method, the Ruth Well Formation is intruded by the Karratha Granodiorite, samples of which have been dated between 3270 Ma and 3261 Ma. The stratigraphic correlation with the Sulphur Springs Group implies a maximum depositional age of c. 3280 Ma.

6.3.2.1 Ruth Well Formation

Ultramafic lithologies in the Ruth Well Formation include metamorphosed komatiite with well-developed olivine spinifex textures (Fig. 6.2), tremolite–chlorite schist, serpentinite, and talc–chlorite schist, metaperidotite and metapyroxenite. Mafic lithologies include amphibolite to upper greenschist facies metabasalt, basaltic schist, metabasalt containing minor chert units, variolitic basalt and komatiitic basalt and metagabbro. Chert lithologies include massive or weakly layered grey and white chert (Fig. 6.3), bedded green chert, ferruginous chert, and BIF, bedded black chert, massive metachert and grey- and white-layered metachert. Most chert units are interpreted to be silicified sedimentary rocks that were originally deposited as shale or fine-grained volcaniclastic beds between lava flows. Metamorphosed sandstone and siltstone outcrop in the Lower Nickol area, and at Mount Wangee, northeast of Roebourne, quartzite is interbedded with chert.

An absence of pillow structures in basalts of the Ruth Well Formation does not suggest marine deposition. Additionally, there is the more general consideration that Archean komatiites were products of deep, high-temperature mantle melting (Arndt et al. 1998). Komatiites are absent from Phanerozoic oceanic crust that was formed at spreading centres or in arc successions derived from melting of subducted oceanic crust. The basaltic succession of the Ruth Well Formation does not have an oceanic-like geochemical composition. For example, the trace element compositions of tholeiitic basalt, komatiitic basalt, and komatiite in the Ruth Well Formation (Fig. 6.4) suggest significant interaction with continental crust. Concentrations of Th in the basalts are up to 42 times greater than primitive mantle values and the light rare earth elements (LREE) are strongly fractionated, with La at least 20 times primitive mantle values. La/Nb, La/Sm, and La/Yb ratios are high and negative Nb anomalies are strongly developed (Smithies et al. 2007). Enrichments in Th, U, and LREE clearly distinguish these basalts from those of the MORB-like Regal Formation (Ohta et al. 1996; Sun and Hickman 1999; Smithies et al. 2007). The trace element composition of the Ruth Well Formation shows greater LREE enrichment than for any of the Paleoarchean basaltic formations in the East Pilbara (data in

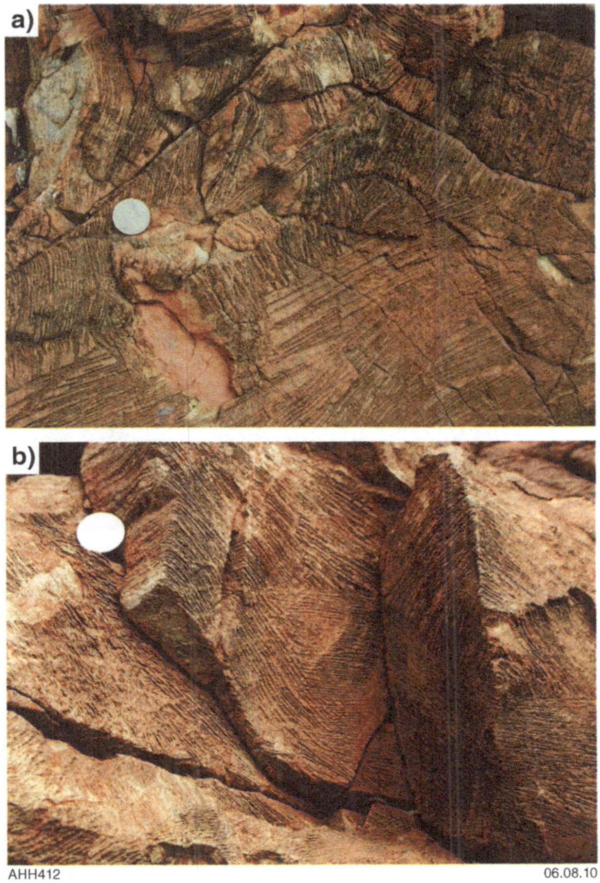

Fig. 6.2 Bladed olivine spinifex texture in a komatiite flow of the Ruth Well Formation at Mount Hall: (**a**) platy olivine crystals, pseudomorphed by serpentine, tremolite, and chlorite, up to 50 cm long; (**b**) close-up of interlocking olivine crystals. Scale in both pictures, 3 cm diameter coin (Zone 50, MGA 520770 E, 7701337N). (From Hickman et al. 2010; with Geological Survey of Western Australia permission)

Smithies et al. 2007), all of which were deposited on continental crust (Smithies et al. 2003, 2007; Hickman 2004, 2012; Van Kranendonk et al. 2007, 2015).

Sm–Nd two-stage depleted mantle model ages ($T_{DM}{}^{2}$) of komatiite, komatiitic basalt, and tholeiitic basalt of the Ruth Well Formation (Arndt et al. 2001; Smithies et al. 2007) vary between c. 3480 Ma and c. 3400 Ma (Hickman 2016). These results are inconsistent with the Ruth Well Formation being juvenile oceanic crust, but instead suggest melting of, or contamination by, much older crust or lithospheric mantle. ε_{Nd} values between −1.92 and + 0.88 (data in Hickman 2016) suggest magma derivation from older continental crust rather than direct derivation from a depleted c. 3300 Ma mantle.

Fig. 6.3 Banded grey-white chert in the Ruth Well Formation west of the Karratha Granodiorite. The chert is interpreted to be a unit of silicified carbonaceous shale between flows of komatiite and basalt (Zone 50, MGA 476560E, 7,696,400 N). (From Hickman et al. 2010; with Geological Survey of Western Australia permission)

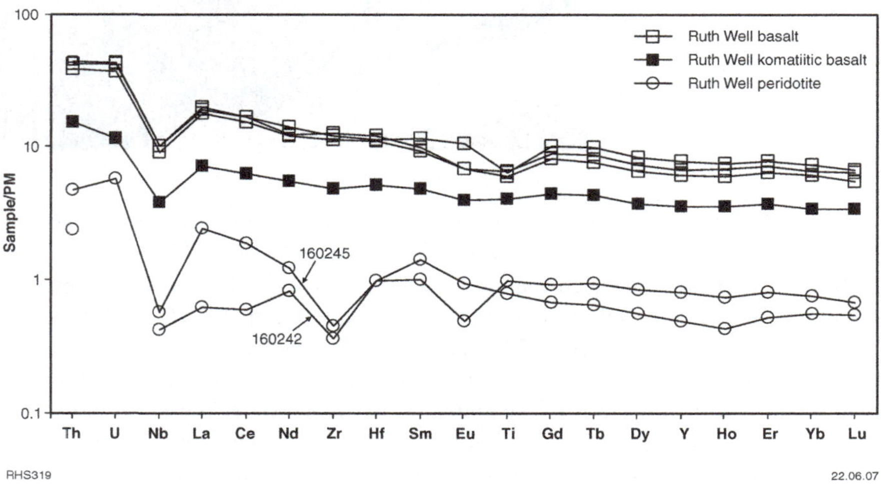

Fig. 6.4 Trace element plots normalized to primitive mantle for komatiites and komatiitic basalts of the Ruth Well Formation (From Smithies et al. 2007; with Geological Survey of Western Australia permission)

6.3.2.2 Weerianna Basalt

The Weerianna Basalt is composed of metamorphosed tholeiitic basalt and fine-grained komatiitic basalt. The formation conformably overlies the Ruth Well Formation and is intruded by sills and dykes of gabbro and dolerite, most of which are interpreted to belong to the Andover Intrusion. In areas near intrusive contacts the metabasalt occurs as xenoliths within the gabbro. Brecciation and gabbroic veining of the basalt is most common in the area immediately southwest of Roebourne. In this area, the stratigraphic thickness of the Weerianna Basalt is up to 1 km and the formation outcrops over a strike length of 20 km. Pillow structures are locally present: for example, 2.5 km east of Carlow Castle. Flow top breccias are exposed west of Roebourne, approximately 1.5 km south of the Weerianna gold mine. The Weerianna Basalt is distinguished from the underlying Ruth Well Formation by an absence of komatiite and chert units, although chert is present at the top of the formation southeast and south of the Weerianna gold mine.

6.3.3 Cleland Supersuite

Emplacement of granitic intrusions of the Cleland Supersuite resulted from partial melting of the crust of the Paleoarchean volcanic plateau during the last mantle plume event in the evolution of the Pilbara Craton. Whereas there is currently no evidence that older supersuites of the EPT were intruded in other terranes of the craton, the Cleland Supersuite was also intruded in the KT and KUT. The variable compositions of the crust over such a large area explain the wide range of granitic compositions in this supersuite, from diorite and tonalite to monzogranite and syenogranite.

The Cleland Supersuite comprises twelve named granitic intrusions and various unnamed intrusions (Hickman 2021). Crystallization ages range between c. 3274 and 3223 Ma. One of the youngest intrusions of the Supersuite, the c. 3240 Ma Strelley Monzogranite, includes comagmatic dolerite. Because the supersuite was intruded after the last significant Paleoarchean deformation event (Emu Pool Event), the granitic rocks lack well-developed tectonic foliations.

In the Northwest Pilbara, the supersuite is represented by two tonalite–trondhjemite–granodiorite (TTG) intrusions: the c. 3270–3261 Ma Karratha Granodiorite (Figs. 6.5 and 6.6) and the c. 3236 Ma Tarlwa Pool Tonalite. In the EPT, the supersuite is mainly composed of granodiorite and monzogranite. Most intrusions of the Cleland Supersuite in the east Pilbara were emplaced close to the granitic centres of the Paleoarchean domes. Although the supersuite outcrops in most domes of the EPT, associated felsic volcanic rocks are restricted to the Kangaroo Caves Formation of the Sulphur Springs Group. This formation is underlain by a subvolcanic laccolith, the c. 3240 Ma Strelley Monzogranite.

AHH389 23.07.10

Fig. 6.5 Sampling site (Geological Survey of Western Australia sample142433) in the Karratha Granodiorite near Mount Regal. The crystallization age of the granodiorite, using the U–Pb zircon method, was calculated as c. 3270 Ma and the Sm–Nd T_{DM}^2 model age was calculated as c. 3480 Ma (data in Smithies et al. 2007) (From Hickman et al. 2010; with Geological Survey of Western Australia permission)

In common with the Emu Pool Supersuite, the Cleland Supersuite is composed of high- and low-Al granites (Champion and Smithies 2007). High-Al granites have elevated Al_2O_3, Na_2O, and Sr. They also have high light rare earth element: heavy rare earth element ratios (LREE/HREE) and low FeO (total), HREE, Y and $(Gd/Yb)_N$ compared to low-Al granites. Geochemical data suggest that the high-Al granites were mainly derived by partial melting of mafic crust. The less common low-Al granites have relatively high SiO_2 and K_2O contents, high FeO (total) and high HREE contents. The low-Al granites also have pronounced negative Eu anomalies (Champion and Smithies 2007) indicative of fractional crystallization.

ε_{Nd} values for the supersuite are mainly negative and Sm–Nd two-stage depleted mantle model ages (T_{DM}^2) range between c. 3560 and 3460 Ma, consistent with melting of older crust (data reviewed in Hickman 2021).

6.4 Continental Breakup

Increasing crustal extension and rifting during the deposition of the Sulphur Springs Group (Vearncombe et al. 1995, 1998; Brauhart et al. 1998; Van Kranendonk 2004a, b) led to the development of at least two rift basins: the Regal Basin and

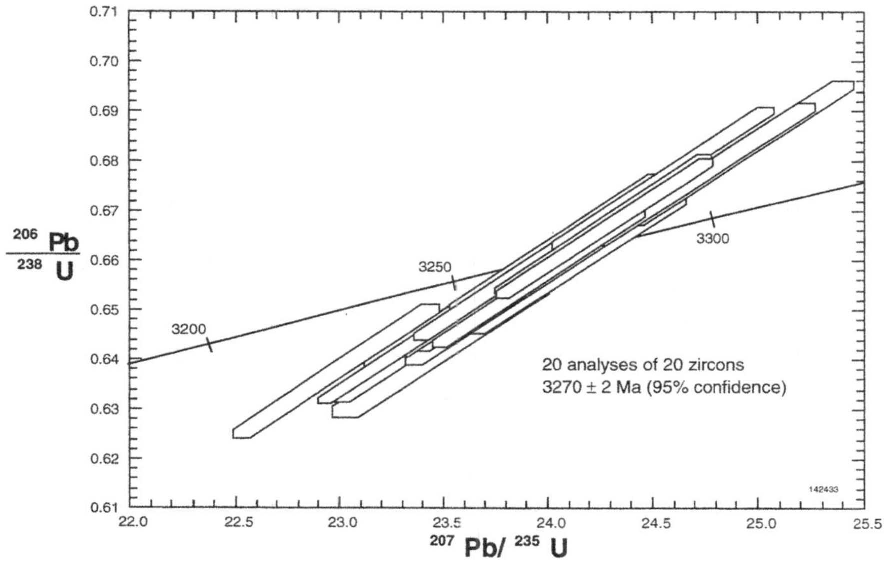

Fig. 6.6 Concordia plot of U–Pb zircon data for a sample of tonalite (Geological Survey of Western Australia sample 142433) from the Karratha Granodiorite near Mount Regal (From Nelson 1998; with Geological Survey of Western Australia permission)

the Mosquito Creek Basin. The Regal Basin developed between the EPT and KT, whereas the Mosquito Creek Basin began to form between the EPT and the KUT. Both rift basins were bounded by major normal faults; the Regal Basin by faults in the positions of the current Tabba Tabba and Sholl Shear Zones (Fig. 1.3), and the Mosquito Creek Basin by faults that later evolved into the Kurrana and Nullagine River Shear Zones. At c. 3220 Ma, continuing extension split the continental crust of both basins, and each began to evolve as proto-oceanic basins containing juvenile, MORB-like basaltic crust. In the case of the Regal Basin, this evolution lasted until c. 3160 Ma when plate separation was halted by the KT colliding with another plate converging from the northwest (Hickman 2016).

6.4.1 *Evidence*

Separation of the KT from the EPT, forming the Regal Basin, is the most direct evidence for the c. 3220 Ma continental breakup of the Paleoarchean Pilbara Craton. Related evidence has come from the recognition that post-3220 Ma clastic sedimentary successions such as the lower part of the Soanesville Group, the Budjan Creek Formation, and the Nickol River Formation were deposited in separate passive margin basins (Chap. 7). Other evidence is provided by the geological histories of major faults such as the Tabba Tabba and Sholl Shear Zones, and by a post-3220 Ma divergence of the geological evolutions of the Northwest and East Pilbara Craton.

6.4.1.1 Recognition of the Regal Basin

Evidence leading to the recognition of the Regal Basin was obtained from the 1994–2005 Pilbara Craton Mapping Project (Chap. 1). The area of the northwest Pilbara underlain by the Regal Basin was described as the Central Pilbara Tectonic Zone (CPTZ) (Hickman 2001, 2004; Hickman et al. 2001; Van Kranendonk et al. 2002). Hickman (2001) interpreted the CPTZ (Fig. 1.3) as a northeast–southwest trending zone of Mesoarchean deformation between the EPT and the KT that originated as an extensional basin containing the c. 3125 Ma Whundo Group. The same zone was subsequently the depositional site of the Mallina Basin. Hickman et al. (2001) referred to Sm–Nd isotope data (Sun and Hickman 1998) indicating that the CPTZ is not underlain by crust older than c. 3300 Ma, in contrast to early Paleoarchean crust in the EPT and KT. Hickman et al. (2001) concluded that the basin was formed by post-3300 Ma rifting of Paleoarchean crust.

The <3300 Ma Nd model ages in the CPTZ and > 3500 Ma Nd model ages in the EPT and KT have been confirmed by later work (Smithies et al. 2004, 2007; Van Kranendonk et al. 2007; Champion 2013; Gardiner et al. 2018). Combined with evidence that the exposed Paleoarchean geology in the KT (Roebourne Group and Karratha Granodiorite) correlates with the late Paleoarchean stratigraphy of the EPT (Sulphur Springs Group and Cleland Supersuite) (Sun and Hickman 1998; Hickman 2001; Hickman et al. 2006; Van Kranendonk et al. 2006), these data support the interpretation that the Regal Basin originated from a c. 3200 Ma separation of the terranes.

The basaltic crust of the Regal Basin, the c. 3200–3165 Ma Regal Formation (Chap. 7), was obducted onto the KT between 3160 and 3070 Ma (Chap. 7). Geochemical analyses have shown the basalts of the Regal Formation to have MORB-like compositions (Glikson et al. 1986; Ohta et al. 1996; Sun and Hickman 1998, 1999; Shibuya et al. 2007; Smithies et al. 2007). Data from one sample of the Regal Formation indicated a Nd model age (T_{DM}^2) of c. 3160 Ma and a ε_{Nd} value of +3.5, close to the theoretical depleted mantle value at c. 3200 Ma (Smithies et al. 2007). These data are consistent with eruption of the Regal Formation from an oceanic-like spreading centre (Fig. 1.5), although evolution of the Regal Basin was interrupted by a plate collision with the result that it never attained the dimensions of an ocean.

6.4.1.2 Passive Margin Successions

The stratigraphy and interpreted depositional settings of c. 3200 Ma sedimentary units such as the Soanesville Group, Budjan Creek Formation, and Nickol River Formation (Chap. 7) indicate deposition in passive margin basins. In the cases of the

Soanesville Group and Budjan Creek Formation, this interpretation was first reached by Eriksson (1981). The distribution of these successions is consistent with the c. 3200 Ma existence of rifted margins on the northwest and southeast sides of the EPT and on the southeast side of the KT.

6.4.1.3 Evidence from Major Faults

The CPTZ is bounded by two major shear zones: along its northwest side by the Sholl Shear Zone and along its southeast margin by the Tabba Tabba Shear Zone. Based on detailed investigations of both these shear zones (Smith et al. 1998; Beintema et al. 2001, 2003; Beintema 2003), they originated during the EPTRE.

6.4.1.3.1 Tabba Tabba Shear Zone

The Tabba Tabba Shear Zone (TTSZ) is a 250 km-long zone of intense faulting and shearing that defines the southeast boundary of the CPTZ and separates the EPT from the Mallina Basin and the Northwest Pilbara Carton (Figs. 1.3 and 6.7). The TTSZ is up to 3 km wide and contains strongly foliated to mylonitic units derived from almost all lithologies in the EPT and the Mallina Basin. Following the acquisition of Landsat imagery in the early 1980s, the structure was identified as one of several major lineaments that extend across the northern Pilbara Craton (Krapež and Barley 1987; Krapež 1993; Barley 1997; Krapež and Eisenlohr 1998). However, the geological significance of these lineaments was questioned for several years (Smithies et al. 1999; Hickman et al. 2000; Hickman 2001, 2004; Van Kranendonk et al. 2002).

The origins, geological history, and tectonic significance of the TTSZ are closely linked to the evolution of the CPTZ (Hickman 2016). Evidence that the TTSZ originated as a c. 3235 Ma extensional fault includes the presence of c. 3235 Ma gabbroic intrusions within the shear zone (Beintema 2003; Beintema et al. 2003). On the southeast side of the TTSZ, the c. 3235 Ma gabbros intruded c. 3257 and 3250 Ma granodiorite and granite (Beintema 2003; Beintema et al. 2003; GSWA 160745, Nelson 2001) and 3253 ± 4 Ma felsic volcanic rock (GSWA 160258, Wingate et al. 2010). The granitic rocks are part of the Cleland Supersuite, and the felsic volcanic rocks are correlated with the Sulphur Springs Group.

Xenocrystic zircon in the c. 3250 Ma granitic rocks of the TTSZ have ages consistent with derivation from Paleoarchean sources in the EPT (GSWA 160745, Nelson 2001; KB 263, 312, 770, 779, Beintema 2003). By contrast, granitic intrusions in the Mallina Basin immediately northwest of the TTSZ contain no xenocrystic zircons older than c. 3246 Ma (GSWA 160727, Nelson 2001), and the single 3246 Ma grain detected was most likely inherited from sedimentary rocks of the Mallina Formation rather than from basement rocks. Thus, the TTSZ marks the northwestern edge of the EPT, and a major change in the composition of the Pilbara Craton crust.

Fig. 6.7 Interpreted bedrock geology in the central section of the Tabba Tabba Shear Zone, showing a complete mismatch of the geology between the East Pilbara Terrane in the southeast and the Central Pilbara Tectonic Zone in the northwest. Reference in 6.7b. (From Smithies et al. 2001b; with Geological Survey of Western Australia permission)

The second episode of deformation along the TTSZ occurred after c. 3160 Ma when separation of the EPT and KT was reversed, and these terranes reconverged until 3070 Ma. Based on ages of xenocrystic zircon within granitic rocks (KB 746, 770, Beintema 2003), the TTSZ records 3160–3070 Ma movement and associated igneous activity. Beintema et al. (2001, 2003) described evidence of early dextral movement on the shear zone and suggested this occurred at or before c. 3115 Ma.

Structural studies of the TTSZ have reported evidence of major c. 2940 Ma strike-slip movement (Beintema et al. 2001, 2003; Hickman et al. 2001; Smithies and Champion 2002; Beintema 2003). Mesoscopic structures, such as C – S fabrics and

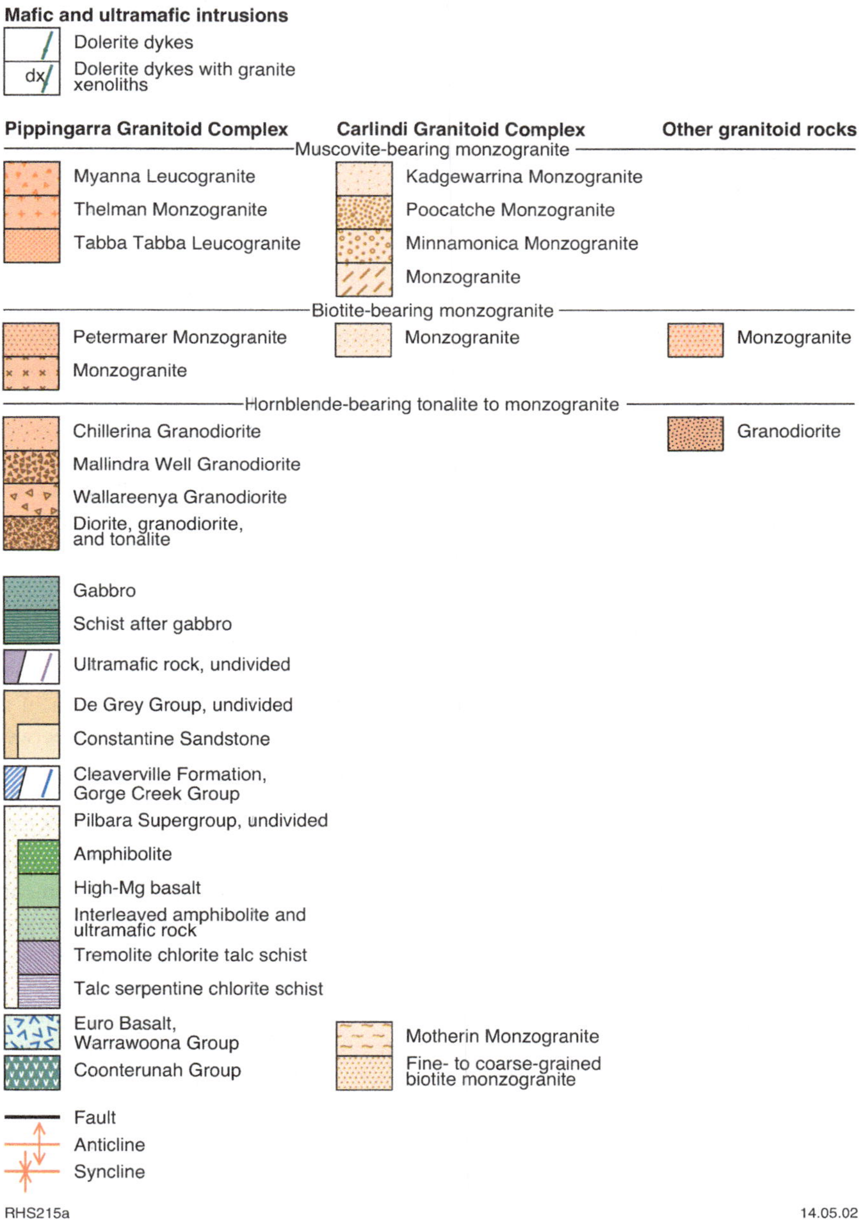

Fig. 6.7 (continued)

rotated feldspar phenocrysts, indicate a sinistral component of displacement. Additionally, northwest-plunging mineral and stretching lineations indicate normal movement consistent with juxtaposition of the Croydon Group in the Mallina Basin against Paleoarchean rocks of the EPT to the southeast. The intrusion of gabbro and high-Mg diorite along the TTSZ reveals that it was a deep crustal structure controlling the migration and emplacement of mantle-derived magmas (Smithies and Champion 2000).

6.4.1.3.2 Sholl Shear Zone

The Sholl Shear Zone (SSZ) is a major sinistral strike-slip fault that separates Paleoarchean crust of the Karratha Terrane (KT) from Mesoarchean crust of the Sholl Terrane (ST) (Fig. 1.3). Geophysical evidence indicates that the onshore extent of the SSZ is c. 350 km (Hickman et al. 2001). Its total length, including sections concealed by the Fortescue Basin in the southwest and offshore beneath the Northern Carnarvon Basin to the northeast, is likely to be c. 600 km (Hickman 2016). In areas where the shear zone is well exposed, as in the Nickol River area 20 km southeast of Karratha (Fig. 6.8a-c), it is a sub-vertical unit of mylonite and schist between 1 and 2 km wide (Hickman 1997, 2004; Hickman et al. 1998).

In view of the major differences between the KT and the ST, and the interpretation that the Roebourne Group was displaced from the northern section of the EPT, Sun and Hickman et al. (1998) suggested major sinistral strike-slip movement on the SSZ. This movement must have occurred at some time between deposition of the Whundo Group at c. 3130 Ma and deposition of the Gorge Creek Group between 3067 and 3015 Ma because the Cleaverville Formation was deposited on both sides of the SSZ. Small intrusions of the Railway Supersuite, including one dated at 3123 ± 2 Ma (sample N4413, Smith 2003), intrude the southern side of the SSZ at Twin Table Hills 17 km south of Karratha. This suggests that a fault already existed in this position during eruption of the Whundo Group. Most sinistral displacement is likely to have occurred during plate collision at c. 3070 Ma (Prinsep Orogeny, Chap. 7).

Earlier deformation along the SSZ is suggested by a $^{40}Ar/^{39}Ar$ age of 3144 ± 35 Ma (KB 34, Beintema 2003) in amphibolite of the Ruth Well Formation immediately north of the Sholl Shear Zone. Beintema et al. (2003) interpreted this to be a cooling age after amphibolite facies metamorphism. This metamorphism is likely to have coincided with reactivation of the SSZ during the Karratha Event when separation of the KT and EPT was reversed (Hickman 2016). Much rifting during the c. 3220–3165 Ma evolution of the Regal Basin is likely to be concealed by the Regal Formation. However, the earliest rifting should now be at the margins of the basin. In this regard, it is significant that the abrupt change in Nd model ages from the CPTZ to the KT is located at either the SSZ or the Loudens Fault (Hickman 2016). The Loudens Fault is a major branch of the SSZ that separates the Whim Creek greenstone belt from the Mallina Basin (overlying the Regal Basin) in the CPTZ.

Fig. 6.8 Mylonite in the Sholl Shear Zone at Nickol River (MGA Zone 50, 494950E, 7689700N): (**a**) extensive outcrop in the Nickol River showing vertically inclined layers of felsic and mafic mylonite. Felsic mylonite (pale colour) was derived from intensely sheared granitic rocks and mafic mylonite originating from similarly sheared greenstones; (**b**) view of mylonite layering from above, showing minor folding of some layers indicating dextral shear sense (2940–2920 Ma). Scale: lens cap, 5 cm diameter; (**c**) isoclinal folding of mylonite fabric. Scale: coin, 2 cm diameter. (From Hickman 2016; with Geological Survey of Western Australia permission)

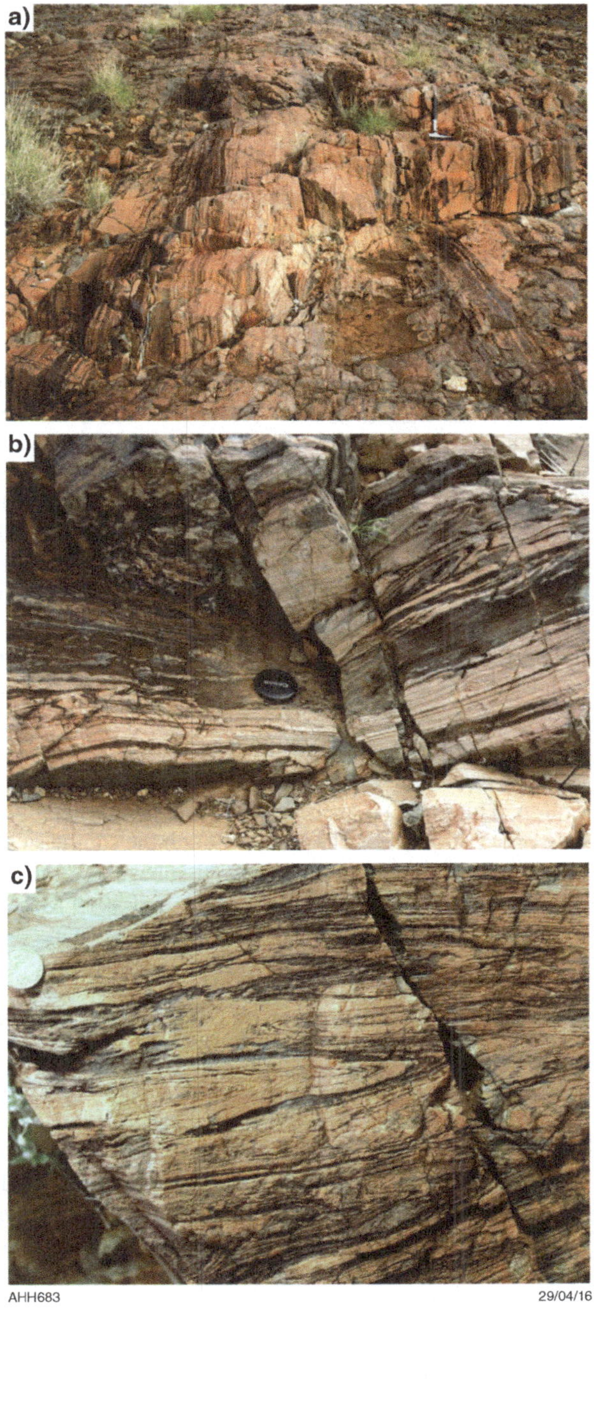

It is concluded that the SSZ originated as a c. 3220 Ma rift fault along what is now the northwest boundary of the CPTZ.

6.4.1.4 Mesoarchean Differences between Northwest and East Pilbara

Once U–Pb zircon geochronology began to be more widely applied in the northern Pilbara Craton, it became evident that the Northwest Pilbara Craton contains stratigraphic units such as the 3130–3110 Ma Whundo Group and 3130–3093 Ma Railway Supersuite that are absent from the East Pilbara Craton. Whereas the Northwest Pilbara was tectonically and magmatically active from c. 3165 to 3070 Ma (Hickman 1997, 2001, 2004, 2016; Hickman et al. 2001; Van Kranendonk et al. 2006, 2007; Smithies et al. 2005), no additions were made to the stratigraphy of the East Pilbara during this period, and additionally the EPT experienced no deformation. However, there are stratigraphic correlations between the Northwest and East Pilbara before 3220 Ma (e.g. Cleland Supersuite) and after 3070 Ma (e.g. Gorge Creek Group). The absence of correlations between 3220 and 3070 Ma is explained by the interpretation that during this period the newly formed oceanic-like crust of the Regal Basin (Fig. 1.5) effectively isolated the EPT from the effects of the convergence between the KT and a northwest plate (Chaps. 7 and 8). Compression and failure of the thin crust in the Regal Basin, including the formation of an internal subduction zone, absorbed all the effects of convergence until the KT and EPT collided at c. 3070 Ma, closing the basin, and producing the Prinsep Orogeny (Chap. 7).

With closure of the Regal Basin, ongoing compression of the Northwest Pilbara caused the relocation of plate subduction to the northwest margin of the Pilbara Craton. This led to a succession of continental magmatic arcs and retro-arc basins between c. 3067 and 2919 Ma. Although the southeast migrating magmatism did not reach the EPT until c. 2950 Ma, the first basin (Gorge Creek Basin, Chap. 8) formed across the northern EPT between 3067 and 3015 Ma.

6.4.2 *Other Fragments of the Paleoarchean Plateau*

The geology of the Northern Pilbara Craton has provided the evidence for the extension and breakup of the Paleoarchean volcanic plateau. This continental plateau has been interpreted to have included the Eastern Kaapvaal Craton (Zegers et al. 1998; Hickman 2021), and it might also have included the Singhbhum and Bastar Cratons of eastern India. Geochronological data suggest that all these cratons had closely contemporaneous magmatic events, and it can be argued that this suggests a common ancestry on a single continent rather than parallel evolutions on different continents. There are remarkable similarities between the Paleoarchean–Paleoproterozoic stratigraphic successions of the Pilbara and Kaapvaal Cratons (Fig. 1.10), although most published comparisons have so far been made between the Neoarchean–Paleoproterozoic successions (Cheney et al. 1988; Cheney 1996; Martin et al. 1998; Eriksson et al. 2002; Bleeker 2003; Strik et al. 2003; De Kock et al. 2009, 2012; Huston et al. 2012; Smirnov et al. 2013). However, Nelson et al. (2014) noted close similarities between the Paleoarchean geology of the Pilbara and Singhbhum Cratons, and closer comparison of these cratons might be informative. Already, there is evidence that the cratons contain a common dyke swarm dated at c. 2765 Ma (Kumar et al. 2017).

Most reconstructions of Precambrian supercontinents place Western Australia, South Africa, and India close together and it was not until the breakup of Gondwana in the Jurassic to Cretaceous that they became widely separated.

References

Arndt N, Ginibre C, Chauvel C, Albarède F, Cheadle M, Herzberg C, Jenner G, Lahaye Y (1998) Were komatiites wet? Geology 26:739–742

Arndt N, Bruzak G, Reischmann T (2001) The oldest continental and oceanic plateaus: geochemistry of basalts and komatiites of the Pilbara Craton, Australia. In: Ernst RE, Buchan K (eds) Mantle plumes: their identification through time. Geological Society of America, pp 359–387

Bagas L, Farrell TR, Nelson DR (2004) The age and provenance of the Mosquito Creek Formation. Geological Survey of Western Australia Annual Review 2003–04:62–70

Bagas L (2005) Geology of the Nullagine 1:100 000 sheet. 1:100 000 geological series explanatory notes. Geological Survey of Western Australia

Barley ME (1997) The Pilbara Craton. In: De Wit MJ, Ashwel L (eds) Greenstone belts. Oxford University Press, Oxford, UK, pp 657–664

Beintema KA (2003) Geodynamic evolution of the west and central Pilbara Craton: A mid-Archean active continental margin. PhD, Geologica Ultraiectina, Utrecht University

Beintema KA, de Leeuw GAM, White SH, Hein KAA (2001) Tabba shear: a crustal-scale structure in the Pilbara Craton, WA. In: Cassidy KF, Dunphy JM, Van Kranendonk MJ (eds) Extended Abstracts–4th International Archaean Symposium, Perth, 24–28 September 2001. AGSO (Geoscience Australia), pp 285–287

Beintema KA, Mason PRD, Nelson DR, White SH, Wijbrans JR (2003) New constraints on the timing of tectonic activity in the Archaean central Pilbara Craton, Western Australia. J Virtual Explor 13. https://doi.org/10.3809/jvirtex.vol.2003.013

Bleeker W (2003) The late Archean record: a puzzle in ca. 35 pieces. Lithos 71:99–134

Blewett RS, Champion DC (2005) Geology of the Wodgina 1:100 000 sheet. 1:100 000 Geological Series Explanatory Notes. Geological Survey of Western Australia

Brauhart C (1999) Regional alteration systems associated with Archean volcanogenic massive sulphide deposits at Panorama, Pilbara, Western Australia. PhD, The University of Western Australia

Brauhart C, Groves DI, Morant P (1998) Regional alteration systems associated with volcanogenic massive sulphide mineralization at Panorama, Pilbara, Western Australia. Econ Geol 93:292–302

Buick R, Brauhart CW, Morant P, Thornett JR, Maniw JG, Archibald NJ, Doepel MG, Fletcher IR, Pickard AL, Smith JB, Barley ME, McNaughton NJ, Groves DI (2002) Geochronology and stratigraphic relationships of the Sulphur Springs Group and Strelley granite: a temporally distinct igneous province in the Archean Pilbara Craton, Australia. Precambrian Res 114:87–120

Champion DC (2013) Neodymium depleted mantle model age map of Australia: explanatory notes and user guide. Record, 2013/44. Geoscience Australia

Champion DC, Huston DL (2016) Radiogenic isotopes, ore deposits and metallogenic terranes: novel approaches based on regional isotopic maps and the mineral systems concept. Ore Geol Rev 76:229–256. https://doi.org/10.1016/j.oregeorev.2015.09.025

Champion DC, Smithies RH (2007) Geochemistry of Paleoarchean granites of the East Pilbara Terrane, Pilbara Craton, Western Australia: implications for early Archean crustal growth. In: Van Kranendonk MJ, Bennett VC, Smithies RH (eds) Earth's oldest rocks. Elsevier B.V, Burlington, Massachusetts, USA, pp 369–410

Cheney ES (1996) Sequence stratigraphy and plate tectonic significance of the Transvaal succession of southern Africa and its equivalent in Western Australia. Precambrian Res 79:3–24

Cheney ES, Roering C, Stettler E (1988) Valbara. In: Geocongress '88: extended abstracts. Geological Society of South Africa, pp 85–88

De Kock MO, Beukes NJ, Armstrong RA (2012) New SHRIMP U–Pb zircon ages from the Hartswater Group, South Africa: implications for correlations of the Neoarchean Ventersdorp Supergroup on the Kaapvaal Craton and with the Fortescue Group on the Pilbara Craton. Precambrian Res 204–205:66–74

De Kock MO, Evans DAD, Beukes NJ (2009) Validating the existence of Vaalbara in the Neoarchean. Precambrian Res 174:145–154

Eriksson KA (1981) Archaean platform-to-trough sedimentation, East Pilbara Block, Australia. In: Glover JE, Groves DI (eds) Archaean geology: second International Symposium, Perth 1980. Geological Society of Australia, pp 235–244

Eriksson KA (1982) Geometry and internal characteristics of Archaean submarine channel deposits, Pilbara Block, Western Australia. J Sediment Res 52:383–393

Eriksson PG, Condie KC, van der Westhuizen W, van der Merwe R, de Bruiyn H, Nelson DR, Altermann W, Catuneanu O, Bumby AJ, Lindsay J (2002) Late Archaean superplume events: a Kaapvaal-Pilbara perspective. J Geodyn 34:207–247

Farrell TR (2006) Geology of the Eastern Creek 1:100 000 sheet. 1:100 000 Geological Series Explanatory Notes. Geological Survey of Western Australia

Gardiner NJ, Hickman AH, Kirkland CL, Lu Y, Johnson TE, Wingate MTD (2018) New Hf isotope insights into the Paleoarchean magmatic evolution of the Mount Edgar Dome, Pilbara Craton: implications for early Earth and crust formation processes. Report, vol 181. Geological Survey of Western Australia

Glikson AY, Davy R, Hickman AH (1986) Geochemical data files of Archaean volcanic rocks, Pilbara Block, Western Australia. Record 1986/014. Bureau Mineral Resources, Canberra, 136p

Glikson AY (2001) The astronomical connection of terrestrial evolution: crustal effects of post-3.8 Ga mega-impact clusters and evidence for major 3.2 Ga bombardment of the earth–moon system. J Geodyn 32:205–229

Glikson AY, Vickers J (2006) The 3.26–3.24 Ga Barberton asteroid impact cluster: tests of tectonic and magmatic consequences, Pilbara Craton, Western Australia. Earth Planet Sci Lett 241:11–20

Glikson AY, Vickers J (2010) Asteroid impact connections of crustal evolution. Aust J Earth Sci 57: 79–95

Glikson AY (2005) Geochemical signatures of Archean to early Proterozoic Maria-scale oceanic impact basins. Geology 33:125–128. https://doi.org/10.1130/G21034.1

Glikson AY (2008) Field evidence of Eros-scale asteroids and impact-forcing of Precambrian geodynamic episodes, Kaapvaal (South Africa) and Pilbara (Western Australia) cratons. Earth Planet Sci Lett 267:558–570

Glikson AY (2014) The Archaean: geological and geochemical windows into the early earth. Springer, new approaches in solid earth sciences 9. Springer, The Netherlands. https://doi.org/10.1007/978-3-319-07908-0_1

Hickman AH (1980) Lithological map and stratigraphic interpretation of the Pilbara Block. Geology if the Pilbara Block and its environs. Geological Survey of Western Australia

Hickman AH (1981) Crustal evolution of the Pilbara Block. In: Glover JE, Groves DI (eds) Archaean geology: second International Symposium, Perth 1980. Geological Society of Australia, pp 57–69
Hickman AH (1983) Yarrie, Western Australia. 1:250 000 Geological Series Explanatory Notes. Geological Survey of Western Australia
Hickman AH (1997) A revision of the stratigraphy of Archaean greenstone successions in the Roebourne–Whundo area, West Pilbara. In: Geological Survey of Western Australia annual review 1996–97. Geological Survey of Western Australia, pp 76–81
Hickman AH, Smithies RH, Huston DL (2000) Archaean geology of the West Pilbara Granite–Greenstone Terrane and the Mallina Basin, Western Australia—a field guide. Record, 2000/9. Geological Survey of Western Australia
Hickman AH (2001) Geology of the Dampier 1:100 000 sheet. 1:100 000 geological series explanatory notes. Geological Survey of Western Australia
Hickman AH (2004) Two contrasting Granite–Greenstones Terranes in the Pilbara Craton, Australia: evidence for vertical and horizontal tectonic regimes prior to 2900 Ma. Precambrian Res 131:153–172
Hickman AH, Huston DL, Van Kranendonk MJ, Smithies RH (2006) Geology and mineralization of the West Pilbara—a field guide. Record, 2006/17. Geological Survey of Western Australia
Hickman AH (2012) Review of the Pilbara Craton and Fortescue Basin, Western Australia: crustal evolution providing environments for early life. Island Arc 21:1–31
Hickman AH (2016) Northwest Pilbara Craton: a record of 450 million years in the growth of Archean continental crust. Report, 160. Geological Survey of Western Australia
Hickman AH (2021) East Pilbara Craton: a record of one billion years in the growth of Archean continental crust. Report 143, Geological Survey of Western Australia
Hickman AH, Lipple SL (1975) Explanatory notes on the Marble Bar 1:250 000 geological sheet, Western Australia. Record, 1974/20. Geological Survey of Western Australia
Hickman AH, Smithies RH, Huston DL (1998) Excursion guide to the geology of the Granite–Greenstone Terrane of the West Pilbara. Geological Survey of Western Australia and Australian Geological Survey Organisation, Perth
Hickman AH, Smithies RH, Pike G, Farrell TR, Beintema KA (2001) Evolution of the West Pilbara Granite–Greenstone Terrane and Mallina Basin, Western Australia–a field guide. Record, 2001/16. Geological Survey of Western Australia
Hickman AH, Smithies RH, Tyler IM (2010) Evolution of active plate margins: West Pilbara Superterrane, De Grey Superbasin, and the Fortescue and Hamersley Basins–a field guide. Record, 2010/3. Geological Survey of Western Australia
Hickman AH, Van Kranendonk MJ (2008) Marble Bar, WA sheet 2855. Geological Survey of Western Australia
Huston DL, Blewett RS, Champion DC (2012) Australia through time: a summary of its tectonic and metallogenic evolution. Episodes 35:23–43. https://doi.org/10.18814/epiiugs/2012/v35i1/004
Kisters AFM, Belcher RW, Poujol M, Dziggel A (2010) Continental growth and convergence-related arc plutonism in the Mesoarchean: evidence from the Barberton granitoid-greenstone terrain, South Africa. Precambrian Res 178:15–26
Krapež B (1993) Sequence stratigraphy of the Archaean supracrustal belts of the Pilbara Block, Western Australia. Precambrian Res 60:1–45
Krapež B, Barley ME (1987) Archaean strike-slip faulting and related ensialic basins: evidence from the Pilbara Block. Australia Geol Mag 124:555–567. https://doi.org/10.1017/S0016756800017386
Krapež B, Eisenlohr B (1998) Tectonic settings of Archaean (3325–2775 Ma) crustal–supracrustal belts in the West Pilbara Block. Precambrian Res 88:173–205
Kumar A, Parashuramula V, Shankar R, Besse J (2017) Evidence for a Neoarchean LIP in the Singhbhum Craton, Eastern India: implications to Vaalbara supercontinent. Precambrian Res 292:163–174

Lowe DR, Byerly GR (1999) Stratigraphy of the west-central part of the Barberton Greenstone Belt, South Africa. In: Lowe DR, Byerly GR (eds) Geologic evolution of the Barberton Greenstone Belt. Geological Society of America, South Africa, pp 1–36

Lowe DR, Byerly GR (2007) An overview of the geology of the Barberton Greenstone Belt and vicinity: implications for early crustal development. In: Van Kranendonk MJ, Bennett VC, Smithies RH (eds) Earth's oldest rocks. Elsevier B.V, Burlington, Massachusetts, USA, pp 481–526

Martin DMB, Clendenin CW, Krapež B, McNaughton NJ (1998) Tectonic and geochronological constraints on late Archaean and Palaeoproterozoic stratigraphic correlation within and between the Kaapvaal and Pilbara cratons. J Geol Soc 155:311–322

Morant P (1995) The Panorama Zn–Cu VMS deposits, Western Australia. In: Ho SE, Amann WJ (eds) Recent developments in base metal geology and exploration: papers presented at an AIG seminar, Perth, 29 November 1995. Australian Institute of Geoscientists, pp 75–84

Morant P (1998) Panorama zinc–copper deposits. In: Berkman DA, Mackenzie DH (eds) Geology of Australian and Papua New Guinean mineral deposits, vol 22. The Australasian Institute of Mining and Metallurgy, Melbourne, pp 287–292

Nelson DR (1997) Compilation of SHRIMP U-Pb zircon geochronology data, 1996. Record, 1997/2. Geological Survey of Western Australia

Nelson DR (1998) Compilation of SHRIMP U-Pb zircon geochronology data, 1997. Record, 1998/2. Geological Survey of Western Australia

Nelson DR (2001) Compilation of Compilation of geochronology data, 2000. Record, 2001/2. Geological Survey of Western Australia

Nelson DR, Bhattacharya HN, Thern ER, Altermann W (2014) Geochemical and ion-microprobe U–Pb zircon constraints on the Archaean evolution of the Singhbhum Craton, Eastern India. Precambrian Res 255:412–432

Ohta H, Maruyama S, Takahashi E, Watanabe Y, Kato Y (1996) Field occurrence, geochemistry and petrogenesis of the Archaean Mid-Oceanic Ridge Basalts (AMORBs) of the Cleaverville area, Pilbara Craton, Western Australia. Lithos 37:199–221

Poujol M, Robb LJ, Anhaeusser CR, Gericke B (2003) A review of the geochronological constraints on the evolution of the Kaapvaal Craton, South Africa. Precambrian Res 127:181–213. https://doi.org/10.1016/S0301-9268(03)00187-6

Rasmussen B, Fletcher IR, Muhling JR (2007) In situ U–Pb dating and element mapping of three generations of monazite: unravelling cryptic tectonothermal events in low-grade terranes. Geochim Cosmochim Acta 71:670–690

Shibuya T, Kitajima K, Komiya T, Terabayashi M, Maruyama S (2007) Middle Archean Ocean ridge hydrothermal metamorphism and alteration recorded in the Cleaverville area, Pilbara Craton. J Meta Geol 25:751–767

Smirnov AV, Evans ADA, Ernst RE, Söderlund U, Li ZX (2013) Trading partners: tectonic ancestry of Southern Africa and Western Australia, in Archean supercratons Vaalbara and Zimgarn. Precambrian Res 224:11–12

Smith JB (2003) The episodic development of intermediate to silicic volcano-plutonic suites in the Archaean West Pilbara, Australia. Chem Geol 194:275–295

Smith JB, Barley ME, Groves DI, Krapež B, McNaughton NJ, Bickle MJ, Chapman HJ (1998) The Sholl Shear Zone, West Pilbara; evidence for a domain boundary structure from integrated tectonostratigraphic analysis, SHRIMP U–Pb dating and isotopic and geochemical data of granitoids. Precambrian Res 88:143–172

Smithies RH, Champion DC, Cassidy KF (2003) Formation of Earth's early Archaean continental crust. Precambrian Res 127:89–101

Smithies RH, Champion DC, Sun S-S (2004) Evidence for early LREE-enriched mantle source regions: diverse magmas from the c. 3.0 Ga Mallina Basin, Pilbara Craton, NW Australia. J Petrol 45:1515–1537

Smithies RH, Van Kranendonk MJ, Champion DC (2005) It started with a plume—early Archaean basaltic proto-continental crust. Earth Planet Sci Lett 238:284–297

Smithies RH, Champion DC (2000) The Archaean high-Mg diorite suite: links to tonalite–trondhjemite–granodiorite magmatism and implications for early Archaean crustal growth. J Petrol 41:1653–1671

Smithies RH, Champion DC (2002) Assembly of a composite granite intrusion at a releasing bend in an active Archean shear zone. In: Geological Survey of Western Australia annual review 2000–01. Geological Survey of Western Australia, Perth, Western Australia, pp 63–68

Smithies RH, Champion DC, Blewett RS (2001a) Wallaringa, WA sheet 2656. Geological Survey of Western Australia

Smithies RH, Hickman AH, Nelson DR (1999) New constraints on the evolution of the Mallina Basin, and their bearing on relationships between the contrasting Eastern and Western Granite–Greenstone Terranes of the Archaean Pilbara Craton, Western Australia. Precambrian Res 94: 11–28

Smithies RH, Nelson DR, Pike G (2001b) Development of the Archaean Mallina Basin, Pilbara Craton, Northwestern Australia; a study of detrital and inherited zircon ages. Sediment Geol 141–142:79–94

Smithies RH, Champion DC, Van Kranendonk MJ, Hickman AH (2007) Geochemistry of volcanic rocks of the Northern Pilbara Craton, Western Australia. Report, vol 104. Geological Survey of Western Australia

Strik G, Blake TS, Zegers TE, White SH, Langereis CG (2003) Palaeomagnetism of flood basalts in the Pilbara Craton, Western Australia: late Archaean continental drift and the oldest known reversal of the geomagnetic field. J Geophys Res 108. https://doi.org/10.1029/2003JB002475

Sun S-S, Hickman AH (1998) New Nd-isotopic and geochemical data from the West Pilbara–implications for Archaean crustal accretion and shear zone development. Australian Geological Survey Organisation, Research Newsletter:25–29

Sun S-S, Hickman AH (1999) Geochemical characteristics of ca 3.0-Ga Cleaverville greenstones and later mafic dykes, West Pilbara: implication for Archaean crustal accretion. Australian Geological Survey Organisation, Research Newsletter:23–29

Van Kranendonk MJ (1997) Results of field mapping, 1994–1996, in the North Shaw and Tambourah 1:100 000 sheet areas, Eastern Pilbara Craton, Northwestern Australia. Record, 1997/23. Australian Geological Survey Organisation (Geoscience Australia)

Van Kranendonk MJ (1999) North Shaw, WA sheet 2755. Geological Survey of Western Australia

Van Kranendonk MJ (2003) Geology of the Tambourah 1:100 000 sheet. 1:100 000 Geological Series Explanatory Notes. Geological Survey of Western Australia

Van Kranendonk MJ (2004a) Carlindie, WA Sheet 2756. Geological Survey of Western Australia

Van Kranendonk MJ (2004b) Geology of the Carlindie 1:100 000 sheet. 1:100 000 geological series explanatory notes. Geological Survey of Western Australia

Van Kranendonk MJ (2006) Volcanic degassing, hydrothermal circulation and the flourishing of early life on earth: a review of the evidence from c. 3490–3240 Ma rocks of the Pilbara Supergroup, Pilbara Craton. Western Australia Earth-Science Reviews 74:197–240

Van Kranendonk MJ (2008) Structural geology of the central part of the Lalla Rookh–Western Shaw structural corridor, Pilbara Craton, Western Australia. Report 103. Geological Survey of Western Australia

Van Kranendonk MJ, Hickman AH, Smithies RH, Nelson DN, Pike G (2002) Geology and tectonic evolution of the Archaean North Pilbara Terrain, Pilbara Craton, Western Australia. Econ Geol 97:695–732. https://doi.org/10.2113/gsecongeo.97.4.695

Van Kranendonk MJ, Hickman AH, Smithies RH, Williams IR, Bagas L, Farrell TR (2006) Revised lithostratigraphy of Archean supracrustal and intrusive rocks in the Northern Pilbara Craton, Western Australia. Record, 2006/15. Geological Survey of Western Australia

Van Kranendonk MJ, Kröner A, Hoffman JE, Nagel T, Anhaeusser CR (2014) Just another drip: re-analysis of a proposed Mesoarchean suture from the Barberton Mountain land, South Africa. Precambrian Res 254:19–35. https://doi.org/10.1016/j.precamres.2014.07.022

Van Kranendonk MJ, Morant P (1998) Revised Archaean stratigraphy of the North Shaw 1:100 000 sheet, Pilbara Craton. In: Geological Survey of Western Australia Annual Review 1997–98. Geological Survey of Western Australia, pp 55–62

Van Kranendonk MJ, Pawley MJ (2002) Tambourah, WA sheet 2754. Geological Survey of Western Australia

Van Kranendonk MJ, Smithies RH, Griffin WL, Huston DL, Hickman AH, Champion DC, Anhaeusser CR, Pirajno F (2015) Making it thick: a volcanic plateau origin of Paleoarchean continental lithosphere of the Pilbara and Kaapvaal cratons. Geol Soc Lond, Spec Publ 389:89–111. https://doi.org/10.1144/SP389.12

Van Kranendonk MJ, Smithies RH, Hickman AH, Champion DC (2007) Review: secular tectonic evolution of Archean continental crust: interplay between horizontal and vertical processes in the formation of the Pilbara Craton, Australia. Terra Nova 19:1–38

Van Kranendonk MJ, Smithies RH, Hickman AH, Wingate MTD, Bodorkos S (2010) Evidence for Mesoarchean (~3.2 Ga) rifting of the Pilbara Craton: the missing link in an early Precambrian Wilson cycle. Precambrian Res 177:145–161

Van Ruitenbeek FJA, Cudahy TJ, Van der Meer FD, Hale M (2012) Characterization of the hydrothermal systems associated with Archean VMS-mineralization at Panorama, Western Australia, using hyperspectral, geochemical and geothermometric data. Ore Geol Rev 45:33–46

Vearncombe S, Barley ME, Groves DI, McNaughton NJ, Mikucki EJ, Vearncombe JR (1995) 3.26 Ga black smoker-type mineralization in the Strelley Belt, Pilbara Craton, Western Australia. J Geol Soc Lond 152:587–590

Vearncombe S, Kerrich R (1999) Geochemistry and geodynamic setting of volcanic and plutonic rocks associated with early Archaean volcanogenic massive sulphide mineralization, Pilbara Craton. Precambrian Res 98:243–270

Vearncombe S, Vearncombe JR, Barley ME (1998) Fault and stratigraphic controls on volcanogenic massive sulphide deposits in the Strelley belt, Pilbara Craton, Western Australia. Precambrian Res 88:67–82

Wilhelmij HR, Dunlop JSR (1984) A genetic stratigraphic investigation of the Gorge Creek Group in the Pilgangoora syncline. In: Muhling JR, Groves DI, Blake TS (eds) Archaean and Proterozoic basins of the Pilbara, Western Australia: evolution and mineralization potential. The University of Western Australia, Geology Department and University Extension, pp. 68–88

Wingate MTD, Kirkland CL, Bodorkos S, Hickman AH (2010) 160258: felsic metavolcanic rock. Orchard Well

Zegers TE (1996) Structural, kinematic, and metallogenic evolution of selected domains of the Pilbara Granite–Greenstone Terrain. PhD, Geologica Ultraiectina. Utrecht University

Zegers TE, Wijbrans JR, White SH (1999) $^{40}Ar/^{39}Ar$ age constraints on tectonothermal events in the Shaw area of the Eastern Pilbara Granite–Greenstone Terrain (Western Australia): 700 Ma of Archaean tectonic evolution. Tectonophysics 311:45–81

Zegers TE, de Wit MJ, Dann J, White SH (1998) Vaalbara, Earth's oldest assembled continent? A combined structural, geochronological, and palaeomagnetic test. Terra Nova 10:250–259

Chapter 7
Mesoarchean Rift and Marginal Basins of the Pilbara Craton

Abstract The c. 3220 Ma breakup of the Paleoarchean Pilbara Craton was followed by a 60-million-year period of northwest–southeast plate separation. Three continental microplates (East Pilbara, Karratha, and Kurrana Terranes) were separated by two expanding rift basins. Between the Karratha Terrane (KT) and the East Pilbara Terrane (EPT), the Regal Basin evolved from a spreading centre to form a belt of MORB-like basaltic crust. The early Mesoarchean Mosquito Creek Basin formed with separation of the EPT and Kurrana Terrane (KUT) but is now almost entirely concealed beneath the late Mesoarchean Mosquito Creek Formation.

Continental breakup of the Pilbara Craton marked the end of Paleoarchean vertical deformation and the beginning of Mesoarchean plate tectonic processes in the Pilbara. Paleoarchean melts derived from crustal recycling of much older crust were succeeded by Mesoarchean juvenile, mantle-derived melts. Passive margin successions such as the 3223–3165 Ma Soanesville Group and the 3220–3160 Ma Nickol River Formation were deposited along the margins of the continental microplates. Separation of the EPT and KT ended at c. 3160 Ma due to the collision of the KT with another plate converging from the northwest.

Keywords Plate tectonics · Continental breakup · Plate separation · Collision · Obduction · Passive margin

7.1 Introduction

Continental rifting and breakup of the Pilbara Craton at c. 3220 Ma (Hickman 2001, 2004, 2016; Van Kranendonk et al. 2002, 2006, 2007b, 2010; Hickman and Van Kranendonk 2008, 2012), referred to as the East Pilbara Terrane Rifting Event (Chap. 6), marked the beginning of a 150-million-year period during which the Eoarchean–Paleoarchean craton was separated into at least three continental microplates, the EPT, KT, and KUT. It is likely that additional continental microplates were produced by the breakup, probably including the eastern Kaapvaal Craton of South Africa and the Singhbhum and Bastar Cratons of eastern India, but if so these apparently remained separated from the Pilbara Craton throughout the

A. H. Hickman, *Archean Evolution of the Pilbara Craton and Fortescue Basin*, Modern Approaches in Solid Earth Sciences 24,
https://doi.org/10.1007/978-3-031-18007-1_7

Mesoarchean. The eastern Kaapvaal Craton has very strong stratigraphic and structural similarities to the EPT (Zegers et al. 1998; Hickman 2012, 2016, 2021a; Van Kranendonk et al. 2015), yet the course of its Mesoarchean evolution diverged following the breakup. This is evident in a comparison of the stratigraphy of the two cratons (Fig. 1.10). As in the Pilbara Craton, a major change in the crustal evolution of the east Kaapvaal Craton is recorded between 3230 and 3215 Ma (Byerly et al. 2019) when volcanism and vertical tectonic processes of the previous 300 million years ended with deformation events, including crustal extension, followed by the development of large clastic sedimentary basins such as that of the Moodies Group. Initially, lithofacies and depositional environments of the Moodies Group were very similar to those of the Soanesville Group and Nickol River Formation in the Pilbara Craton.

As the continental microplates moved apart, they became separated by basaltic rift basins, including the Regal Basin and the early Mosquito Creek Basin. This explains a key feature of Pilbara Craton geology: that apart from the common development of c. 3200 Ma passive margin basin successions, no stratigraphic or structural correlations are possible between the Northwest and East Pilbara Craton between 3220 and 3070 Ma (Chap. 6). Between c. 3165 Ma and 3070 Ma, the Northwest Pilbara was subjected to deformation and metamorphism at c. 3160 Ma, subduction and magmatism between 3130 and 3093 Ma, and obduction of part of the Regal Basin onto the KT prior to 3070 Ma. The activity in the Northwest Pilbara formed two new terranes: the Sholl Terrane (ST) comprising the volcanic Whundo Group and the mainly granitic Railway Supersuite, and the Regal Terrane (RT) composed of ophiolites obducted from the Regal Basin (Fig. 7.1). Between about 3100 and 3070 Ma, northwest–southeast compression of the Northwest Pilbara juxtaposed the ST with the KT and RT along the Sholl Shear Zone to form the West Pilbara Superterrane (Van Kranendonk et al. 2006, 2007b).

Deformation and magmatic activity in the Northwest Pilbara commenced from about 3160 Ma following a regional metamorphic event (Karratha Event) in the KT (Kiyokawa 1993; Smith et al. 1998; Beintema 2003). This event coincided with a change from separation of the KT and EPT, either side of a spreading centre in the Regal Basin, to KT–EPT convergence. Compression of the Regal Basin indicated by the formation of a subduction zone (Smithies et al. 2005, 2007; Van Kranendonk et al. 2006, 2007a, b) is now interpreted to have resulted from a collision between the KT with another plate converging from the northwest (Hickman 2016). The identity of the northwest plate is unknown, and it might have been a plate or basin that separated from the Pilbara Craton following the breakup. From about 3160 Ma onwards the Regal Basin began to close as the northwest plate pushed the KT back towards the EPT. Less is known about the development of the basaltic rift basin between the EPT and KUT (Mosquito Creek Basin), although 3199–3178 Ma intrusion of granodiorite along the basin margin of the KUT suggests that crustal separation lasted little more than 20 Ma (Hickman 2021a).

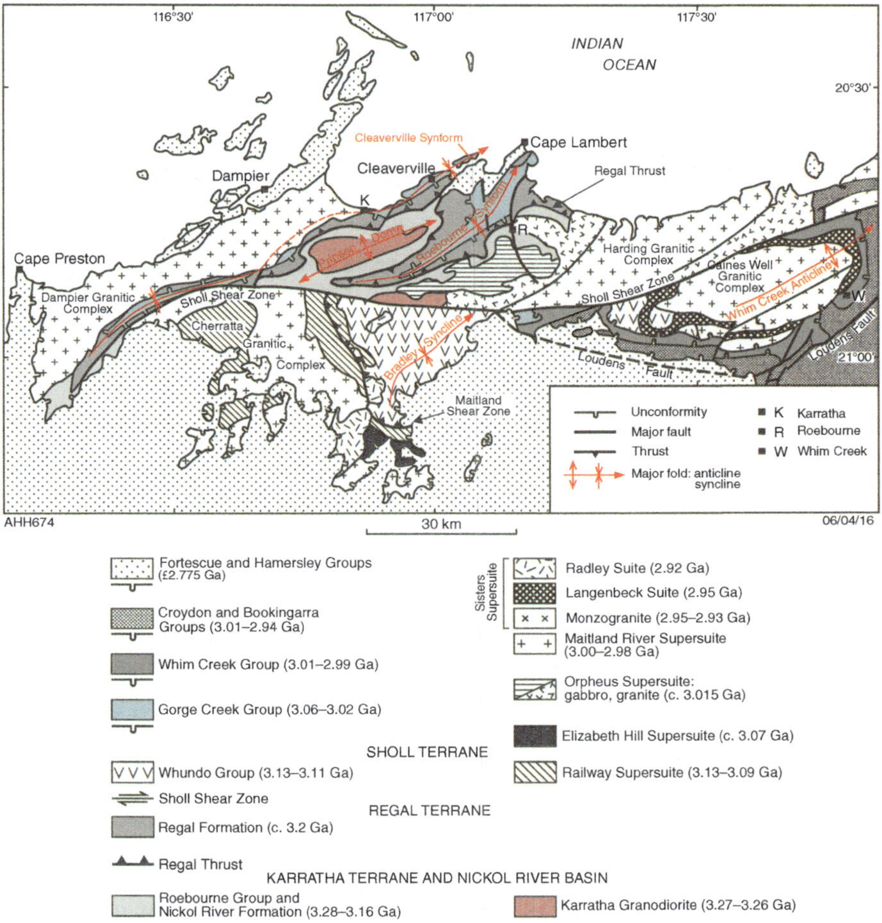

Fig. 7.1 Simplified geological map of the Northwest Pilbara Craton between Cape Preston and Whim Creek, showing lithostratigraphy, tectonic units, and major structures (From Hickman 2016; with Geological Survey of Western Australia permission)

7.2 Basaltic Rift Basins

Between the EPT and KT, MORB-like basaltic crust of the Regal Formation is interpreted to have been erupted from a spreading centre in the Regal Basin. Large sections of the Regal Formation were subsequently obducted onto the KT on the northwest margin of the Regal Basin (Hickman 2001, 2004, 2012, 2016; Hickman et al. 2001, 2010; Hickman and Van Kranendonk 2012). Geochemical data (Glikson et al. 1986; Ohta et al. 1996; Sun and Hickman 1998, 1999; Shibuya et al. 2007; Smithies et al. 2007) and isotopic data (Sun and Hickman 1998, 1999; Smithies et al. 2007) indicate that the Regal Formation is composed of juvenile basaltic crust.

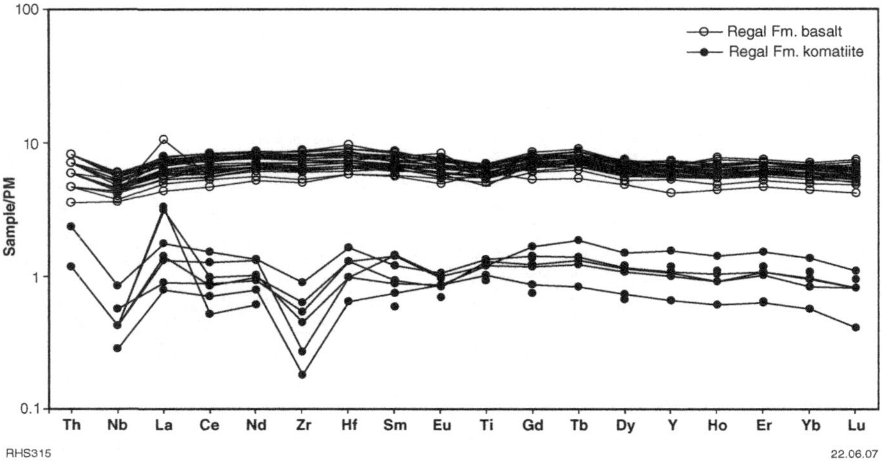

Fig. 7.2 Trace element plots normalized to primitive mantle for komatiites and komatiitic basalts of the Regal Formation (From Smithies et al. 2007; with Geological Survey of Western Australia permission)

Smithies et al. (2007) analysed 38 samples of the Regal Formation collected from four localities in the Northwest Pilbara. Six of the samples were olivine spinifex-textured komatiites. The non-komatiitic samples were all tholeiitic basalt and all had very similar major and trace element compositions. The basalts have relatively flat normalized trace element patterns (Fig. 7.2) with similar compositions to MORB except for slight LREE enrichment. Smithies et al. (2007) attributed this enrichment to crustal contamination. This is thought to be consistent with small size of the Regal Basin compared to modern oceans and the likely presence of some underlying crust and basal sediments at an early stage in plate separation. Mapping has established that the Regal Formation was thrust across the continental crust of the KT and passive margin units of the Nickol River Formation over an area of at least 3000 km^2 (Hickman 1997b, 2000, 2001, 2002; Hickman and Strong 2001, 2003; Hickman et al. 2010).

The stratigraphically lowest formation in the Mosquito Creek Basin, the Coondamar Formation, is composed of metamorphosed ultramafic and mafic volcanic rocks, mafic volcaniclastic sedimentary rocks, chlorite–actinolite schist, metasandstone, pelite, and metachert. The depositional ages of the protoliths of this complex assemblage are inferred to be between 3220 and 3175 Ma which matches the timing of Paleoarchean rifting and breakup of the Pilbara Craton. Ultramafic and mafic sills along the northern and southern margins of the basin are undated but interpreted to have been emplaced during rifting and are assigned to the c. 3185 Ma Dalton Suite (Hickman 2012, 2021b). The 3199–3178 Ma TTG of the Golden Eagle Orthogneiss (KUT) was most likely derived by partial melting of recently formed mafic crust, the only potential known source being the Coondamar

Formation. Sm–Nd isotope data indicate that the orthogneiss was derived from crust far more juvenile than any components of the EPT.

7.2.1 Regal Basin

The preserved c. 500 km southwest–northeast length of the Regal Basin is almost entirely concealed by the 3015–2930 Ma Mallina Basin. The length of the basin prior to Neoarchean breakup of the Pilbara Craton between 2755 and 2501 Ma would have been considerably greater than 500 km. The present maximum width of the basin is 100 km although, inferring modern spreading rates, its original width prior to post-3160 Ma compression might have been at least 300 km. The apparent displacement of the Paleoarchean Sulphur Springs–Roebourne Group succession across the basin suggests that separation of the EPT and KT included about 200 km of sinistral strike-slip movement (Sun and Hickman 1998), although the timing of this movement is uncertain.

The Regal Basin is mainly composed of the 3200–3160 Ma Regal Formation, which is a c. 3 km-thick basaltic succession metamorphosed to lower amphibolite facies. In the Cleaverville area, on the Northwest Pilbara coast, the basin also includes the greenschist-facies basaltic succession of the Port Robinson Basalt (Kiyokawa et al. 2012). Almost all information on the stratigraphy of the Regal Basin has been obtained from the RT, an obducted slice of the Regal Formation that overlies the KT across the 3160–3070 Ma Regal Thrust. Excellently exposed sections of the Regal Formation are exposed at Mount Regal, immediately South of Karratha, Southeast of Cleaverville, and Southeast of Wickham. At Mount Regal and South of Karratha, the lowest stratigraphic units of the Regal Formation are locally preserved spinifex-textured komatiites.

7.2.1.1 Regal Formation

The Regal Formation is composed of pillowed and massive tholeiitic basalt, minor komatiite and komatiitic basalt, and rare chert units. Sills of dolerite and gabbro intrude the formation and a metamorphosed swarm of parallel dolerite dykes (sheeted dykes?) is present in the ridge immediately South of Karratha. Well-preserved pillow structures are exposed in many basalt outcrops (Fig. 7.3) but in areas of higher strain, such as the ridge South of Karratha, the pillows are flattened and stretched. The lowest preserved stratigraphic sections of the formation immediately overlying the Regal Thrust are locally composed of spinifex-textured komatiite and komatiitic basalt. Good examples of komatiite are exposed north of Mount Regal and in an east–west striking section of the Roebourne greenstone belt 2 km South of Karratha.

Chemical analyses of the Regal Formation komatiites have revealed MgO contents of 27.0–36.9 wt. %, Cr contents of 1720–5229 ppm, and Ni contents of

Fig. 7.3 Pillow structures in basalt of the Regal Formation exposed on a wave-cut platform near Cleaverville (Zone 50, MGA 503290E, 7716644N). The local succession of pillow basalt flows is 1 km thick, with most pillow structures being between 1.0 and 1.5 m wide in cross section. Convex pillow tops and cuspate tail structures developed above adjoining underlying pillows indicate stratigraphic way-up. (From Hickman 2016; with Geological Survey of Western Australia permission)

1489–2487 ppm (Smithies et al. 2007). The trace element composition of the komatiites (Fig. 7.2) suggests a degree of interaction with either continental crust or underlying clastic sedimentary rocks. Concentrations of Th and La are relatively high, and the komatiites show negative Nb anomalies. Local crustal contamination of the Regal Formation is also indicated by variable Nd model ages ranging between 3380 and 3160 Ma (Smithies et al. 2007; Champion 2013). Based on Sm–Nd data, Smithies et al. (2007) interpreted the depositional age of the Regal Formation to be c. 3200 Ma.

The minimum depositional age of the Regal Formation is indicated by metamorphism in the KT at c. 3160 Ma (Kiyokawa 1993; Kiyokawa and Taira 1998; Smith et al. 1998). This c. 3160 Ma metamorphic event is significant in the tectonic evolution of the Northwest Pilbara because it coincides with a change from separation of the KT and EPT to convergence. This is inferred to be due to collision of the KT with another plate northwest of the Pilbara Craton. The maximum age of the Port Robinson Basalt, interpreted to correlate with the Regal Formation at Cleaverville, is suggested by a date of 3195 ± 15 Ma on the underlying Dixon Island Formation (Kiyokawa et al. 2012). The theoretical maximum depositional age of the Regal Formation is constrained by the timing of the East Pilbara Terrane Rifting Event at c. 3220 Ma. The lithological composition (pillow basalt and rare thin chert units) and geochemistry (MORB-like) of the Regal Formation are suggestive of oceanic crust (Ohta et al. 1996; Sun and Hickman 1998; Kiyokawa and Taira 1998; Beintema 2003; Shibuya et al. 2007). Interpretation of the Regal Formation as a slice of oceanic crust (ophiolite) is consistent with flat REE patterns and ε_{Nd} values up to +3.5, which is close to the depleted mantle value (+3.2) at 3200 Ma (Smithies et al. 2007). Across 3000 km^2 of the Northwest Pilbara, the Regal Formation is sandwiched between the underlying continental crust of the KT, and the overlying Cleaverville Formation that was deposited at c. 3020 Ma.

7.2.1.2 Regal Thrust

The Regal Thrust was discovered during GSWA mapping of the Dampier, Karratha, and Roebourne areas of the Northwest Pilbara Craton (Hickman 1997b, 2000). This almost horizontal thrust forms a tectonic contact between the Regal Formation and the Ruth Well Formation over an area of 3000 km^2 (Fig. 7.1) (Hickman 2001, 2004, 2016). Because the Regal Formation is MORB-like basaltic crust and the Ruth Well Formation is part of the continental crust of the Karratha Terrane, the Regal Thrust was formed by obduction. In detail, the Regal Thrust is a 1 km-thick belt of thrusting incorporating slices of the overlying Regal Formation and the underlying Nickol River Formation (this chapter) and Ruth Well Formation. The uppermost thrust plane immediately underlies tectonically stretched and flattened basaltic rocks of the Regal Formation and includes mylonite units several metres thick (Fig. 7.4). Below this level, the zone of tectonic lensing includes silicic mylonites derived from extreme shearing of the Nickol River Formation. These mylonites locally exhibit refolded isoclinal folds and sheaf folds indicating multiple phases of deformation (Fig. 7.5). The age of the Regal Thrust is between 3160 and 3070 Ma because obduction of the Regal Formation occurred during the compression of the Regal Basin. After c. 3015 Ma, and most probably at c. 2950 Ma, the Regal Thrust was folded by major upright folds, causing the thrust and the overlying Regal Formation outcrop on three-fold limbs (Fig. 7.1).

7.3 Early Mesoarchean Passive Margins

Early Mesoarchean passive margin successions were deposited in three basins (Nickol River, Soanesville, and Mosquito Creek) on the continental margins of the KT, EPT, and KUT (Hickman et al. 2010; Van Kranendonk et al. 2010; Hickman 2012, 2016). Outcrops of the basins are shown on Fig. 1.3. The Soanesville Basin unconformably overlies the volcano-sedimentary basin of the Sulphur Springs Group (Chap. 6) in the eastern half of the EPT. Until c. 3185 Ma deposition of the Soanesville Group was entirely sedimentary, consisting of conglomerate, sandstone, shale, and banded iron-formation (BIF) but, with deeper rifting northwest towards the Regal Basin, became entirely volcanic. On the southeast margin of the KT,

Fig. 7.4 Outcrops of basaltic mylonite in the Regal Thrust southeast of Mount Regal: (**a**) mylonite dipping northwest under the Regal Formation (top right); (**b**) close-up of the mylonite showing its strongly planar tectonic foliation (MGA Zone 50, 474800E 7698290N). (From Hickman et al. 2010; with Geological Survey of Western Australia permission)

deposition of the Nickol River Formation commenced with conglomerate and sandstone, passing upwards into shale and BIF. The top of the Nickol River Formation is missing due to tectonic removal by the Regal Thrust, but there is evidence that the clastic sedimentary rocks might have been overlain by basalt that passed laterally southeast into the Regal Formation. On the southeast margin of the EPT, the Budjan Creek Formation is interpreted to have passed eastwards into the rift related Coondamar Formation of the Mosquito Creek Basin. The Budjan Creek Formation is lithologically similar to the lower Soanesville Group and the Nickol River Formation, whereas the Coondamar Formation is composed of sedimentary and mafic volcanic rocks.

Fig. 7.5 Structures in silicic mylonite of the Regal Thrust 14 km southeast of Karratha: (**a**) finely laminated silicic mylonite deformed by isoclinal folds (field of view 1 m across); (**b**) close-up of a refolded isocline (lens cap 5 cm diameter); (**c**) sheaf folds with a parallel mineral lineation (MGA Zone 50, 492000E 7696800N). (From Hickman et al. 2010; with Geological Survey of Western Australia permission)

7.3.1 Soanesville Basin

In ascending stratigraphic order the succession of the Soanesville Basin, the Soanesville Group, comprises the following formations: Cardinal Formation (shale and BIF, up to 1 km thick); Corboy Formation (sandstone, conglomerate, wacke, and shale, 2.5 km thick); Paddy Market Formation (BIF and Fe-rich clastic sedimentary rocks, 1 km thick); Honeyeater Basalt (komatiitic and tholeiitic basalt, 1 km thick); Pyramid Hill Formation (BIF and chert, up to 900 m thick); Hong Kong Chert (chert, basalt, and siliciclastic sedimentary rocks, approximately 150 m thick); and Empress Formation (komatiitic basalt, komatiite, and thin chert units, between 1.5 and 2.0 km thick). The c. 3185 Ma Dalton Suite, consisting of numerous ultramafic–mafic sills and dykes intruding the lower Soanesville Group, was genetically related to the Honeyeater Basalt (Van Kranendonk et al. 2010).

Sections of the Soanesville Group are preserved in the Soanesville, Pincunah, Abydos, Tambina, Emerald Mine, and Western Shaw greenstone belts, and the Honeyeater Basalt is interpreted to form parts of the Wodgina, Pilbara Well, and Cheearra greenstone belts (Fig. 1.7) (Van Kranendonk et al. 2010).

Representing a major change in the crustal evolution of the Pilbara Craton, the unconformity between the Soanesville Group and the underlying Sulphur Springs Group is arguably more important than any of the other unconformities in the Pilbara Craton. In the first regional stratigraphic interpretation of the eastern part of the craton (Lipple 1975; Hickman and Lipple 1975, 1978), the greenstone succession was divided into two main stratigraphic units, the Warrawoona Group and the Gorge Creek Group, and these were separated by what is now the contact between the Sulphur Springs and Soanesville Groups. Even so, the nature of the contact has been controversial. Most workers have interpreted it to be an unconformity (Wilhelmij and Dunlop 1984; Morant 1995, 1998; Glikson 2001; Buick et al. 2002; Glikson and Vickers 2006, 2010; Rasmussen et al. 2007). However, Van Kranendonk (1997, 2000) interpreted the contact to vary from conformable to disconformable, locally involving an onlap of the Soanesville Group across the Sulphur Springs Group. Van Kranendonk et al. (2006) described the contact as a disconformity and expressed a view that the onset of deposition of the Soanesville Group directly followed deposition of the Sulphur Springs Group. This led to the Soanesville Group being included in the Pilbara Supergroup (Van Kranendonk et al. 2006, 2007a, b, 2010), although this was inconsistent with the Sulphur Springs and Soanesville Groups being separated by the breakup of the Pilbara Craton at 3220 Ma (Van Kranendonk et al. 2010). For this and other reasons (e.g. different tectonic settings), the Soanesville Group was removed from the Pilbara Supergroup (Hickman 2011).

7.3.1.1 Cardinal Formation

The Cardinal Formation (Van Kranendonk et al. 2006) is a unit of red-weathering grey shale, siltstone, chert, BIF, and minor felsic volcaniclastic rocks, up to 1 km thick, which forms the base of the Soanesville Group in several greenstone belts. The shale was previously part of the 'Pincunah Hill Formation' which included the 800–1000 m-thick BIF unit at the base of the group in the Pincunah greenstone belt (Van Kranendonk and Morant 1998). Sedimentological studies (Eriksson 1981; Wilhelmij and Dunlop 1984) indicated that the BIF and shale are facies variations of the same formation. Wilhelmij and Dunlop (1984) observed that several shale units are included within the BIF in the eastern section of the Pincunah greenstone belt, and they interpreted a west to east change from BIF to shale before the onset of submarine-fan deposition (Corboy Formation). The BIF was named as the Pincunah Banded-Iron Member of the Cardinal Formation (Hickman 2021a).

In the Pincunah greenstone belt (Fig. 1.7), the basal 10–40 m-thick succession of the Cardinal Formation is locally composed of metamorphosed and partly mylonitized quartz sandstone. Wilhelmij and Dunlop (1984) named this unit the 'Tank Pool quartzite' (informal). The unit was mapped by Hickman and Lipple (1978) as an unnamed quartzite extending laterally along a strike length of about 15 km in the vicinity of Tank Pool. Blewett and Champion (2005) described it as a mylonitized quartzite including units of semipelitic schist and BIF. Wilhelmij and Dunlop (1984) observed trough cross-bedding and cross-lamination in the metasandstone, and noted that is intercalated with, and grades laterally into, the 800 m-thick BIF (Pincunah Banded-Iron Member). Wilhelmij and Dunlop (1984) and Blewett and Champion (2005) identified a lithologically similar unit including sandstone and conglomerate underlying the Corboy Formation near the northern boundary of the Pincunah greenstone belt. This suggests that the unit is a tectonically attenuated fluviatile sandstone that was locally deposited at the base of the Soanesville Group.

7.3.1.1.1 Pincunah Banded-Iron Member

The Pincunah Banded-Iron Member is 800–1000 m thick in the Pincunah greenstone belt (Fig. 1.7) and mainly composed of red and black, thinly bedded BIF, with minor purplish-red to purplish-grey shale, chert, siltstone, and sandstone. The BIF is highly magnetic and contains a magnetite resource in the eastern Pincunah greenstone belt (Iron Bridge magnetite project). The member partly overlies, and is partly laterally transitional into, a 10- to 40 m-thick unit of sheared sandstone. The sandstone preserves trough cross-bedding and cross-lamination, suggesting deposition from currents carrying detritus eroded from the EPT. The Pincunah Banded-Iron Member is interpreted to have been a submarine-fan to basin-plain deposit. Contemporaneous basaltic volcanism in the oceanic-like spreading centre of the Regal Basin is likely to have increased levels of Fe in the ocean during deposition of the Soanesville Group.

Apart from its presence in the Pincunah greenstone belt, the Pincunah Banded-Iron Member forms part of the Cardinal Formation in the Western Shaw, Emerald Mine, Tambina, Panorama, and North Shaw greenstone belts (Fig. 1.7). Additionally, thin BIF units in the Cardinal Formation of the Soanesville greenstone belt are likely to be related to the member. In the southern section of the Western Shaw greenstone belt, BIF of the Pincunah Banded-Iron Member is composed of millimetre- to centimetre-thick layers of red-brown iron formation and microcrystalline quartz (Van Kranendonk 2003). This BIF is overlain by light grey siltstone and quartz sandstone passing up into red and grey shales and interbedded grey shales and siltstones of the Cardinal Formation.

7.3.1.2 Corboy Formation

The Corboy Formation is the thickest (2.5 km) and most widespread formation of the Soanesville Group. The formation was deposited in tectonically active basins during crustal extension and is mainly composed of submarine-fan deposits (Eriksson 1981). In the Pincunah greenstone belt, Wilhelmij and Dunlop (1984) described the succession of the Corboy Formation, above the Pincunah Banded-Iron Member of Cardinal Formation, as comprising shales, siltstones, and sandstones, and multiple thin units of chert and BIF. The succession was interpreted to be dominated by turbidites and pelagic shales. Eriksson (1981) described lenticular channel deposits containing conglomerate and sandstone locally up to 500 m thick and 1500 m wide and separated from the turbidites by levees of laminated shale and siltstone. In the southern Pincunah greenstone belt, the Pincunah Banded-Iron Member (Cardinal Formation) is overlain by a 100 m-thick unit of quartzite, siltstone, wacke, and conglomerate (Van Kranendonk 2000). Above this basal unit of the Corboy Formation is a homogeneous unit of brown-weathering lithic sandstone or greywacke almost 1 km thick. An overlying thin unit of shale and polymictic conglomerate is succeeded by an upper succession of siltstone, shale, and sandstone of uncertain thickness (Van Kranendonk 2000). In the western Pincunah greenstone belt, the Corboy Formation is overlain by a c. 200 m-thick unit of quartz-sandstone and sandstone assigned to the Farrel Quartzite (Hickman 2010).

In the Sulphur Springs area of the Soanesville greenstone belt, the Corboy Formation is the basal formation of the Soanesville group in the Sulphur Springs area of the northern Soanesville greenstone belt, although in the south of this greenstone belt, and in the Pincunah, Tambina, and Emerald Mine greenstone belts, it is underlain by the Cardinal Formation. At Sulphur Springs, the Corboy Formation has a maximum thickness of 475 m but this thickness varies across a mass of breccia and intrusive rhyodacite above the 'marker chert' at the top of the Sulphur Springs Group. This breccia and rhyodacite formed a topographic high during deposition of the Corboy Formation (Van Kranendonk 2000). Basal polymictic conglomerate includes large pebbles of black chert derived from erosion of the 'marker chert' (Van Kranendonk 2000). Above the conglomerate, the main part of the formation is an upward-fining sequence of graded beds of pebbly sandstone and sandstone overlain by graded sandstone–siltstone beds. The depositional setting was probably a series of shallow-marine deltas and fans (Wilhelmij 1986), which contrasts with the deep-water submarine-fan setting of the Corboy Formation in the Pincunah greenstone belt. The Corboy Formation also forms parts of the Tambina and Emerald Mine greenstone belts (Fig. 1.7) although most of the Soanesville

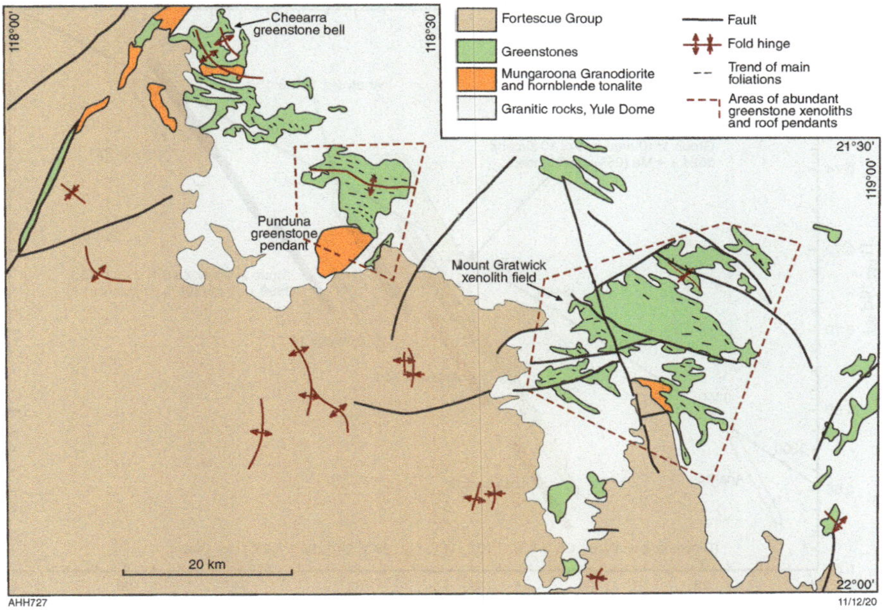

Fig. 7.6 Simplified structural geology of the southern Yule Dome showing outcrops of high-Mg diorite and granodiorite, and hornblende tonalite. The east-southeast trend of fold structures, foliations, greenstone xenoliths, and amphibolite facies metamorphism is at a high angle to structural trends farther north in the Yule Dome and transects the dome-and-keel architecture of the East Pilbara Terrane. (Modified from Smithies 2003; with Geological Survey of Western Australia permission)

Group in these areas is shale and iron formation facies assigned to the Cardinal and Paddy Market Formations.

Quartzite and metasandstone of the Cheearra greenstone belt, in the western Yule Dome (Fig. 7.6), and preserved in isolated outliers and enclaves across the granitic core of the Yule Dome, are tentatively assigned to the Corboy Formation, although no stratigraphic or sedimentological studies have been undertaken. Most detrital zircons in the metasandstones have Eoarchean to early Paleoarchean U–Pb ages (Figs. 7.7 and 7.8) (GSWA 169013, Nelson 2004; GSWA 178045, Nelson 2005b; Kemp et al. 2015a, b). This indicates that detritus comprising the Corboy Formation was substantially derived from erosion of the EPT and Eoarchean rocks of the 3800–3530 Ma crust (Chap. 2).

7.3.1.3 Paddy Market Formation

The Paddy Market Formation conformably overlies the Corboy Formation in the Soanesville greenstone belt and is distinguished from that formation by being mainly composed of units of BIF intercalated with ferruginous clastic sedimentary rocks.

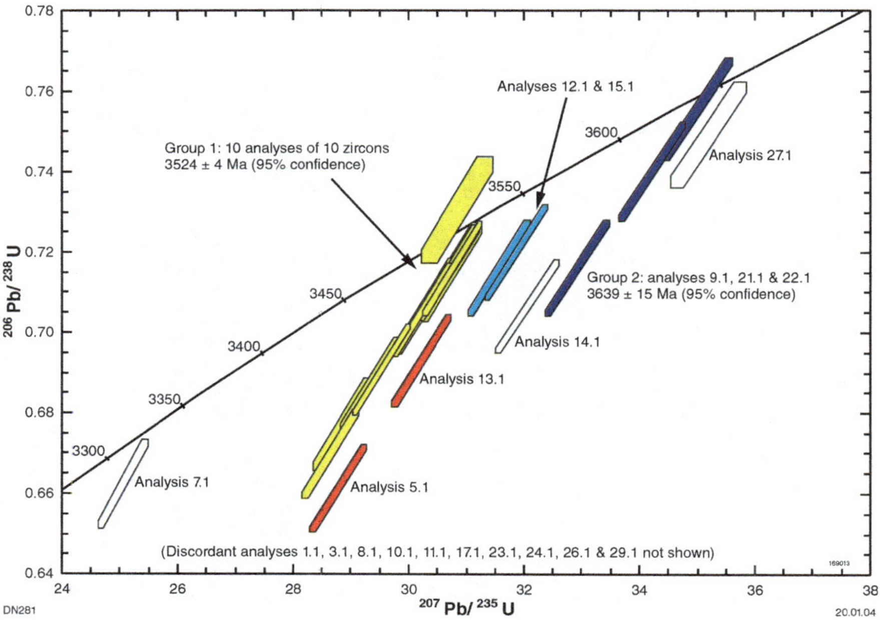

Fig. 7.7 Concordia plot of U–Pb zircon data for a sample of metasandstone (Geological Survey of Western Australia 169013) from the Corboy Formation in the Cheearra greenstone belt, southwest Yule granitic complex. Many detrital zircons pre-date the Pilbara Supergroup. (From Nelson 2004; with Geological Survey of Western Australia permission)

Lipple (1975) estimated the total thickness of the formation as 1 km and described it as comprising units of BIF, ferruginous shale, and ferruginous siltstone and sandstone. Van Kranendonk (2000, 2008) reported the distribution of the Paddy Market Formation as extending through the Soanesville, East Strelley, Panorama, North Shaw, Emerald Mine, and Tambina greenstone belts (Fig. 1.7). Southwest from Shaw Gorge in the North Shaw greenstone belt, the base of the formation is composed of finely layered chert overlain by thinly bedded BIF, red and green shale, black, white, and brown banded chert (after shale), up to 10 m of sandstone, up to 100 m of jaspilite, and chert. Van Kranendonk (2000) interpreted the stratigraphically highest chert to contain felsic volcanic breccia and silicified, well-bedded felsic volcaniclastic tuff overlain by 200 m of massive to spherulitic dacite. These felsic rocks have not been dated and it is uncertain if the dacite is extrusive or intrusive. Cenozoic silicification of ferruginous shale and siltstone has formed secondary chert and BIF on the tops of ridges. Van Kranendonk (2003) recorded similar silicification of ferruginous shale in the Emerald Mine greenstone belt (Fig. 1.7). The minimum depositional age of the Paddy Market Formation, and the two underlying formations of the Soanesville Group, is c. 3185 Ma because it is intruded by a gabbro sill of the Dalton Suite dated at 3185 ± 2 Ma (Fig. 7.9) (GSWA 178185, Wingate et al. 2009a; Van Kranendonk et al. 2010).

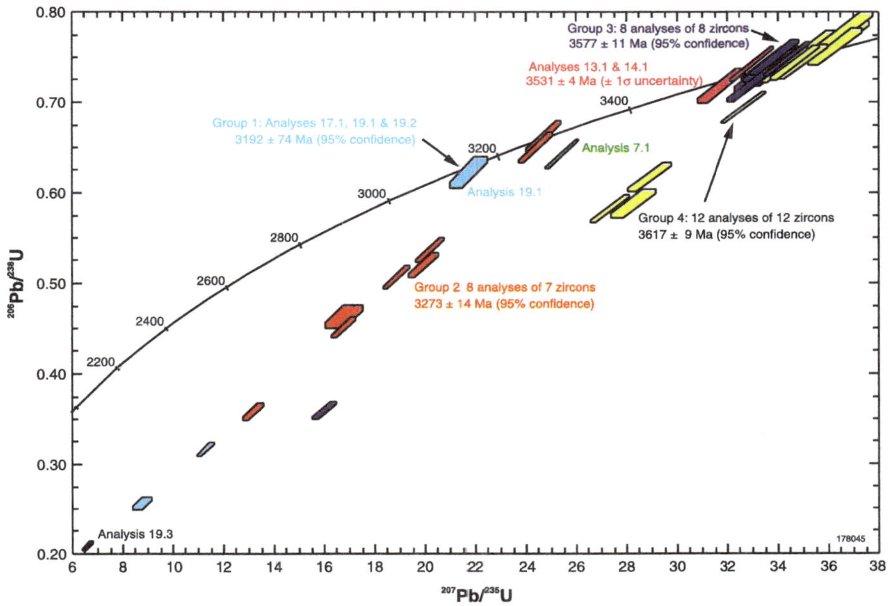

Fig. 7.8 Concordia plot of U–Pb zircon data for a sample of sandstone (Geological Survey of Western Australia sample 178045) from the Corboy Formation near Quininya Well in the northeast Yule granitic complex. Most detrital zircons pre-date the Pilbara Supergroup. (From Nelson 2005b; with Geological Survey of Western Australia permission)

7.3.1.4 Honeyeater Basalt

The Honeyeater Basalt was first recognized and defined in the Soanesville greenstone belt (Lipple 1975) where it is up to 1 km thick and is composed of komatiitic and tholeiitic basalt (Glikson and Hickman 1981a, b; Van Kranendonk 2000; Van Kranendonk et al. 2006; Smithies et al. 2007) and conformably or disconformably overlies the Paddy Market Formation. The depositional age of the basalt in this greenstone belt is indicated by the c. 3185 Ma date from gabbro of underlying the Dalton Suite (Fig. 7.9). The Dalton Suite is interpreted to be composed of magma that crystallized in conduits that had fed magma in the Honeyeater Basalt. In the Pilbara Well greenstone belt a welded rhyolite tuff within the Honeyeater Basalt (Fig. 7.10) was dated at 3176 ± 3 Ma (Fig. 7.11) (GSWA 180098, Wingate et al. 2009b; Van Kranendonk et al. 2010).

A stratigraphic section through the Honeyeater Basalt in the southern Soanesville greenstone belt (Glikson and Hickman 1981a) indicates that komatiitic basalt is confined to the lower half of the formation, pillow basalts are restricted to the central and upper parts of the formation, and dolerite and gabbro sills are present throughout the entire formation. Basalt formations that have been correlated with the Honeyeater Basalt of the Soanesville greenstone belt are exposed in the North Shaw greenstone belt (Van Kranendonk 2000) and in the Wodgina and Pilbara Well

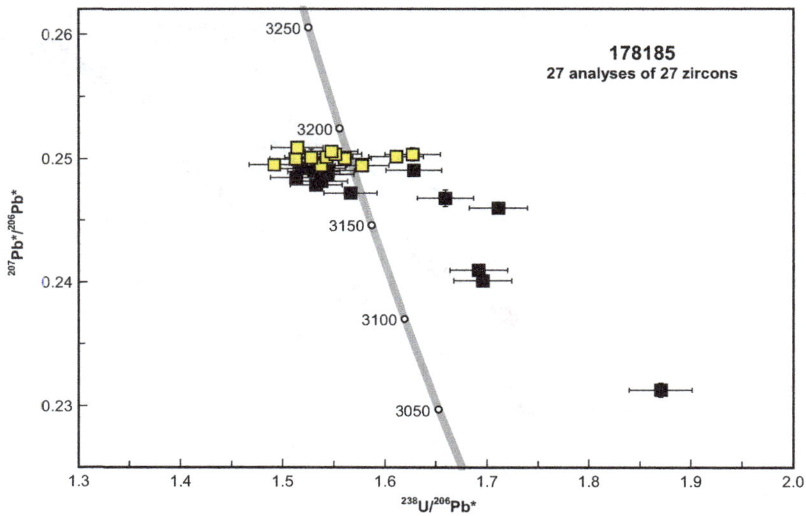

Fig. 7.9 U–Pb analytical data for sample Geological Survey of Western Australia sample 178185 from a gabbro sill of the Dalton Suite at Sulphur Springs. Yellow squares indicate Group I (magmatic zircons); black squares indicate Group P (radiogenic Pb loss). (From Wingate et al. 2009a; with Geological Survey of Western Australia permission)

Fig. 7.10 Boulder of welded rhyolitic tuff in the Honeyeater Basalt in the Pilbara Well greenstone belt (MGA Zone 50, 632662E 7653025N). (Modified from Van Kranendonk et al. 2010; with Geological Survey of Western Australia permission)

greenstone belts (Hickman 2010; Van Kranendonk et al. 2010). Van Kranendonk et al. (2010) also assigned tholeiitic pillow basalt in the far western East Strelley greenstone belt (Fig. 1.7) to the Soanesville Group. Alternatively, since this basalt unit is not underlain by the Sulphur Springs Group or lower formations of the Soanesville Group it might be part of the Euro Basalt (Hickman 2010). Limited Sm–Nd isotope data from basaltic rocks of the Wodgina and Pilbara Well greenstone

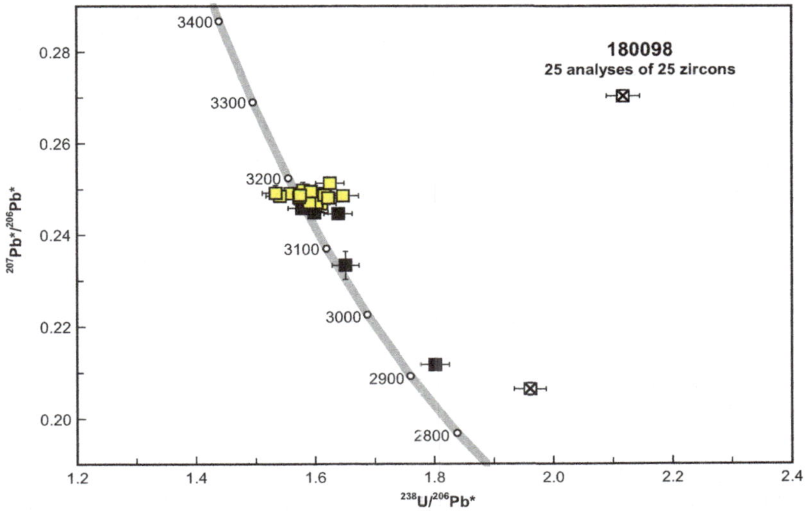

Fig. 7.11 U–Pb analytical data from a unit of welded rhyolitic tuff in the Honeyeater Basalt (Geological Survey of Western Australia sample 180098, boulder shown in Fig. 7.10). Yellow squares indicate Group I magmatic zircons; black squares indicate Group P radiogenic Pb loss; crossed squares indicate Group D (discordance >5%) (From Wingate et al. 2009b; with Geological Survey of Western Australia permission)

belts suggest magma derivation from evolved and juvenile sources, making the amount of Honeyeater Basalt in these areas uncertain. Geochronology in the Pilbara Well greenstone belt (Van Kranendonk et al. 2010) indicates a thick basalt succession of approximately the same age as the Honeyeater Basalt of the Soanesville greenstone belt, and the basalts of both areas are stratigraphically overlain by the Gorge Creek Group.

7.3.1.5 Pyramid Hill Formation

The Pyramid Hill Formation is composed of red to black shale, grey and white banded chert, and red to brown BIF. The formation outcrops in the Soanesville Syncline in the southern Soanesville greenstone belt, and on Agrippa Ridge between the Panorama and North Shaw greenstone belts. In the Soanesville Syncline, the Pyramid Hill Formation conformably overlies the c. 3185 Ma Honeyeater Basalt and is unconformably overlain by the Mount Roe Basalt of the Fortescue Group. At Agrippa Ridge, BIF, chert and shale correlated with the Pyramid Hill Formation conformably overlie the Honeyeater Basalt and unconformably underlie the c. 2950 Ma Lalla Rookh Sandstone. However, there are no geochronological data from any units at Agrippa Ridge to test stratigraphic correlations with the Soanesville Group.

In the Soanesville Syncline, the Pyramid Hill Formation is apparently 100 to 200 m thick, although the top of the formation is not exposed. Evidence that the maximum depositional age of the Pyramid Hill Formation is c. 3185 Ma is provided by dating of the gabbro sill interpreted to be a subvolcanic feeder to the underlying Honeyeater Basalt. In the Soanesville area, the Pyramid Hill Formation conformably overlies the Honeyeater Basalt and comprises red to black shale, grey and white banded chert, and red to brown BIF. The BIF is composed of alternating cm-thick layers of hematite-rich chert and silica-rich chert.

At the southern end of Sunset Ridge (Zone 50 K, MGA 735200E 7653600 N), 3 km north of Honeyeater Creek, BIF and shale underlying the Lalla Rookh Sandstone were previously interpreted to be parts of the Pyramid Hill Formation (Van Kranendonk and Morant 1998; Van Kranendonk 1999, 2000). This correlation is no longer preferred because aeromagnetic imagery, combined with geological mapping to the north and west, indicates that the Lalla Rookh Sandstone is underlain by the Cleaverville Formation throughout the Lalla Rookh Synclinorium. The Sunset Ridge outcrop of BIF is separated from the Honeyeater Basalt by a large fault, preventing an interpretation that this BIF is conformable with the basalt.

Conformably overlying the Honeyeater Basalt, and apparently restricted in its regional distribution, the Pyramid Hill Formation might be a relatively proximal hydrothermal deposit rather than a sedimentary BIF. An alternative possibility is that the formation is laterally equivalent to the Hong Kong Chert that overlies the Honeyeater Basalt in the Pilbara Well greenstone belt.

7.3.1.6 Hong Kong Chert

The Hong Kong Chert (Fitton et al. 1975) is apparently restricted to the Pilbara Well greenstone belt although it occupies the same stratigraphic position, immediately overlying the Honeyeater Basalt, as the Pyramid Hill Formation in the Soanesville greenstone belt. The formation is included in the Soanesville Group (Hickman 2021a) based on dating of the underlying Honeyeater Basalt at c. 3176 Ma and a minimum depositional age of c. 3165 Ma for the overlying Empress Formation (see below).

The Hong Kong Chert comprises two units of grey- and white-layered chert separated by silicified basalt, komatiitic basalt, gabbro and clastic sedimentary rocks including conglomerate, breccia, and sandstone. When first named (Fitton et al. 1975), the formation was interpreted it to be a single chert formation folded by a major northeast plunging isoclinal anticline. This structure was mapped in the core of a larger anticline that formed the northwest part of a fold-pair that dominated the structural geology of the Pilbara Well greenstone belt. Southeast of the anticline, Fitton et al. (1975) interpreted a more open northeast plunging syncline (John Bull Syncline). Following new geological mapping (Smithies and Farrell 2000; Van Kranendonk et al. 2010), a revised interpretation is that the axis of the northwest anticline does not fold the Hong Kong Chert but is actually located along a major northeast-striking fault that bisects the Pilbara Well greenstone belt (Smithies and

Fig. 7.12 Basaltic breccia at the base of the Hong Kong Chert in the Annie Gap area of the Pilbara Well greenstone belt. Note the cuspate fragments in the bottom left corner of the photograph (MGA Zone 50, 637000E 7659000N). (From Smithies and Farrell 2000; with Geological Survey of Western Australia permission)

Farrell 2000; Van Kranendonk et al. 2010). This fault is interpreted to be a major thrust referred to as the Kangaroo Flat Thrust (Hickman 2021a). This structure replaced the northwest limb of the John Bull Syncline, transporting the upper Honeyeater Basalt, the Hong Kong Chert, and the Empress Formation, over older greenstone units to the southeast. Pillow structures in the Honeyeater Basalt on both sides of the thrust indicate younging to the northwest; moreover, Van Kranendonk et al. (2010) emphasized that the metamorphic grade of greenstones in the succession northwest of the thrust is lower greenschist facies, whereas the greenstone succession to the southeast is metamorphosed to amphibolite and upper greenschist facies.

This revised structural interpretation is stratigraphically important because the greenstone succession northwest of the Kangaroo Flat Thrust is younger than the succession to the southeast. In describing the lithology of the Hong Kong Chert, Smithies and Farrell (2000) combined evidence from the Hong Kong Chert near Hong Kong mine northwest of the thrust with observations on a chert and sandstone unit North of Pilbara Well on the southeast limb of the John Bull Syncline. This southeast unit is now correlated with the 3426–3350 Ma Strelley Pool Formation. Detrital zircon data from sandstone of the southeast unit are very similar to data from the Strelley Pool Formation at the type locality in the East Strelley greenstone belt. Additionally, the basalt overlying the Strelley Pool Formation includes komatiite units with peridotite compositions, which is a characteristic feature of the Euro Basalt in other greenstone belts.

In the vicinity of the Hong Kong mine, the Hong Kong Chert includes a silicified basal sequence of mafic volcaniclastic breccia (Fig. 7.12), conglomerate, and sandstone (Smithies and Farrell 2000). The breccia contains angular and cuspate fragments of altered basalt in a tuffaceous sandstone matrix. The breccia matrix has relics of shard-like fragments up to 2 mm long, some of which are cuspate or have partial vesicle walls. These are interpreted to be altered fragments of basaltic glass. The unit is poorly sorted and appears to be thick bedded, suggesting a mass flow deposit. Smithies and Farrell (2000) described the overlying succession as comprising a lower chert 'zone', an intervening succession of basalt, silicified metagabbro, and strongly silicified mafic and sedimentary rocks, and an upper chert 'zone'. The chert zones rank as members of the formation, although there is currently insufficient detail to name and define them. The colour of the chert varies from off-white to black, and the degree of layering is variable. Layers are between 0.1 and 40 mm thick, and most of the thicker layers have fine-scale internal laminae or 'microbanding'. The upper chert member contains alternating dark grey and cream or pale-grey layers, which are up to about 700 mm thick.

The formation has a stratigraphic thickness of 150 m and outcrops northeast–southwest over a strike length of 13 km, being unconformably overlain by younger formations at both ends of this exposure.

7.3.1.7 Empress Formation

The Empress Formation (Fitton et al. 1975) outcrops along the northwest side of the Pilbara Well greenstone belt over a total strike length of 60 km, although between Hong Kong mine and Nunyerry Gap it is largely concealed by the Fortescue Group. The formation is 1.5–2.0 km thick and comprises komatiitic basalt and komatiite with sills or flows of peridotite and numerous thin chert units. In the Northern Pilbara Well greenstone belt, the Empress Formation conformably overlies the Hong Kong Chert and is unconformably overlain by the Gorge Creek and Croydon Groups. Inclusion of the Empress Formation within the Soanesville Group is based on apparently conformable contacts with the underlying Hong Kong Chert and on its depositional age being greater than c. 3165 Ma. Near Nunyerry Gap, the Empress Formation is intruded by the c. 3165 Ma Flat Rocks Tonalite, constraining its minimum depositional age. The maximum age of the formation is constrained by the 3176 Ma date on the underlying Honeyeater Basalt.

The geochemical and isotope compositions of the juvenile basalt of the Empress Formation (Smithies et al. 2007; Van Kranendonk et al. 2010), combined with its deformation by recumbent folding and thrusting (Blewett 2002; Blewett and Champion 2005; Van Kranendonk et al. 2010), suggest that it might be part of an ophiolite succession obducted southeast from the c. 3200 Ma Regal Basin. If such obduction occurred, it is likely to have been between c. 3165 and 3070 Ma during convergence of the East Pilbara and KTs, and before the Prinsep Orogeny. This scenario is analogous to the northwest obduction of the Regal Formation onto the KT (Hickman 2001, 2004, 2016; Hickman et al. 2010). The 3066–3015 Ma Gorge Creek Group, deposited following plate collision (Prinsep Orogeny), unconformably overlies the obducted juvenile crust on both sides of the Mallina Basin.

7.3.1.8 Tectonic Setting

The first detailed sedimentological investigation of the early Mesoarchean basins was by Eriksson (1981, 1982), who interpreted an initial 'cratonic rift stage' in which sedimentation was governed by crustal extension and localized in grabens. This resulted in successions of sandstone, siltstone, shale, and BIF. A subsequent study of Cardinal and Corboy Formations the Soanesville Group in the Pincunah greenstone belt (Wilhelmij and Dunlop 1984) indicated a tectonically active basin, with deposition in fault-bounded sub-basins. Relatively thin, and laterally discontinuous, sandstone, siltstone, and shale units at the base of the succession were interpreted as wave- or tide-dominated, transgressive shelf deposits. With deepening of the basin, the lower sandstone, shale, and BIF were overlain by coarse sandstone and turbidites that were deposited in submarine fans with migrating proximal channels. Wilhelmij and Dunlop (1984) interpreted the basin to have been rapidly subsiding and deepening to the southwest. Coarse-grained sediment was deposited in submarine fans, with aggradational facies stacking in an uplapping fan-wedge system. The tectonic setting was interpreted to be an unstable continental margin. The Soanesville Group is now interpreted to be a passive margin succession that was deposited during the plate separation stage of the East Pilbara Terrane Rifting Event. Rifting across the EPT is likely to have commenced at different times in different areas.

7.3.1.9 Geochemistry

Geochemical studies of the Soanesville Group have been restricted to volcanic rocks of the Honeyeater Basalt and the Empress Formation (Glikson and Hickman 1981a; Glikson et al. 1986; Smithies et al. 2007). Glikson and Hickman (1981a) analysed 28 volcanic samples from the Honeyeater Basalt in the Soanesville Syncline of the southern Soanesville greenstone belt, and Glikson et al. (1986) provided more accurate REE and high field-strength element (HFSE) analyses for four of these samples. Smithies et al. (2007) provided analyses of five samples of the Honeyeater Basalt from the northern Soanesville greenstone belt. These five samples were all low-Ti basalts with strongly LREE-depleted normalized trace element patterns. Glikson and Hickman (1981a) reported low-Ti basalts with relatively high K_2O (>0.5 wt. %), Ba (400 ppm) and Rb (20 ppm). Data from Glikson et al. (1986) reveal LREE enrichment, probably resulting from either crustal contamination or a crustal component in the source. Such LREE enrichment was recorded by Van Kranendonk et al. (2010) from c. 3170 Ma basalts in the lower Pilbara Well greenstone belt succession. In contrast, samples from the Empress Formation, and from the Wodgina greenstone belt (Fig. 1.7), have depleted LREE profiles with no evidence of crustal contamination.

Whole-rock Nd model ages and ε_{Nd} values from basalt samples from the Pilbara Well greenstone belt (Smithies et al. 2007) support geochemical evidence that the

Honeyeater Basalt and Empress Formation were deposited in different tectonic settings. Two basalt samples from the c. 3180 Ma Honeyeater Basalt yielded Nd model ages of c. 3540 and 3470 Ma and ε_{Nd} values of -1.52 and -0.57, suggesting Paleoarchean bulk sources. However, a Nd model age of c. 3320 Ma and a ε_{Nd} value of $+1.32$ from basalt of the Empress Formation indicate more juvenile sources (Van Kranendonk et al. 2010). The Nd data from this sample are similar to data from juvenile basaltic crust of the Regal Formation, which underlies the Mallina Basin northwest of the Pilbara Well greenstone belt. Two basalt samples collected from the Wodgina greenstone belt are isotopically similar to the Empress Formation.

7.3.1.10 Geochronology

In the northern Soanesville greenstone belt, diagenetic xenotime in the Corboy Formation was dated at 3190 ± 10 Ma (Rasmussen et al. 2007), indicating a minimum depositional age. In the same area, Wingate et al. (2009a) applied the U–Pb zircon method to date a gabbro sill of the Dalton Suite at 3185 ± 2 Ma. Dalton Suite intrusions are interpreted to have been feeders to the overlying Honeyeater Basalt (Van Kranendonk et al. 2010). About 90 km to the west, in the Pilbara Well greenstone belt, Van Kranendonk et al. (2010) reported a date of 3176 ± 2 Ma from a thin rhyolite unit within basalt correlated with the Honeyeater Basalt (GSWA 180098, Wingate et al. 2009b). The Empress Formation has not been dated, although in the Nunyerry area of the Southwest Pilbara Well greenstone belt it was intruded by the c. 3165 Ma Flat Rocks Tonalite (GSWA 142946, Nelson 2000; GSWA 142948, Nelson 2000), suggesting an eruptive age similar to that of the Honeyeater Basalt.

The maximum depositional age of the Soanesville Group is constrained by two lines of evidence. Firstly, in the Soanesville greenstone belt, the underlying Kangaroo Caves Formation of the Sulphur Springs Group is well dated at 3253–3235 Ma, with most volcanism occurring between c. 3238 and 3235 Ma (Buick et al. 2002). This timing coincides with that of the underlying Strelley Monzogranite dated at c. 3239 Ma (Buick et al. 2002). Buick et al. (2002) concluded that the Strelley Monzogranite was consanguineous with the felsic volcanic rocks of the Kangaroo Caves Formation and described the monzogranite as a subvolcanic intrusion emplaced within its own volcanic edifice. Considered in isolation, it might therefore be interpreted that the maximum depositional age of the Soanesville Group might be as old as 3235 Ma. However, a second constraint on the maximum depositional age of the Soanesville Group is that regional stratigraphic and sedimentological evidence, indicates that it was deposited in a subsiding extensional basin after rifting and breakup of the Paleoarchean Pilbara Craton (Van Kranendonk et al. 2010; Hickman 2012; Hickman and Van Kranendonk 2012). Breakup and plate separation, resulting in marginal-basin successions, followed the final magmatic event in the EPT that was responsible for intrusion of the 3270–3223 Ma Cleland Supersuite (Van Kranendonk et al. 2006, 2007b). This suggests a more robust maximum depositional age of c. 3223 Ma for the Soanesville Group. Accordingly, currently available geochronology indicates that first deposition of the Soanesville Group occurred between 3223 Ma and 3190 Ma.

7.3.2 *Nickol River Basin*

The Nickol River Basin unconformably overlies the southeast margin of the KT in the Northwest Pilbara Craton (Hickman 2012, 2016), and is exposed in the Roebourne and Devil Creek greenstone belts (Fig. 7.13). This basin evolved along the northwest margin of the Regal Basin, whereas the Soanesville Basin evolved on the southeast margin. Because most of the KT is concealed by the Indian Ocean (Hickman 2004), the depositional extent of the Nickol River Basin is interpreted to have been considerably greater than is evident on land. The basin is entirely composed of metasedimentary rocks of the Nickol River Formation that outcrop

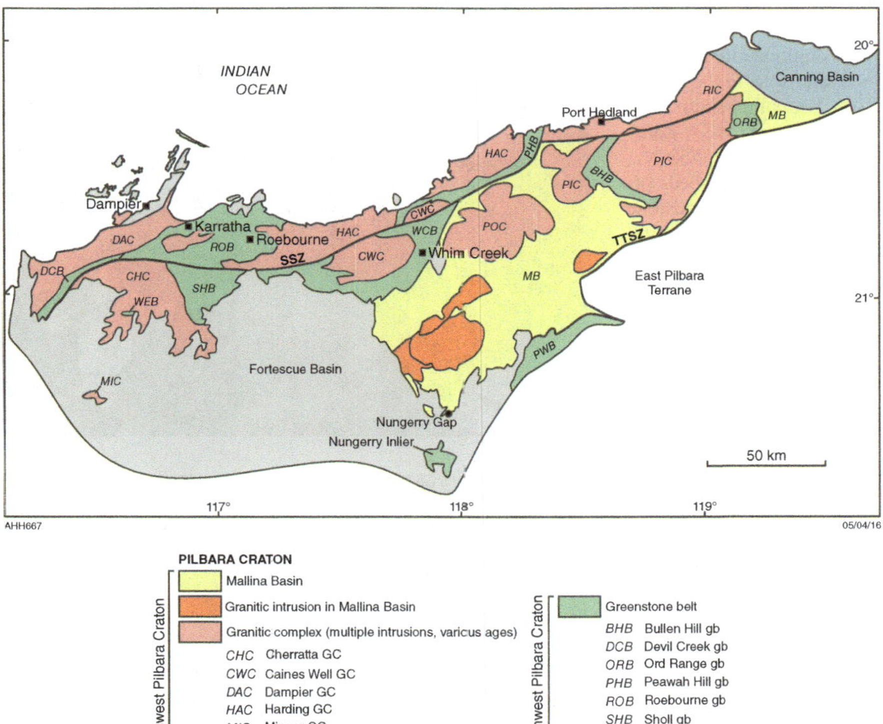

Fig. 7.13 Greenstone belts and granitic complexes of the Northwest Pilbara Craton. Each of the granitic complexes contains more than one supersuite (From Hickman 2016; with Geological Survey of Western Australia permission)

over a northeast–southwest strike length of 100 km. The most complete stratigraphic section through the formation is exposed southeast of Mount Regal, and a traverse was described by Hickman et al. (2010).

7.3.2.1 Nickol River Formation

The Nickol River Formation is a 1 km-thick, upward-fining clastic sedimentary formation comprising basal conglomerate and sandstone overlain by black shale, chert, carbonate rocks, and banded iron-formation (BIF). The conglomerate contains rounded to angular boulders, pebbles, and smaller clasts of chert (black, banded, and green) (Figs. 7.14 and 7.15) and altered volcanic rocks derived by erosion of the underlying Roebourne Group. The preserved thickness of the formation is likely to under-represent its depositional thickness by at least 1 km because upper sections

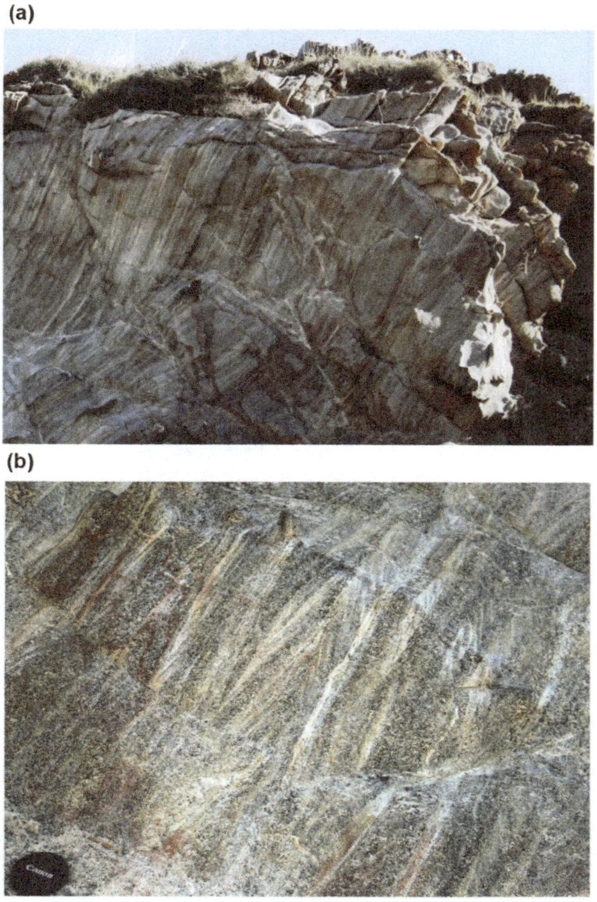

Fig. 7.14 Breakaway outcrop of metaconglomerate and metasandstone in the Nickol River Formation south of Port Robinson (MGA Zone 50, 502725 E, 7713758 N): (**a**) Steeply dipping beds of cross-bedded metasandstone exposed in a 4 m-high cliff; (**b**) close-up of cross-bedding indicating variable paleocurrent directions. Most dark grains in the metasandstone are fragments of black chert. Scale: lens cap, 5 cm diameter (Modified from Hickman 2016; with Geological Survey of Western Australia permission)

Fig. 7.15 Outcrop of strongly sheared matrix-supported conglomerate in the Nickol River Formation 2 km east of Lydia gold mine (MGA Zone 50, 501505E, 7707865N): (**a**) acutely stretched clasts of grey chert and pelitic schist within mylonitized sandstone; (**b**) small cobble of grey chert, largely undeformed, within sheared and lineated metasandstone matrix; (**c**) stretched boulders of grey chert and fine-grained metasedimentary rock within strongly sheared metasandstone; (**d**) numerous clasts of grey chert define a strongly oriented lineation in the metasandstone. Scale: lens cap, 5 cm diameter. (From Hickman 2016; with Geological Survey of Western Australia permission)

were tectonically removed by the regionally extensive Regal Thrust (Hickman 2001). This sub-horizontal thrust obducted the c. 3200 Ma Regal Formation (MORB-like basalt) over the Nickol River Formation, and the underlying KT, across an area of at least 3000 km^2 (Hickman 2004; Hickman et al. 2010). Obduction of the Regal Formation was a response to compression of the Regal Basin during 3160–3070 Ma convergence of the KT with the EPT (Sun and Hickman 1999; Hickman et al. 2001, 2010; Hickman 2001, 2004, 2016; Van Kranendonk et al. 2002, 2006).

The basal contact of the Nickol River Formation is an unconformity with the KT (Kiyokawa et al. 2002), whereas its upper contact with metabasalt of the Regal Formation is tectonic (Hickman 1997a, 2001). Beneath the Regal Thrust, the Nickol River Formation is variably deformed by subsidiary thrusts and isoclinal folding, and between Roebourne and Carlow Castle the formation is replaced by a zone of siliceous mylonites. In most outcrops, the formation is strongly deformed by shearing and the tectonic elongation of smaller clasts defines a stretching lineation.

Granitic clasts have not been observed, but detrital zircon grains in a quartzite form a single age group dated at 3269 ± 2 Ma (GSWA 136819, Nelson 1998), consistent with derivation through erosion of the underlying 3270–3261 Ma Karratha Granodiorite (Chap. 6).

Upper sections of the Nickol River Formation are composed of well-bedded black and ferruginous chert, metamorphosed carbonate rocks, quartz-sericite schist, ferruginous and carbonaceous shale, and BIF. In the Cleaverville area, the Regal Formation is underlain by black shale and chert of the Dixon Island Formation that is correlated with the Nickol River Formation (Hickman 2016). Kiyokawa et al. (2002) dated felsic tuff in the Dixon Island Formation at 3195 ± 15 Ma.

7.3.2.2 Tectonic Setting

Unconformably overlying the Roebourne Group (correlated with the Sulphur Springs Group in the EPT) and the Karratha Granodiorite (correlated with the Cleland Supersuite), the Nickol River Formation occupies a similar stratigraphic position to the Soanesville Group in the East Pilbara. Based on an early investigation of the Cleaverville area, Kiyokawa et al. (2002) referred to the Nickol River Formation as the 'Lydia Mine complex' (informal name) and interpreted it to be a continental shelf succession. Lithologies in the upper part of the Nickol River Formation suggest deepening of the depositional basin with increased plate separation. Kiyokawa et al. (2012) interpreted deposition of the Dixon Island Formation to have taken place during normal faulting of the sea floor and Kiyokawa et al. (2014) described evidence for hydrothermal activity, including numerous dykes of black chert. Dykes of black chert underlying bedded chert elsewhere in the Northern Pilbara Craton were formed by circulation of hydrothermal fluids in normal faults under conditions of crustal extension (Nijman et al. 1998; Van Kranendonk 2006; Hickman 2008).

7.3.2.3 Geochronology

The Nickol River Formation is undated, but it unconformably overlies the 3270–3261 Ma Karratha Granodiorite (Cleland Supersuite), indicating a maximum depositional age of c. 3261 Ma. Another intrusion of the Cleland Supersuite in the Northwest Pilbara Craton, the Tarlwa Pool Tonalite, has been dated at 3236 ± 3 Ma (Nelson 1998, GSWA 142535), which implies a younger maximum depositional age. The fact that the formation unconformably overlies the Roebourne Group, which is correlated with the 3280–3235 Ma Sulphur Springs Group of the EPT, also indicates a c. 3235 Ma maximum depositional age. A more precise depositional age is suggested by correlation of the black shale in the formation with the Dixon Island Formation, dated at 3195 ± 15 Ma (Kiyokawa et al. 2002). Finally, the interpretation that the Nickol River Formation was deposited on the southeast margin of the KT after the breakup of the Pilbara Craton implies a c. 3220 maximum depositional age.

7.3.3 *Early Mosquito Creek Basin*

The Mosquito Creek Basin is one of the most distinctive tectonic units of the Pilbara Craton, occupying an east–west trending belt of strongly deformed sedimentary rocks (mainly schistose sandstone, wacke, and pelite) east of Nullagine (Fig. 1.7). The exposed east–west length of the basin is 60 km, and it is 30 km wide. However, because the eastern and western exposures are unconformably overlain by the Neoarchean Fortescue Group, most of the basin length is concealed (Hickman 2004). The main sedimentary fill of the basin is the late Mesoarchean Mosquito Creek Formation (Hickman 1975a; Lipple 1975) but this overlies a much older formation, the late Paleoarchean to early Mesoarchean Coondamar Formation. This formation was deposited along the northern and southern margins of the c. 3220 Ma Mosquito Creek rift basin between the EPT and the KUT. It is inferred that mafic volcanic rocks were erupted into the centre of the Mosquito Creek Basin from c. 3220 Ma onwards, and that the Coondamar Formation represents two passive margin successions deposited unconformably on the margins of the EPT and KUT.

The northern succession of the Coondamar Formation is interpreted to be laterally equivalent to a mainly sedimentary passive margin succession, the Budjan Creek Formation, outcropping 15 km west in the Kelly greenstone belt of the EPT. Owing to cover by the Neoarchean Fortescue Group, there are no exposed contacts between the Budjan Creek Formation, on the southeast side of the Corunna Downs Dome, and the Mosquito Creek Basin. However, gravity data (Blewett et al. 2000; Hickman 2004; Bagas et al. 2008; Nijman et al. 2010) indicate that the Budjan Creek Formation was deposited on the northwest margin of the Mosquito Creek Basin, and therefore records part of the earliest stage in its evolution. Detrital zircon ages (Nelson 2001, GSWA 168908) and paleocurrent data (Eriksson 1981; Bagas et al. 2004b) from the Budjan Creek Formation indicate sediment transport from the Corunna Downs Dome southeast into the Mosquito Creek Basin. The upper part of the Budjan Creek Formation contains pillowed komatiitic and tholeiitic basalt (Bagas 2005; Williams and Bagas 2007) similar to lithologies in the Coondamar Formation.

7.3.3.1 Coondamar Formation

Prior to mapping by Bagas (2005) and Farrell (2006), the entire fill of the Mosquito Creek Basin was assigned to the Mosquito Creek Formation (Hickman 1975a, 1983; Lipple 1975). Although the Coondamar Formation was separated from the Mosquito Creek Formation (Van Kranendonk et al. 2006), the stratigraphy of the formation is poorly understood due to thrusting and folding within the Mosquito Creek Basin. In the eastern part of the basin, the formation underlies the Mosquito Creek Formation, and commonly outcrops in the cores of anticlines (Farrell 2006). Hickman (2021b) interpreted the formation to unconformably underlie the Mosquito Creek Formation along the northern margin of the Mosquito Creek Basin from an area 7 km South of Lionel to South of the Cooke Creek mining area, a west to east distance of 40 km. As noted above, Nijman et al. (2010) had reached a similar conclusion, although they did not directly equate their 'Cooke Creek unit' with the Coondamar Formation.

A previous interpretation that the Coondamar and Mosquito Creek Formations are conformable (Bagas et al. 2004a, 2008; Van Kranendonk et al. 2006) cannot be sustained in view of the reported unconformity south of the Middle Creek mining area (Blewett 2002) and the interpreted c. 200-million-year depositional age difference between the formations (Hickman 2012, 2021a). Detrital zircon ages indicate that parts of the Mosquito Creek Formation were deposited at c. 2930 Ma (Bagas et al. 2004a, 2008; Van Kranendonk et al. 2006), whereas the tectonic evolution of the EPT suggests a depositional age between c. 3220 and c. 3175 Ma.

The Coondamar Formation comprises a moderately metamorphosed assemblage of sandstone, mafic volcaniclastic sandstone, chloritic metasedimentary rocks, chert, and metamorphosed komatiitic and tholeiitic basalt (Blewett 2002; Bagas 2005; Farrell 2006; Van Kranendonk et al. 2006; Nijman et al. 2010). On the northern side of the Mosquito Creek Basin, the Mosquito Creek Formation overlies a 1 km-wide, 40 km-long, east–west-striking belt of metamorphosed sedimentary rocks, komatiitic basalt, tholeiitic basalt, and ultramafic and mafic intrusions. This assemblage, complicated by shearing, was assigned to the Coondamar Formation by Hickman (2016). Alternatively, since the komatiitic and tholeiitic basalt are lithologically similar to the upper part of the Budjan Creek Formation, and interpreted to be approximately the same age, the Budjan Creek Formation might east of its previously mapped outcrop to extend along the northern margin of the Mosquito Creek Basin, a possibility requiring testing by geochemistry and geochronology. Thrusting in the east–west-striking belt locally overrides southerly dipping conglomerate along 8 km of the contact with the EPT (Bagas et al. 2004a). The conglomerate contains angular to sub-rounded clasts of basalt, chert, and vein quartz in a mafic matrix (Bagas 2005); that is, its composition is lithologically like sandstone and conglomerate of the Coondamar Formation to the east and southeast.

The Coondamar Formation has not been dated isotopically although there is indirect evidence that its depositional age is between 3220 and 3175 Ma. One line of evidence is provided by data from the Golden Eagle Orthogneiss which forms the KUT in tectonic contact with the Mosquito Creek Basin. U–Pb zircon dating indicates that the intrusive age of the orthogneiss is between c. 3199 (GSWA 178013, Nelson 2005a) and 3178 Ma (GSWA 178012, Nelson 2004). Sm–Nd isotope data indicate a Nd model age of c. 3369 Ma and a ε_{Nd} value of +1.75 (sample 76,338, Champion 2013). The positive ε_{Nd} value and the tonalite–trondhjemite–granodiorite (TTG) composition of the orthogneiss suggest it was partly derived by melting of juvenile basaltic crust, but such crust is absent from the EPT and neither has any been recorded in the KUT. However, pre-3200 Ma juvenile basaltic crust in the interpreted rift basin (early Mosquito Creek Basin) would potentially have provided a source for the Golden Eagle Orthogneiss. Another consideration in relation to the Golden Eagle Orthogneiss is that there is no equivalent unit in the EPT along the northern margin of the Mosquito Creek Basin. If the basin had formed 'after' intrusion of the TTG in the KUT, the same TTG might also be expected in the adjacent EPT.

Mapping of the KUT has revealed that the Golden Eagle Orthogneiss contains units of chlorite–actinolite schist, quartz–muscovite schist, amphibolite schist, ultramafic schist, and pelitic schist lithologically similar to units in the adjacent Coondamar Formation (Bagas 2005; Farrell 2006). In most instances, field evidence has been insufficient to determine if these units originated as xenoliths or were tectonically interleaved with the granitic rocks. An additional line of evidence might be available from large mafic and ultramafic sills in the Coondamar Formation. These were emplaced along and close to the tectonic contact with the KUT and have been tentatively correlated with the c. 3185 Ma Dalton Suite (Hickman 2012, 2016). The correlation is based on the location of the intrusions close to the margins of the basin, which suggests a relationship to rifting. There are no similar intrusions in the Mosquito Creek Formation which suggests they were emplaced relatively early in the evolution of the basin. The EPT does not contain any younger suite of mafic and ultramafic intrusions until c. 2950 Ma when scattered intrusions of the Sisters Supersuite were emplaced. This suggests that the interpreted c. 3185 Ma age of the intrusions is correct, and that this places an additional constraint on the minimum age of the Coondamar Formation.

7.3.3.2 Budjan Creek Formation

The Budjan Creek Formation (Noldart and Wyatt 1962; Lipple 1975) is well exposed along a 60 km southwest–northeast strike length in the eastern Kelly greenstone belt (Fig. 1.7). The formation is an upward-fining succession of conglomerate, sandstone, siltstone, shale, and felsic volcaniclastic rocks, but includes pillowed komatiitic and tholeiitic basalt towards the northeast end of the outcrop (Bagas et al. 2004b; Bagas 2005; Williams and Bagas 2007). The formation unconformably overlies the 3350–3315 Ma Kelly Group and is unconformably overlain by the Gorge Creek Group. In the southwest of this linear outcrop, the basal part of the formation is a 1.2 km-thick succession starting with a matrix-supported boulder conglomerate (Fig. 7.16). The conglomerate fines upward into a succession of pebble conglomerate interbedded with arkosic sandstone, siltstone, and shale. Clasts in the conglomerate include vein quartz and chert with rare felsic volcanic rocks and are consistent with derivation from the underlying Kelly Group. The arkosic composition of the sandstone suggests derivation from granitic rocks, most likely those of the Emu Pool Supersuite in the centre of the Corunna Downs Dome. Sandstone beds in the basal unit contain planar cross-beds and rare trough cross-beds that indicate paleocurrents trending towards the south and southeast (Bagas et al. 2004b). This basal clastic unit is conformably overlain by a 600 m-thick unit of lithic wacke, siltstone, minor conglomerate, felsic tuff and fine-grained volcaniclastic sandstone

Fig. 7.16 Conglomerate at the base of the Budjan Creek Formation (MGA Zone 50, 791200E 7578400N). The boulder in the bottom right-hand corner of the photograph is 0.5 m across. (From Bagas et al. 2004b; with Geological Survey of Western Australia permission)

(Bagas et al. 2004b). The stratigraphically highest units in the formation are felsic volcaniclastic rocks derived from erosion of the Wyman Formation and Boobina Porphyry. Towards the northeast end of the 60 km outcrop, the formation includes pillowed komatiitic and tholeiitic basalt (Bagas 2005; Williams and Bagas 2007).

7.3.3.2.1 Tectonic Setting

Deposition of the Budjan Creek Formation commenced with intracontinental rifting and produced horst and graben structures (Eriksson 1981). The formation was initially deposited along the southeast margin of the EPT during rifting, and subsequently during separation of the KUT from the EPT (Hickman 2012, 2021a). Paleocurrent data and facies changes indicate that the depositional basin became deeper to the south or south-southeast (Eriksson 1981; Bagas et al. 2004a), suggesting deposition on the northwest margin of the Mosquito Creek Basin (Fig. 1.3). The Budjan Creek Formation included a basal continental sedimentary succession overlain by a passive margin succession on the southeast side of the EPT.

7.3.3.2.2 Geochronology

A fine-grained felsic volcaniclastic unit sampled near Copper Hills contained a dominant zircon age component at 3308 ± 5 Ma (Nelson 2001, GSWA 168908). This age does not indicate the depositional age of the Budjan Creek Formation because the zircons are detrital and derived from erosion of the Boobina Porphyry, Wyman Formation, and the Emu Pool Supersuite. GSWA 168908 also contained a single near-concordant younger zircon dated at 3228 ± 6 Ma (1σ). Assuming this age is not the result of Pb loss, this zircon was most likely derived from erosion of the Cleland Supersuite. The youngest granitic intrusion of the Cleland Supersuite in the EPT was dated at 3223 ± 3 Ma (Bodorkos et al. 2006, GSWA 178076). Because intrusion of the Cleland Supersuite pre-dated breakup and plate separation of the EPT, the c. 3223 Ma date suggests a maximum depositional age for the Budjan Creek Formation.

7.3.3.3 Tectonic Setting

Previous tectonic interpretations of the Mosquito Creek Basin have been reviewed by Bagas et al. (2008) and Nijman et al. (2010). Nijman et al. (2010) interpreted the basin to be the upper sedimentary part of a greenstone belt, with deformation resulting from north–south compression. A previous interpretation, derived from regional studies of the entire Northern Pilbara Craton (Van Kranendonk et al. 2006), was that the Mosquito Creek Basin originated through extension and rifting of the EPT at c. 3220 Ma, and initially evolved with central basaltic volcanism. The high proportion of metamorphosed basaltic rocks in the Coondamar Formation is consistent with the eruption of juvenile magmas in a c. 3200 Ma rift basin (Hickman 2012), although the geochemistry, geochronology, and isotopic composition of the basaltic rocks remain to be tested.

The present 30 km width of the Mosquito Creek Basin is the result of major south–north compression of the succession during the 2930–2900 Ma Mosquito Creek Orogeny. This timing is c. 30 Ma later than that of the 2955–2919 Ma North Pilbara Orogeny in the Mallina Basin. Accordingly, the Mosquito Creek Orogeny is interpreted to have been a delayed response to the same collision between the Northwest Pilbara Craton and a plate converging from the northwest (Hickman 2016). It is interpreted that until c. 2930 Ma tectonic effects of the collision were initially absorbed by compression within the Mallina Basin but, with final closure of the Mallina Basin, deformation in the Pilbara Craton moved to the relatively thin crust of the Mosquito Creek Basin. From c. 2930 Ma the EPT and KUT were driven together and the sedimentary succession of the Mosquito Creek Basin, probably >100 km wide at c. 2930 Ma, was progressively compressed into a 30–35 km-wide synclinorium. Basin inversion and reactivation of c. 3200 Ma extensional faults into c. 2905 Ma thrusts probably evolved along the lines of experimental laboratory models testing the transformation of graben into basin inversions (Del Ventisette et al. 2006).

The uniform width of the basin along its exposed length of 60 km suggests an original east–west length far exceeding 100 km. Gravity data indicate that south and southwest of Nullagine the basin strikes southwest for another 100 km beneath the unconformably overlying Fortescue Group (Bagas et al. 2004a; Hickman 2004). Evidence for an even greater lateral extent was provided by Bagas et al. (2008) who noted that a number of detrital zircon age components in the Mosquito Creek Formation have no apparent provenance in the northern Pilbara Craton. Since there is no evidence that the Mosquito Creek Basin extended into the southern area of the Pilbara Craton, it is possible that the basin extended west, rather than southwest, under the area now occupied by the large east–west trending valley of the Fortescue River.

7.3.3.4 Kurrana Shear Zone

A major structural feature of the Mosquito Creek Basin is the Kurrana Shear Zone between the Coondamar Formation and the KUT (Fig. 1.7). This important zone of shearing along the northern margin of the Golden eagle Orthogneiss has been briefly mentioned in most studies of the Mosquito Creek Basin (Hickman 1975a, b, 1978; Tyler et al. 1992; Blewett 2002; Bagas et al. 2004a, 2008; Bagas 2005; Farrell 2006; Nijman et al. 2010), although no detailed structural description has been published. Hickman (1975b) interpreted the shear zone to be a tectonic slide developed on the attenuated southern limb of the 'Mosquito Creek Synclinorium'. Tyler et al. (1992) interpreted the Kurrana Shear Zone to be a suture resulting from c. 3000–2760 Ma accretion of the EPT and KUT (previously unrelated according to these authors). Interpretations that the Mosquito Creek Basin originated as an extensional basin (Bagas et al. 2008; Hickman 2012; Hickman and Van Kranendonk 2012) imply that the Kurrana Shear Zone originated as a normal fault at c. 3220 Ma and was reactivated during the Mosquito Creek Orogeny. The present c. 60° northerly dip of the shear zone (Hickman 1975b; Bagas et al. 2004a; Farrell 2006) suggests southward thrusting of the Coondamar Formation over the KUT during north–south compression. Thrusts on the northern margin of the basin are inclined south indicating northerly thrusting of the Coondamar and Mosquito Creek Formations onto the EPT. A diagrammatic section across the basin consistent with this interpretation was provided by Farrell (2006).

References

Bagas L (2005) Geology of the nullagine 1:100 000 sheet. 1:100 000 geological series explanatory notes. Geological Survey of Western Australia

Bagas L, Bierlein FP, Bodorkos S, Nelson DR (2008) Tectonic setting, evolution, and orogenic gold potential of the late Mesoarchaean Mosquito Creek Basin, North Pilbara Craton, Western Australia. Precambrian Res 160:237–244

Bagas L, Farrell TR, Nelson DR (2004a) The age and provenance of the Mosquito Creek Formation. Geological Survey of Western Australia Annual Review 2003–04:62–70

Bagas L, Van Kranendonk MJ, Pawley MJ (2004b) Geology of the split rock 1:100 000 sheet. 1:100 000 geological series explanatory notes. Geological Survey of Western Australia

Beintema KA (2003) Geodynamic evolution of the West and Central Pilbara Craton: a mid-Archean active continental margin. PhD, Geologica Ultraiectina, Utrecht University

Blewett RS, Champion DC (2005) Geology of the Wodgina 1:100 000 sheet. 1:100 000 geological series explanatory notes. Geological Survey of Western Australia

Blewett RS, Wellman P, Ratajkoski M, Huston DL (2000) Atlas of North Pilbara, geology and geophysics, 1:1.5 million scale. Record, 2000/2. Australian Geological Survey Organisation

Blewett RS (2002) Archaean tectonic processes: a case for horizontal shortening in the North Pilbara Granite–Greenstone Terrane, Western Australia. Precambrian Res 113:87–120

Bodorkos S, Love GJ, Nelson DR, Wingate MTD (2006) 178076: biotite monzogranite, No. 7 Well; Geochronology Record 626. Geological Survey of Western Australia

Buick R, Brauhart CW, Morant P, Thornett JR, Maniw JG, Archibald NJ, Doepel MG, Fletcher IR, Pickard AL, Smith JB, Barley ME, McNaughton NJ, Groves DI (2002) Geochronology and stratigraphic relationships of the Sulphur Springs Group and Strelley Granite: a temporally distinct igneous province in the Archean Pilbara Craton, Australia. Precambrian Res 114:87–120

Byerly GR, Lowe DR, Heubeck C (2019) Chapter 24–geologic evolution of the Barberton Greenstone Belt–a unique record of crustal development, surface processes, and early life 3.55–3.20Ga. In: Van Kranendonk MJ, Bennett VC, Hoffmann JE (eds) Earth's oldest rocks, 2nd edn. Elsevier B.V, pp 569–613

Champion DC (2013) Neodymium depleted mantle model age map of Australia: explanatory notes and user guide. Record, 2013/44. Geoscience Australia

Del Ventisette C, Montanari D, Sani F, Bonini M (2006) Basin inversion and reactivation in laboratory experiments. J Struct Geol 28:2067–2083

Eriksson KA (1981) Archaean platform-to-trough sedimentation, East Pilbara Block, Australia. In: Glover JE, Groves DI (eds) Archaean Geology: Second International Symposium, Perth 1980. Geological Society of Australia, pp 235–244

Eriksson KA (1982) Geometry and internal characteristics of Archaean submarine channel deposits, Pilbara Block, Western Australia. J Sediment Res 52:383–393

Farrell TR (2006) Geology of the Eastern Creek 1:100 000 sheet. 1:100 000 Geological Series Explanatory Notes. Geological Survey of Western Australia

Fitton MJ, Horwitz RC, Sylvester G (1975) Stratigraphy of the early Precambrian in the West Pilbara. Western Australia, Report, FP, p 11

Glikson AY (2001) The astronomical connection of terrestrial evolution: crustal effects of post-3.8 Ga mega-impact clusters and evidence for major 3.2 Ga bombardment of the Earth–Moon system. J Geodyn 32:205–229

Glikson AY, Davy R, Hickman AH (1986) Geochemical data files of Archaean volcanic rocks, Pilbara Block, Western Australia. BMR (Geoscience Australia) Record, 1986/014

Glikson AY, Hickman AH (1981a) Geochemistry of Archaean volcanic successions, Eastern Pilbara Block, Western Australia. BMR (Geoscience Australia) Record, 1981/036

Glikson AY, Hickman AH (1981b) Geochemical stratigraphy and petrogenesis of Archaean basic–ultrabasic volcanic units, Eastern Pilbara Block, Western Australia. In: Glover JE, Groves DI (eds) Archaean Geology: Second International Symposium, Perth 1980, vol 7. Geological Society of Australia, pp 287–300

Glikson AY, Vickers J (2006) The 3.26–3.24 Ga Barberton asteroid impact cluster: tests of tectonic and magmatic consequences, Pilbara Craton, Western Australia. Earth Planet Sci Lett 241:11–20

Glikson AY, Vickers J (2010) Asteroid impact connections of crustal evolution. Aust J Earth Sci 57: 79–95

Hickman AH (1975a) Explanatory notes on the Nullagine 1:250 000 geological sheet. Record, 1975/5. Geological Survey of Western Australia

Hickman AH (1975b) Precambrian structural geology of part of the Pilbara Region. In: Annual report for the year 1974. Geological Survey of Western Australia, pp 68–73

Hickman AH (1978) Nullagine, Western Australia, 2nd edn. 1:250 000 Geological Series Explanatory Notes. Geological Survey of Western Australia

Hickman AH (1997a) A revision of the stratigraphy of Archaean greenstone successions in the Roebourne–Whundo area, West Pilbara. In: Geological Survey of Western Australia Annual Review 1996–97. Geological Survey of Western Australia, pp 76–81
Hickman AH (1997b) Dampier, WA Sheet 2256. Geological Survey of Western Australia
Hickman AH (2000) Roebourne, WA Sheet 2356. Geological Survey of Western Australia
Hickman AH (2001) Geology of the Dampier 1:100 000 sheet. 1:100 000 Geological Series Explanatory Notes. Geological Survey of Western Australia
Hickman AH (2002) Geology of the Roebourne 1:100 000 sheet. 1:100 000 Geological Series Explanatory Notes. Geological Survey of Western Australia
Hickman AH (2004) Two contrasting Granite–Greenstones Terranes in the Pilbara Craton, Australia: evidence for vertical and horizontal tectonic regimes prior to 2900 Ma. Precambrian Res 131:153–172
Hickman AH (2008) Regional review of the 3426–3350 Ma Strelley Pool Formation, Pilbara Craton, Western Australia. Record, 2008/15. Geological Survey of Western Australia
Hickman AH (2010) Marble Bar, WA sheet SF50–8, 3rd edn. Geological Survey of Western Australia
Hickman AH (2011) Pilbara Supergroup of the East Pilbara Terrane, Pilbara Craton: Updated lithostratigraphy and comments on the influence of vertical tectonics. In: Geological Survey of Western Australia Annual Review 2009–10. Geological Survey of Western Australia, pp. 50–59
Hickman AH (2012) Review of the Pilbara Craton and Fortescue Basin, Western Australia: crustal evolution providing environments for early life. Island Arc 21:1–31
Hickman AH (2016) Northwest Pilbara Craton: a record of 450 million years in the growth of Archean continental crust. Report, 160. Geological Survey of Western Australia
Hickman AH (2021a) East Pilbara Craton: a record of one billion years in the growth of Archean continental crust. Report 143, Geological Survey of Western Australia
Hickman AH (2021b) Interpreted bedrock geology of the East Pilbara Craton (version 2). East Pilbara Craton: a record of one billion years in the growth of Archean continental crust. Geological Survey of Western Australia
Hickman AH, Lipple SL (1975) Explanatory notes on the Marble Bar 1:250 000 geological sheet, Western Australia. Record, 1974/20. Geological Survey of Western Australia
Hickman AH, Lipple SL (1978) Marble Bar, WA sheet SF50–8, 2nd edn. Geological Survey of Western Australia
Hickman AH, Smithies RH, Pike G, Farrell TR, Beintema KA (2001) Evolution of the West Pilbara Granite–Greenstone Terrane and Mallina Basin, Western Australia–a field guide. Record, 2001/16. Geological Survey of Western Australia
Hickman AH, Smithies RH, Tyler IM (2010) Evolution of active plate margins: West Pilbara Superterrane, De Grey Superbasin, and the Fortescue and Hamersley Basins–a field guide. Record, 2010/3. Geological Survey of Western Australia
Hickman AH, Strong CA (2001) Dampier–Barrow Island, WA sheet SF50–2 and part of sheet SF 50–1, 2nd edn. Geological Survey of Western Australia
Hickman AH, Strong CA (2003) Dampier–Barrow Island, Western Australia. 1:250 000 Geological Series Explanatory Notes. Geological Survey of Western Australia
Hickman AH, Van Kranendonk MJ (2008) Archean crustal evolution and mineralization of the Northern Pilbara Craton–a field guide. Record, 2008/13. Geological Survey of Western Australia
Hickman AH, Van Kranendonk MJ (2012) Early earth evolution: evidence from the 3.5–1.8 Ga geological history of the Pilbara region of Western Australia. Episodes 35:283–297. https://doi.org/10.18814/epiiugs/2012/v35i1/028
Kemp AIS, Hickman AH, Kirkland CL (2015a) Early evolution of the Pilbara Craton, Western Australia, from hafnium isotopes in detrital and inherited zircons. Report, vol 151. Geological Survey of Western Australia

Kemp AIS, Hickman AH, Kirkland CL, Vervoort JD (2015b) Hf isotopes in detrital and inherited zircons of the Pilbara Craton provide no evidence for hadean continents. Precambrian Res 261: 112–126

Kiyokawa S (1993) Stratigraphy and structural evolution of a Middle Archean greenstone belt, Northwestern Pilbara Craton, Australia. PhD, University of Tokyo

Kiyokawa S, Ito T, Ikehara M, Yamaguchi KE, Koge S, Sakamoto R (2012) Lateral variations in the lithology and organic chemistry of a black shale sequence on the Mesoarchean seafloor affected by hydrothermal processes: the Dixon Island Formation of the coastal Pilbara Terrane, Western Australia. Island Arc 21:118–147

Kiyokawa S, Koge S, Ito T, Ikehara M (2014) An ocean-floor carbonaceous sedimentary sequence in the 3.2 Ga Dixon Island Formation, coastal Pilbara Terrane. Western Australia Precambr Res 255:124–143

Kiyokawa S, Taira A (1998) The Cleaverville Group in the West Pilbara coastal Granite–Greenstone Terrane of Western Australia: an example of a mid-Archaean immature oceanic island-arc succession. Precambrian Res 88:102–142

Kiyokawa S, Taira A, Byrne T, Bowring S, Sano Y (2002) Structural evolution of the middle Archean coastal Pilbara Terrane, Western Australia. Tectonics 21:1–24. https://doi.org/10.1029/2001TC001296

Lipple SL (1975) Definitions of new and revised stratigraphic units of the Eastern Pilbara Region. In: Annual report for the year 1974. Geological Survey of Western Australia, pp. 58–63

Morant P (1995) The Panorama Zn–Cu VMS deposits, Western Australia. In: Ho SE, Amann WJ (eds) Recent developments in base metal geology and exploration: Papers presented at an AIG Seminar, Perth, 29 November 1995. Australian Institute of Geoscientists, pp 75–84

Morant P (1998) Panorama zinc–copper deposits. In: Berkman DA, Mackenzie DH (eds) Geology of Australian and Papua new Guinean mineral deposits, The Australasian Institute of Mining and Metallurgy, vol 22. Melbourne, Australia, pp 287–292

Nelson DR (1998) Compilation of SHRIMP U-Pb zircon geochronology data, 1997. Record, 1998/2. Geological Survey of Western Australia

Nelson DR (2000) Compilation of geochronology data, 1999. Record, 2000/2. Geological Survey of Western Australia

Nelson DR (2001) Compilation of geochronology data, 2000. Record, 2001/2. Geological Survey of Western Australia

Nelson DR (2004) Compilation of geochronology data, 2003. Record, 2004/2. Geological Survey of Western Australia

Nelson DR (2005a) 178013: quartz diorite gneiss, Quartz Hill; Geochronology Record 550; Geological Survey of Western Australia

Nelson DR (2005b) 178045: sandstone, Quininya Well; Geochronology Record 567. Geological Survey of Western Australia

Nijman W, Clevis Q, De Vries ST (2010) The waning stage of a greenstone belt: the Mesoarchean Mosquito Creek Basin of the East Pilbara, Western Australia. Precambrian Res 180:251–271

Nijman W, de Bruijne KCH, Valkering ME (1998) Growth fault control of early Archaean cherts, barite mounds and chert–barite veins, North Pole Dome, Eastern Pilbara, Western Australia. Precambrian Res 88:25–52

Noldart AJ, Wyatt JD (1962) The geology of portion of the Pilbara Goldfield, covering the Marble Bar and Nullagine 4-mile map sheets. Bulletin, vol 115. Geological Survey of Western Australia

Ohta H, Maruyama S, Takahashi E, Watanabe Y, Kato Y (1996) Field occurrence, geochemistry and petrogenesis of the Archaean Mid-Oceanic Ridge Basalts (AMORBs) of the Cleaverville area, Pilbara Craton, Western Australia. Lithos 37:199–221

Rasmussen B, Fletcher IR, Muhling JR (2007) In situ U–Pb dating and element mapping of three generations of monazite: unravelling cryptic tectonothermal events in low-grade terranes. Geochim Cosmochim Acta 71:670–690

Shibuya T, Kitajima K, Komiya T, Terabayashi M, Maruyama S (2007) Middle Archean Ocean ridge hydrothermal metamorphism and alteration recorded in the Cleaverville area, Pilbara Craton. J Meta Geol 25:751–767

Smith JB, Barley ME, Groves DI, Krapež B, McNaughton NJ, Bickle MJ, Chapman HJ (1998) The Sholl Shear Zone, West Pilbara; evidence for a domain boundary structure from integrated tectonostratigraphic analysis, SHRIMP U–Pb dating and isotopic and geochemical data of granitoids. Precambrian Res 88:143–172

Smithies RH (2003) Geology of the White Springs 1:100 000 sheet. 1:100 000 geological series explanatory notes. Geological Survey of Western Australia

Smithies RH, Champion DC, Van Kranendonk MJ, Hickman AH (2007) Geochemistry of volcanic rocks of the Northern Pilbara Craton, Western Australia. Report, vol 104. Geological Survey of Western Australia

Smithies RH, Champion DC, Van Kranendonk MJ, Howard HM, Hickman AH (2005) Modern-style subduction processes in the Mesoarchean: geochemical evidence from the 3.12 Ga Whundo intra-oceanic arc. Earth Planet Sci Lett 231:221–237

Smithies RH, Farrell TR (2000) Geology of the Satirist 1:100 000 sheet. 1:100 000 Geological Series Explanatory Notes. Geological Survey of Western Australia

Sun S-S, Hickman AH (1998) New Nd-isotopic and geochemical data from the West Pilbara–implications for Archaean crustal accretion and shear zone development. Australian Geological Survey Organisation, Research Newsletter:25–29

Sun S-S, Hickman AH (1999) Geochemical characteristics of ca 3.0-Ga Cleaverville greenstones and later mafic dykes, West Pilbara: implication for Archaean crustal accretion. Australian Geological Survey Organisation, Research Newsletter:23–29

Tyler IM, Fletcher IR, de Laeter JR, Williams IR, Libby WG (1992) Isotope and rare earth element evidence for a late Archaean terrane boundary in the Southeastern Pilbara Craton, Western Australia. Precambrian Res 54:211–229

Van Kranendonk MJ (1997) Results of field mapping, 1994–1996, in the North Shaw and Tambourah 1:100 000 sheet areas, Eastern Pilbara Craton, Northwestern Australia. Record, 1997/23. Australian Geological Survey Organisation (Geoscience Australia)

Van Kranendonk MJ (2000) Geology of the North Shaw 1:100 000 sheet. 1:100 000 Geological Series Explanatory Notes. Geological Survey of Western Australia

Van Kranendonk MJ (2003) Geology of the Tambourah 1:100 000 sheet. 1:100 000 Geological Series Explanatory Notes. Geological Survey of Western Australia

Van Kranendonk MJ (2008) Structural geology of the central part of the Lalla Rookh–Western Shaw structural corridor, Pilbara Craton, Western Australia. Report 103. Geological Survey of Western Australia

Van Kranendonk MJ, Hickman AH, Smithies RH, Champion DC (2007a) Paleoarchean development of a continental nucleus: the East Pilbara Terrane of the Pilbara Craton, Western Australia. In: Van Kranendonk MJ, Bennett VC, Smithies RH (eds) Earth's oldest rocks. Elsevier B.V, Burlington, Massachusetts, USA, pp 307–337

Van Kranendonk MJ, Hickman AH, Smithies RH, Nelson DN, Pike G (2002) Geology and tectonic evolution of the Archaean North Pilbara Terrain, Pilbara Craton, Western Australia. Econ Geol 97:695–732. https://doi.org/10.2113/gsecongeo.97.4.695

Van Kranendonk MJ, Hickman AH, Smithies RH, Williams IR, Bagas L, Farrell TR (2006) Revised lithostratigraphy of Archean supracrustal and intrusive rocks in the Northern Pilbara Craton, Western Australia. Record, 2006/15. Geological Survey of Western Australia

Van Kranendonk MJ, Morant P (1998) Revised Archaean stratigraphy of the North Shaw 1:100 000 sheet, Pilbara Craton. In: Geological Survey of Western Australia Annual Review 1997–98. Geological Survey of Western Australia, pp 55–62

Van Kranendonk MJ, Smithies RH, Griffin WL, Huston DL, Hickman AH, Champion DC, Anhaeusser CR, Pirajno F (2015) Making it thick: a volcanic plateau origin of Paleoarchean continental lithosphere of the Pilbara and Kaapvaal Cratons. Geol Soc Lond, Spec Publ 389:89–111. https://doi.org/10.1144/SP389.12

Van Kranendonk MJ, Smithies RH, Hickman AH, Champion DC (2007b) Review: secular tectonic evolution of Archean continental crust: interplay between horizontal and vertical processes in the formation of the Pilbara Craton, Australia. Terra Nova 19:1–38

Van Kranendonk MJ, Smithies RH, Hickman AH, Wingate MTD, Bodorkos S (2010) Evidence for Mesoarchean (~3.2 Ga) rifting of the Pilbara Craton: the missing link in an early Precambrian Wilson cycle. Precambrian Res 177:145–161

Wilhelmij HR (1986) Depositional history of the middle Archaean sedimentary sequences in the Pilbara Block, Western Australia–a genetic stratigraphic analysis of the terrigenous rocks of the Pilgangoora syncline. PhD, The University of Western Australia

Wilhelmij HR, Dunlop JSR (1984) A genetic stratigraphic investigation of the Gorge Creek Group in the Pilgangoora syncline. In: Muhling JR, Groves DI, Blake TS (eds) Archaean and Proterozoic basins of the Pilbara, Western Australia: evolution and mineralization potential. The University of Western Australia, Geology Department and University Extension, pp 68–88

Williams IR, Bagas L (2007) Geology of the Mount Edgar 1:100 000 sheet. 1:100 000 Geological Series Explanatory Notes. Geological Survey of Western Australia

Wingate MTD, Bodorkos S, Van Kranendonk MJ (2009a) 178185: gabbro sill, Sulphur Springs, geochronology record 804. Geological Survey of Western Australia

Wingate MTD, Bodorkos S, Van Kranendonk MJ (2009b) 180098: welded rhyolitic tuff, Hong Kong mine, geochronology record 814. Geological Survey of Western Australia

Zegers TE, de Wit MJ, Dann J, White SH (1998) Vaalbara, Earth's oldest assembled continent? A combined structural, geochronological, and palaeomagnetic test. Terra Nova 10:250–259

Chapter 8
Mesoarchean Subduction in the Pilbara Craton

Abstract Two episodes of Mesoarchean subduction in the Pilbara Craton are attributed to interaction of the craton with another plate along its northwest margin. Northwest–southeast compression of the craton resulting from plate convergence is interpreted to explain the tectonic and magmatic evolution of the Northwest Pilbara Craton between 3160 and 2913 Ma. The first consequence of this convergence was the compression of the MORB-like basaltic crust of the Regal Basin as the adjacent Karratha and East Pilbara Terranes were forced closer together. Resulting failure of the thin crust of the basin established a subduction zone above which the volcanic arc of the 3130–3110 Ma Whundo Group was formed. Another consequence of the basin compression was obduction of a large section of its basaltic crust (Regal Formation) onto the Karratha Terrane. Eventual closure of the basin, marked by the 3070 Ma Prinsep Orogeny, ended the first episode of subduction. With ongoing convergence, the northwest plate began to be subducted under the northwest margin of the Pilbara Craton. This second episode of subduction lasted from about 3067 to 2913 Ma and resulted in the evolution of a series of continental magmatic arcs and retro-arc basins.

Keywords Regal Basin · Whundo Group · Prinsep Orogeny · Subduction · Obduction · Magmatic arcs

8.1 Introduction

Following the 3165–3140 Ma Karratha Event (Chap. 7), convergence of the Karratha Terrane (KT) and East Pilbara Terrane (EPT) resulted in northwest–southeast compression of the intervening Regal Basin. Development of a subduction zone within the thin, MORB-like basaltic crust of the basin led to the evolution of a volcanic arc (Smithies et al. 2005, 2007a; Van Kranendonk et al. 2006, 2007). Over 20 million years, a 10-km-thick volcanic succession (3130–3110 Ma Whundo Group) included tholeiitic basalt, boninite, adakite, calc-alkaline volcanics, and felsic pyroclastic units. Contemporaneous with the volcanism, intrusive rocks of the 3130–3093 Ma Railway Supersuite accumulated beneath the arc and together

A. H. Hickman, *Archean Evolution of the Pilbara Craton and Fortescue Basin*, Modern Approaches in Solid Earth Sciences 24,
https://doi.org/10.1007/978-3-031-18007-1_8

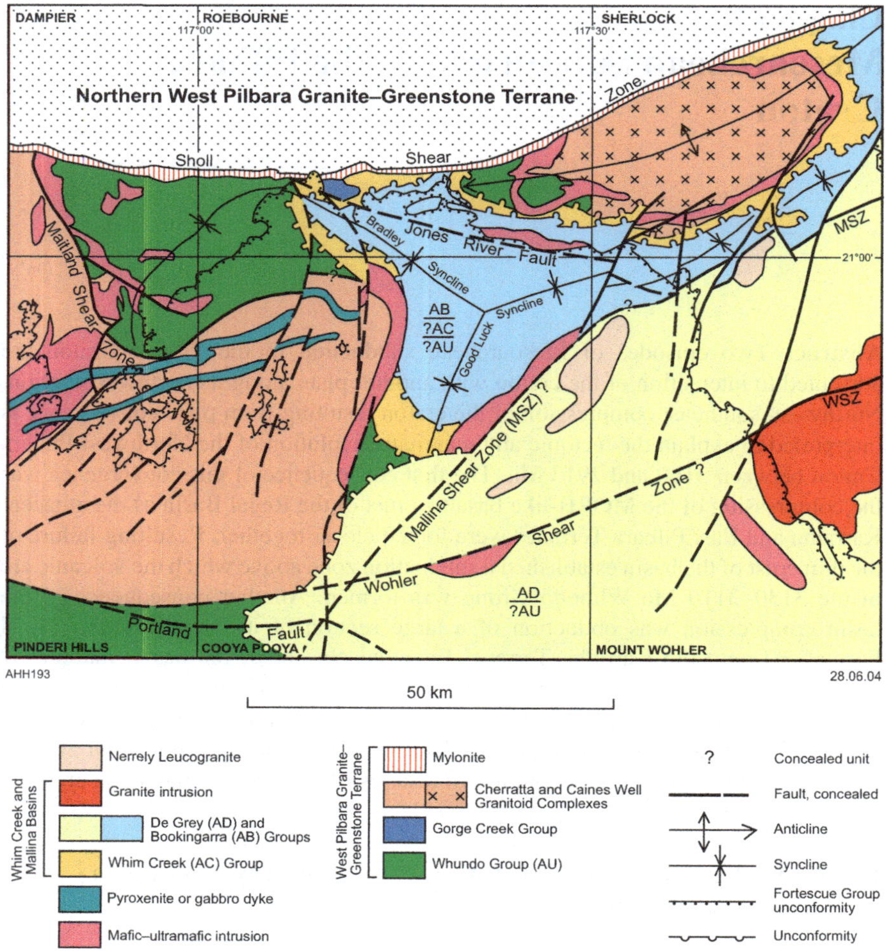

Fig. 8.1 Interpreted pre-Fortescue Group outcrops of the Whundo Group and adjacent stratigraphic units based on geophysical data. The Whundo Group is interpreted to underlie the Mallina Basin (from Hickman 2004a; with Geological Survey of Western Australia permission)

with the Whundo Group evolved to form the Sholl Terrane (ST) (Fig. 1.3). Figure 8.1 interprets the geological extent of the Whundo Group beneath the cover of the Fortescue Group.

8.2 Sholl Terrane

The Sholl Terrane (ST) outcrops over an area of 4000 km^2 south of the Sholl Shear Zone (SSZ) (Fig. 1.7) and is interpreted to unconformably underlie much of the Central Pilbara Tectonic Zone (Fig. 1.3). The volcanic component of the terrane, the 3130–3110 Ma Whundo Group, is exposed in two main areas of the Northwest Pilbara: around Mount Sholl, adjacent to the Cherratta granitic complex, and near Mount Fisher on the western margin of the Caines Well granitic complex (Figs. 7.13 and 8.1). The depositional extent of the Whundo Group is likely to have been considerably greater than this because the underlying subduction zone would have extended at least several hundred kilometres northeast–southwest along the length of the Regal Basin.

Emplaced beneath the Whundo Group, the 3130–3093 Ma Railway Supersuite is composed of tonalite–trondhjemite–granodiorite (TTG), minor monzogranite, and at least one large ultramafic–mafic intrusion (Bullock Hide Intrusion). Many parts of the Railway Supersuite consist of banded orthogneiss, containing greenstone enclaves. The bulk of the supersuite is exposed in the Cherratta granitic complex, but it also forms part of the Caines Well granitic complex (Fig. 7.13). Small felsic intrusions of the supersuite that intrude the basal Nallana Formation of the Whundo Group, and dated at c. 3120 Ma (Smith et al. 1998; Smith 2003), very likely represent feeders to overlying felsic volcanics of the Tozer and Woodbrook Formations.

Between the Sholl greenstone belt and the Cherratta granitic complex, the Whundo Group is separated from most parts of the Railway Supersuite by the Maitland Shear Zone (MSZ) (Figs. 1.9 and 8.1). However, dated intrusions of the Railway Supersuite also intrude the Whundo Group within the Sholl greenstone belt. Likewise, the Whundo Group is intruded by the Railway Supersuite in the Cherratta and Caines Well granitic complexes. The MSZ is up to 1 km wide and includes thick zones of mylonite tectonically interleaved with amphibolite and granitic gneiss. Although the MSZ is now steeply dipping, it is interpreted to have been formed a low-angle thrust when the Whundo Group was thrust southwards across the Railway Supersuite during the 3070 Ma Prinsep Orogeny (Hickman 2001; Hickman and Kojan 2003).

8.2.1 Whundo Group

The 3130–3110 Ma Whundo Group is a 10-km-thick volcanic succession comprising four lithostratigraphic formations: from base to top, Nallana Formation, Tozer Formation, Bradley Basalt, and Woodbrook Formation (Hickman 1997a). South of the SSZ, the metamorphic grade of the Whundo Group is lower greenschist facies but southwards and eastwards the grade increases to amphibolite facies. Based on geochemical traverses in the northern part of the Whundo greenstone belt, Smithies et al. (2005) recognized three chemically distinct packages: lower, middle, and upper

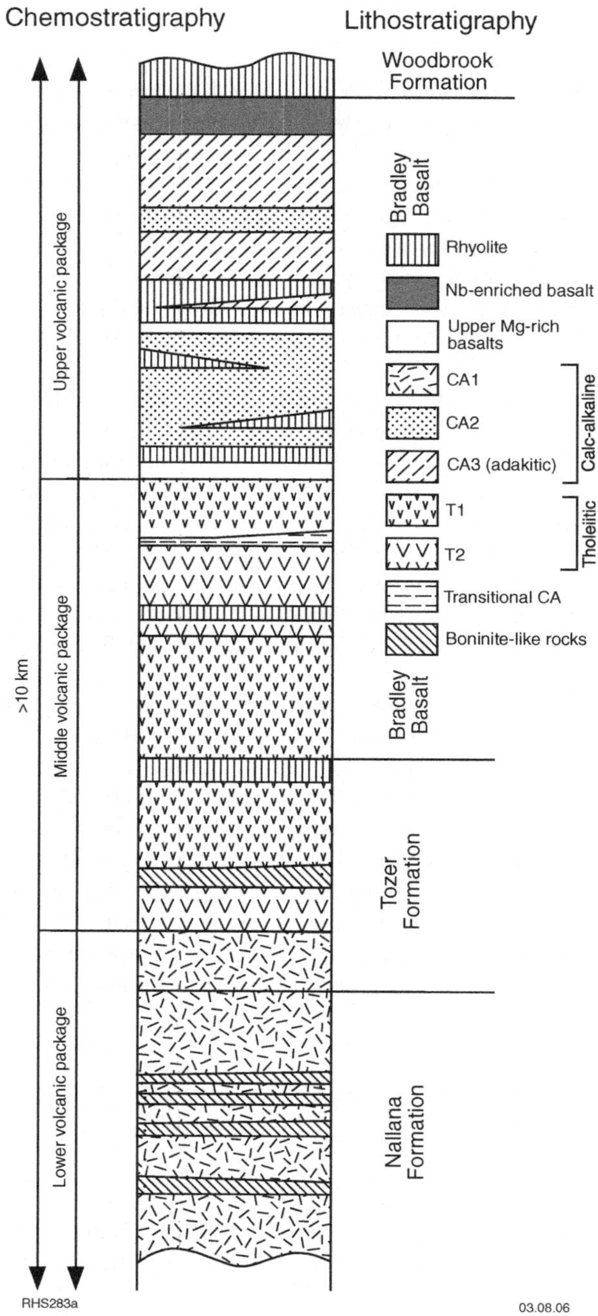

Fig. 8.2 Chemostratigraphic column of the Whundo Group, with lithostratigraphic formation boundaries shown for comparison (modified from Smithies et al. 2005; with Geological Survey of Western Australia permission)

(Fig. 8.2). They interpreted the depositional environment of the Whundo Group to have been like that of an intra-oceanic arc, but they made no reference to the associated granites of the Railway Supersuite.

8.2.1.1 Nallana Formation

The Nallana Formation is the basal formation of the Whundo Group and comprises a > 2-km-thick succession of metamorphosed basaltic rocks, local ultramafic rocks, isolated felsic rocks, and sills of pyroxenite, gabbro, and dolerite. Volcanic lithologies include pillow basalt, metabasalt of amphibolite to upper greenschist facies, metamorphosed ultramafic volcanic rock, ultramafic schist derived from volcanic rock, dacitic to rhyolitic volcaniclastic and tuffaceous rocks, and felsic schist of rhyolite to andesite composition. Sedimentary rocks form only a very minor part of the formation but include quartzofeldspathic schist derived from felsic igneous rocks or clastic sedimentary rocks, calc-silicate schist, and metachert. Lithologies forming sills within the formation include metapyroxenite, dolerite, gabbro, and leucogabbro. It is uncertain which of the sills are subvolcanic intrusions and which were intruded later.

A north–south geochemical traverse through the Nallana Formation (Smithies et al. 2005) indicated that the formation is dominated by calc-alkaline basaltic to andesitic lavas but that it also includes minor boninite-like lavas. These lavas are vesicular, extremely fine-grained, glass-rich rocks containing phenocrysts of subhedral olivine (now chlorite and serpentine) and acicular pyroxene (now actinolite). Depositional features in the Nallana Formation include hyaloclastite units, pillow structures, and flow-top breccias.

The lithological composition of the Nallana Formation is consistent with early stages in the growth of a volcanic arc in which basaltic magma was derived from melting of the mantle or older basaltic crust. Certain geochemical features of the calc-alkaline basaltic to andesitic lavas (high Th and La) are consistent with crustal contamination, although not necessarily involving felsic crust. Because the Regal Basin is likely to have been no more than 300 km wide, it is likely that fine-grained, basin floor sediments were subducted along with the MORB-like crust of the Regal Formation. Smithies et al. (2005) noted that the calc-alkaline basalts have lower initial ε_{Nd} values than the tholeiites, which suggests a greater crustal component.

Field exposures of the formation has not been studied in sufficient detail to determine water depths, although the lack of evidence for erosion surfaces within the >2-km-thick succession, and the apparent absence of clastic sediments, suggests moderately deep water. Small units of chert 6 km southeast of Whundo mine, and quartzofeldspathic schist 4.5 km west of Whundo mine, might be metamorphosed clastic sedimentary rocks near the base of the formation. In both instances, the protoliths are uncertain due to alteration, and, as noted above, fine-grained clastic sedimentary rocks might have been present at the base of the Nallana Formation.

Some of the volcanic units in the Nallana Formation are so extensively silicified that primary compositions are difficult to determine. Metamorphosed ultramafic rocks, some of which might be intrusive, are mainly restricted to the area south of

Whundo mine. An altered felsic unit near the top of the formation north of Mount Sholl was dated at 3125 ± 4 Ma (Nelson 1996, GSWA 114350), although the possibility of Pb loss suggests this date might understate the true depositional age. This possibility is supported by two slightly older U–Pb zircon dates from the overlying Tozer Formation. Near the SSZ, the formation is intruded by a rhyolite dyke dated at 3123 ± 2 Ma (Smith 2003), and by the ultramafic–mafic Bullock Hide Intrusion, dated at c. 3122 Ma (Wingate and Hickman 2009).

8.2.1.2 Tozer Formation

The c. 3120 Ma Tozer Formation is 2.5 km thick and composed of metamorphosed basalt, andesite, dacite, rhyolite, and thin units of sedimentary rocks, including volcaniclastic turbidite deposits, quartz-rich volcaniclastic sandstone, chert, and BIF. Felsic volcanic rocks, including thick massive deposits of rhyolite breccia, bedded volcaniclastic rocks, and flow-banded rhyolite, make up approximately 30% of the Tozer Formation. The felsic volcanic piles, although up to 1 km thick, are of very variable thickness and most lens out laterally over distances of 5 to10 km. The wedge-shaped felsic units are separated by basalt and basaltic andesite. More distal fringes of the felsic volcanoes, as exposed in outcrops immediately north of Whundo mine, show well-developed, graded bedding in reworked tuffaceous deposits.

U–Pb zircon dating of the felsic volcanic rocks has revealed no evidence of older felsic crust, as might be expected if the depositional setting of the formation was deposited adjacent to significantly older crust, as in a continental rift basin. Apart from one c. 3449 Ma zircon grain of unknown origin (in sample GSWA 114350, Nelson 1996), no zircon grains older than 3160 Ma were detected amongst 123 grains from 7 samples. Likewise, the oldest zircon grain of 84 grains from 4 samples of the Railway Supersuite was c. 3151 Ma. Very uncommon occurrences of siliciclastic sedimentary rocks are locally present within the formation but have not been dated. It is likely that these were derived from erosion of the underlying felsic volcanics.

Fine-grained felsic units near the stratigraphic base of the Tozer Formation contain relatively small deposits of VMS copper and zinc mineralization that have been mined at Whundo, West Whundo, and Yannery Hill.

8.2.1.3 Bradley Basalt

The Bradley Basalt is 4 km thick and mainly composed of weakly metamorphosed tholeiitic pillow basalt with thin units of felsic volcaniclastic rocks and dolerite sills. In many areas there is a well-defined stratigraphic boundary between the felsic volcanic rocks at the top of the Tozer Formation and massive and pillow basalt at the base of the Bradley Basalt. The boundary is locally marked by quartz-rich volcaniclastic sandstone, suggesting erosion and reworking of the Tozer Formation prior to mafic volcanism. In some areas the lowermost flows of the Bradley Basalt contain pyroxene spinifex texture, indicating a change in magma sources. Thin units

Fig. 8.3 Pillow lava in the Bradley Basalt 2.5 km northwest of Harding Dam (MGA Zone 50, 508900E, 7682000N). The pillow structures are 1.5 to 2 m in diameter, and their morphology is exceptionally well revealed due to the weathering and removal of inter-pillow material. Scale provided by geological hammer (modified from Hickman 2002; with Geological Survey of Western Australia permission)

of felsic tuffaceous material close to the base of the formation indicate either erosion of the underlying Tozer Formation or a continuation of bimodal volcanism. However, data from a geochemical traverse showed no significant lithological or chemical change at the mapped contact between the two formations (Smithies et al. 2005). This might be explained by the variable lithological composition of the Tozer Formation along strike. Pillow basalt is common throughout the Bradley Basalt, and some excellent examples are exposed west of Harding Dam (Fig. 8.3).

Geochemical data indicate that in the upper 1 km of the Bradley Basalt is composed of calc-alkaline volcanic rocks including basalt and andesite. Nb-enriched basalts and adakites were erupted prior to deposition of the c. 1.5-km-thick rhyolitic breccia and volcaniclastic rocks of the Woodbrook Formation at the top of the Whundo Group. In the Bradley Syncline (Fig. 7.1), approximately 7 km northwest of Harding Dam, the change to calc-alkaline volcanics coincides with several thin units of felsic volcaniclastic rocks within the otherwise basaltic succession. Where fine grained, these felsic rocks include exceptionally well-preserved sedimentary structures, such as graded bedding, cross-bedding, flame structures, and slump folds (Fig. 8.4). The lithology of these units suggests distal deposition, possibly from turbidity currents, during erosion of one or more felsic

Fig. 8.4 Sedimentary structures exposed in an outcrop of a 20-m-thick felsic volcaniclastic unit within the Bradley Basalt (MGA Zone 50, 505426E, 7684197 N): (**a**) view of the outcrop showing well-developed bedding; (**b**) fine-scale cross-bedding revealed in a weathered vertical section through eroded ripples; (**c**) slump folding and local diapiric injection of fine-grained felsic volcaniclastic sediment into overlying coarser-grained units (basal parts of upward-fining graded beds); (**d**) well-developed flame structures (centre and right) and syndepositional slump folding (left). Scale, b–d: coin, 2 cm diameter. (From Hickman 2016; with Geological Survey of Western Australia permission)

volcanic centres. However, it is unknown if these distant felsic sources were contemporaneous with the Bradley Basalt or merely exposed areas of the Tozer Formation.

Lower basalts of the formation locally exhibit pyroxene spinifex texture, indicating more mafic compositions than higher in the formation. Even so, thin units of felsic tuffaceous material are also locally present near the base of the formation. Unless these felsic volcaniclastic lenses were derived from erosion of the Tozer Formation, this suggests a continuation of the bimodal volcanism into the Bradley Basalt. On one traverse (Smithies et al. 2005), geochemistry defines an important volcanic change midway through the formation: tholeiites, like those of the underlying Tozer Formation, give way to calc-alkaline volcanics, including adakite-like rocks (CA3 in Fig. 8.2).

8.2.1.4 Woodbrook Formation

The Woodbrook Formation is composed of calc-alkaline volcanic rocks similar to those in the Bradley Basalt, but it is distinguished by a predominantly felsic composition. In the type area west of Woodbrook, the formation comprises a 1000-m-thick succession of metamorphosed rhyolite tuff and agglomerate. Fragments within the tuff include pumice, flow-banded lava, and porphyritic and spherulitic lava. This rock was dated at 3117 ± 3 Ma (GSWA 127378, Nelson 1998), which is a very similar age to most dated samples from the Bradley Basalt and Tozer Formation. This suggests that the middle and upper parts of the Whundo Group were deposited within five million years. The lower part of the Woodbrook Formation includes minor basalt, and a single thin BIF unit was mapped near the top of the formation (Hickman 2000).

8.2.1.5 Geochemistry

Early geochemical data from basalts of the Whundo Group (Glikson et al. 1986) were consistent with juvenile crust generated in a subduction zone environment (Sun and Hickman 1998). However, Smithies et al. (2004a) were first to liken the light rare earth element (LREE)-enriched basalts of the Whundo Group to modern boninites. Boninites are high-Mg basaltic to andesitic rocks, typically with low Ti concentrations, high large-ion lithophile element (LILE) concentrations, and with very high Al_2O_3/TiO_2, low Gd/Yb, and high La/Gd ratios compared to primitive mantle. Phanerozoic boninites are confined to convergent margins (Crawford et al. 1989), typically in fore-arc settings during the early stages of subduction. The Whundo Group boninite-like basalts have SiO_2 contents ranging from 47.3 to 53.2 wt.%, and MgO contents varying between 8.1 and 9.9 wt.%. Mg numbers range between 61 and 66. Smithies et al. (2004a) noted that high Al_2O_3/TiO_2 ratios (35–58) and high CaO/TiO_2 ratios (18–29) distinguish the boninite-like basalts of the Whundo Group from Paleoarchean basalts of the EPT and KT.

Additional geochemical evidence for subduction was provided by detailed geochemical traverses described by Smithies et al. (2005). These revealed an arc-like assemblage of lithologies, including boninite-like rocks, interlayered tholeiitic and calc-alkaline volcanics, Nb-enriched basalts, adakites, and rhyolites (Fig. 8.5). The inclusion of adakites in the Whundo Group provides further evidence for subduction. Adakites are felsic igneous rocks characterized by high La/Yb and Sr/Y and low heavy rare earth elements (HREE). Phanerozoic adakites are typically associated with arcs above subduction zones and are the product of partial melting of subducting hydrous oceanic crust (oceanic slab) or of the mantle metasomatized by slab melt.

Geochemical and Sm–Nd isotope data from the tholeiitic volcanic rocks of the Whundo Group suggest minor crustal contamination. These data include La/Nb and Th/Nb ratios slightly higher than N-MORB values and Sm–Nd two-stage depleted

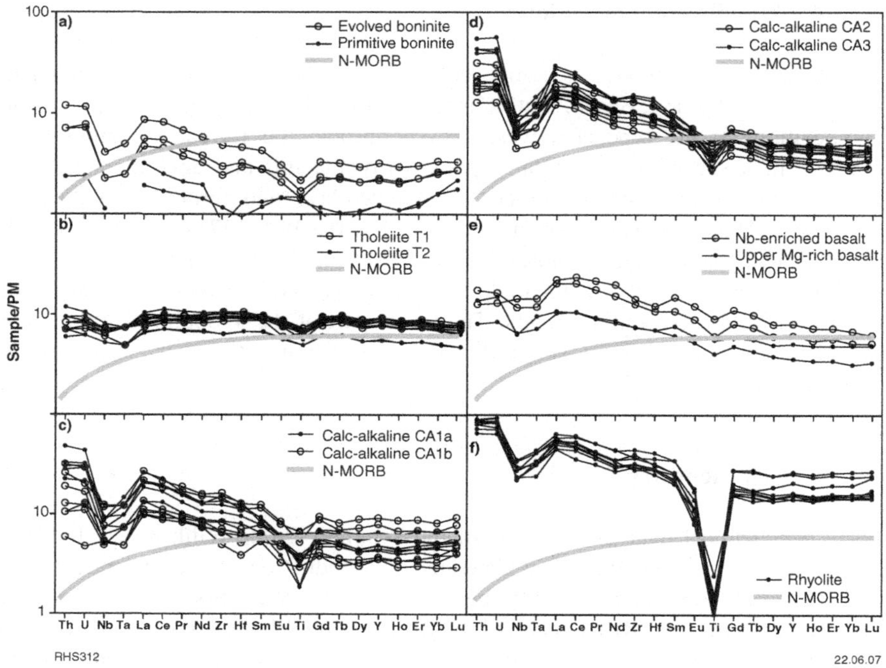

Fig. 8.5 Trace element plots normalized to primitive mantle for various volcanic rocks of the Whundo Group (From Smithies et al. 2007a; with Geological Survey of Western Australia permission)

mantle model ages (T_{DM}^2) up to 100 million years older than depositional ages. The evidence for some crustal input questions the initial interpretation that the depositional setting of the Whundo Group was an intra-oceanic island arc (Smithies et al. 2005). Alternatively, some crustal input is consistent with subduction within the Regal Basin, which was relatively narrow and during its evolution is likely to have been up to 400 km wide (Hickman 2016, 2021). During the early evolution of a rift basin, clastic sediments derived from erosion of the adjacent continental plates would have formed part of the basin fill.

Compared to the tholeiites, the calc-alkaline rocks of the Whundo Group are depleted in HREE and distinctly enriched in incompatible trace elements, with higher La/Nb (Fig. 8.5c, d). Based on geochemistry and lithological association, Smithies et al. (2005, 2007a) divided the calc-alkaline rocks into four groups: CA1a, CA1b, CA2, and CA3 (Figs. 8.2 and 8.5), which they assigned to 'lower' and 'upper' volcanic packages. The trace element plots (Figs 8.5c, d) indicate a large degree of compositional overlap across the four groups. In line with lithological compositions, the CA1a andesites have lower V (and Sc) and higher Th/Yb, Th/Nb, Th/La, and La/Sm ratios, and generally higher Th, HFSE, and REE concentrations, than the CA1b basalts. The CA2 group, including andesites and basalts, has trace element patterns similar to those of the andesites and basalts in CA1. However, the CA2 lavas are interbedded with locally abundant units of felsic lava, whereas CA1 lavas are interbedded throughout with boninite-like rocks. The CA3 group, in the upper 1 km of the Bradley Basalt, is composed of andesitic to dacitic lavas with similar compositions to Phanerozoic adakites. However, although the subducting crust of the Regal Basin was MORB-like in composition, it was not oceanic. The adakite-like rocks of group CA3 are interbedded with felsic volcanic and volcaniclastic rocks and unusually Nb-rich basalt.

Nb-enriched basalt interbedded with felsic and CA3 lavas at the top of the Bradley Basalt (Fig. 8.2) are geochemically transitional between the tholeiites and the calc-alkaline rocks. They are more Nb enriched (Nb, 8.6–10.5 ppm) than all the tholeiites and most of the calc-alkaline rocks (Smithies et al. 2007a). The composition of these basalts might reflect assimilation of felsic material during intrusion through felsic volcanic rocks of the underlying Whundo Group and the granitic rocks of the Railway Supersuite. Available geochemical data indicate that felsic volcanic rocks of the Whundo Group are mainly low- to medium-K_2O rhyolites characterized by low Al_2O_3 and high Na_2O/K_2O. They have high HREE (Y from 60 to 130 ppm), Zr (300–500 ppm), and Nb (16–25 ppm) concentrations combined with low to moderate LILE concentrations (Smithies et al. 2007a).

8.2.1.6 Sm–Nd Isotope Data

Sm–Nd isotope data indicate that lavas of the Whundo Group were derived from juvenile sources not much older than the group (Sun and Hickman 1998; Smithies et al. 2005). Sun and Hickman et al. (1998) observed that the Railway Supersuite has similar Nd model ages to those from the Whundo Group and contrasted the available 3250–3150 Ma model ages from the group to 3480–3430 Ma model ages from c. 3270 Ma rocks of the KT. Average ε_{Nd} values of +2.0 for the Whundo tholeiites, compared to the theoretical depleted mantle ε_{Nd} value of +3.2 at 3120 Ma, suggest a minor amount of crustal input. ε_{Nd} values of the felsic volcanic rocks of the Whundo Group are also positive, in the range of +1.3 to +3.6. Data from the calc-alkaline rocks indicate more crustal input than in the tholeiites.

8.2.1.7 Geochronology

U–Pb zircon dating on formations of the Whundo Group has provided a range of dates between 3128 ± 7 Ma (Smith 2003, N 4325) and 3112 ± 6 Ma (Horwitz and Pidgeon 1993). The group was erupted during the intrusion of the 3130 to 2093 Ma Railway Supersuite (Van Kranendonk et al. 2006; Hickman 2016), suggesting that its maximum depositional age is c. 3130 Ma. The stratigraphic base of the group, overlying the 3200 to 3160 Ma basaltic crust of the Regal Formation, is not preserved.

The stratigraphic top of the Whundo Group is not exposed, and dating of the upper two formations of the group has given slightly inconsistent ages. A felsic tuff in the lower part of the Bradley Basalt in the Bradley Syncline was dated at 3115 ± 5 Ma (Nelson 1996, GSWA 114305). However, two samples from the overlying Woodbrook Formation were dated at 3117 ± 3 Ma (Nelson 1998, GSWA 127378) and 3118 ± 2 Ma (Nelson 1998, GSWA 144256). Either the Bradley Basalt is older than c. 3115 Ma or the dates from the volcaniclastic units of the Woodbrook Formation were calculated on zircon analyses that included xenocrystic material. The date of 3112 ± 6 Ma (Horwitz and Pidgeon 1993) was obtained on an altered rock within the Tozer Formation and was based on only five zircon analyses. Two granitic samples from the Railway Supersuite were individually dated at 3114 ± 5 Ma (Smith et al. 1998, JS 20, JS 33). From the available geochronology, it is inferred that the Whundo Group was deposited between c. 3130 and 3110 Ma.

8.2.2 Railway Supersuite

The second major component of the ST, the Railway Supersuite, is mainly composed of metamorphosed tonalite–trondhjemite–granodiorite (TTG), and minor monzogranite, dated between 3130 and 3093 Ma (Van Kranendonk et al. 2006). The supersuite forms large parts of the Cherratta granitic complex but also forms intrusions within the Whundo Group. In the Caines Well granitic complex, it is represented by unnamed 3111–3093 Ma banded biotite–trondhjemite gneiss (Nelson 1997; Smithies and Champion 1998). Intrusions within the Whundo Group include the 25 km^2 Bullock Hide Intrusion, dated at 3122 ± 3 Ma (Wingate and Hickman 2009), and monzogranite and rhyolite dykes of the Twin Table Monzogranite dated at 3123 ± 2 Ma (Smith 2003). Granitic gneiss boulders in conglomerate at the base of the 3000–2990 Ma Whim Creek Group near Red Hill were probably eroded from the Railway Supersuite in the Caines Well granitic complex.

8.2.2.1 Pinnacle Hill Gneiss

The Pinnacle Hill Gneiss now occupies a substantial part of the Cherratta Granitic complex (Hickman et al. 2006), although the precise extent remains to be tested by more detailed mapping, geochemistry, and geochronology. The composition of the gneiss is mainly tonalite to trondhjemite (Smithies and Champion 1998), and it is typically banded with elongate greenstone xenoliths, some of which are up to several hundred metres in length. In view of the juvenile chemistry of the gneiss and the lack of contamination by significantly older crust, the greenstones were most likely derived from the lower section of the Whundo Group, either by igneous injection or by tectonic interleaving during thrusting related to formation of the MSZ. Excellent exposures of the Pinnacle Hill Gneiss are present in the bed of the

Maitland River 6 km south-southwest of Whundo copper mine and in outcrops immediately beneath the unconformity of the Fortescue Group between Mount Leopold and Byong Creek.

8.2.2.2 Bullock Hide Intrusion

The Bullock Hide Intrusion (Hickman 1997b) intrudes the base of the Nallana Formation over an area of 25 km^2 area south of the SSZ. The intrusion is composed of layered units of serpentinite (metaperidotite) and metamorphosed gabbro, dolerite, and leucogabbro. Northern outcrops of the intrusion are mainly composed of gabbro, containing numerous angular blocks of metabasalt. This suggests initial fragmentation and net-veining of the basalt followed by its incorporation as xenoliths within the gabbro. The sample dated at 3122 $\pm$ 3 Ma (Wingate and Hickman 2009, GSWA 178164) was a coarse-grained leucogabbro from the upper part of the intrusion.

8.2.2.3 Twin Table Monzogranite

The Twin Table Monzogranite intrudes the Nallana Formation of the Whundo Group forming small monzogranite to granodiorite stocks but also locally exposed as rhyolite dykes. Monzogranite of the formation has been dated at c. 3114 Ma, whereas rhyolite has been dated at c. 3123 Ma. The Twin Table Monzogranite is interpreted to be genetically related to felsic volcanic units of the Whundo Group, most likely the 3128–3116 Ma Tozer Formation. This formation forms a large felsic volcanic centre in the central part of the Whundo Group.

8.2.2.4 Geochemistry

Smith (2003) described the rare earth element (REE) geochemistry of two samples from the supersuite, noting LREE enrichment, negative Eu anomalies, flat HREE patterns, and high levels of high field strength elements (HFSE). This REE pattern is similar to that for felsic volcanic rocks of the Whundo Group, and Smith (2003) concluded that the intrusive and extrusive rocks had a similar source.

8.2.2.5 Sm–Nd Isotope Data

Sm–Nd two-stage depleted mantle model ages (${T_{DM}}^2$) for the supersuite range between 3310 and 3210 Ma (Smith et al. 1998; Sun and Hickman 1998; Smithies et al. 2007a). Smithies and Champion (1998) reported that the Railway Supersuite of the Cherratta and Caines Well Granitic complexes has high ε_{Nd} between +2.5 and + 2.8 and Nd model ages very similar to the crystallization ages. They, and

Sun and Hickman et al. (1998), concluded that these granitic rocks were derived from juvenile crust, and this is now interpreted to have been the 3200–3165 Ma Regal Formation. Two additional samples of the supersuite (GSWA 142534 and 142,536) reported by Smithies et al. (2007a) have moderately high ε_{Nd} values (+1.63 and + 1.38) and Nd model ages of 3220 to 3240 Ma.

8.2.2.6 Geochronology

The maximum age of the Railway Supersuite is inferred to be c. 3130 Ma based on a date of 3130 ± 4 Ma from the Pinnacle Hill Gneiss (Nelson 1998, GSWA 142835). However, this sample also contained a minor zircon age component dated at 3149 ± 15 Ma that is interpreted to be from an earlier intrusive event. Another constraint on the maximum age is that the subduction event that led to the formation of the ST did not commence until after the c. 3165–3144 Ma Karratha Event when the Regal Basin began to be compressed by convergence of the KT and EPT (Hickman 2016, 2021).

The minimum age of the Railway Supersuite is inferred to be c. 3093 Ma based on a date of 3093 ± 4 Ma from gneiss of the Caines Well granitic complex (Nelson 1997, GSWA 118965). A more general constraint is that the supersuite is older than the c. 3070 Ma Prinsep Orogeny that marked the end of subduction in the Regal Basin (Hickman 2016).

8.3 Ophiolite (3220–3165 Ma Regal Formation)

Compression of the Regal Basin between 3160 and 3070 Ma not only resulted in the formation of an intra-basin subduction zone and a Mesoarchean magmatic arc but also caused parts of the basaltic crust of the basin floor (Regal Formation) to be obducted onto the adjacent KT (Sun and Hickman 1999; Hickman et al. 2000; Hickman 2001, 2004b). The obducted greenstone succession of the Regal Formation arguably constitutes the world's best documented example of a Mesoarchean ophiolite.

The 3-km-thick MORB-like basaltic succession of the Regal Formation overlies >30-km-thick Paleoarchean continental crust of the KT along a tectonic contact that extends across at least 3000 km^2 of the Northwest Pilbara Craton (Hickman 1997a, b, 2001; Hickman et al. 1998). The tectonic contact is a 1-km-thick layer-parallel zone of mylonite and schist named the Regal Thrust (Chap. 7). Where the Regal Formation overlies the KT and the Nickol River Formation tectonically, it is referred to as the Regal Terrane (RT) (Van Kranendonk et al. 2006).

Geochemical evidence indicates that the Regal Formation is entirely composed of MORB-like basaltic crust (Ohta et al. 1996; Kato et al. 1998; Kiyokawa and Taira 1998; Sun and Hickman 1998, 1999; Hickman 2001, 2004b, 2012, 2016; Smithies et al. 2007a; Hickman and Van Kranendonk 2012). This conclusion is supported by

Sm–Nd isotope data indicating that the basalt is juvenile with only minor evidence of crustal contamination (Sun and Hickman 1998, 1999; Smithies et al. 2005, 2007a; Hickman and Van Kranendonk 2012).

Although originally almost horizontal, the Regal Thrust was subsequently deformed by major anticlines and synclines (Fig. 7.1) during the 2955–2919 Ma North Pilbara Orogeny. In consequence, it is now exposed on the southeast and northwest limbs of the Roebourne Syncline and, farther northwest, on the southeast and northwest limbs of the Cleaverville Syncline (Hickman et al. 2010). Structures within the Regal Thrust include mylonite zones that separate tectonically deformed slices of the Nickol River Formation, mesoscopic recumbent isoclines in the Nickol River Formation, and smaller-scale isoclinal folds and sheaf folds developed in the mylonite layering (Fig. 7.5). The mylonite contains a strongly developed stretching lineation, and clasts in the Nickol River Formation are tectonically elongated (Fig. 7.15).

Thrusting of the MORB-like basaltic crust of the Regal Formation across the adjacent continental crust of the KT occurred during the 3160–3070 Ma convergence of the KT and EPT (Hickman 2001; Van Kranendonk et al. 2002). The minimum age of this obduction is constrained by the timing of the Prinsep Orogeny at c. 3070 Ma, when the EPT collided with the KT and ST (Van Kranendonk et al. 2006, 2010; Hickman et al. 2010; Hickman 2012, 2016), effectively closing the Regal Basin. During the same convergence, but within the Regal Basin, the Whundo Group was thrust over the Railway Supersuite across the MSZ (Hickman 2001; Hickman and Kojan 2003).

8.4 Prinsep Orogeny and Elizabeth Hill Supersuite

Plate convergence, commencing with the Karratha Event at c. 3165 Ma, eventually led to collision of the EPT and KT resulting in the c. 3070 Ma Prinsep Orogeny (Van Kranendonk et al. 2006). At the same time, 3068–3066 Ma tonalite–trondhjemite–granodiorite (TTG) intrusions of the Elizabeth Hill Supersuite were emplaced along the western side of the EPT and beneath the MSZ in the Northwest Pilbara Craton. The crystallization ages of intrusions in the Elizabeth Hill Supersuite are inferred to closely date the main orogenic event, although zircon age components in younger igneous and sedimentary rocks suggest that the orogeny might have spanned c. 3080 Ma to 3060 Ma. Thrusting on the MSZ and the Regal Thrust and major sinistral strike–slip movements on the SSZ are interpreted to have culminated during the Prinsep Orogeny. Deformation during the Prinsep Orogeny produced structural styles typical of horizontal tectonics, in contrast to structures related to the Paleoarchean vertical deformation of the EPT. In addition to the Regal Thrust and the SSZ and MSZ within the Central Pilbara Tectonic Zone, large-scale thrusts close to the western margin of the EPT, in the Pilbara Well and East Strelley greenstone belts, are also attributed to the c. 3070 Ma collision event (Van Kranendonk et al. 2010).

8.4.1 Elizabeth Hill Supersuite

The Elizabeth Hill Supersuite was first identified by geochronology in the Northwest Pilbara Craton where the Cliff Pool Tonalite was dated at 3068 ± 4 Ma (Nelson 1998). Subsequent mapping and geochronology in the Yule Dome of the EPT (Fig. 1.7) has indicated the widespread intrusion of another TTG intrusion, the Cockeraga Leucogranite. Samples from this intrusion were dated at 3066 ± 4 Ma and 3068 ± 22 Ma (Nelson 2002). The Cliff Pool Tonalite includes xenocrystic zircons of Railway Supersuite age, but no zircons older than c. 3169 Ma. A Sm–Nd two-stage depleted mantle model age of c. 3220 Ma and a positive ε_{Nd} value of +1.15 (Smithies et al. 2007a) also suggest little or no involvement of crust older than the Regal Formation. The Cliff Pool Tonalite intruded the Central Pilbara Tectonic Zone, and its composition suggests partial melting of juvenile mafic crust.

The chemistry and physical features of the Cockeraga Leucogranite were interpreted in terms of disequilibrium melting of a range of Paleoarchean protoliths (Smithies 2003). The intrusion includes numerous undated greenstone and metasedimentary xenoliths of unknown origin. The Cockeraga Leucogranite includes biotite (−hornblende) tonalite and granodiorite with minor monzogranite (Smithies 2003). The interpreted extent of the Cockeraga Leucogranite is >1500 km^2, although this remains to be confirmed by additional geochronology and geochemistry. Smithies (2003) interpreted the source of melting for the Cockeraga Leucogranite to be underlying Paleoarchean crust. Sm–Nd two-stage depleted mantle model ages between 3560 and 3520 Ma and negative ε_{Nd} values (Smithies et al. 2007a; Hickman 2021) indicate inclusion of old crustal sources, consistent with intrusion within the EPT.

8.5 Magmatic Arcs of the De Grey Superbasin

Some workers have interpreted a subduction zone along the northwest margin of the Pilbara Craton from about 3265 Ma onwards (Krapež 1993; Krapež and Eisenlohr 1998; Smith et al. 1998; Beintema 2003). However, there is no evidence that the present northwest margin of the craton existed before the c. 3220 Ma breakup. Although it is possible that subduction commenced along the northwest margin between 3165 and 3070 Ma, there is no stratigraphic evidence of it northwest of the SSZ. Following collision of the KT and EPT at c. 3070 Ma, continued convergence of the northwest plate resulted in subduction moving from the Regal Basin to the northwest margin of the craton. Post-3070 Ma subduction led to the evolution of a series of continental magmatic arcs and retro-arc basins between 3066 and 2913 Ma. From about 3000 Ma onwards, the magmatic arcs migrated southeast, entering the EPT at about 2950 Ma (Hickman 2016, 2021).

8.5.1 Orpheus Supersuite

The oldest preserved magmatic arc in the Northwest Pilbara is that of the Orpheus Supersuite. Older magmatic arcs might be concealed on the northwest shelf that extends about 150 km into the Indian Ocean (Hickman 2004b). The Orpheus Supersuite comprises nine named formations and four informal formations that outcrop within 70 km of the Northwest Pilbara coast between Karratha and Balla Balla Landing. The supersuite evolved in a c. 3023 to 3007 Ma magmatic arc above a subduction zone along the northwest margin of the Pilbara Craton. Intrusive and extrusive formations of the supersuite are distributed along an east-northeast–west-southwest trending belt, but these formations represent only remnants of the original arc. This arc lay on the northwest side of the retro-arc depositional basin of the Gorge Creek Group which is far more completely preserved. This relationship suggests that the original east-northeast–west-southwest extent of the supersuite was at least 500 km.

Outcrops of the supersuite are mainly north of the SSZ where it is composed of intrusive and extrusive units varying in composition from peridotite to granite and rhyolite; rhyolite is assigned to the Rea Dacite and Mount Wangee Dacite. Two units south of the shear zone establish that it post-dates major strike-slip movement on this major fault. No other igneous rocks of this age have been identified more than 70 km inland. Major intrusions north of the shear zone include the Andover Intrusion, a layered ultramafic–mafic intrusion, and the Black Hill Well Monzogranite, Mount Gregory Monzodiorite, and Forestier Bay Granodiorite. South of the shear zone, the only dated intrusions are the South Whundo Monzogranite and the very local Stone Yard Granophyre. Additional geochronology is required to establish the age of several layered ultramafic–mafic intrusions south of the shear zone, including the Maitland, Mount Sholl, and North Whundo Intrusions.

Volcanic and subvolcanic formations of the supersuite include the Rea and Mount Wangee Dacites that have been dated between c. 3023 and 3014 Ma. Geochronology on felsic volcaniclastic sandstones of the Cleaverville Formation (Gorge Creek Group) is consistent with volcanism during sedimentary deposition within the retro-arc basin.

8.5.1.1 Andover Intrusion

The Andover Intrusion is the largest layered ultramafic–mafic intrusion of the Northern Pilbara Craton and occupies an area of about 200 km^2 southeast and southwest of Roebourne. The intrusion is a complex body formed during at least two stages of intrusion (Wallace 1992a, b; Hoatson et al. 1992; Hickman 2002; Hoatson and Sun 2002). The lower ultramafic zone of the Andover Intrusion consists of rhythmically layered, peridotite, lherzolite, olivine websterite, clinopyroxenite, olivine orthopyroxenite, orthopyroxenite, and websterite with minor anorthosite, leucogabbro, and gabbro. Above this, there is gabbroic zone composed of gabbro, leucogabbro, and anorthosite. GSWA mapping suggests that the intrusion is a lopolith or a funnel-shaped body, up to 3 km thick, in which the layering of the lower ultramafic zone is partly obscured by late-stage discordant and irregular intrusion of gabbro and leucogabbro. Upper levels of the intrusion, mainly composed of gabbro and dolerite, are exposed in the northwest outcrops south of the Carlow Castle copper mine, whereas the lower ultramafic zones outcrop southeast of Roebourne.

8.5.1.2 Black Hill Well Monzogranite

The Black Hill Monzogranite is one of the largest intrusions of the Orpheus Supersuite and has been dated at c. 3018 Ma. The monzogranite intrudes the south and northeast margins of the Andover Intrusion and is interpreted to underlie an area of approximately 200 km^2 in the western part of the Harding granitic complex. In the type area, 3.5 km southeast of Black Hill Well, the rock is a weakly foliated monzogranite containing leucocratic veins. In this area, the Black Hill Monzogranite is bounded by the SSZ in the south and by the Black Hill Shear Zone in the north.

8.5.1.3 Mount Gregory Monzodiorite

The Mount Gregory Monzodiorite forms several small intrusive stocks into the basal, southeast section of the Andover Intrusion. The largest stock, dated at c. 3016 Ma, is composed of medium- to coarse-grained monzodiorite, pegmatitic diorite, and granodiorite. The variable composition is inferred to be due to partial assimilation of mafic rocks of the Andover Intrusion, although multiple intrusive phases are possible.

East of Mount Gregory, the monzodiorite outcrops along the contact between the Andover Intrusion and the c. 3018 Ma Black Hill Well Monzogranite. All three intrusions form part of the Orpheus Supersuite that was emplaced in a subduction-related magmatic arc close to the northwest margin of the Pilbara Craton.

8.5.1.4 Forestier Bay Granodiorite

The Forestier Bay Granodiorite is an intrusion of variably foliated porphyritic granodiorite that intrudes compositionally layered granitic gneiss north of the SSZ. The porphyritic granodiorite is less foliated than the gneissic granitic rock and has been dated at c. 3014 Ma. The locally undated gneissic unit is interpreted to be part of the c. 3093 Ma gneiss that was dated in the Caines Well granitic complex 16 km to the southwest.

8.5.1.5 South Whundo Monzogranite

The South Whundo Monzogranite forms several intrusions into the c. 3130 to 3110 Ma Whundo Group. The monzogranite is a pale grey, non-foliated, and relatively coarse-grained (4 to 5 mm) rock. Numerous granitic dykes extend outwards from the South Whundo Monzogranite into the Whundo Group and the undated Maitland Intrusion. A felsic sill that intrudes the Whundo Group south of the largest intrusion was dated at c. 3012 Ma.

8.5.1.6 Geochemistry

Geochemical data from the Andover Intrusion (Sun et al. 1991; Wallace 1992a, b; Hoatson et al. 1992) reveal minor differences from the compositions of layered ultramafic–mafic intrusions of the c. 2930–2913 Ma Radley Suite (Sisters Supersuite). The petrology and geochemistry of the Andover Intrusion suggest that it assimilated more felsic crustal material that the other intrusions, all of which are located within the Central Pilbara Tectonic Zone south of the SSZ.

8.5.1.7 Sm–Nd Isotope Data

Sm–Nd isotope data from the ultramafic–mafic Andover Intrusion (Smithies et al. 2007a) indicate a two-stage depleted mantle model ages ($T_{DM}{}^2$) between 3330 and 3220 Ma with ε_{Nd} values ranging between −0.83 and + 0.65 ($n = 6$). These ε_{Nd} values indicate far more evolved sources than those from which the ST was derived. The data are consistent with magma derivation through melting of late Paleoarchean to early Mesoarchean mafic crust subducted beneath the northwest margin of the Pilbara Craton. Sm–Nd isotope data from the Mount Gregory Monzodiorite indicate an Nd model age ($T_{DM}{}^2$) of c. 3250 Ma and an ε_{Nd} value of +0.33, similar to data from the Andover Intrusion.

Three felsic igneous rocks from the Orpheus Supersuite have provided Nd model ages ($T_{DM}{}^2$) between 3400 and 3300 Ma and ε_{Nd} values ranging between −1.65 and − 0.42 (Smithies et al. 2007a). Compared to the mafic rocks of the supersuite, these data suggest more melting of Paleoarchean felsic crust.

8.5.1.8 Geochronology

The oldest dated unit of the Orpheus Supersuite is a mylonitized granitic rock in the SSZ dated at 3024 ± 4 Ma (Smith et al. 1998, JS25). Additionally, a sample of the Rea Dacite was dated at 3023 ± 9 Ma (Nelson 1997, GSWA 118976). However, the maximum age of the supersuite is otherwise poorly constrained by being younger than the c. 3067 Ma age of the Elizabeth Hill Supersuite that was emplaced during the Prinsep Orogeny.

The minimum age of the supersuite might be indicated by a date of c. 3007 Ma on a leucogabbro in the upper section of the Andover Intrusion. However, the error on this date is relatively large allowing for an older minimum age. Other dates on units of the supersuite are between c. 3018 and 3012 Ma, and c. 3012 Ma is the preferred minimum age here.

8.5.2 Maitland River Supersuite

The 2999–2982 Ma Maitland River Supersuite forms a 50-km-wide belt of tonalite, granodiorite, and monzogranite intrusions underlying the coastal plain of the Northwest Pilbara. Geophysical data indicate that the supersuite extends about 250 km northeast along the Pilbara coast and northwest towards the northwest margin of the Pilbara Craton concealed by the Indian Ocean. The supersuite comprises ten intrusions, four of which are granodiorites or tonalites, with the other six being mainly composed of monzogranite and minor syenogranite. The linear distribution of the supersuite and its genetic and spatial relationships to the Whim Creek and Mallina Basins suggest it comprises the remnants of a continental magmatic arc (Hickman 2016, 2021). Relatively young Sm–Nd two-stage depleted mantle model ages are consistent with magma derivation by partial melting of subducted mafic crust.

North of the SSZ, the supersuite is the main constituent of the Dampier and Harding granitic complexes, whereas south of the SSZ, it forms much of the Cherratta and Caines Well granitic complexes. North of the SSZ, the Maitland River Supersuite intrudes all parts of the West Pilbara Superterrane, including the 3280–3260 Ma KT. South of the SSZ, it intrudes the Whundo Group and the Railway Supersuite of the 3130–3093 Ma ST and locally intrudes the c. 3067 Ma Elizabeth Hill Supersuite.

8.5.2.1 Granodiorite and Tonalite Intrusions

South of the SSZ, tonalite–trondhjemite–granodiorite (TTG) intrusions in the Cherratta granitic complex include the 2995 Ma Toorare Tonalite, 2988 Ma Jean Well Granodiorite, and the 2994 Ma Whyjabby Granodiorite. The Toorare Tonalite is mainly composed of foliated hornblende- and biotite-rich tonalite with local granodiorite. Monzogranite is a minor constituent but is undated and likely to be derived from the 2954–2913 Ma Sisters Supersuite. The Whyjabby Granodiorite is a foliated to gneissic microcline-porphyritic hornblende granodiorite. Hornblende is present in most exposures, making up between 5 and 10% of the rock. The Jean Well Granodiorite comprises foliated and weakly banded granodiorite and monzogranite.

8.5.2.2 Monzogranite Intrusions

Three monzogranite intrusions (2999 Ma Hearson Cove Monzogranite, 2997 Ma Miaree Granite, and 2982 Ma Eramurra Monzogranite) make up the Dampier granitic complex north of the SSZ. Three other monzonites (undated Brill Monzogranite, 2988 Ma Waloo Waloo Monzogranite, and 2990 Ma Caines Monzogranite) represent the supersuite south of the shear zone. All these intrusions are lithologically similar, comprising weakly foliated, variably K-feldspar porphyritic monzogranite, minor granodiorite, and local syenogranite. Intrusive into the Roebourne and Devil Creek greenstone belts, the Miaree and Eramurra intrusions contain greenstone xenoliths and enclaves of banded gneiss. Likewise, the Caines Monzogranite, located north of the Loudens Fault (similar structure to the SSZ), contains large greenstone xenoliths from the Whundo Group and banded gneiss from the Railway Supersuite. The Caines Monzogranite underlies the 3000–2990 Ma Whim Creek Group and is interpreted to have formed in a subvolcanic magma chamber that fed the Red Hill Volcanics (Hickman 2016).

8.5.2.3 Geochemistry

Geochemical data from the Maitland River Supersuite are very limited. Smith (2003) reported the composition of a sample from the Toorare Tonalite (N3132) in the Cherratta granitic complex. The sample contained high contents of SiO_2 (75.10 wt.%), Na_2O (5.88 wt.%), Sr (895 ppm), LREE and LILE, and low Y (13) and HREE. This composition is similar to that of high-SiO_2 adakite as described by Martin et al. (2005). Phanerozoic adakites are associated with arcs above subduction zones and are the product of partial melting of subducting mafic crust. The relatively low contents of MgO, Cr, and Ni reported by Smith (2003) might be evidence of Archean flat or low-angle subduction, a process discussed by Smithies et al. (2003, 2007a) and Martin et al. (2005).

8.5.2.4 Sm–Nd Isotope Data

North of the SSZ, the supersuite was intruded through evolved Paleoarchean crust of the KT and into Mesoarchean granitic rocks of the Orpheus Supersuite and the Gorge Creek Group. South of the shear zone, it was intruded through 3200–3165 Ma juvenile mafic crust of the Regal Formation and through mixed juvenile mafic and felsic crust of the 3130–3093 Ma ST. These differences are reflected in Sm–Nd two-stage depleted mantle model ages (Smith et al. 1998; Sun and Hickman 1998; Smithies et al. 2005, 2007a; Hickman 2016).

Samples of the Miaree Granite (GSWA 136844) and Eramurra Monzogranite (GSWA 142893) in the Dampier granitic complex north of the SSZ gave Nd model ages of 3360 and 3330 Ma and ε_{Nd} values of -1.44 and -1.14. Two samples from the Caines Monzogranite (GSWA 160502 and 160501), in the Caines Well granitic complex north of the Loudens Fault, gave Nd model ages of 3350 and 3310 Ma and ε_{Nd} values of −1.34 and − 0.81. By contrast, samples from the Toorare Tonalite (GSWA 136862) and Jean Well Granodiorite (JS 44 and 84,770,083), in the Cherratta granitic complex south of the SSZ, gave Nd model ages of 3190, 3220, and 3240 Ma, and ε_{Nd} values of +0.86, +0.33, and + 0.19. The results from the TTG rocks of the Cherratta granitic complex are consistent with partial melting of relatively juvenile mafic crust, either from the subducted slab or from the Regal Formation or ST.

8.5.2.5 Geochronology

The maximum age of the supersuite is indicated by a date of 2999 ± 10 Ma (Wingate and Hickman 2009, GSWA 178148) on the Hearson Cove Monzogranite. This is supported by the interpreted genetic relationship between the supersuite and the volcanic Whim Creek Group interpreted to have an age range of c. 3000 to c. 2990 Ma (Chap. 9). The youngest intrusion in the supersuite is the Eramurra Monzogranite, dated at c. 2982 Ma. Accordingly, the age range of the Maitland River Supersuite is interpreted to be 2999–2982 Ma.

8.5.3 Sisters Supersuite

The Sisters Supersuite was initially intruded into the Mallina Basin where it also includes alkaline granite, hornblende-granodiorite, high-Mg diorite (sanukitoid), and ultramafic−mafic layered intrusions. Geochronology (Fig. 8.6) has established that from 2954 to 2913 Ma magmatic activity migrated southeast across the Tabba Tabba Shear Zone (TTSZ) into the centre of the EPT where it is mainly composed of leucocratic high-K monzogranite substantially derived from partial melting of Paleoarchean rocks. Unlike intrusions in the Mallina Basin, many intrusions in the EPT include early Paleoarchean zircon xenocrysts (Hickman 2016).

Sm–Nd isotope data indicate that sources within the central trough of the Mallina Basin were younger than c. 3300 Ma (Van Kranendonk et al. 2006, 2007a, b) and likely included the 3130–3093 Ma ST. Sm–Nd two-stage depleted mantle model ages (T_{DM}^2) <3300 Ma from the northwest sides of the Carlindi and Yule granitic complexes were once interpreted to indicate that the northwest margin of the EPT does not coincide with the Tabba Tabba Shear Zone (TTSZ) but lies somewhere within these complexes (Van Kranendonk et al. 2004, 2006, 2007; Smithies et al. 2007b). However, these young Nd model ages cannot define the northwest limit of Paleoarchean crust because other geochronological data establish the presence of this crust within and adjacent to the TTSZ (Hickman 2016, 2021). An alternative explanation of the <3300 Ma Nd model ages in the northwest granitic complexes is

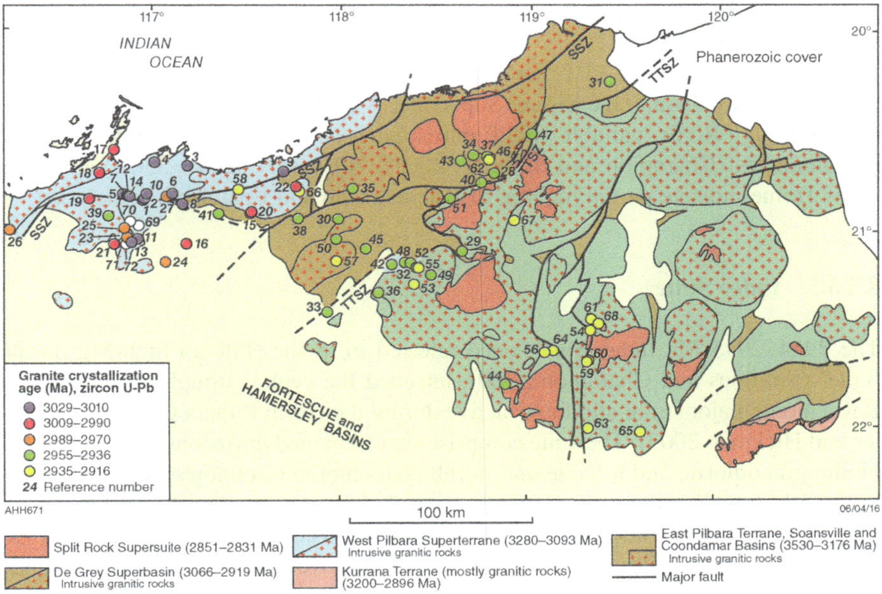

No.	Sample	Sample details	Age (Ma)	Reference
1	JS25	tonalite	3024 ± 4	Smith (2003)
2	118976	porphyritic dacite	3023 ± 9	Nelson (1997h)
3	144224	porphyritic dacite	3021 ± 3	Nelson (1999j)
4	127327	porphyritic dacite	3018 ± 2	Nelson (1998a)
5	N4028	granodiorite	3017 ± 6	Smith (2003)
6	168936	monzodiorite	3016 ± 4	Nelson (2001j)
7	N3188	granodiorite	3016 ± 2	Smith (2003)
8	127320	quartz granophyre	3014 ± 6	Nelson (1997j)
9	118966	granodioritic gneiss	3014 ± 3	Nelson (1997c)
10	118979	quartz-feldspar porphyry	3014 ± 2	Nelson (1997i)
11	N4350	monzogranite	3013 ± 4	Smith (2003)
12	N4097	granodiorite	3013 ± 8	Smith (2003)
13	142832	intermediate rock	3012 ± 4	Nelson (2000a)
14	N4128	granodiorite	3012 ± 3	Smith (2003)
15	141936	welded tuff	3009 ± 4	Nelson (1998g)
16	168932	porphyritic granodiorite	3006 ± 12	Nelson (2001h)
17	178148	biotite granite	2999 ± 10	Wingate, Hickman (2009a)
18	136844	granite	2997 ± 3	Nelson (1998e)
19	118974	porphyritic granodiorite	2994 ± 2	Nelson (1997f)
20	UWA	porphyritic dacite	2991 ± 12	Barley et al. (1994)
21	142657	granodiorite	2990 ± 3	Nelson (1999d)
22	142950	porphyritic monzogranite	2990 ± 5	Nelson (2000n)
23	142438	granodiorite	2988 ± 4	Nelson (1999c)
24	168934	biotite monzogranite	2988 ± 7	Nelson (2001i)
25	N3132	granodiorite	2985 ± 7	Smith (2003)
26	142893	monzogranite	2982 ± 5	Nelson (1999i)
27	142430	monzogranite	2970 ± 5	Nelson (1999b)
28	142935	hornblende granodiorite	2954 ± 4	Nelson (2000e)
29	180038	metamonzodiorite	2952 ±7	Wingate et al. (2009a)
30	118967	tonalite	2948 ± 5	Nelson (1997d)
31	169025	rhyolite	2948 ± 3	Nelson (2002e)
32	142941	porphyritic dacite	2946 ± 4	Nelson (2000g)
33	142945	porphyritic andesite	2946 ± 20	Nelson (2000j)
34	160730	biotite granodiorite	2946 ± 3	Nelson (2001d)
35	142889	alkali granite	2946 ± 6	Nelson (1999g)
36	142938	tonalite	2945 ± 5	Nelson (2000f)
37	160744	granodiorite	2945 ± 2	Nelson (2001e)
38	160498	granodiorite	2945 ± 6	Nelson (2000b)
39	136826	tonalite gneiss	2944 ± 5	Nelson (1997k)
40	KB351	aplite	2944 ± 7	Beintema (2003)
41	144261	rhyolite	2943 ± 7	Nelson (1998u)
42	180097	dacite dyke	2942 ± 8	Wingate et al. (2009d)
43	142934	biotite monzogranite	2941 ± 4	Nelson (2000b)
44	169018	biotite monzogranite	2941 ± 5	Nelson (2002c)
45	142892	porphyritic rhyolite	2941 ± 4	Nelson (1999h)
46	160728	biotite monzogranite	2940 ± 2	Nelson (2001c)
47	160745	biotite granodiorite	2940 ± 3	Nelson (2001g)
48	KB746	granite	2939 ± 12	Beintema (2003)
49	142176	biotite monzogranite	2938 ± 3	Nelson (1999a)
50	141973	biotite monzogranite	2938 ± 4	Nelson (1998h)
51	KB770	granite	2938 ± 15	Beintema (2003)
52	142936	leucocratic monzogranite	2937 ± 7	Nelson (2000d)
53	142937	leucocratic monzogranite	2935 ± 3	Nelson (2000e)
54	T94/31	granite	2934 ± 2	Zegers (1996)
55	160442	biotite monzogranite	2933 ± 4	Nelson (2000r)
56	142884	biotite syenogranite	2933 ± 3	Nelson (1998q)
57	141977	granite	2931 ± 5	Nelson (1998i)
58	169080	quartz diorite, drillcore	2931 ± 8	Nelson (2004)
59	142965	monzogranite	2929 ± 4	Nelson (2000p)
60	142967	biotite monzogranite	2929 ± 5	Nelson (2000q)
61	142882	biotite monzogranite	2928 ± 2	Nelson (1998o)
62	160727	biotite granodiorite	2928 ± 6	Nelson (2001c)
63	178047	monzogranite	2927 ± 19	Wingate (in prep.)
64	142885	biotite monzogranite	2927 ± 3	Nelson (1998r)
65	178049	biotite monzogranite	2926 ± 6	Nelson (2005)
66	118964	foliated granite	2925 ± 4	Nelson (1997a)
67	169021	leucocratic syenogranite	2923 ± 3	Nelson (2002d)
68	142883	porphyritic syenogranite	2919 ± 3	Nelson (1998p)
69	N3162	granite, Radley Suite	2930 ± 4	Smith (2003)
70	N4450	granite, Radley Suite	2929 ± 13	Smith (2003)
71	103227	pegmatite, Radley Suite	2925 ± 16	Arndt et al. (1991)
72	142436	monzonite, Radley Suite	2924 ± 5	Nelson (1998k)

Fig. 8.6 Southeast migration of granitic intrusion in the Northwest Pilbara from c. 3024 Ma to c. 2919 Ma: (**a**) granite crystallization ages from all available SHRIMP U–Pb zircon geochronology; (**b**) summary of geochronological data used in (**a**). (From Hickman 2016; with Geological Survey of Western Australia permission)

that from c. 2950 onwards Mesoarchean crust was being subducted and melted beneath the northwest EPT.

In the main part of the EPT, particularly in the eastern section of the Yule Dome, and in the Shaw Dome, the Sisters Supersuite is characterized by high K_2O/Na_2O, generally high Rb/Sr, variable HREE enrichment, and moderate to strong negative Eu anomalies (Champion and Smithies 2001). Nd model ages >3500 Ma in the Shaw Dome indicate partial melting of Paleoarchean and Eoarchean felsic crust.

8.5.3.1 Indee Suite

The 2954–2945 Ma Indee Suite, once referred to as the 'Pilbara high-Mg diorite Suite' (Smithies and Champion 2000), intruded the central trough of the Mallina Basin along major strike-slip faults active during the North Pilbara Orogeny (Smithies and Hickman 2003). The suite comprises seven named intrusions of hornblende–biotite granodiorite and tonalite and hornblende–biotite (−clinopyroxene) granodiorite. Additionally, there are several concealed intrusions likely to be high-Mg diorites that have not been named. Most of the intrusions are relatively small (less than 20 km^2), but the Peawah Granodiorite outcrops over an area of 180 km^2, and the poorly exposed Jallagnoonina Granodiorite occupies an area of 160 km^2. The best documented intrusions are distributed along a narrow, 150-km-long linear trend within the basin, parallel to the interpreted northwest margin of the Pilbara Craton during c. 3066–2913 Ma subduction. The rocks are equigranular to porphyritic, with plagioclase phenocrysts up to 1 cm long (Fig. 8.7). Fine-grained clinopyroxene- and orthopyroxene-bearing diorite forms chilled margins to some intrusions.

Named intrusions of the suite are the Wallareenya Granodiorite, Mallindra Well Granodiorite, Peawah Granodiorite, Toweranna Porphyry, Jallagnoonina Granodiorite, Jones Well Granodiorite, and Geemas Granodiorite (Smithies and Champion 2000). The Peawah Granodiorite is the largest intrusion, underlying an area of 180 km^2, and is a composite body that can be geochemically divided into eastern and western parts (Smithies and Champion 2000). Geochronology indicates that the Indee Suite was coeval with the alkaline rocks of the Portree Suite.

8.5.3.1.1 Geochemistry

The distinguishing feature of the Indee Suite is its high-Mg diorite (sanukitoid) geochemistry that sets it apart from average tonalite–trondhjemite–granodiorite (TTG). Dioritic to granodioritic rocks of the suite typically contain 59–70% wt.% SiO_2; 2–5 wt.% MgO; >100 ppm Cr; >50 ppm Ni; 10–20 ppm Th; high LILE, LREE, and LREE/HREE; low HFSE; and high Mg numbers (molecular $Mg/(Mg + Fe^{2+})$ of c. 60 (data in Smithies and Champion 2000). Magma derivation has been attributed to melting of a mantle source combined with a Th-, Zr-, and LREE-enriched crustal component (Smithies and Champion 2000; Smithies 2002; Smithies et al. 2004a, b). Mantle enrichment might have occurred during a previous

Fig. 8.7 Tonalite of the Indee Suite containing numerous megacrysts of hornblende, 3 km south of Mallindra Well on the Wallaringa 1:100,000 map sheet area. (From Smithies et al. 2002; with Geological Survey of Western Australia permission)

subduction event: either the c. 3130 Ma subduction responsible for the Whundo Group, the c. 3020 Ma Orpheus subduction event, or the 2999–2982 Ma subduction that produced the Maitland River Supersuite and Whim Creek Group (Blewett 2002; Smith 2003; Hickman 2004b, 2012; Pike et al. 2006). Sanukitoids are directly linked to slab melting during subduction (Martin et al. 2005), and involvement of an enriched mantle source at c. 2950 Ma is consistent with the subduction history of the Northwest Pilbara Craton.

8.5.3.1.2 Sm–Nd Isotope Data

Sm–Nd isotope data (Smithies et al. 2005) support the interpretation that the Indee Suite was derived from sources more evolved than a depleted c. 2950 Ma mantle. Sm–Nd two-stage depleted mantle model ages ($T_{DM}{}^2$) of five samples range between c. 3220 and 3160 Ma. ε_{Nd} values (2950 Ma) vary between +0.8 and − 0.1, which contrasts with the theoretical depleted mantle ε_{Nd} value of about +4 at 2950 Ma (Champion 2013). The model ages are consistent with magma derivation through melting of early Mesoarchean mafic crust. The ε_{Nd} values at c. 2950 Ma indicate relatively evolved sources or mixing of juvenile material with older crust or sedimentary rocks.

8.5.3.2 Portree Suite

The c. 2946 Ma Portree Suite comprises a cluster of six genetically related alkaline granites in the northern Mallina Basin. Intrusions of the Portree Suite are poorly exposed and restricted to a 1000 km^2 area northeast of Whim Creek. They are composed of leucocratic alkali granite consisting of quartz, microcline or microcline–microperthite, sodic plagioclase, sodic clinopyroxene, and minor amounts of biotite. Blue alkali amphibole is a late magmatic to sub-solidus replacement of clinopyroxene. The northern part of the suite is composed of coarse-grained alkali granite. The best exposed intrusion is the Portree Granite, a high-temperature alkaline granite interpreted to have been derived from partial melting of metasomatized basaltic crust. This melting was most likely related to subduction along the northwest margin of the Pilbara Craton.

8.5.3.3 Langenbeck Suite

The Langenbeck Suite comprises numerous c. 2955–2945 Ma layered ultramafic–mafic sills in the Central Pilbara Tectonic Zone (Fig. 1.3). The Sherlock and Opaline Well Intrusions are major sills within the Whim Creek greenstone belt, whereas the name Millindinna Intrusion is applied to many separate sills within the central part of the Mallina Basin. Most of the intrusions are interpreted to be intrusive equivalents of the Louden and Mount Negri Volcanics in the Whim Creek greenstone belt, although none of these units have been dated. The Sherlock and Opaline Well Intrusions were folded at c. 2940 Ma, and some sills of the Millindinna Intrusion are within the limbs of folds formed at or before c. 2955 Ma.

Prior to a stratigraphic revision of the Northern Pilbara Craton in 2006, the Langenbeck Suite was informally referred to as 'LREE-enriched gabbros' and, together with boninite-like basaltic rocks, was included is the 'Mallina mafic suite' (informal name). Apart from light rare earth element (LREE) enrichment, the characteristic geochemical features of the Langenbeck Suite include high $(La/Nb)_{PM}$, high $(La/Yb)_{PM}$, and negative ε_{Nd} values (2950 Ma) indicating a significant crustal component within the magmas. Being part of the Sisters Supersuite, the Langenbeck Suite was intruded during subduction along the northwest margin of the Pilbara Craton. Magmas were most likely derived by melting of older Mesoarchean mafic crust in the subducted plate together with sedimentary material.

8.5.3.3.1 Sherlock Intrusion

The Sherlock Intrusion is a layered sill, locally over 1 km thick that was intruded along the stratigraphic contact of the Caines Well granitic complex with the overlying Whim Creek Group. Prior to 2940 Ma folding and erosion, the Sherlock Intrusion is estimated to have extended beneath an area of at least 1000 km^2. The intrusion underlies the ultramafic–volcanic succession of the c. 2950 Ma Bookingarra Group, suggesting it is a subvolcanic intrusion.

8.5.3.3.2 Opaline Well Intrusion

The Opaline Well Intrusion forms sills up to 100 m thick within the Cistern Formation of the Bookingarra Group and also occurs as dykes up to 400 m wide in the underlying Whim Creek Group (Pike et al. 2006). Metamorphism is low grade and rock types include peridotite, peridotitic gabbro, olivine gabbro, gabbro, dolerite, and basalt. The basalt and dolerite typically contain acicular clinopyroxene suggesting quenching of a crystal-rich magma, and many of the rocks are petrographically similar to mafic and ultramafic components of the overlying Louden Volcanics of the Bookingarra Group.

8.5.3.3.3 Millindinna Intrusion

Sills of the Millindinna Intrusion are widespread in the lower part of the Croydon Group within the central trough of the Mallina Basin. Lithologies range from dunite to peridotite, pyroxenite, or gabbro. In the type area, within a radius of 10 km from Millindinna Hill, the intrusion forms a sill up to 400 m thick (Smithies and Farrell 2000). A basal layer of serpentized lherzolite is overlain by a central unit of metapyroxenite (now actinolite–chlorite rock) and a thin upper layer of metamorphosed melanogabbro.

8.5.3.4 Radley Suite

Intrusions of the c. 2930 Ma Radley Suite were emplaced along a 180-km-long north-northeast trending zone immediately south of the SSZ. The alignment coincides with gravity and magnetic lineaments and with several exposed faults. This suggests control by a deep structural feature, and this might have originated as an early branch of the SSZ (Hickman 2016). The intrusions are either lopoliths or funnel-shaped bodies (Hoatson et al. 1992; Hoatson 1998; Hoatson and Sun 2002), the largest being the 5.5 km-thick Munni Munni Intrusion that underlies an area of 300 km^2. Most intrusions of the suite comprise a lower section of ultramafic layers (dunite, peridotite, and pyroxenite) overlain by layered units of gabbro, leucogabbro, norite, and more rarely anorthosite and granophyre.

Most of the intrusions were described by Hoatson et al. (1992), Wallace (1992a), and Hoatson and Sun (2002). However, geochronology has revealed that some, once thought to be the same age as the well-studied c. 2930–2925 Ma Munni Munni Intrusion, are actually considerable older: examples include the c. 3122 Ma Bullock Hide and Dingo Intrusions, the c. 3023 to 3007 Ma Andover Intrusion, and the undated Maitland Intrusion which is intruded by granitic dykes apparently related to the c. 3012 Ma South Whundo Monzogranite. The range of intrusive ages, and the fact that the intrusions cluster along the north-northeast trending zone, strongly indicates a deep structural control.

Three intrusions of the Radley Suite (Munni Munni, Radio Hill, and Mount Sholl) contain significant Ni-Cu and PGE mineralization. Underground mining at Radio Hill produced substantial tonnages of Ni and Cu, along with Co, Pd, Pt, and Ag as by-products (Hickman 2016). The Munni Munni Intrusion contains one of the largest PGE deposits (23.6 Mt. at 2.9 g/t PGE + Au) in Australia, although it has not been mined. PGE mineralization occurs in a 30–80-m-thick layer of plagioclase websterite orthocumulate that separates the lower, 1.85-km-thick ultramafic zone from the upper, and 3.63-km-thick gabbroic zone. Hoatson and Sun (2002) interpreted the porphyritic websterite to be a product of mixing between the ultramafic and gabbroic zones. The Mount Sholl Intrusion contains Ni-Cu mineralization with inferred resources of about 6 Mt. at 0.5% Ni and 0.6% Cu.

8.5.3.4.1 Munni Munni Intrusion

The Munni Munni Intrusion is c. 5.5 km thick and outcrops over an area of 150 km^2. Gravity data indicate an additional southwest section of similar areal extent that is concealed by the Neoarchean Fortescue Group. There are more than 40 cyclic units in the lower ultramafic zone, individual layers having a basal peridotite unit that grades upwards into clinopyroxenite. Fine-scale layering in the clinopyroxenite consists of olivine-rich layers alternating with pyroxenite at intervals of 2 to 200 mm.

8.5.3.4.2 Radio Hill Intrusion

The Radio Hill Intrusion consists of a basal ultramafic zone comprised of lherzolite, dunite, and websterite, and an overlying gabbroic zone including quartz gabbro, gabbro, and gabbronorite. The basal layer of the ultramafic zone contains massive, disseminated, veined, and brecciated pyrrhotite and pentlandite–chalcopyrite–magnetite at or near the contact with the Nallana Formation. The gabbroic zone is overlain by granitic rocks dated at 2929 ± 13 Ma (Smith 2003). In the nearby North Whundo Intrusion, the upper granitic component, the Yannery Granite, was dated at 2930 ± 4 Ma (Smith 2003).

8.5.3.4.3 Geochemistry

The composition of the Radley Suite indicates siliceous high-Mg, Al-depleted basalt magmas with 9 to 12 percent MgO, low Al_2O_3/TiO_2, and enrichment of Th and light rare earth elements (LREE) (Hoatson and Sun 2002). Hoatson and Sun (2002) considered that magma generation had involved a pre-3000 Ma subduction event that enriched the lithospheric mantle. Smithies et al. (2004a) recorded siliceous high

Mg in other formations of the Mallina Basin such as the c. 2950 Ma Louden Volcanics and Mount Negri Volcanics of the Bookingarra Group, although they did not mention the slightly younger Radley Suite. Features of the c. 2950 Ma Mallina Basin units include high Th, Zr, and LREE concentrations and high Th/HFSE and LREE/HFSE ratios, consistent with significant crustal input. Smithies et al. (2004a) considered that pre-3000 Ma subduction of oceanic crust and sediment had established an enriched mantle source region from which magma was intruded into Mallina Basin in an intracontinental rift setting (Smithies et al. 2001). However, the revised tectonic model for the evolution of the Northwest Pilbara Craton interprets ongoing subduction from 3066 to 2019 Ma (Hickman 2016, 2021). Accordingly, although strike-slip faulting and rifting occurred during the evolution of the Mallina Basin, intrusion of the Sisters Supersuite was directly related to contemporaneous subduction.

8.5.3.4.4 Geochronology

The maximum age of the Radley Suite is interpreted to be c. 2930 Ma based on a date of 2930 ± 4 Ma (N3162, Smith 2003) obtained from the Yannery Granite that is part of the North Whundo Intrusion. This age is consistent with a date of 2929 ± 13 Ma (N4450, Smith 2003) from a similar granite at the top of the adjacent Radio Hill Intrusion. Various units of the Munni Munni Intrusion have been dated between 2925 and 2922 Ma (data in Hickman 2016). Pooled Sm–Nd data from clinopyroxene, plagioclase, and orthopyroxene in porphyritic websterite and gabbro of the Munni Munni Intrusion gave a weighted mean mineral isochron age of 2927 ± 13 Ma (2σ) (Hoatson and Sun 2002). The Mount Sholl and Maitland Intrusions have not been dated. These data from the layered intrusions of the Radley Suite indicate an age range of c. 2930 to 2922 Ma.

The minimum age of the Munni Munni Intrusion is indicated by unpublished baddeleyite dates on the cross-cutting Zebra Hill Dolerite. A date of 2914 ± 7 Ma was obtained on this dolerite 1 km east of the Munni Munni Intrusion (A. Gumsley 2021, written communication), and a date 2911 ± 8 Ma was obtained on the same dolerite 10 km east-northeast of the Munni Munni Intrusion (A. Gumsley 2021, written communication). These data indicate that the age of the Zebra Hill Dolerite is c. 2913 Ma, and this age constrains the minimum age of the Munni Munni Intrusion and the Radley Suite. However, the Zebra Hill Dolerite is not interpreted to be part of the Radley Suite because it comprises east–west and east-northeast–west-southwest trending dykes in contrast to the large, layered intrusions of the Radley Suite that occupy a north-northeast trending zone. The Radley Suite was intruded during west-northwest–east-southeast extension, whereas the Zebra Hill Dolerite was intruded during north–south to north-northwest–south-southeast extension, consistent with crustal relaxation after the 2955–2919 Ma North Pilbara Orogeny that mainly involved north–south to north-northwest–south-southeast compression.

References

Beintema KA (2003) Geodynamic evolution of the West and Central Pilbara Craton: a mid-Archean active continental margin. Geologica Ultraiectina, Utrecht University, PhD

Blewett RS (2002) Archaean tectonic processes: a case for horizontal shortening in the North Pilbara Granite–Greenstone Terrane, Western Australia. Precambrian Res 113:87–120

Champion DC, Smithies RH (2001) The geochemistry of the Yule Granitoid Complex, East Pilbara Granite-Greenstone Terrane: evidence for early felsic crust. In: Geological Survey of Western Australia Annual Review 1999–2000. Geological Survey of Western Australia, Perth, pp 42–48

Champion DC (2013) Neodymium depleted mantle model age map of Australia: explanatory notes and user guide. Record, 2013/44. Geoscience Australia

Crawford AJ, Falloon TJ, Green DH (1989) Classification, petrogenesis and tectonic setting of boninites. In: Crawford AJ (ed) Boninites. Unwin-Hyman, London, pp 1–49

Glikson AY, Davy R, Hickman AH (1986) Geochemical data files of Archaean volcanic rocks, Pilbara Block, Western Australia. BMR (Geoscience Australia) Record, 1986/014

Hickman AH (1997a) A revision of the stratigraphy of Archaean greenstone successions in the Roebourne–Whundo area, West Pilbara. In: Geological Survey of Western Australia annual review 1996–97. Geological Survey of Western Australia, pp 76–81

Hickman AH (1997b) Dampier, WA Sheet 2256. Geological Survey of Western Australia

Hickman AH (2000) Roebourne, WA sheet 2356. Geological Survey of Western Australia

Hickman AH (2001) Geology of the Dampier 1:100 000 sheet. 1:100 000 Geological Series Explanatory Notes. Geological Survey of Western Australia

Hickman AH (2002) Geology of the Roebourne 1:100 000 sheet. 1:100 000 Geological Series Explanatory Notes. Geological Survey of Western Australia

Hickman AH (2004a) Geology of the Cooya Pooya 1:100 000 sheet. 1:100 000 Geological Series Explanatory Notes. Geological Survey of Western Australia

Hickman AH (2004b) Two contrasting Granite–Greenstones Terranes in the Pilbara Craton, Australia: evidence for vertical and horizontal tectonic regimes prior to 2900 Ma. Precambrian Res 131:153–172

Hickman AH (2012) Review of the Pilbara Craton and Fortescue Basin, Western Australia: crustal evolution providing environments for early life. Island Arc 21:1–31

Hickman AH (2016) Northwest Pilbara Craton: a record of 450 million years in the growth of Archean continental crust. Report, 160. Geological Survey of Western Australia

Hickman AH (2021) East Pilbara Craton: a record of one billion years in the growth of Archean continental crust. Report 143, Geological Survey of Western Australia

Hickman AH, Huston DL, Van Kranendonk MJ, Smithies RH (2006) Geology and mineralization of the West Pilbara–a field guide. Record, 2006/17. Geological Survey of Western Australia

Hickman AH, Kojan CJ (2003) Geology of the Pinderi Hills 1:100 000 sheet. 1:100 000 Geological Series Explanatory Notes. Geological Survey of Western Australia

Hickman AH, Smithies RH, Huston DL (1998) Excursion guide to the geology of the Granite–Greenstone Terrane of the West Pilbara. Geological Survey of Western Australia and Australian Geological Survey Organisation, Perth

Hickman AH, Smithies RH, Huston DL (2000) Archaean geology of the West Pilbara Granite–Greenstone Terrane and the Mallina Basin, Western Australia–a field guide. Record, 2000/9. Geological Survey of Western Australia

Hickman AH, Smithies RH, Tyler IM (2010) Evolution of active plate margins: West Pilbara Superterrane, De Grey Superbasin, and the Fortescue and Hamersley Basins–a field guide. Record, 2010/3. Geological Survey of Western Australia

Hickman AH, Van Kranendonk MJ (2012) Early earth evolution: evidence from the 3.5–1.8 Ga geological history of the Pilbara region of Western Australia. Episodes 35:283–297. https://doi.org/10.18814/epiiugs/2012/v35i1/028

Hoatson DM, Sun S-S (2002) Archean layered mafic-ultramafic intrusions in the West Pilbara Craton, Western Australia: a synthesis of some of the oldest orthomagmatic mineralizing systems in the world. Econ Geol 97:847–872
Hoatson DM, Wallace DA, Sun S-S, Macias LF, Simpson CJ, Keays RR (1992) Petrology and platinum-group element geochemistry of Archaean layered mafic–ultramafic intrusions, West Pilbara Block, Western Australia. Bulletin 242. Australian Geological Survey Organisation (Geoscience Australia)
Hoatson DM (1998) Platinum-group element mineralisation in Australian Precambrian layered mafic-ultramafic intrusions. AGSO Journal of Australian Geology & Geophysics 17:139–151
Horwitz R, Pidgeon RT (1993) 3.1 Ga tuff from the Sholl Belt in the West Pilbara: further evidence for diachronous volcanism in the Pilbara Craton. Precambrian Res 60:175–183
Kato Y, Ohta I, Tsunematsu T, Watanabe E, Isozaki Y, Maruyama S, Imai N (1998) Rare earth element variations in mid-Archean banded iron formations: implications for the chemistry of ocean and continent plate tectonics. Geochim Cosmochim Acta 62:3475–3497
Kiyokawa S, Taira A (1998) The Cleaverville Group in the West Pilbara coastal Granite–Greenstone Terrane of Western Australia: an example of a mid-Archaean immature oceanic island-arc succession. Precambrian Res 88:102–142
Krapež B (1993) Sequence stratigraphy of the Archaean supracrustal belts of the Pilbara Block, Western Australia. Precambrian Res 60:1–45
Krapež B, Eisenlohr B (1998) Tectonic settings of Archaean (3325–2775 Ma) crustal–supracrustal belts in the West Pilbara Block. Precambrian Res 88:173–205
Martin H, Smithies RH, Rapp R, Moyen J-F, Champion DC (2005) An overview of adakite, tonalite–trondhjemite–granodiorite (TTG), and sanukitoid: relationships and some implications for crustal evolution. Lithos 79:1–24. https://doi.org/10.1016/j.lithos.2004.04.048
Nelson DR (1998) Compilation of SHRIMP U-Pb zircon geochronology data, 1997. Record, 1998/2. Geological Survey of Western Australia
Nelson DR (2002) Compilation of geochronology data, 2001. Record, 2002/2. Geological Survey of Western Australia
Nelson DR (1996) Compilation of SHRIMP U-Pb zircon geochronology data, 1995. Record, 1996/5. Geological Survey of Western Australia
Nelson DR (1997) Compilation of SHRIMP U-Pb zircon geochronology data, 1996. Record, 1997/2. Geological Survey of Western Australia
Ohta H, Maruyama S, Takahashi E, Watanabe Y, Kato Y (1996) Field occurrence, geochemistry and petrogenesis of the Archaean Mid-Oceanic Ridge Basalts (AMORBs) of the Cleaverville area, Pilbara Craton, Western Australia. Lithos 37:199–221
Pike G, Cas RAF, Hickman AH (2006) Archean volcanic and sedimentary rocks of the Whim Creek greenstone belt, Pilbara Craton, Western Australia. Report, vol 101. Geological Survey of Western Australia
Smith JB (2003) The episodic development of intermediate to silicic volcano-plutonic suites in the Archaean West Pilbara, Australia. Chem Geol 194:275–295
Smith JB, Barley ME, Groves DI, Krapež B, McNaughton NJ, Bickle MJ, Chapman HJ (1998) The Sholl Shear Zone, West Pilbara; evidence for a domain boundary structure from integrated tectonostratigraphic analysis, SHRIMP U–Pb dating and isotopic and geochemical data of granitoids. Precambrian Res 88:143–172
Smithies RH (2002) Archaean boninite-like rocks in an intracratonic setting. Earth Planet Sci Lett 197:19–34
Smithies RH (2003) Geology of the White Springs 1:100 000 sheet. 1:100 000 geological series explanatory notes. Geological Survey of Western Australia
Smithies RH, Champion DC (1998) Secular compositional changes in Archaean granitoid rocks of the West Pilbara. In: Geological Survey of Western Australia Annual Review 1997–98. Geological Survey of Western Australia, pp 71–76

Smithies RH, Champion DC (2000) The Archaean high-Mg diorite suite: links to tonalite–trondhjemite–granodiorite magmatism and implications for early Archaean crustal growth. J Petrol 41:1653–1671

Smithies RH, Champion DC, Blewett RS (2002) Geology of the Wallaringa 1:100 000 sheet. 1:100 000 Geological Series Explanatory Notes. Geological Survey of Western Australia

Smithies RH, Champion DC, Cassidy KF (2003) Formation of Earth's early Archaean continental crust. Precambrian Res 127:89–101

Smithies RH, Champion DC, Sun S-S (2004a) Evidence for early LREE-enriched mantle source regions: diverse magmas from the c. 3.0 Ga Mallina Basin, Pilbara Craton, NW Australia. J Petrol 45:1515–1537

Smithies RH, Champion DC, Sun S-S (2004b) The case for Archaean boninites. Contrib Mineral Petrol 147:705–725. https://doi.org/10.1007/s00410-004-0579-x

Smithies RH, Champion DC, Van Kranendonk MJ, Howard HM, Hickman AH (2005) Modern-style subduction processes in the Mesoarchean: geochemical evidence from the 3.12 Ga Whundo intra-oceanic arc. Earth Planet Sci Lett 231:221–237

Smithies RH, Farrell TR (2000) Geology of the Satirist 1:100 000 sheet. 1:100 000 Geological Series Explanatory Notes. Geological Survey of Western Australia

Smithies RH, Hickman AH (2003) Archaean geology of the Mallina and Whim Creek Basins (1: 250 000 scale). Geological Survey of Western Australia

Smithies RH, Nelson DR, Pike G (2001) Development of the Archaean Mallina Basin, Pilbara Craton, Northwestern Australia; a study of detrital and inherited zircon ages. Sediment Geol 141–142:79–94

Smithies RH, Champion DC, Van Kranendonk MJ, Hickman AH (2007a) Geochemistry of volcanic rocks of the Northern Pilbara Craton, Western Australia. Report data package, vol 104. Geological Survey of Western Australia

Smithies RH, Van Kranendonk MJ, Champion DC (2007b) The Mesoarchean emergence of modern-style subduction. In: Maruyama S, Santosh M (eds) Island arcs: past and present, pp 50–68

Sun S-S, Hickman AH (1998) New Nd-isotopic and geochemical data from the West Pilbara–implications for Archaean crustal accretion and shear zone development. Australian Geological Survey Organisation, Research Newsletter: 25–29

Sun S-S, Hickman AH (1999) Geochemical characteristics of ca 3.0-Ga Cleaverville greenstones and later mafic dykes, West Pilbara: implication for Archaean crustal accretion. Australian Geological Survey Organisation, Research Newsletter:23–29

Sun S-S, Wallace DA, Hoatson DM, Glikson AY, Keays RR (1991) Use of geochemistry as a guide to platinum group element potential of mafic-ultramafic rocks: examples from the West Pilbara Block and Halls Creek Mobile Zone, Western Australia. Precambrian Res 50:1–35

Van Kranendonk MJ, Hickman AH, Smithies RH, Champion DC (2007a) Paleoarchean development of a continental nucleus: the East Pilbara Terrane of the Pilbara Craton, Western Australia. In: Van Kranendonk MJ, Bennett VC, Smithies RH (eds) Earth's oldest rocks. Elsevier B.V, Burlington, Massachusetts, USA, pp 307–337

Van Kranendonk MJ, Hickman AH, Smithies RH, Nelson DN, Pike G (2002) Geology and tectonic evolution of the Archaean north Pilbara Terrain, Pilbara Craton, Western Australia. Econ Geol 97:695–732. https://doi.org/10.2113/gsecongeo.97.4.695

Van Kranendonk MJ, Hickman AH, Smithies RH, Williams IR, Bagas L, Farrell TR (2006) Revised lithostratigraphy of Archean supracrustal and intrusive rocks in the Northern Pilbara Craton, Western Australia. Record, 2006/15. Geological Survey of Western Australia

Van Kranendonk MJ, Smithies RH, Hickman AH, Bagas L, Williams IR, Farrell TR (2004) Event stratigraphy applied to 700 million years of Archean crustal evolution, Pilbara Craton, Western Australia. Geological Survey of Western Australia Annual Review 2003–04:49–61

Van Kranendonk MJ, Smithies RH, Hickman AH, Champion DC (2007b) Review: secular tectonic evolution of Archean continental crust: interplay between horizontal and vertical processes in the formation of the Pilbara Craton, Australia. Terra Nova 19:1–38

Van Kranendonk MJ, Smithies RH, Hickman AH, Wingate MTD, Bodorkos S (2010) Evidence for Mesoarchean (~3.2 Ga) rifting of the Pilbara Craton: the missing link in an early Precambrian Wilson cycle. Precambrian Res 177:145–161

Wallace DA (1992a) Andover mafic–ultramafic complex, West Pilbara, Western Australia. Record, 1992/13. Bureau of Mineral Resources, Geology and Geophysics (Geoscience Australia)

Wallace DA (1992b) Field relationships, petrography, and geochemistry of the Andover Complex. In: Hoatson DM, Wallace DA, Sun S-S, Macias LF, Simpson CJ, Keays RR, 'Petrology and platinum group element geochemistry of layered Archaean mafic-ultramafic intrusions, West Pilbara Block, Western Australia'. Bureau of Mineral Resources, Geology and Geophysics, Bulletin 242

Wingate MTD, Hickman AH (2009) 178164: leucogabbro, bottom bore; geochronology record 803. Geological Survey of Western Australia

Chapter 9
Mesoarchean Basin Evolution Inland of Magmatic Arcs

Abstract Tectonic, magmatic, and depositional processes from the c. 3070 Ma Prinsep Orogeny to the end of the North Pilbara Orogeny at c. 2919 Ma were directly related to ongoing northwest–southeast compression of the Pilbara Craton. This compression is interpreted to have resulted from c. 3160 Ma collision and subsequent interaction with an exotic plate northwest of the craton. Following closure of the Regal Basin at c. 3070 Ma (Chap. 7), geochronological evidence indicates the development of a subduction zone on the northwest margin of the Pilbara Craton resulting in a series of southeast-migrating continental magmatic arcs (Chap. 8). Deposition to the southeast of these arcs led to the evolution of the De Grey Supergroup, comprising the 3066–3015 Ma sedimentary Gorge Creek Group in a retro-arc basin related to the Orpheus arc; the 3015–2931 Ma, mainly sedimentary Croydon Group, overlying the Gorge Creek Group and related to the Orpheus, Maitland River, and Sisters arcs; the 3000–2990 Ma volcano-sedimentary Whim Creek Group on the southeast margin of the Maitland River arc; and the 2955–2945 Ma Bookingarra Group related partly to the Sisters arc and associated rifting.

The Gorge Creek and Croydon Groups were deposited in large retro-arc basins that extended across most of the presently exposed Northern Pilbara Craton. Prior to Neoarchean breakup of the Pilbara Craton at c. 2710 Ma (Chap. 12), the northeast–southwest extents of these basins were probably between 500 and 1000 km. Current exposures, and allowances for deformation, indicate northwest–southeast basin widths of at least 200 km.

Keywords Subduction · Retro-arc basins · Gorge Creek Group · Croydon Group, Whim Creek Group · Bookingarra Group

9.1 Introduction

Following the c. 3070 Ma Prinsep Orogeny (Chap. 8), Mesoarchean sedimentation and volcanism across the Northern Pilbara Craton occurred within three stacked retro-arc foreland basins that evolved behind three successive magmatic arcs.

A. H. Hickman, *Archean Evolution of the Pilbara Craton and Fortescue Basin*, Modern Approaches in Solid Earth Sciences 24,
https://doi.org/10.1007/978-3-031-18007-1_9

Collectively making up the 3067–2919 Ma De Grey Superbasin (Fig. 1.3), these basins now occupy more than 30,000 km^2 of the Northern Pilbara, although prior to a Neoarchean breakup of the Pilbara Craton at c. 2710 Ma (Chap. 12) they would have covered a far greater area. The stratigraphy of the De Grey Superbasin is assigned to the De Grey Supergroup (Tables 1.1, 3.1).

Deposition in the oldest basin, the c. 3067–3015 Ma Gorge Creek Basin accumulated a 2-km-thick succession of sandstone, BIF, and shale that makes up the Gorge Creek Group. This basin overlay thick continental crust formed by the tectonic amalgamation of the West Pilbara Superterrane (WPS) and the East Pilbara Terrane (EPT). Overlying the Gorge Creek Basin, the c. 3015–2931 Ma Mallina Basin evolved in a more tectonically active setting during alternating compression and extension and igneous intrusion. Deposition in this basin included conglomerate, sandstone, wacke, and shale and the intrusion and eruption of ultramafic and mafic rocks. The succession of the Mallina Basin is assigned to the c. 3015–2931 Ma Croydon Group (mainly sedimentary) and the c. 2955–2945 Ma Bookingarra Group (predominantly volcanic). A third basin, the mainly volcanic c. 3000–2990 Ma Whim Creek Basin, contains the 1-km-thick Whim Creek Group. The Whim Creek Group was deposited in an intra-arc setting between the Mallina Basin and a c. 2999–2982 magmatic arc that is now preserved as the Maitland River Supersuite. A major crustal fracture, the Loudens Fault, separates the Croydon Group from the Whim Creek and Bookingarra Groups (Hickman 2016).

The De Grey Supergroup also includes a local basaltic formation in the Shay Gap area of the Northeast Pilbara Craton, the undated Coonieena Basalt. This basaltic formation, geochemically similar to basaltic members of the Croydon and Bookingarra Groups (Van Kranendonk et al. 2006), was erupted between deposition of the Gorge Greek and Croydon Groups.

The Croydon and Bookingarra Groups were deposited in three northeast–southwest trending zones (Hickman 2016):

1. A central deep-water trough in which the Constantine Sandstone and Mallina Formation of the Croydon Group were deposited between c. 3015 and c. 2940 Ma.
2. A southeast shallow-water shelf, on which the Cattle Well Formation and Lalla Rookh Sandstone of the Croydon Group were deposited between 3015 and 2931 Ma.
3. A northwest shallow-water shelf on which five formations of the Bookingarra Group were deposited between c. 2955 and c. 2931 Ma.

When first defined, the De Grey Superbasin was interpreted to include a fourth basin in the far eastern part of the Northern Pilbara, the Mosquito Creek Basin (Van Kranendonk et al. 2006). This basin is now excluded because it originated as a rift basin during the c. 3220 Ma breakup of the Pilbara Craton (Chap. 6), that is, about 160 Ma before formation of the De Grey Superbasin (Hickman 2021a).

9.2 De Grey Supergroup

Comprising three groups, the Mesoarchean De Grey Supergroup was deposited across most of the Northern Pilbara Craton and marks the reconnection of the EPT and the Karratha Terrane (KT) that were previously separated by the Paleoarchean breakup of the craton (Chap. 6). Correlations between the stratigraphic successions of the East and Northwest Pilbara, before the c. 3220 Ma breakup (Chap. 6) and after the c. 3070 Ma collision (Chap. 8 and this chapter), are key to the Mesoarchean evolution of the Pilbara Craton.

9.2.1 Gorge Creek Group

The Gorge Creek Group comprises three formations, in ascending stratigraphic order: the 3067–3022 Ma Farrel Quartzite, up to 1.5 km thick, and composed of metamorphosed sandstone and conglomerate; the 3022 to 3015 Ma Cleaverville Formation, generally no more than 1 km thick and comprising banded iron formation (BIF), chert, carbonaceous shale, siltstone, and sandstone; and the Cundaline Formation (age uncertain), between 0.8 and 1.1 km thick, and composed of conglomerate, sandstone, wacke, and ferruginous shale.

In the northwest part of the Pilbara Craton, the Gorge Creek Group is mainly represented by BIF, chert, ferruginous clastic rocks, and shale and accordingly is assigned to the Cleaverville Formation. However, lower and upper parts of the northwest succession include relatively thin lithological units that might be parts of the Farrell Quartzite or the Cundaline Formation. For example, in areas between Wickham and Roebourne, the base of the Cleaverville Formation is locally composed of conglomerate, sandstone, siltstone, shale, or chert. Collectively, these units constitute a succession lithologically like the Farrell Quartzite. Near Cleaverville, BIF of the Cleaverville Formation is separated from the underlying Regal Formation by a unit of sandstone and conglomerate up to 100 m thick. Kiyokawa et al. (2012) referred to this unit using an informal name, 'Sixty-six Hill member', and recognized an unconformable contact with the underlying Regal Formation. They provided geochronological evidence that the maximum depositional age of the sandstone is c. 3060 Ma. Previously, the same sandstone was dated at between c. 3058 and c. 3022 Ma (Nelson 1998, GSWA 127330). Both results are consistent with the interpreted depositional age range of the Farrel Quartzite. Formations of the Gorge Creek Group, in particular the Cleaverville Formation which is composed mainly of BIF, outcrop in areas as far apart as the Devil Creek greenstone belt in the Northwest Pilbara (Hickman and Strong 2003; Hickman 2021b), the northern Isabella Range in the Northeast Pilbara (Williams 2001), and Budjan Creek area in the East Pilbara (Bagas et al. 2002; Hickman 2012). Stratigraphic evidence indicates that the Isabella Range and Devil Creek successions were not deposited close to margins of the basin. This is especially evident in the Northeast Pilbara where one of the thickest sections

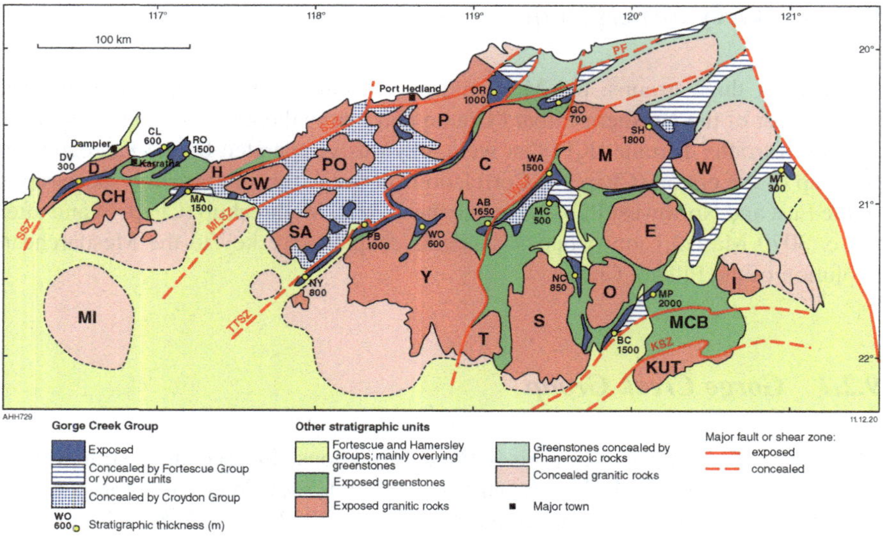

Fig. 9.1 Simplified geological map of the Northern Pilbara Craton showing outcrops and exposed thicknesses of the Gorge Creek Group. Interpreted concealed sections of the group (from magnetic imagery) illustrate the minimum original extent of its depositional basin. Thickness data indicate that the basin extended well beyond the present outcrop areas. In the East Pilbara, the group was folded around the domes, establishing a late Mesoarchean reactivated stage of doming. Domes and igneous complexes: *C* Carlindi Dome; *CH* Cherratta Igneous Complex; *CW* Caines Well Igneous Complex; *D* Dampier Igneous Complex; *E* Mount Edgar Dome; *H* Harding Igneous Complex; *I* Yilgalong Dome; *KUT* Kurrana Igneous Complex; *M* Muccan Dome; *MI* Mingar Igneous Complex; *O* Corunna Downs Dome; *P* Pippingarra Igneous Complex; *PO* Portree, Igneous Complex; *S* Shaw Dome; *SA* Satirist Granite; *T* Tambourah Dome; *W* Warrawagine Dome; *Y* Yule Dome. Major faults and shear zones: *KSZ* Kurrana Shear Zone; *LWF* Lalla Rookh–Western Shaw Fault; *MSZ* Maitland Shear Zone; *Pf* Pardoo Fault; *SSZ* Sholl Shear Zone; *TTSZ*, Tabba Tabba Shear Zone. Section locations: *AB* Abydos; *BC* Budjan Creek; *CL* Cleaverville; *DV* Devil Creek; *GO* Goldsworthy; *MA* Mount Ada; *MC* Miralga Creek; *MP* McPhee Creek; *MT* Mount Cecelia; *NC* North Coongan; *NY* Nunyerry Gap; *OR* Ord Range; *PB* Pilbara Well; *RO* Roebourne; *SH* Shay Gap; *WA* Warralong; *WO* Wodgina. (From Hickman 2021a, b; with Geological Survey of Western Australia permission)

of the Gorge Creek Group is exposed in the Shay Gap greenstone belt (SH on Fig. 9.1). The present northeast limit of the Pilbara Craton is a faulted margin dating from the Neoarchean breakup of the craton (Chap. 1). Therefore, the original southwest–northeast length of the Gorge Creek Basin might have been twice its currently preserved 500 km length. The northwest–southeast width of the basin is at least 200 km, the southeast margin being an unconformable contact with the EPT.

Most greenstone belts of the Northern Pilbara Craton include parts of the Gorge Creek Group, exceptions being the Mount Elsie and McPhee greenstone belts in the southeast, the Western Shaw, Tambina and Emerald Mine greenstone belts in the south, and apparently the poorly exposed Peawah Hill greenstone belt (Fig. 7.13) in the north. Successions overlying the EPT are remarkably similar except that the upper formation of the group, the Cundaline Formation, has been recognized in only the northeast greenstone belts (Williams 1999). Stratigraphic correlations between the eastern greenstone belts and those of the Northwest Pilbara (Hickman 1980b, 1983) were questioned by some workers (Williams 1999; Van Kranendonk et al. 2002) but subsequently reaffirmed by Van Kranendonk et al. (2006). Stratigraphic differences between the east and northwest successions reflect differences in the underlying crust. In the East Pilbara, the Gorge Creek Group was deposited on deeply eroded Paleoarchean stratigraphy (greenstones and granites), resulting in basal erosional unconformities overlain by thick units of the Farrel Quartzite. In areas of the Northwest Pilbara north of the Sholl Shear Zone (SSZ) (Fig. 7.1), the group was deposited on Mesoarchean basaltic crust of the 3220–3165 Ma Regal Formation without any angular unconformities and only thin basal clastic units. Here, the group is only rarely exposed south of the SSZ where its basalt contact with the 3130–3110 Ma Whundo Group is concealed.

9.2.1.1 Farrel Quartzite

In the East Pilbara, the Farrel Quartzite overlies a regional unconformity at the base of the De Grey Supergroup. Until a major lithostratigraphic revision of the northern part of the Pilbara Craton arising from results of the GSWA–GA Pilbara Craton Mapping Project (Van Kranendonk et al. 2006), all conglomerate–sandstone units immediately overlying this unconformity were assigned to the Corboy Formation. This interpretation followed the first formal definition of the Gorge Creek Group (Lipple 1975; Hickman and Lipple 1975, 1978). Accordingly, many descriptions and interpretations of the Corboy Formation that were published prior to 2006 now apply to the Farrel Quartzite. Exceptions are descriptions of the Corboy Formation in the Soanesville and Pincunah greenstone belts.

9.2.1.1.1 Depositional Environment

Important isotopic evidence on the exposed geology in the East Pilbara Craton has come from U–Pb zircon dating of detrital zircons in the Farrel Quartzite. Fig. 2.2c, d, e show that the ages of individual detrital zircon grains are mainly 3600–3340 Ma indicating widespread exposures of early Paleoarchean crust. The almost complete absence of 3340–3070 Ma zircons suggests that younger felsic crust such as granitic rocks of the Emu Pool, Cleland, and Mount Billroth Supersuites and felsic volcanic formations including the Wyman and Kangaroo Caves Formations were not exposed. A likely explanation is that much of the EPT was submerged between 3070 and 3015 Ma, with islands being dominated by pre-3400 Ma crust. In rare cases where the Farrel Quartzite unconformably overlies granitic crust, as in the Shay Gap and Ord Range greenstone belts (SH and OR on Fig. 9.1), the formation is very thin. This suggests that the partly eroded granitic cores of the EPT domes probably formed most of the islands and larger upland areas that were never covered by the formation. Most sedimentological evidence on the depositional environment of the Farrel Quartzite has been obtained from the Goldsworthy, Warralong, and East Strelley greenstone belts (Fig. 1.7). However, additional evidence on the overall paleogeography of the basin has been obtained from other greenstone belts in the west and southeast parts of the EPT. In the Northwest Pilbara Craton, the Cleaverville Formation is interpreted to underlie most of the Croydon Group in the Mallina Basin, although thicknesses of the Farrel Quartzite there are expected to be minor. This is based on the exposed stratigraphy of the group farther west, both north and south of the SSZ.

In the Goldsworthy greenstone belt (GO on Fig. 9.1), depositional environments of the Farrel Quartzite were fluvial to coastal plain, with evaporite deposition in coastal salinas (Sugitani et al. 2003, 2006, 2007). Microprobe analyses and crystallographic studies have indicated evaporitic gypsum, nahcolite ($NaHCO_3$), and barite, later pseudomorphed by silica. One bed of black chert containing microfossils (Sugitani et al. 2007, 2009a, b), and associated with evaporite minerals, was traced laterally for 7 km. Sugahara et al. (2010) analysed the microfossiliferous black chert and detected positive La anomalies, HREE enrichment, slightly positive europium anomalies, and chondritic to slightly super-chondritic Y/Ho ratios. They interpreted these features as evidence of freshwater deposition, possibly influenced by geothermal water, but not high-temperature hydrothermal solutions. Retallack et al. (2016) supported a coastal plain setting that included evaporites and interpreted other features as evidence of sandy and silty paleosols.

Farther south in the Warralong greenstone belt (WA on Fig. 9.1), deposition was initially in nearshore, high-energy beach environments (Van Kranendonk 2010). The lower part of the Farrel Quartzite includes beds of polymictic conglomerate up to 20 m thick with numerous pebbles and boulders up to 30 cm in diameter (Fig. 9.2). Higher in the formation, quartz sandstone contains ripple marks and cross-bedding, whereas near the top of the formation interbedded quartz-rich sandstone and siltstone indicate an upward-fining transition into deeper-water facies characteristic of the overlying Cleaverville Formation. The depositional environment was tectonically active, with syndepositional crustal extension producing horst and graben structures within the formation. Normal growth faults extend to the top of the formation, but not into the overlying Cleaverville Formation (Van Kranendonk 2004, 2010; Van Kranendonk et al. 2006). The growth faults separate panels of similar lithology but different sedimentary thicknesses (Van Kranendonk 2004).

Still farther south, in the East Strelley greenstone belt (AB on Fig. 9.1), the Farrel Quartzite is approximately 500–650 m thick and is a fining-upward succession commencing with coarse- to medium-grained sandstone overlying an erosional unconformity developed on the Kelly and Sulphur Springs Groups. Wilhelmij and Dunlop (1984) divided the succession into lower and upper parts, with a distinctive unit of siltstone, shale, and beds of tuffaceous green chert 30–60 m above the base of the formation. They interpreted this part of the formation, with soft-sediment deformation, symmetrical ripple marks, climbing ripple cross-lamination, and lenticular and flaser bedding, as a wave- or tide-dominated, transgressive shallow shelf deposit. The 600-m-thick upper division comprises alternating units of regularly

Fig. 9.2 Outcrops of the Farrel Quartzite in the Warralong greenstone belt: (**a**) polymictic boulder conglomerate overlying an angular unconformity on the Euro Basalt (Zone 50, MGA 755640E 7690200N); (**b**) graded conglomerate–sandstone–mudstone beds from near the top the basal polymictic conglomerate of the Farrel Quartzite (MGA Zone 50, 755700E 7690250N). (From Van Kranendonk 2004; with Geological Survey of Western Australia permission)

bedded sandstone, siltstone, and shale with minor chert that Wilhelmij and Dunlop (1984) interpreted as thin-bedded and proximal turbidites. The thin-bedded turbidites contain mid- and upper sections of the typical Bouma sequence, whereas the proximal turbidites are represented by >10-cm-thick graded sandstone layers stacked or separated by thin shale partings. The depositional environment of the upper division was interpreted as a submarine fan with migrating proximal channels. The Farrel Quartzite grades up into the Cleaverville Formation via a sedimentary transition succession of siltstone, shale, chert, and BIF.

West of the East Strelley greenstone belt, in the Wodgina greenstone belt (WO on Fig. 9.1), the Farrel Quartzite is a succession of sandstone and siltstone that conformably underlies the Cleaverville Formation (Van Kranendonk et al. 2010; Duuring et al. 2016). Duuring et al. (2016) described units here correlated with the Farrel Quartzite as comprising a basal unit of volcanogenic sedimentary rocks (possibly tuffaceous), overlain by siltstone and chert and overlain by quartz arenite and quartzite.

Farther west in the Pilbara Well greenstone belt (PB on Fig. 9.1), the Farrel Quartzite is between 200 and 400 m thick and consists of basal conglomerate, sandstone, felsic volcaniclastic rocks, siltstone, shale, and minor felsic lava (Smithies and Farrell 2000). Smithies and Farrell (2000) reported trachyte and trachytic fragmental rocks in the sandstone, and trachytic volcaniclastic material is included in the sandstone at Nunyerry Gap (Nelson 1998). The increased content of felsic volcaniclastic rocks in the Farrel Quartzite of this greenstone belt is interpreted to reflect its greater proximity to the Elizabeth Hill Supersuite and the Orpheus magmatic arc of the Northwest Pilbara. The greater content of fine-grained clastic rocks compared to succession of the formation farther east and northeast suggests a westerly deepening of the depositional environment to submarine fan settings.

Most successions of the Farrel Quartzite overlying the southeast part of the EPT have not been documented in sufficient detail to interpret any lateral changes in depositional environments. In the southeast Kelly greenstone belt (BC on Fig. 9.1), between Copper Hills and Cookindina Pool, polymictic conglomerate, sandstone, siltstone, and shale, correlated with the Farrel Quartzite (Hickman 2012), unconformably overlie the Budjan Creek Formation (Bagas et al. 2004). The conglomerate contains cobble- to pebble-sized clasts of chert, porphyritic dacite or rhyolite, and rare mafic volcanic rocks, indicating derivation from the underlying Kelly Group. This clastic succession fines upwards and is conformably overlain by BIF, carbonaceous shale, and chert of the Cleaverville Formation. In the eastern part of the Kelly greenstone belt (MP on Fig. 9.1), the Farrel Quartzite is 1.5 km thick (Williams and Bagas 2004) and is composed of metamorphosed sandstone, polymictic conglomerate, and grey, black, and Fe-rich chert. Clasts in the conglomerate include quartzite, chert, and vein quartz derived from erosion of the EPT.

9.2.1.2 Cleaverville Formation

The Cleaverville Formation is composed of BIF (Fig. 9.3), chert, shale, and siltstone with minor volcaniclastic rocks and sandstone. Near-surface silicification of Fe-rich shales has locally formed cherty BIF-like cappings that have been mapped as BIF. Such silicification is evident in some cliff exposures or in drilling below depths of near-surface alteration. Some units of BIF and chert include ooids (Fig. 9.4) (Van Kranendonk 2004) suggesting silicification of carbonate rocks, although

Fig. 9.3 Folded jaspilitic iron formation in the Cleaverville Formation of the Warralong greenstone belt (MGA Zone 50, 756120E 7689400N) (From Van Kranendonk 2004; with Geological Survey of Western Australia permission)

Fig. 9.4 A bed containing densely packed ooids in the Cleaverville Formation of the Warralong greenstone belt (MGA Zone 50, 751870E 7678350N) (From Van Kranendonk 2004; with Geological Survey of Western Australia permission)

hydrothermal processes can also form ooids. Other outcrops show layering of iron formation and chert at 1 cm intervals (Fig. 9.5), similar to 'mesobanding' in the Paleoproterozoic Dales Gorge Member of the Brockman Iron Formation (Trendall and Blockley 1970). Another interesting sedimentary structure in the BIF is intraformational breccia, or 'tabular clast conglomerate' (Fig. 9.6) that forms well-defined beds up to 80 cm thick and consists of tabular rip-up clasts of laminated chert (? a product of secondary silicification) in a fine-grained matrix. The same structure is present in the Cleaverville Formation at Cleaverville Beach. Very similar breccias occur in the c. 2720 Ma sedimentary carbonates of the Meentheena Member of the Tumbiana Formation (Chap. 12) where they are referred to a 'flat-pebble

RHS265 12.02.04

Fig. 9.5 Mesobanding in BIF of the Cleaverville Formation in the Ord Range (MGA Zone 50, 722495E 7754034N) (Modified from Smithies 2004; with Geological Survey of Western Australia permission)

RHS269 12.02.04

Fig. 9.6 Flat pebble conglomerate/breccia within BIF of the Cleaverville Formation in the Ord Range (MGA Zone 50, 722495E 7754034N) (From Smithies 2004; with Geological Survey of Western Australia permission)

conglomerates'. In similar carbonates of the slightly older Mopoke Member of the Kylena Formation (Chap. 12), these breccia units are termed 'tabular clast edgewise conglomerates' and are interpreted to have been produced by wave-driven hydrodynamic reworking in shallow water. Hickman (1983), Van Kranendonk et al. (2006), and Olivier et al. (2012) have illustrated the same structure in the c. 3460 Ma Marble Bar Chert Member, suggesting that an interpretation that part of this chert unit originated as sedimentary carbonate.

The name 'Nimingarra Iron Formation' was briefly applied to the Cleaverville Formation in the Northeast Pilbara (Williams 1999) before a previous correlation with the Cleaverville Formation of the Northwest Pilbara was confirmed (Van Kranendonk et al. 2006). This followed recognition that the Cleaverville Formation of both areas was deposited after the c. 3070 Ma Prinsep Orogeny. The stratigraphic thickness of the Cleaverville Formation is relatively constant across the Northern Pilbara Craton, varying between 500 and 1000 m (Hickman 1983, 2021a; Williams 1999; Bagas et al. 2004; Duuring et al. 2016, 2017). In most areas, over 50% of the formation thickness is composed of BIF with an average composition of 57 wt.% Fe_2O_3 and 41 wt.% SiO_2 (Duuring et al. 2016). By comparison, BIF of the c. 2490 Ma Dales Gorge Member of the Brockman Iron Formation averages about 42 wt.% Fe_2O_3 and 50 wt.% SiO_2 (data in Trendall and Blockley 1970).

9.2.1.2.1 Depositional Environment

Evidence on the depositional environments of the Cleaverville Formation is partly stratigraphic and partly geochemical. The stratigraphic evidence indicates a regionally extensive depositional basin across the Northern Pilbara that progressively subsided during deposition of the Gorge Creek Group. The total stratigraphic thickness of the Gorge Creek Group in the east Pilbara is approximately 2 km, suggesting this was the minimum amount of basin subsidence. By analogy with deposition of the Neoarchean Fortescue Group, it is likely that some gravity-driven reactivation of the EPT domes occurred during deposition of the Cleaverville Formation; in this case, basin deepening would have been greatest above underlying greenstone belts. Turbidites in the Cleaverville Formation (Duuring et al. 2016) suggest syndepositional tectonic activity; also, the Cundaline Formation is locally unconformable on the Cleaverville Formation (Williams 1999) indicating uplift and erosion.

In the northern part of the Gorge Creek Basin, relatively close to the magmatic arc of the Orpheus Supersuite, there is evidence of hydrothermal activity during deposition of the Cleaverville Formation. Smithies et al. (2004) reported that the Cleaverville Formation of the Goldsworthy greenstone belt contains layers of dark grey fine-grained silica up to 0.5 m thick that were fed from fracture zones containing hydrothermal brecciation. Sugahara et al. (2010) analysed black chert and BIF from the Cleaverville Formation of the Goldsworthy greenstone belt and reported a vertical trend of increasing Y/Ho with HREE enrichment and positive La anomalies. These results were interpreted to indicate a change from non-marine deposition in the Farrel Quartzite to marine deposition of the Cleaverville Formation. They also noted increasingly positive Eu anomalies upwards into the Cleaverville Formation and interpreted this as evidence of increasing syndepositional hydrothermal activity. This trend is consistent with southeast migration of the Orpheus magmatic arc (Hickman 2016).

9.2.1.3 Cundaline Formation

The Cundaline Formation stratigraphically overlies the Cleaverville Formation across a contact that varies from a sedimentary transition to an erosional unconformity. As defined by Williams (1999), the Cundaline Formation is approximately 800 m thick and consists of reddish Fe-rich shale, grey-green shale, siltstone, wacke, sandstone, immature pebbly sandstone, and pebble conglomerate. Lithic clasts are mainly chert, jasper, and minor devitrified felsic volcanic rock. Thick, coarse-grained clastic units locally contain graded bedding. Intercalated thin units of komatiitic basaltic rock are present in the Shay Gap greenstone belt. Rather than being volcanic units, these might be subvolcanic sills related to the overlying Coonieena Basalt. In the northeast part of the Marble Bar greenstone belt, the Cundaline Formation unconformably overlies the Cleaverville Formation and includes basal conglomerate and breccia derived from the BIF and chert (Williams 1999). Erosion of the Cleaverville Formation during deposition of the Cundaline Formation suggests areas of uplift in the Gorge Creek Basin, and it is notable that the Cundaline Formation is unusual in that it was deposited where the Gorge Creek Group overlies granitic rocks. The Cundaline Formation was deposited across the Muccan and Warrawagine Domes and overlies BIF of the Cleaverville Formation at Mount Cecilia (Fig. 9.1).

9.2.2 Regional Stratigraphic Continuity

Some workers have questioned the regional stratigraphic continuity of the Cleaverville Formation (Williams 1999; Van Kranendonk 2000; Van Kranendonk et al. 2002; Sheppard et al. 2017). Figure 9.1 provides estimates of the total thickness of the Gorge Creek Group (Farrel Quartzite plus Cleaverville Formation) at many localities across the northern Pilbara. These thicknesses establish that the succession of the Gorge Creek Group is present in almost all greenstone belts of the Northern Pilbara Craton; exceptions are present where this stratigraphic level is not exposed.

The depositional extent of the Gorge Creek Group (Fig. 9.1) is consistent with the interpretation that it was deposited in a single, regionally extensive retro-arc foreland basin (Hickman 2021a). An important consideration is that the depositional setting of BIF in the Cleaverville Formation is very different from BIF within volcanic successions, referred to by many authors as Algoma-type iron formation (Gross 1980). Volcanic-hosted BIF is locally present in the Warrawoona Group: for example, thin BIF units in the c. 3515 Ma Coucal Formation, local BIF units in the 3530–3490 Ma North Star Basalt, and BIF in the c. 3477 Ma McPhee Formation. These volcanic-hosted BIF units are no more than 20 m thick and have very limited lateral extents. Bekker et al. (2010, 2014) commented that Algoma-type iron formations elsewhere in the world rarely have lateral extents greater than 10 km and are typically less than 50 m thick, whereas BIF units in extensive sedimentary basins extend for hundreds of kilometres and are typically hundreds of metres thick. Konhauser et al. (2017) noted that dissolved hydrothermal Fe(II) can be transported more than 2000 km from seafloor-hydrothermal systems, such as at mid-ocean ridges or in arc- or back-arc settings, before it is precipitated by various oxidative mechanisms in relatively shallow-water photic zones.

Thick BIF units in Archean and Proterozoic sedimentary basins are likely to have been deposited over many hundreds of kilometres, that is like the depositional extent of the Gorge Creek Basin. Trendall (1968) described three examples: the Hamersley Basin of Western Australia, the Animikie Basin of North America, and the Transvaal Basin of South Africa. As noted above, the thickness and Fe content of the Cleaverville Formation is comparable to, if not greater than, the thickness and Fe content of BIF in the Brockman Iron Formation of the Hamersley Basin. The evidence that the Gorge Creek Basin is a retro-arc basin linked to a magmatic arc and subduction zone on the northwest margin of the Pilbara Craton (Hickman 2016, 2021a) gives some indication of its original depositional extent. This tectonic interpretation is similar to previous back-arc or retro-arc basin interpretations for the overlying Mallina Basin (Krapež 1993; Krapež and Eisenlohr 1998; Smith et al. 1998; Beintema 2003; Smith 2003; Pike et al. 2006; Hickman 2012, 2016). Present exposures of the basin, even after breakup of the Pilbara Craton at about 2710 Ma (Chap. 12), establish that its minimum southwest–northeast extent was 500 km and its northwest–southeast width was at least 200 km. Prior to the Neoarchean breakup event, the length of the basin might have exceeded 1000 km.

It is useful to consider the stratigraphic evidence gained from regional mapping in the Northern Pilbara. The following summary of thickness variations of the Cleaverville Formation from the northwest Pilbara to the east Pilbara confirms the regional continuity of the formation. Greenstone belts and areas referred to in the following text are identified on Fig. 9.1. Local erosion of the Cleaverville Formation prior to deposition overlying stratigraphy explains some of the regional thickness variations.

In the Devil Creek greenstone belt (area D), the Cleaverville Formation is up to 300 m thick, although measurement of stratigraphic thickness is complicated by folding and shearing adjacent to the Sholl Shear Zone. At Cleaverville (area CL), the formation is between 300 and 600 m thick (Kiyokawa et al. 2012; Ryan and Kriewaldt 1964), whereas south of Roebourne (area MA) the Cleaverville Formation is about 1.5 km thick (Hickman 1997, 2001b). At Nunyerry Gap (area NY), the Cleaverville Formation makes up 400 m of the 800-m-thick Gorge Creek Group, the remainder being a clastic succession of the Farrel Quartzite (Smithies 1998a). Here, and farther northeast on the northwest margin of the Pilbara Well greenstone belt, the BIF was partly eroded prior to deposition of the Constantine Sandstone of the Croydon Group (Fitton et al. 1975; Smithies and Farrell 2000). In the Pilbara Well greenstone belt (area PB), the Cleaverville Formation is 800 m thick and conformably underlain by a 200-m-thick succession of the Farrel Quartzite. In the Ord Range (area OR), the total thickness of the Cleaverville Formation is 1 km, with the largest individual BIF unit being 500 m thick (Duuring et al. 2016).

Duuring et al. (2016) reported that the thickness of the Cleaverville Formation in the Wodgina greenstone belt (area WO) is 600 m, and in the Abydos area of the East Strelley greenstone belt (area AB), they estimated a thickness of 1 km. In the northeast Panorama greenstone belt at Miralga Creek, mapping indicates that BIF of the Cleaverville Formation is approximately 500 m thick (Van Kranendonk 1999; Hickman 2013), and the formation maintains this thickness northwards into the Warralong and Doolena Gap greenstone belts. In the Warralong greenstone belt (area WA), the Cleaverville Formation is up to 500 m thick. In the northern section of the Coongan greenstone belt (area NC), the Cleaverville Formation is 750 m thick (Duuring et al. 2016). At Goldsworthy (area GO), the Cleaverville Formation is over 400 m thick and in the Shay Gap area (area SH) of the Muccan Dome the Cleaverville Formation is 1 km thick, with the overlying Cundaline Formation being 800 m thick (Williams 2000). Aeromagnetic data indicate that BIF of the Cleaverville Formation underlies Proterozoic and younger rocks up to 30 km northeast of Shay Gap, and possibly as far as the faulted contact between the Pilbara Craton and the Paterson Orogen (Fig. 9.1).

North of Nullagine, in the McPhee Creek area (area MP), the formation is at least 500 m thick (Bagas et al. 2004), and the same thickness extends south to the Copper Hills–Budjan Creek area (area BC). East of Nullagine, in the Mount Elsie greenstone belt (Fig. 1.7), the Cleaverville Formation is not preserved, apparently as a result of erosion to the level of the Kelly Group. In the far northeast of the East Pilbara, at Mount Cecilia (area MT) in the northern part of the Isabella Range, the Cleaverville Formation is faulted and tightly folded, but outcrop width suggests a minimum stratigraphic thickness of 300 m. Aeromagnetic data indicate not only that this outcrop is an eastern extension of the Cleaverville Formation between the Mount Edgar and Warrawagine Domes (Williams 2003) but also that the Cleaverville Formation is thickly developed on the northeast side of the EPT.

9.2.2.1 Stratigraphic Continuity Between Greenstone Belts

Stratigraphic continuity of the Cleaverville Formation, from the type area on the Cleaverville Peninsula southeast across the Karratha and Sholl Terranes and the Mallina Basin to the northwest boundary of the EPT (Fig. 1.3), is revealed by inliers of the formation within the Mallina Basin (Smithies and Farrell 2000; Hickman 2016). Accordingly, the Cleaverville Formation is interpreted to underlie the Croydon Group in all parts of the Mallina Basin except in areas of granitic intrusion (Fig. 9.1). Eastward continuation of the formation onto the EPT is established by a succession of BIF and underlying clastic rocks on the western margin of the Pilbara Well greenstone belt (Fitton et al. 1975; Smithies and Farrell 2000). The Cleaverville Formation in the Pilbara Well greenstone belt is continuous with BIF and chert in the Wodgina greenstone belt, and via the Kangan Syncline area, the BIF continues northwest along the Tabba Tabba Shear Zone towards the Ord Range.

The stratigraphic successions of the Ord Range, Goldsworthy, and Shay Gap greenstone belts are extremely similar, consisting of a thin basal unit of quartzite and chert (Farrel Quartzite) overlain by a thick BIF–shale unit (Cleaverville Formation). Despite the 100 km separation of the Ord Range and Shay Gap greenstone belts, most workers have correlated the successions (Dawes et al. 1995a, b; Williams 1999, 2000; Smithies 2004; Van Kranendonk and Smithies 2006; Sheppard et al. 2017). The Ord Range and Goldsworthy outcrops were mostly likely adjacent prior to c. 2940 Ma sinistral displacement along the TTSZ and its eastern continuation, the Pardoo Fault (Fig. 1.3). Stratigraphic correlation of the quartzite–BIF–shale successions in these three greenstone belts provides an important link between the Gorge Creek Group of the Northeast Pilbara and successions that underlie the Mallina Basin in the Northwest Pilbara. This is evidence that the Gorge Creek Group was deposited in a single basin that extended from the Northwest Pilbara to the Northeast Pilbara.

Geological mapping and regional aeromagnetic imagery obtained during the PCMP (Chap. 1) establish that the BIF of the Shay Gap greenstone belt extends 50 km south into the BIF of the northeast Marble Bar greenstone belt (Williams 1998). From there, the BIF extends west and southwest for 120 km, through the Marble Bar, Warralong, and East Strelley greenstone belts, and south into the Coongan and Kelly greenstone belts. Details of the successions of these greenstone belts were provided by Hickman (2021a).

9.2.3 Conclusions regarding the Gorge Creek Basin

The Gorge Creek Group was deposited in a single, laterally extensive basin across the Northern Pilbara. Formation of the basin followed collision of the EPT and WPS at c. 3070 Ma, and the subsequent development of a subduction zone along the northwest margin of the craton. Previously published evidence (Van Kranendonk et al. 2006, 2007) indicates that from c. 3067 to 2919 Ma, a succession of magmatic arcs migrated southeast across the Northern Pilbara (Fig. 8.6) (Hickman 2016). The oldest arc exposed on land is the 3024–3012 Ma Orpheus Supersuite although remnants of 3067–3024 Ma arcs might be preserved within a 150–200-km-wide offshore section of the Pilbara Craton that is concealed by the Northern Carnarvon Basin (Fig. 1.2). Therefore, the Gorge Creek Basin is interpreted as a retro-arc basin southeast of one or more subduction-related magmatic arcs. A back-arc or retro-arc setting has long been interpreted for the Mallina Basin (Krapež 1993; Krapež and Eisenlohr 1998; Smith et al. 1998; Beintema 2003; Smith 2003; Pike et al. 2006; Hickman 2012, 2016), and the Mallina Basin evolved from the underlying Gorge Creek Basin. Stratigraphic evidence (Smithies 1998a; Hickman 2016) has revealed that the Gorge Creek Group has a transitional contact with the Croydon Group in the Mallina Basin. Thus, the retro-arc tectonic setting and the subduction related to both basins extended from c. 3070 to c. 2919 Ma.

Throughout the East Pilbara, the earliest sedimentation in the Gorge Creek Basin was that of sandstone and conglomerate of the 3067–3022 Ma Farrel Quartzite. In some areas, the Farrel Quartzite is over 1000 m thick. By contrast, only thin units of sandstone and siltstone underlie the Cleaverville Formation in the Northwest Pilbara. This is consistent with the absence of a basal angular unconformity in the Northwest Pilbara (Hickman 2016). Deposition of the Farrel Quartzite commenced in subaerial to shallow-water environments including coastal plain, river channel, coastal salina, delta, supratidal and tidal flats, and shallow shelf. Although there have been no paleocurrent studies, it is likely that the topography of the depositional basin was complicated by shallow horst and graben structures and islands composed of mainly Paleoarchean rocks. Some sections through the Farrel Quartzite have provided evidence of progressive basin subsidence, including upward-fining sequences, an introduction of turbidites, and transitions into the Cleaverville Formation with increases in the amounts of siltstone, shale, chert, and BIF (Wilhelmij and Dunlop 1984).

Based on investigations of BIFs elsewhere (Rasmussen et al. 2014, 2016; Gauger et al. 2015; Smith 2015; Konhauser et al. 2017), the principal protoliths of BIF in the Cleaverville Formation are likely to have been iron-silicate mud, Fe-rich clays such as greenalite, ferric oxyhydroxides, and Fe-rich carbonate precipitates. The thicknesses of the BIF-shale successions in the Cleaverville Formation, locally exceeding 1000 m, testify to deposition in a large basin, which is consistent with the retro-arc foreland basin environment interpreted here. Prior to the Neoarchean breakup of the Pilbara Craton (Blake and Barley 1992; Martin et al. 1998a, b; Müller et al. 2005), the Gorge Creek Basin extended southwest–northeast for more than 500 km, and southeast for at least 200 km. In contrast to the Farrel Quartzite, most of the Cleaverville Formation was probably deposited in moderately deep-water remote from significant terrigenous influx. However, in the Northwest Pilbara, where the Cleaverville Formation was deposited adjacent to felsic volcanism of the Orpheus Supersuite, the formation includes felsic tuff and volcaniclastic sandstone.

9.2.4 *Geochronology*

The maximum depositional age of the Cleaverville Formation in the Northwest Pilbara is indicated by three dates between 3022 and 3015 Ma (Nelson 1998). An important source of the detrital zircons in the dated samples was the Orpheus Supersuite which outcrops along the northwest margin of the Gorge Creek Basin. Thirteen samples of the Orpheus Supersuite from the Northwest Pilbara were dated between 3024 and 3012 Ma (Nelson 1997, 1998, 1999, 2001; Smith et al. 1998; Smith 2003). Two samples of the Gorge Creek Group overlying the EPT, at Nunyerry Gap and Shay Gap, have also provided data indicating a Mesoarchean depositional age.

9.2.4.1 Geochronology at Nunyerry Gap

At Nunyerry Gap (NY on Fig. 9.1), in the Pilbara Well greenstone belt close to the western margin of the EPT, U–Pb zircon dating of felsic volcaniclastic sandstone immediately underlying BIF and chert of the Cleaverville Formation indicated a maximum depositional age of 3016 ± 13 Ma (Nelson 1998, GSWA 142842). All the dated zircons were interpreted as detrital and separated into five age components: c. 3016 Ma, most likely derived from the Orpheus Supersuite in the Northwest Pilbara; c. 3068 Ma, derived from either the 3068–3066 Ma Elizabeth Hill Supersuite or a pre-Orpheus magmatic arc close to the northwest margin of the Pilbara Craton; c. 3106 Ma, consistent with derivation from the 3130–3093 Ma Sholl Terrane in the adjacent Regal Basin; c. 3141 Ma, possibly derived from felsic magmatic activity related to the 3165–3140 Ma Karratha Event; and a large zircon age component at dated c. 3171 Ma, consistent with derivation from underlying formations of the 3185–3165 Ma Soanesville Group in the Pilbara Well greenstone belt.

Along strike, in the northern part of the Pilbara Well greenstone belt, approximately 50 km northeast of Nunyerry Gap, a similar BIF and chert unit is also underlain by felsic volcaniclastic sandstone (PB locality on Fig. 9.1). Notably, the volcaniclastic sandstones at Nunyerry Gap and Pilbara Well each contain abundant clasts of trachytic lava (Nelson 1998; Smithies and Farrell 2000). Trachyte has not been recorded elsewhere in Archean rocks of the Northern Pilbara, strengthening the case for correlation of the Nunyerry and Pilbara Well successions. Aeromagnetic imagery between Nunyerry Gap and the northern Pilbara Well greenstone belt suggests continuity of the BIF–chert units.

All previous descriptions of the Pilbara Well greenstone belt (Fitton et al. 1975; Hickman 1980a; Horwitz 1990; Smithies and Farrell 2000) have correlated the BIF and chert at Pilbara Well with the Cleaverville Formation, and the c. 3016 Ma date supports correlation with the Cleaverville Formation of the northwest Pilbara. This correlation has wider implications in the East Pilbara because the BIF and chert at Pilbara Well are laterally contiguous with BIF and chert in the Wodgina greenstone belt, and with BIF and chert in the Tabba Tabba Shear Zone (TTSZ) along the northwest margin of the EPT. The TTSZ unit comes close to connecting with thick BIF and chert in the Ord Ranges in the Goldsworthy greenstone belt. Duuring et al. (2016) reported stratigraphic and geochemical evidence that the BIFs of the Ord Range and Wodgina greenstone belts are parts of the same formation.

9.2.4.2 Geochronology at Shay Gap

At Shay Gap (SH on Fig. 9.1), a c. 1800-m-thick unit of BIF, chert, and shale has been correlated with the Cleaverville Formation of the Northwest Pilbara (Hickman 1980b, 1990, 2021b; Horwitz 1990; Dawes et al. 1995a, b). During geological mapping in the late 1990s, it was briefly renamed as the 'Nimingarra Iron Formation'

(Williams (1999, 2004) pending more evidence on its depositional age and stratigraphic relationships to other Pilbara BIF units. The name Cleaverville Formation was reapplied by Van Kranendonk et al. (2006) based on regional stratigraphic evidence and a new tectonic interpretation of the Archean crustal evolution of the Northern Pilbara. Van Kranendonk et al. (2006) made the name 'Nimingarra Iron Formation' obsolete.

Sheppard et al. (2017) reported 78 SHRIMP $^{207}Pb/^{206}Pb$ analyses of zircon grains from a shale unit within the BIF and chert at Shay Gap. As at Nunyerry Gap, several zircon age components were identified in the Shay Gap sample, which indicates detrital zircons in the rock. Sheppard et al. (2017) used 14 of the least discordant analyses (± 5% discordance) to interpret a pooled age of 3104 ± 16 Ma, and they concluded that this represents the actual depositional age of the BIF. This interpretation conflicts with evidence in this chapter that the BIF at Shay Gap is the Cleaverville Formation. It also conflicts with the previous important interpretation that the BIF at Shay Gap and therefore most, if not all, of the BIF of the East Pilbara were deposited after the c. 3070 Ma Prinsep Orogeny when the Regal Basin had been closed by collision of the EPT with the West Pilbara Superterrane (Van Kranendonk et al. 2006, 2007; Hickman and Van Kranendonk 2008, 2012; Hickman 2012, 2016). Until the c. 3070 Ma collision, these terranes had evolved in isolation, separated by the Regal Basin. This raises another problem for the interpretation by Sheppard et al. (2017) because there is no known source of c. 3104 Ma zircons within the EPT; the EPT contains no granitic intrusions or volcanic rocks formed between 3164 and 3068 Ma.

An alternative interpretation of the geochronological data reported by Sheppard et al. (2017) is that the maximum depositional age of the BIF and shale at Shay Gap is best defined by the age of the youngest concordant detrital zircon grain dated, and on that basis, the maximum depositional age is 3027 ± 32 Ma. This maximum depositional age would be consistent with the depositional age of Cleaverville Formation elsewhere in the northern Pilbara, as reviewed by Hickman (2016). Four main lines of evidence support this interpretation:

1. The spread of zircon $^{207}Pb/^{206}Pb$ ages within the 14 analyses that Sheppard et al. (2017) used to calculate the 3104 ± 16 Ma age is 3146–3027 Ma. This c. 120 Ma range is inconsistent with a single volcanic source, and it is difficult to reconcile with ash-fall zircons in a felsic tuff, the lithology interpreted by Sheppard et al. (2017). The wide variation of zircon ages is far more characteristic of a mixed detrital origin. The total age range of the 47 least altered zircons in the sample is 3479–2964 Ma, confirming a high content of detrital zircons.
2. The zircons dated by Sheppard et al. (2017) were contained in graded beds <1 mm thick. Fine-scale graded bedding in the shale is consistent with deposition from turbidity currents. Duuring et al. (2016) reported turbidites in the Cleaverville Formation at Wodgina (WO on Fig. 9.1), and turbidites are components of other Pilbara BIFs, for example, in the Hamersley Basin (Krapež et al. 2003; Pickard et al. 2004). Limited abrasion of the zircon grains in the Shay Gap

shale might be a consequence of rapid transport in suspension by turbidity currents.

3. Sheppard et al. (2017) correlated the BIF at Shay Gap with an equally thick BIF in the Ord Range. The same correlation has been made by previous workers (Hickman 1980b; Hickman and Gibson 1981; Horwitz 1990; Williams 1999, 2004; Smithies 2002, 2004; Van Kranendonk and Smithies 2006). The BIF in the Ord Range (OR on Fig. 9.1) immediately underlies the 3015–2931 Ma Croydon Group in the Mallina Basin (Van Kranendonk and Smithies 2006), and inliers of this BIF are present farther southwest in the Mallina Basin (Fig. 9.1). Throughout the Mallina Basin, this BIF has been correlated with the 3022–3016 Ma Cleaverville Formation (Smithies 1998a, b; Smithies and Farrell 2000; Smithies and Hickman 2003, 2004; Hickman 2016). The same 3022–3016 Ma BIF is exposed at Nunyerry Gap where it extends from the Mallina Basin onto the northwest margin of the EPT, and where volcaniclastic sandstone conformably underlying it has an interpreted maximum depositional age of 3016 ± 13 Ma (Nelson 1998, GSWA 142842). Sheppard et al. (2017) agreed that the BIF at Nunyerry Gap is the same unit as that in the Mallina Basin.
4. The Gorge Creek and Mallina Basins are now interpreted to have formed as retro-arc basins southeast of magmatic arcs, the formation of which was triggered by collision of the EPT and WPS at c. 3070 Ma. The Gorge Creek Basin underlies, and is transitional into, the Mallina Basin (Hickman 2016), for which most tectonic interpretations favour a back-arc or retro-arc setting (Krapež and Eisenlohr 1998; Smith et al. 1998; Blewett 2002; Beintema 2003; Smith 2003; Pike et al. 2006; Hickman 2012, 2016). A retro-arc foreland basin setting for the Gorge Creek Basin is consistent with stratigraphic evidence that the basin extended across most of the Northern Pilbara Craton.

9.2.5 Coonieena Basalt

Outcrops of the Coonieena Basalt (Williams 1999) are restricted to the Shay Gap and Marble Bar greenstone belts where the formation comprises a c. 1500-m-thick succession of pillowed komatiitic and tholeiitic basalt lava flows intruded by dolerite sills. The formation unconformably to disconformably overlies the Gorge Creek Group and is disconformably overlain by the Cattle Well Formation of the Croydon Group (Williams 1999). Williams (1999, 2003) observed that the Gorge Creek Group underlying the Coonieena Basalt is intruded by numerous mafic and ultramafic dykes and sills that were feeders to the basalt; none of the dykes continue into the overlying Cattle Well Formation. This suggests crustal extension during eruption of the Coonieena Basalt, consistent with contemporaneous rifting in the Mallina Basin (Smithies et al. 1999, 2001b; Van Kranendonk et al. 2002).

The Coonieena Basalt is intruded by the Shay Intrusion, although $^{207}Pb/^{206}Pb$ ages of zircon grains from a granophyre at the top of the intrusion indicate inheritance from intrusion of the Cattle Well Formation (GSWA 180056, Wingate et al.

2009a). The Cattle Well Formation of the Croydon Group has a maximum depositional age of c. 2988 Ma (GSWA 180048, Wingate et al. 2009b), although because it is conformable with the overlying Lalla Rookh Sandstone its depositional age might be closer to c. 2950 Ma. Therefore, geochronological data from the Shay Gap greenstone belt does not constrain the eruptive age of the Coonieena Basalt except to suggest an age between c. 3015 and c. 2950 Ma.

Geochemical data from the Coonieena Basalt and mafic volcanic and intrusive rocks in the central trough of the Mallina Basin suggest they were derived from very similar LREE-enriched mantle sources (Smithies et al. 2004, 2005, 2007a; Van Kranendonk et al. 2006, 2007). Additionally, the c. 2950 Ma Louden Volcanics and Mount Negri Volcanics of the Bookingarra Group (Hickman 2016) have similar compositions (Smithies et al. 2004). On the basis of these geochemical similarities, Van Kranendonk et al. (2006) assigned the Coonieena Basalt to the Croydon Group, despite outcrops of the Coonieena Basalt and the Louden and Mount Negri Volcanics being separated by more than 200 km. Smithies et al. (2007a) noted that other komatiitic basalt units overlying the Gorge Creek Group in the Goldsworthy and Ord Range greenstone belts (Fig. 1.7) are also chemically very similar to the Coonieena Basalt. Like the Louden Volcanics and Mount Negri Volcanics, the approximate eruptive age of these basalts is 2950 Ma (Hickman 2016). ε_{Nd} values of c. $-$ 2.00 and Nd $T_{DM}{}^2$ model ages of c. 3400 Ma for the Coonieena Basalt and the Louden Volcanics and Mount Negri Volcanics (Smithies et al. 2007a) also favour the stratigraphic correlation suggested by Van Kranendonk et al. (2006). Arndt et al. (2001) argued that magmas for the Louden and Mount Negri Volcanics were derived by melting of continental lithosphere due to a mantle plume, although adopting this interpretation for the Coonieena Basalt would require it to have been eroded from most areas prior to deposition of the Lalla Rookh Sandstone. Smithies et al. (2005) referred to high La/Nb and La/Sm ratios of the Coonieena Basalt as evidence against simple crustal contamination and more consistent with magma derivation from a refractory mantle source that was enriched by a subduction-derived component.

9.2.6 Croydon Group

The Mallina Basin (Fig. 1.3) is elongated southwest–northeast and, as preserved today, is up to 100 km wide from the northwest to the southeast. Prior to orogenic northwest–southeast crustal shortening and closure of the basin during the 2955–2919 Ma North Pilbara Orogeny, the width of the central Mallina Basin is likely to have been at least 200 km. Regional gravity and aeromagnetic data indicate that the total length of the rift basin, including concealed sections in the southwest and northeast, exceeds 600 km (Hickman 2004).

Although the Mallina Basin is mainly composed of sedimentary rocks, the northwest and southeast margins include thick volcanic formations. On the northwest shelf, the upper part of the Bookingarra Group is composed of a komatiitic and tholeiitic basalt succession up to 2 km thick (Louden Volcanics and Mount Negri

Volcanics). On the southeast shelf, the Coonieena Basalt is a 1.5-km-thick formation of komatiitic and tholeiitic basalt underlying the Cattle Well Formation. These basalts are geochemically similar and were once assigned to the same formation (Van Kranendonk et al. 2006; Smithies et al. 2007a), but limited geochronology suggests that the Coonieena Basalt is at least 30 Ma older than the Bookingarra Group, and its inclusion in the Croydon Group is problematic. Lithologically similar basaltic units are locally present within the central trough of the Mallina Basin and include the South Mallina Basalt Member and the Yareweeree Boninite Member. However, these units are much thinner than the shelf basalts and also vary in age (Hickman 2016). The Yareweeree Boninite Member is restricted to the Constantine Sandstone and is folded by structures older than 2955 Ma (Hickman 2016). The South Mallina Basalt Member forms thin, silicified basaltic units at various stratigraphic levels (Van Kranendonk et al. 2006). Another basaltic unit, the Salt Well Member, is now assigned to the Lalla Rookh Sandstone (Hickman 2016). The present interpretation is that the geochemical similarities between all these various basaltic units are due to a common origin, rifting of the basin at various stages during its 3015–2931 Ma evolution.

The stratigraphy of the Mallina Basin is complicated by major faults separating the three zones. Large parts of the central trough are very poorly exposed, and most geochronology has been on granitic intrusions of the Sisters Supersuite (Chap. 8). A U–Pb zircon date of 2994 ± 4 Ma on the Constantine Sandstone collected south of Whim Creek (Fig. 9.7) (Nelson 2000, GSWA 142942) represents either the maximum depositional age or the true depositional age of the rock. Although the dated sample contains several detrital zircons, by far the largest zircon component gives the youngest date of c. 2994 Ma, and these zircons might be ash-fall zircons introduced to the sandstone during eruption of the nearby c. 3000–2990 Ma Red Hill Volcanics (this chapter). The dated sample was not collected from the stratigraphic base of the Constantine Sandstone, supporting other information that the base of the formation is conformable with the underlying Cleaverville Formation (Hickman 2016). Parts of the Mallina Formation might be almost as old as the oldest units in the Constantine Sandstone, but there is geochronology to indicate that upper units of the Mallina Formation were deposited between 2945 and 2931 Ma. The minimum depositional age of the Mallina Formation is indicated by an igneous crystallization age of 2931 ± 5 Ma for the Satirist Monzogranite which crosscuts the formation (Nelson 1998, GSWA 141977). However, deposition of most of the formation apparently occurred before c. 2940 Ma, because it is intruded by >2945 Ma intrusions of the Indee, Portree, and Langenbeck Suites and by the 2948–2941 Ma Kialrah Rhyolite (Hickman 2016).

On the southeast shelf, the maximum age of the Cattle Well Formation is c. 2988 Ma. This formation disconformably overlies the undated Coonieena Basalt and is conformably overlain by the regionally extensive Lalla Rookh Sandstone. Attempts to date the depositional age of the Lalla Rookh Sandstone have faced the same problem encountered in dating the Farrel Quartzite of the underlying Gorge Creek Group. Detritus in both formations was almost entirely derived from the Paleoarchean EPT and Eoarchean older crust. Where the Lalla Rookh Sandstone

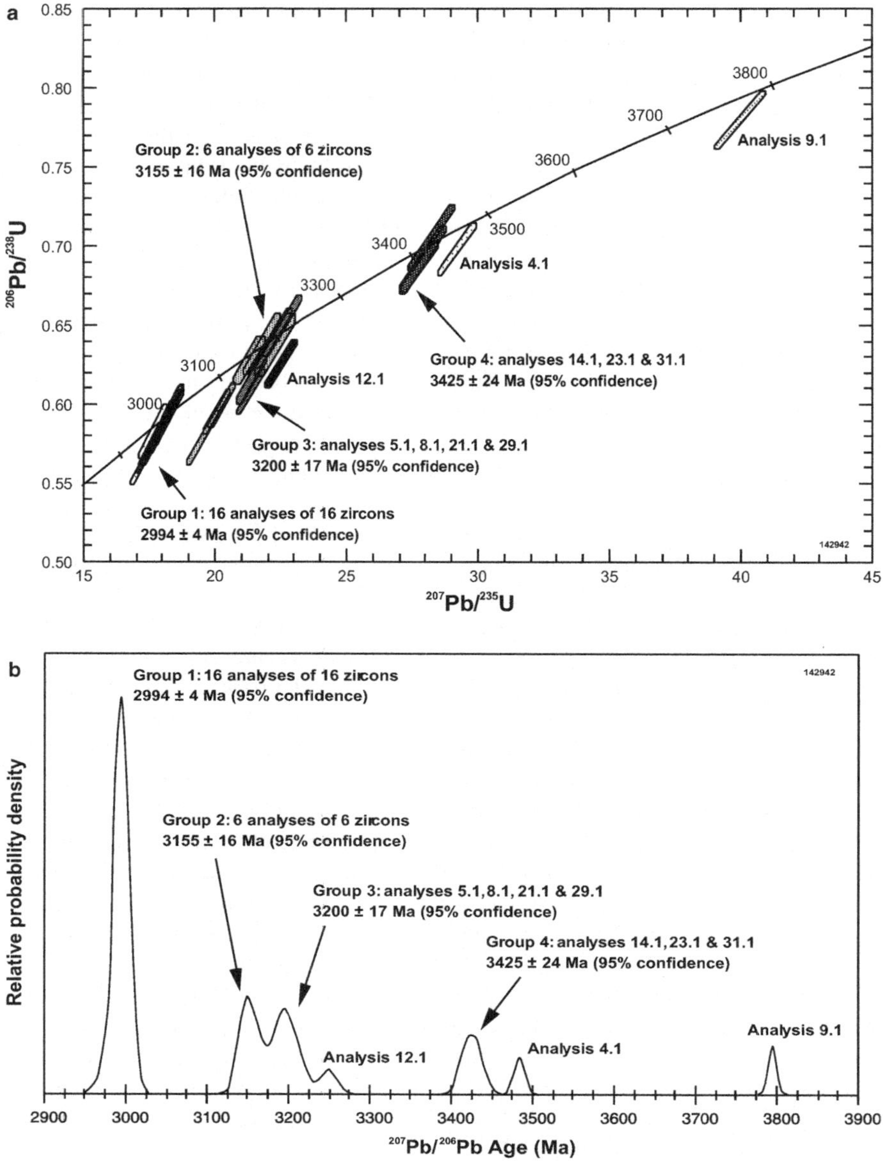

Fig. 9.7 U–Pb zircon data from the Constantine Sandstone (sample Geological Survey of Western Australia 142942), Croydon Well: (**a**) concordia plot showing most zircon analyses clustered at c. 2994 Ma; (**b**) Gaussian summation probability density plot. (From Nelson 2000; with Geological Survey of Western Australia permission)

overlies the Cattle Well Formation (Shay Gap greenstone belt), its depositional age is inferred to be less than c. 2988 Ma, but in other areas, it unconformably overlies the 3022–3015 Ma Cleaverville Formation. The minimum depositional age of the Lalla Rookh Sandstone is between 2940 and 2930 Ma based on the observation that it is cut by major faults of the Lalla Rookh–Western Shaw Structural Corridor (Fig. 1.7). Therefore, the Lalla Rookh Sandstone is laterally equivalent to the Constantine Sandstone and the Mallina Formation.

The Bookingarra Group of the northwest shelf overlies a c. 2955 Ma unconformity (Pike and Cas 2002; Pike et al. 2006; Hickman 2016) and is separated from the central trough by the Loudens Fault. Although the group has the same depositional age as the upper part of the Croydon Group, the only good stratigraphic correlation across the Loudens Fault involves the c. 2945 Kialrah Rhyolite which is present in the upper parts of both successions.

9.2.6.1 Cattle Well Formation

The Cattle Well Formation is a mixed epiclastic and volcaniclastic succession interpreted by Williams (1999) to have been deposited in a shallow-marine basin or shelf. In the type area around Cattle Well (Shay Gap greenstone belt), the formation comprises a lower unit of thinly bedded sandstone, feldspathic sandstone, and lithic and feldspathic wacke, locally intercalated with siltstone and shale. Middle and upper units of the formation contain layers of felsic tuffaceous and volcaniclastic material, most likely from volcanic activity in the c. 3000–2990 Ma Whim Creek Group. Bedded blue-grey tuff, welded tuff and re-sedimented tuffaceous sandstone, siltstone, wacke, and chert are interbedded with the epiclastic sedimentary rocks. Some tuffs carry abundant shards of devitrified volcanic glass (Williams 1999). Thinly bedded, grey-white dolomitic carbonate and calcareous shale are present near the base of the formation. An isolated outcrop of blue and white chert and chert breccia, interpreted to be a silicified carbonate unit, contains poorly preserved columnar stromatolites up to 10 cm high and 5 cm wide (Williams 2000).

Zircon U–Pb dating of a quartzofeldspathic sandstone revealed detrital zircons with most $^{207}Pb/^{206}Pb$ dates ranging from 3583 to 3289 Ma (GSWA 142867, Nelson 1999). Three other analyses (<5% discordant) returned $^{207}Pb/^{206}Pb$ dates of 3053, 3048, and 3038 Ma. Dating of a similar sandstone from the same formation indicated a maximum depositional age of 2988 ± 5 Ma (Wingate et al. 2009a, b, GSWA 180048). Zircon analyses (<10% discordant) from this sample indicated ages up to c. 3529 Ma.

9.2.6.2 Lalla Rookh Sandstone

The Lalla Rookh Sandstone is the most important stratigraphic unit of the Croydon Group on the southeast shelf of the Mallina Basin. Deposition of this sandstone formation was widespread across the East Pilbara, reaching thicknesses up to 3 km.

Thick conglomerate–sandstone units are preserved at the same stratigraphic level in nine greenstone belts (East Strelley, Panorama, North Shaw, Tambina, Doolena Gap, Coongan, Kelly, Goldsworthy, and Shay Gap). Late Mesoarchean doming of the EPT has left the formation preserved in the cores of synclines (Hickman 1983), whereas it was eroded from areas across the centres of the domes. Because deposition was occurring during doming, there is local evidence of paleocurrents into sinking basins between the rising and eroding domes (Krapež 1984).

In the type area east of Lalla Rookh mine, the Lalla Rookh Sandstone (Lipple 1975) is a c. 3-km-thick succession of feldspathic sandstone, conglomerate, and shale (Hickman and Lipple 1975; Hickman 1983; Krapež 1984; Krapež and Barley 1987; Van Kranendonk and Collins 1998; Van Kranendonk 2000). In this area, the formation is strongly deformed by tight to isoclinal upright folds and faults in a synclinal structure that has been referred to as the Lalla Rookh Syncline (Hickman and Lipple 1975; Eriksson et al. 1988) or Lalla Rookh Synclinorium (Van Kranendonk et al. 2002). The northwest margin of this northeast–southwest trending syncline is partly replaced by the Lalla Rookh–Western Shaw Fault (Fig. 1.4). Sinistral strike-slip movement on this c. 2940 Ma fault post-dates deposition of the Lalla Rookh Sandstone because the formation is also exposed northwest of the fault in the East Strelley greenstone belt (Hickman 2010, 2016).

East of the Lalla Rookh Syncline, the Lalla Rookh Sandstone is preserved in the core of a major east–west trending syncline between the North Pole and Muccan Domes (Fig. 1.7). Farther east, the formation is mainly concealed by the Fortescue Group, but it continues into the core of the syncline between the Shaw and Mount Edgar Domes and is exposed within this syncline in the northern Coongan greenstone belt (Hickman 1980b, 2010, 2016; Hickman and Van Kranendonk 2008). On the northern side of the Shaw Dome, the formation is preserved in a faulted syncline between the Shaw and North Pole Domes. This outcrop of the Lalla Rookh Sandstone is 10 km long and up to 2 km wide. The succession is at least 500 m thick and is composed of coarse-grained sandstone and pebble to boulder conglomerate (Fig. 9.8) (Van Kranendonk (1999, 2000). On the western side of the Shaw Dome, the Lalla Rookh Sandstone outcrops at the southern end of the Keep It Dark Synclinorium (Van Kranendonk 2003). The succession here is dominated by polymictic conglomerate interbedded with pebbly to coarse-grained feldspathic sandstone.

The stratigraphic thickness of the Lalla Rookh Sandstone in the East Strelley greenstone belt is c. 1.5 km (Wilhelmij and Dunlop 1984), and the formation overlies the Cleaverville Formation across an erosional unconformity. The lower 0.6–1.3 km of the formation comprises intercalated packages of conglomerate and trough cross-bedded sandstone and pebbly sandstone (Wilhelmij and Dunlop 1984). These authors reported that this lower part of the formation was deposited in alluvial fans of upward-fining sandstones in which detritus was derived from the underlying Cleaverville Formation, quartzite (Farrel Quartzite), and greenstones. Higher in the succession, increasingly sericite-rich sandstones were interpreted to indicate a change to granitic sources. This observation is consistent with progressively

Fig. 9.8 Conglomerate and breccia at the base of the Lalla Rookh Sandstone in the Gorge Range, consisting of angular to rounded clasts of the underlying Farrel Quartzite in a silicified fine-grained clastic matrix (MGA Zone 50, 761000E 7691200N). (From Van Kranendonk 2010; with Geological Survey of Western Australia permission)

increasing exposure of the granitic cores of rising domes during deposition of the Lalla Rookh Sandstone.

In the Goldsworthy and Ord Range greenstone belts, the Lalla Rookh Sandstone is composed of fine- to coarse-grained sandstone and conglomerate with local mafic volcanic and volcaniclastic units (Smithies 2004). Clastic rocks are mainly of fluvial origin, but subaqueous debris flow deposits are abundant near the base of the formation. In the Ord Range, the basal contact of the Lalla Rookh Sandstone with the underlying Cleaverville Formation (Gorge Creek Group) is an angular unconformity, whereas in the Goldsworthy greenstone belt, it is a disconformity. Units of vesicular and pillowed basalt, mafic volcaniclastic debris flow deposits, mafic breccia, and dolerite sills are assigned to the Salt Well Member. The upper part of the formation is composed of medium- to coarse-grained, trough cross-bedded sandstone, arkosic sandstone, and polymictic conglomerate.

In the Shay Gap greenstone belt, Williams (2003) recorded that the 'Cooragoora Formation' (now Lalla Rookh Sandstone) includes units of polymictic conglomerate containing cobbles and pebbles derived from the underlying Cleaverville Formation, and he interpreted these as channel deposits. The sandstones in the upper part of the formation exhibit trough cross-bedding and both upward-fining and upward-coarsening sequences. Williams (2003) interpreted the formation to be a high-energy, fluvial–deltaic deposit prograding into a shallow-marine or lacustrine basin.

Detrital zircon ages indicate that detritus for feldspathic sandstone in the upper part of the Lalla Rookh Sandstone came from relatively distal sources. U–Pb zircon dating of a sandstone in the Lalla Rookh Syncline revealed a wide range of zircon

$^{207}Pb/^{206}Pb$ ages between 3660 and 3233 Ma (Nelson 2000, GSWA 142951). In particular, four zircon grains older than 3600 Ma have no known sources in the adjacent Carlindi and North Pole Domes. Similar results were obtained from U–Pb zircon dating of sandstone and conglomerate of the formation in the Goldsworthy greenstone belt (Nelson 2005a, b, GSWA 178019 and 178,020).

9.2.6.3 Constantine Sandstone

The Constantine Sandstone is the lower of two clastic sedimentary formations that make up the 3015–2931 Ma Croydon Group in the central trough of the Mallina Basin. Located between the TTSZ to the southeast and the SSZ and Loudens Fault to the northwest, the trough was a deep-water section of the basin in which a > 5-km-thick succession of quartz sandstone, conglomerate, wacke, and turbidites was deposited (Hickman 1983, 2016). The upper formation of the central trough is mainly composed of the Mallina Formation which is characterized by finer-grained lithofacies than within the Constantine Sandstone.

The Constantine Sandstone is up to 3.5 km thick and is predominantly composed of metamorphosed sandstone and coarse-grained wacke. Due to poor exposure across most of the central trough, the detailed stratigraphic relations of the Constantine Sandstone and Mallina Formation remain to be established. Mapping units of the Constantine Sandstone and Mallina Formation has relied on differences between the lithofacies interpreted to be characteristic of the two formations. However, in detail these lithofacies are vertically repeated in the central part of the stratigraphy implying local intercalation of the formations.

The Croydon Group is folded within the trough, and several large anticlinal cores reveal that the thick sandstone and wacke units of the Constantine Sandstone are underlain by intercalated chert, BIF, and clastic sedimentary rocks including conglomerate. This lithological association is interpreted to be a sedimentary transition from the underlying Gorge Creek Group that includes BIF. The largest such exposure of the Gorge Creek Group is in the Ord Ranges about 70 km east of Port Hedland.

9.2.6.4 Mallina Formation

In better exposed areas of the Mallina Basin, the Mallina Formation conformably overlies the Constantine Sandstone although it is likely that some units currently assigned to these formations are intercalated on a regional scale. The Mallina Formation is between 2.5 and 5.0 km thick and is predominantly composed of metamorphosed fine- to medium-grained wacke and shale. The wacke is chloritic and contains abundant plagioclase, suggesting mafic to intermediate sources. The limited exposures available display excellently preserved sedimentary structures of the type found in turbidite deposits, such as graded bedding, cross-bedding, and scour structures. A general regional trend to increasingly fine-grained sedimentary facies in the northern and eastern parts of the Mallina Basin suggests an overall transition to deeper water, basin plain environments (Smithies et al. 1999). The main lithologies are interbedded shale, siltstone, sandstone, and medium- to fine-grained wacke, turbiditic wacke with minor pebble beds and shale, wacke with abundant chert clasts and local conglomerate, laminated shale with minor beds of poorly sorted subarkose, ferruginous siltstone, shale and minor BIF, and siliciclastic sedimentary rocks, containing layers of hornblende-porphyritic dacite.

Shale is metamorphosed and in many outcrops is cleaved or replaced by pelitic schist. The rock is variably ferruginous and contains angular silt-sized grains of chert and quartz; plagioclase is rare. Wacke is typically poorly sorted and contains sand- to silt-sized angular to sub-rounded grains and fining-upward beds. The proportion of quartz exceeds that of feldspar. Lithic fragments are abundant, particularly in coarser grained varieties. The dominant lithic component is grey chert, but fragments of fine-grained and tuffaceous felsic volcanic rocks, shale, and basalt are also present. The matrix is rich in clay minerals and chlorite, with lesser quartz, plagioclase, biotite, epidote, zoisite, and pyrite. The rocks are commonly iron stained, and some are strongly carbonate altered. Coarse-grained wacke is widely exposed in the central and western parts of the Mallina Basin. Southeast and east of the Satirist Monzogranite, the unit locally contains interbeds up to 1 m thick of fine- to medium-grained wacke that might be of volcaniclastic origin (Smithies and Farrell 2000).

Thin units of siliceous komatiitic basalt with pyroxene spinifex textures belong to the South Mallina Basalt Member. The basalt is typically variolitic, commonly pyroxene spinifex-textured, and locally pillowed, and hyaloclastic breccia is locally present. Siliciclastic sedimentary rocks, containing layers of hornblende-porphyritic dacite, form parts of the Mallina Formation close to the Mallina Shear Zone, east of the Peawah Granodiorite, and in the northern section of the trough. Smithies (1998b) described the assemblage as consisting of layers and sills of feldspar(–hornblende) porphyry of andesitic composition, interlayered with shale and wacke. The porphyry consists of subhedral to euhedral phenocrysts of hornblende and plagioclase (up to 3 and 5 mm in size, respectively), in a fine-grained groundmass of quartz and plagioclase.

Sedimentary rocks of the Mallina Formation close to granitic intrusions have been recrystallized by contact metamorphism. Near the Portree Granite, shale and wacke have been converted to hornfels, containing porphyroblasts of andalusite or cordierite or both (Smithies 1999). Adjacent to the Nerrely Monzogranite, contact metamorphism has produced biotite-rich rocks, containing abundant rosettes of sillimanite, and less common porphyroblasts of cordierite (Smithies 1998a). Near to the Peawah Granodiorite, shale of the Mallina Formation has been metamorphosed to quartz–garnet–cordierite–sillimanite–biotite schist (Smithies 1998a).

9.2.7 *Whim Creek Group*

The Whim Creek Group is a succession of mafic and felsic volcanic and intrusive rocks that unconformably overlies the 3022 to 3015 Ma Cleaverville Formation (Gorge Creek Group) and is unconformably overlain by the c. 2955 Ma Cistern Formation (Bookingarra Group). Most outcrops of the group are restricted to the Whim Creek greenstone belt, although more poorly exposed sections of the group are preserved in the Peawah Hill greenstone belt north of the SSZ. The group comprises two mainly volcanic formations—in ascending stratigraphic order: the Warambie Basalt and the Red Hill Volcanics.

Within the 80-km-long Whim Creek greenstone belt, the stratigraphy of the group changes from the southwestern section where the Warambie Basalt is dominant to the northeastern section where only units of the Red Hill Volcanics are exposed. Both formations are exposed in the Red Hill area, where there is minor lateral intercalation between the upper Warambie Basalt and the lower Red Hill Volcanics. The Warambie Basalt is up to 500 m thick and comprises a succession of vesicular and pillowed basalt flows, units of basaltic breccia and basaltic hyaloclastite, and local intercalations of conglomerate, sandstone, and shale. The Red Hill Volcanics is up to 1.3 km thick, although most of this thickness is due to inclusion of the intrusive Mons Cupri Dacite Member. Non-intrusive sections of the Red Hill Volcanics are composed of felsic volcaniclastic rocks, dacite, rhyolite, and volcanic breccia.

Geochemical data are consistent with a continental volcanic intra-arc setting inland from a subduction zone. The group is interpreted to be part of the magmatic arc represented by granitic intrusions of the Maitland River Supersuite (Chap. 8). Isotopic data are consistent with field observations that the group was deposited on continental crust.

9.2.7.1 Warambie Basalt

In the central part of the Whim Creek greenstone belt, the Warambie Basalt unconformably overlies amphibolite facies metabasalt of the 3130–3110 Ma Whundo Group and granitic rocks of the 3130–3093 Ma Railway and 3024–3012 Ma Orpheus Supersuites. The unconformity with the Whundo Group is well exposed between Mount Fraser and Little George River (Fitton et al. 1975; Hickman 1977; Barley 1987; Smithies 1998b; Krapež and Eisenlohr 1998). The base of the formation includes polymictic conglomerate with a fragmental matrix. The conglomerate is clast-supported with granite and basalt boulders up to 1 m in diameter (Fig. 9.9). The basalt was most likely derived from erosion of the underlying Whundo Group. Farther southwest near Mount Oscar, the lower part of the Warambie Basalt includes thick units of conglomerate and sandstone deposited in alluvial fans immediately south of the Sholl Shear Zone. This suggests that this major strike-slip fault included a component of vertical movement and formed a south-facing fault scarp at c. 3000 Ma (Hickman 2002, 2016).

Fig. 9.9 Basal boulder conglomerate of the Warambie Basalt near Red Hill in the Whim Creek greenstone belt. Angular boulders of basalt and granite are set in a poorly sorted sandstone/granulestone matrix (from Hickman et al. 2010; with Geological Survey of Western Australia permission)

Most of the <500 m thickness of the formation is composed of vesicular and pillowed basalt, plagioclase-porphyritic basalt, basaltic tuff, breccia, and hyaloclastite. In the type area of the Warambie Basalt, around Warambie Homestead, the formation is 200 m thick and composed of vesicular, amygdaloidal, and pillowed basalt and basaltic andesite with rare intercalations of chert and thin felsic volcaniclastic rocks. Farther east, near Red Hill, and immediately adjacent to the Caines Well granitic complex, the thickness of the Warambie Basalt varies from 40 to 150 m, and the formation includes minor intercalations of felsic volcanic rocks that probably represent the early stages of deposition of the Red Hill Volcanics (Smithies 1998b; Pike et al. 2006).

Geochemical data from the Warambie Basalt (Pike 2001; Pike et al. 2006; Smithies et al. 2007a) indicate enrichment in K_2O, Rb, Ba, Th, and Ce in comparison to MORB; although this might be attributed to assimilation of continental crust in a rift system, other evidence for an active convergent margin setting (Chap. 8) is consistent with subduction-related basaltic magmas. Smithies et al. (2007a, b) noted that compared with other subduction-related tholeiites, such as those of the Whundo Group, the Warambie Basalt is enriched in Nb and, in particular, Na_2O (up to 4.5 wt. %), supporting magma interaction with the underlying continental crust. Negative ε_{Nd} values (Arndt et al. 2001; Smithies et al. 2007a) suggest contamination by underlying continental crust or epicontinental sediment. Published Sm–Nd two-stage depleted mantle model ages (T_{DM}^2) of 3370 to 3260 Ma (Arndt et al.

2001; Smithies et al. 2007a, b) are significantly older than for the Whundo Group, consistent with underlying Paleoarchean crust (Hickman 2016).

9.2.7.2 Red Hill Volcanics

The Red Hill Volcanics is up to 1.3 km thick and comprises felsic volcaniclastic rocks, dacite and rhyolite lava, volcanic breccia, and intrusive porphyritic rhyolite and dacite. High-level intrusions of rhyolite and dacite assigned to the Mons Cupri Dacite Member are volumetrically the dominant part of the formation, and locally up to 1.1 km thick.

Stratigraphic sections through the formation vary between different areas of the greenstone belt (Pike et al. 2006), with the intrusive units being emplaced at different levels. Exposures of the Red Hill Volcanics are concentrated in a narrow, 50-km-long belt around the southeast and northeast limbs of the Whim Creek greenstone belt where the Whim Creek Group is underlain by contemporaneous granitic rocks of the Maitland River Supersuite.

Volcaniclastic units of the formation include subaqueous pyroclastic flows, debris flows, turbidites, and fluvial deposits. Clastic sedimentary facies include well-bedded sandstone and siltstone, massive tuffaceous sandstone, polymictic conglomerate, and breccia. Near Red Hill, where both formations of the Whim Creek Group are exposed, the Warambie Basalt is conformably overlain by basal sandstone and siltstone of the Red Hill Volcanics, and these clastic units are overlain by rhyodacite pumice breccia. The basal sandstone and siltstone unit is up to 140 m thick, and the pumice breccia is locally 145 m thick.

The most detailed description of volcanic and sedimentary facies within the Red Hill Volcanics was provided by Pike et al. (2006). Medium- to coarse-grained volcaniclastic rocks were assigned to several sedimentary facies associations, viz. juvenile volcaniclastic facies association, grey breccia and sandstone association, and dacite-associated coarse-grained sedimentary facies. The juvenile volcaniclastic facies association comprises various types of pumice breccia, sandstone, and siltstone and was interpreted by Pike et al. (2006) to be evidence of explosive volcanic activity and deposition from debris flows and turbidity currents. The source of the deposits was considered it to be outside the area of the present outcrops. The grey breccia and sandstone association includes volcaniclastic breccia, coarse-grained dacite-rich breccia, pumice and crystal-rich breccia, sandstone, and siltstone. The inclusion of granitic grains indicates that the source region contained granitic rocks in addition to volcanic rocks. Though not mentioned by Pike et al. (2006), this suggests derivation of at least some detritus from the northwest, in areas that included exposures of the Maitland River Supersuite. The dacite-associated, coarse-grained sedimentary facies includes polymictic conglomerate, imbricate conglomerate, and coarse-grained, dacite-rich breccia. Pike et al. (2006) interpreted this facies association to indicate erosion of the Mons Cupri Dacite Member, and deposition in proximal subaqueous fans, fed either fluvial systems or debris flows.

Evidence provided by volcanic and sedimentary facies (Pike et al. 2006) indicates that most units within the Red Hill Volcanics, but particularly the lower stratigraphic units, were deposited from debris flows and turbidity currents. Detritus composition is consistent with erosion of volcanic and granitic rocks in areas northwest of the Whim Creek greenstone belt. The ages of detrital zircons in samples selected for geochronology support this interpretation.

Geochemical data from the Red Hill Volcanics are consistent with volcanic and sedimentary deposition within a continental magmatic arc (Krapež and Eisenlohr 1998; Smith et al. 1998; Blewett 2002; Beintema 2003; Smith 2003; Hickman 2004, 2012, 2016; Pike et al. 2006; Hickman et al. 2010; Hickman and Van Kranendonk 2012). This interpretation implies that the Whim Creek Group was deposited across a much larger area than is now preserved within the Whim Creek greenstone belt. Isotopic data (Smithies et al. 2007a; Hickman 2016) are consistent with a close genetic relationship between 2999 and 2982 Ma monzogranite in the Caines Well granitic complex and the Whim Creek Group. Two Sm–Nd two-stage depleted mantle model ages from the monzogranite are recorded as c. 3350 and 3310 Ma. Epsilon Nd values from the monzogranites were − 0.81 and − 1.34, consistent with relatively evolved sources.

9.2.7.3 Geochronology

Although the Whim Creek Group has been extensively studied, its precise age remains uncertain. One constraint on the maximum depositional age of the Whim Creek Group is that it unconformably overlies the Cleaverville Formation (Hickman 2002, 2016), which has a minimum depositional age of c. 3015 Ma. Samples from the group have been dated from three localities. A volcaniclastic sample from the Red Hill Volcanics near Red Hill was dated at 3009 ± 4 Ma (GSWA 141936, Nelson 1998). However, this date is now interpreted to be too old because five analyses ($^{207}Pb/^{206}Pb$ ages between 3043 and 3017 Ma) used in the calculation were on zircons apparently derived from felsic rocks of the Orpheus Supersuite or older units (Hickman 2016). Numerous samples from the Orpheus Supersuite, which unconformably underlies the Whim Creek Group, have been dated between 3024 and 3012 Ma.

Barley et al. (1994) reported an upper intercept age of c. 2991 ± 12 Ma for porphyritic dacite collected near Red Hill, and the sample is now interpreted to have come from of the Mons Cupri Dacite Member. The date was based on only three discordant analyses and may not be reliable. The geological relations of the Whim Creek Group link it to the 2999–2982 Ma Maitland River Supersuite. Although granitic rocks commonly give slightly younger ages than their volcanic equivalents, it is likely that the maximum age of the Red Hill Volcanics is close to 3000 Ma. The minimum age of the group is interpreted to be greater than c. 2978 Ma because the Cistern Formation of the Bookingarra Group, which unconformably overlies the Whim Creek Group, contains a large detrital zircon component dated at 2978 ± 5 Ma (Nelson 2000, GSWA 142949). Accordingly, based on the youngest c. 2990 Ma date obtained by Barley et al. (1994), the depositional age of the Whim Creek Group is here interpreted to be between c. 3000 and c. 2990 Ma.

9.2.8 Bookingarra Group

The Bookingarra Group was deposited on the northwest side of the Mallina Basin and comprises five sedimentary and volcanic formations unconformably overlying the Whim Creek Group: in ascending stratigraphic order, the Cistern Formation, Rushall Slate, Louden Volcanics, Mount Negri Volcanics, and Kialrah Rhyolite (Table 1.1). By volume, the Bookingarra Group is mainly volcanic, with the combined successions of the basaltic Louden and Mount Negri Volcanics being up to 2 km thick. Available geochronology and regional geological considerations indicate that the age of the unconformity between the groups is c. 2955 Ma (Hickman 2016). This is the same age as the unconformity in the central part of the Mallina Basin, but the two successions are very different lithologically. The most obvious difference is that the Bookingarra Group is mainly composed of volcanic rocks, whereas the Mallina Formation is almost entirely composed of metasedimentary rocks. The only formation recognized in both successions is the Kialrah Rhyolite, units of which has been dated between c. 2948 and 2941 Ma. This overlies or intrudes the Louden Volcanics at the top of the Bookingarra Group but forms flow-banded rhyolite units in the upper Mallina Basin succession. The depositional age of the Louden and Mount Negri Volcanics is c. 2950 Ma, and it is likely that the Loudens Fault, which separates these volcanic formations from the upper Croydon Group, juxtaposed the successions between c. 2950 and 2945 Ma.

Deposited during the North Pilbara Orogeny, formations of the Bookingarra Group are locally separated by unconformities due to periods of uplift and erosion. Deposition of the Cistern Formation resulted from uplift and erosion of the Whim Creek Group at the beginning of the North Pilbara Orogeny. An upward transition from coarse volcaniclastic facies to finer-grained siliciclastic facies suggests a change from relatively shallow-water deposition, adjacent to eroding outcrops of the Whim Creek Group, to deposition in deeper water more distal from sediment supply. Deposition of shale and tuff of the overlying Rushall Slate continued this trend to increasingly deep-water environments, with regular influx of turbidity currents. The observation that pillow structures are rare in the lower part of the overlying Louden Volcanics (Smithies 1998b) suggests uplift of the Cistern–Rushall Slate succession prior to the basaltic volcanism. Such uplift might have occurred above a mantle plume, or because of folding and faulting during the North Pilbara Orogeny. Thin units of sandstone and conglomerate near the base of the Mount Negri Volcanics suggest deposition from locally uplifted areas, possibly near fault scarps. Major lateral thickness and facies changes in the Louden and Mount Negri Volcanics suggest either that deposition was on very uneven topography or that there was substantial syndepositional uplift and erosion (Barley 1987). Volcanism and subvolcanic intrusion of the Kialrah Rhyolite probably occurred during rifting in the

Mallina Basin, and locally in areas of crustal extension adjacent to c. 2945 Ma strike-slip faulting.

9.2.8.1 Cistern Formation

The Cistern Formation is the basal unit of the Bookingarra Group and overlies the Whim Creek Group along an erosional unconformity (Pike and Cas 2002). The formation is up to 600 m thick but is laterally variable in terms of thickness and sedimentary facies (Pike et al. 2006). A thick section of the formation is exposed 5 km southwest of Whim Creek at Mons Cupri where volcanic breccia and coarse conglomerate are associated with a small volcanic centre containing VMS mineralization. Two levels of mineralization are present: a lower, funnel-shaped, Cu-rich disseminated stockwork overlain by a stratiform, 5–10-m-thick Pb–Zn–Ag zone (Low 1963; Miller and Gair 1975; Huston et al. 2000; Huston 2006). The conglomerate at Mons Cupri is poorly sorted with sub-angular to locally rounded clasts and blocks up to 10 m across. The clasts are mainly composed of rhyolite, granite, and lesser basalt. The matrix of the conglomerate contains rhyolite shards, and the conglomerate is intruded by felsic sills, dykes, and domal bodies (Huston et al. 2000) providing evidence of volcanism during conglomerate deposition. The volcanic breccia and conglomerate are overlain by an upward-fining sandstone succession.

Northeast and southwest from Mons Cupri, the volcanic breccia and coarse conglomerate pass laterally into pebble conglomerate and chert breccia overlain by sandstone. In the Good Luck Well area, 10 km southwest of Mons Cupri, the basal conglomerate is about 6 m thick and includes clasts of dacite, basalt, and granite. This is overlain by over 300 m of sandstone, most of which contains clasts of black chert and minor volcanic clasts. Ripple-bedded sandstone is present in the upper part of this succession.

9.2.8.2 Rushall Slate

The Rushall Slate conformably overlies the Cistern Formation via a transition succession, and the contact between the formations is generally difficult to define. The formation reaches a maximum thickness of approximately 300 m and is mainly composed of slate (metamorphosed shale) with minor metamorphosed siltstone and sandstone. Tuffaceous beds and chert locally form minor parts of the formation. Graded bedding in the metamorphosed shale of the formation indicates deposition from distal turbidity currents in relatively deep water (Pike et al. 2002). Isolated outcrops of the formation along the Whim Creek greenstone belt have led to interpretations of deposition in local basins, but there is no direct evidence for this. The Rushall Slate was partly eroded prior to deposition of the Louden and Mount Negri Volcanics (Hickman 1983), and along much of greenstone belt, these thick basaltic formations overlie and conceal the formation. The original depositional extent of the Rushall Slate is likely to have extended far outside the boundaries of the present outcrops.

9.2.8.3 Louden Volcanics

The Louden Volcanics is a 1- to 2-km-thick basaltic formation that outcrops over almost 50% of the Whim Creek greenstone belt. The formation is composed of weakly metamorphosed komatiitic basalt with minor komatiite and flows of massive basalt, pillow basalt, and hyaloclastite near the top of the formation. The upper basaltic section is locally interbedded with medium- to fine-grained sandstone, beds and lenses of shale, and units of polymictic conglomerate. The komatiitic basalt contains coarse pyroxene spinifex textures or randomly orientated, acicular pyroxene phenocrysts, whereas the komatiite consists of plates of skeletal olivine (partly altered to serpentine) set in a groundmass of feathery pyroxene (replaced by amphibole and chlorite). Olivine and pyroxene cumulate zones in the komatiite flows contain up to 32 wt.% MgO, 4500 ppm Cr, and 1150 ppm Ni.

The presence of komatiite and komatiitic basalt in the Louden Volcanics suggests a mantle plume origin; this interpretation was favoured by Arndt et al. (2001) who argued that the composition of the magmas might be explained by plume-related melting of continental lithosphere. However, plume-related basaltic successions typically extend for hundreds of kilometres, whereas the succession is not known extend southeast into the central section of the Mallina Basin. Therefore, if the Louden and Mount Negri Volcanics were products of a mantle plume, the Loudens Fault and Sholl Shear Zone would be expected to have strike-slip displacements greater than 100 km.

Smithies et al. (2007b) suggested an alternative origin for the Louden and Mount Negri Volcanics. On the basis of relatively narrow ranges of La/Nb (2.7 to 3.7), La/Sm (4.8 to 5.8), La/Zr (0.13 to 0.16), and ε_{Nd} (-0.6 to -2.8) in basalts, they suggested that the mafic magmas were derived from a mantle source containing a homogeneous mix of Paleoarchean continental crust (possibly sedimentary rock) and material chemically similar to subduction-related juvenile crust, such as that of c. 3120 Ma Whundo Group. They considered that the limited compositional range of the Louden and Mount Negri Volcanics was unlikely to be simply the result of crustal contamination.

9.2.8.4 Mount Negri Volcanics

The Mount Negri Volcanics is preserved in several small outliers along the length of the Whim Creek greenstone belt. The main outcrops of the formation are at Mount Negri, between Mons Cupri and Hill Well, between Stones Well and Wild Dog Bore, and north and south of Mountain Well. The formation forms the tops of several hills in the Whim Creek greenstone belt but has not been recorded outside this greenstone belt. In contrast to the underlying, more steeply dipping succession of the

Louden Volcanics, flow units of the Mount Negri Volcanics are only weakly folded, and bedding inclinations are typically less than 30 degrees. The formation conformably to unconformably overlies the Louden Volcanics and unconformably overlies the Cistern Formation–Rushall Slate succession in the lower part of the Bookingarra Group and the Red Hill Volcanics of the underlying Whim Creek Group.

The Mount Negri Volcanics is mainly composed of variolitic and vesicular basalt. The variolitic basalt, which forms flows up to 15 m thick, contains small-scale pyroxene spinifex texture within basaltic flows characterized by abundant dark grey-green, pea-sized varioles consisting of acicular clinopyroxene, interstitial plagioclase, and devitrified glass (Smithies 1998b). Euhedral clinopyroxene phenocrysts up to 2 mm in length are distributed randomly throughout varioles and groundmass. Epidote, actinolite, carbonate, and chlorite are common replacement minerals. Smithies (1998b) commented that the combination of rare scoria deposits including bombs up to 20 cm in diameter, a lack of pillow structures, and the presence of hyaloclastite suggest an emergent depositional environment.

The Louden and Mount Negri Volcanics are both characterized by high primitive mantle normalized $(La/Nb)_{PM}$ and $(La/Yb)_{PM}$, negative epsilon Nd values between -1.5 and -4.3, and Sm–Nd two-stage depleted mantle model ages ($T_{DM}{}^2$) between c. 3500 and 3300 Ma (Smithies et al. 2004). These data suggest that the magmas were partly derived from a Paleoarchean crustal component. Smithies et al. (2007a) found that the Mount Negri and Louden Volcanics have very similar trace elements trends to thin basaltic units in the central part of the Mallina Basin and to the Coonieena Basalt in the Shay Gap–Goldsworthy area of the Northeast Pilbara Craton. When compared to the Louden Volcanics, the Mount Negri Volcanics contain consistently higher concentrations of TiO_2, Na_2O, P_2O_5, REE (La, Ce, Pr, Nd, Sm, Eu, Gd, Tb, Dy), HFSE (Nb, Hf, Zr, Y, Th), and the chalcophile elements Cu and Zn but consistently lower MgO and CaO contents and lower Mg#. Plots of Zr vs. Sr, Zr vs. La, Nb vs. La, and Sm vs. La show no overlaps in composition.

9.2.8.5 Kialrah Rhyolite

The c. 2945 Ma Kialrah Rhyolite is the youngest formation of the Bookingarra Group and includes extrusive and intrusive units. The formation outcrops in the Whim Creek greenstone belt and in the Mallina Basin east of the Ord Range and north-northeast of Egina Well in the central trough of the Mallina Basin. This widespread distribution is attributed to its deposition and intrusion after major strike-slip movements on the Loudens Fault and Sholl Shear Zone. These movements juxtaposed the Whim Creek greenstone belt and the central Mallina Basin during intrusion of the Sisters Supersuite.

The largest single outcrop of the Kialrah Rhyolite is located 2 km south of Warambie Homestead where the formation is exposed as a 1-km-thick unit of flow-banded and plagioclase-porphyritic rhyolite stratigraphically overlying the Louden Volcanics. Large sills or flows of the Kialrah Rhyolite 25 km east of the Ord Range intrude felsic volcaniclastic rocks and metasedimentary rocks within the Mallina Formation. The least-deformed rhyolite displays flow banding, and margins to individual units locally show both flow brecciation and hyaloclastite. Approximately 6 km north-northeast of Egina Well, sparsely porphyritic, flow-banded dacite forming low outcrops within the Mallina Formation is correlated with the Kialrah Rhyolite. Quartz-porphyritic rhyolite that intrudes the Cistern Formation at Salt Creek is also assigned to the Kialrah Rhyolite.

Sm–Nd isotope data from the Kialrah Rhyolite are similar to data from the Louden and Mount Negri Volcanics and suggest that the rhyolite was derived by partial melting of Paleoarchean crust. This is consistent with the interpretation that by c. 2945 Ma Paleoarchean crust of the East Pilbara and Karratha Terranes had been amalgamated by convergence and plate collision beneath the Mallina Basin.

9.2.8.6 Geochronology

The maximum depositional age of the Bookingarra Group is c. 2955 Ma because it was deposited on the erosional unconformity that separates the group from the underlying Whim Creek Group (Pike and Cas 2002; Hickman 2016). The age of this unconformity in the Whim Creek greenstone belt is inferred to be the same age as the D6 deformation event in the adjacent Mallina Basin (Hickman 2016). D6 is equivalent to the previously described 'D2' event of Smithies et al. (2002). That event was immediately followed by intrusion of a 2954–2945 Ma high-Mg diorite suite of granites (Smithies and Champion 2000), now referred to as the Indee Suite (Van Kranendonk et al. 2006). Major 'D2' structures in the Mallina Basin, such as the Croydon Anticline (D6 in Hickman 2016), are crosscut by the Indee Suite (Smithies et al. 2001a, 2002).

The minimum depositional age of the Bookingarra Group is constrained by dating of the youngest formation, the Kialrah Rhyolite. South of Warambie Homestead, the Kialrah Rhyolite was initially dated at 2975 ± 4 Ma (Nelson 1998, GSWA 144261). However, based on the presence of a younger zircon group within the sample, Van Kranendonk et al. (2006) reinterpreted the crystallization age as 2943 ± 7 Ma. This younger age closely coincides with dates on similar rhyolite in the Croydon Group of the Mallina Basin. A sample of flow-banded rhyolite within the Mallina Formation east of the Ord Range was dated at 2948 ± 3 Ma (Nelson 2002, GSWA 169025). Rhyolite situated north-northeast of Egina Well was dated at 2941 ± 4 Ma (Nelson 2000, GSWA 142949). All three dated units of rhyolite are interpreted to be parts of the Kialrah Rhyolite. The average age of this formation is therefore estimated at c. 2945 Ma.

9.3 Tectonic Evolution of the De Grey Superbasin

Tectonic, magmatic, and depositional processes between the c. 3070 Ma Prinsep Orogeny and the end of the North Pilbara Orogeny at c. 2919 Ma were directly related to ongoing collisional interactions of the Northwest Pilbara Craton with an exotic plate that had been converging from at least c. 3160 Ma. Subduction along the northwest margin of the Pilbara Craton was inferred in some of the earliest tectonic interpretations of the region (Krapež and Eisenlohr 1998; Smith et al. 1998; Blewett 2002; Beintema 2003; Smith 2003) although associated magmatic arcs had not been identified and there were alternative interpretations that the Mallina and Whim Creek Basins were intracontinental rift basins (Smithies et al. 1999, 2001b; Hickman et al. 2000; Smithies and Champion 2000; Pike and Cas 2002; Pike et al. 2002; Van Kranendonk et al. 2001, 2002). Geochronological evidence of southeast migrating magmatic arcs (Fig. 8.6) following closure of the Regal Basin at c. 3070 Ma has provided a more complete understanding of the evolution of the De Grey Superbasin.

Surviving remnants of the magmatic arcs are preserved as the 3024–3007 Ma Orpheus Supersuite, the 2999–2982 Ma Maitland River Supersuite, and the 2954–2913 Ma Sisters Supersuite. Detrital zircon ages in formations deposited after the Prinsep Orogeny suggest the existence of an earlier, c. 3067 to 3050 Ma supersuite. However, this has not yet been identified by mapping and geochronology, and it might be concealed off the Northwest Pilbara coast. Large retro-arc basins southeast of the magmatic arcs include the 3067–3015 Ma Gorge Creek Basin (sedimentary), the 3000–2990 Ma Whim Creek Basin (volcano-sedimentary), and the 3015–2931 Ma Mallina Basin (mainly sedimentary) (Hickman 2016). Original stratigraphic relations between the Whim Creek and Mallina Basins are uncertain due a major strike-slip fault (Loudens Fault) along the contact; this fault is interpreted to have juxtaposed successions deposited in separate areas of the Northern Pilbara Craton, probably at least 100 km apart. Likewise, the 2955–2941 Ma Bookingarra Group, assigned to the northwest shelf of the Mallina Basin, is in tectonic contact with the central part of the basin. The volcanic formations of this group have a total thickness of 2 km, but there are no volcanic units >100 m thick in the central part of the Mallina Basin.

Northwest–southeast convergence during evolution of the De Grey Superbasin was the cause of most of the complex structural geology of the Central Pilbara Tectonic Zone (Fig. 1.3), but deformation also occurred in the East Pilbara. The earliest structures recognized are syndepositional extensional faults in the Farrel Quartzite (Van Kranendonk 2004, 2010; Van Kranendonk et al. 2006) and Cleaverville Formation of the Gorge Creek Basin. Smithies et al. (2004) reported fractures in the Cleaverville Formation on the Northeast Pilbara through which hydrothermal fluids deposited layers of dark-grey, fine-grained silica up to 0.5 m thick. In the Northwest Pilbara, numerous dolerite, dacite, and granophyre sills and dykes within the Cleaverville Formation are interpreted to have filled extensional fractures during intrusion of the Orpheus Supersuite; Hickman (2016) assigned these extensional structures to D3 in the Northwest Pilbara. Later structures in the Northwest Pilbara, assigned to D4 (Hickman 2016), are upright, tight to isoclinal folds in the Cleaverville Formation. At Mount Ada, 15 km south of Roebourne, east-southeast-trending upright folds deform the Cleaverville Formation and a sill of the c. 3014 Ma Stone Yard Granophyre (Orpheus Supersuite). The minimum age of these structures is constrained by the fact that they are unconformably overlain by the c. 3000 Ma Warambie Basalt (Whim Creek Group). The proximity these folds to the Sholl Shear Zone suggests a 3015–3000 Ma phase of movement along the shear zone due to north–south compression. Locally thick sandstone units in the Warambie Basalt adjacent to the Sholl Shear Zone indicate deposition at the base of a fault scarp (Hickman 2002).

Low-angle thrusts in the Warambie Basalt east of Mount Ada (Hickman 2002), likely to be local structures produced by c. 3000 Ma movement along the SSZ, suggest ongoing north–south compression of the Northwest Pilbara. No thrusting has been recognized in the 3000–2990 Ma Red Hill Volcanics overlying the Warambie Basalt. The deformation of the Warambie Basalt was assigned to D5 by Hickman (2016), and isoclinal folds in the lower part of the Croydon Group in the Mallina Basin might also be D5 structures. Smithies (1998a) recorded early east–west trending folds in the Mallina Basin that deform the Constantine Sandstone and underlying units of the Gorge Creek Group, and which are folded by later north-trending folds such as the D6 Powereena Anticline. The D5 folds are isoclinal and were possibly formed as recumbent folds. North- to north-northeast-trending tight, upright, and locally overturned folds (D2 of Smithies 1998a; D6 of Hickman 2016) in the central part of the Mallina Basin include the Powereena Anticline, Croydon Anticline, and large folds northwest of the Croydon Anticline and south of the Satirist Monzogranite. A steep axial plane schistosity (S6) crosscuts refolded D5 folds. The c. 2948 Ma Peawah Granodiorite cuts across the D6 folds and the structures are inferred to have formed at c. 2955 Ma. D6 and subsequent deformation, metamorphism, and igneous intrusion in the Northwest Pilbara formed part of the 2955–2919 Ma North Pilbara Orogeny.

In the East Pilbara, the main deformation event between c. 3015 and 2955 Ma was a major reactivation of the dome–and–keel structures that had formed between 450 and 200 million years earlier in the EPT east of the Lalla Rookh–Western Shaw Structural Corridor. It is evident that the Mesoarchean doming occurred after deposition of the 3022–3015 Ma Cleaverville Formation because this is strongly deformed around the margins of many domes. Evidence that deformation of the Cleaverville Formation commenced shortly after its deposition is provided by an erosional unconformity separating it from the c. 3015 Ma Cundaline Formation in the Marble Bar greenstone belt (Williams 1999). The minimum age of the doming is poorly constrained and may have differed between domes. In some greenstone belts, the Cleaverville Formation is inclined steeply away from the centres of domes, but in others it is subvertical within narrow, deep graben-like structures of the type described by Hickman (2001a). An excellent example of this is at Coppin Gap in the Marble Bar greenstone belt where subvertical and tight to isoclinally folded beds of the Cleaverville Formation (Fig. 9.10) occupy the faulted boundary between the

Fig. 9.10 Isoclinally folded jaspilitic BIF of the Cleaverville Formation at Coppin Gap (previously unpublished photograph; with Geological Survey of Western Australia permission)

Mount Edgar and Muccan Domes. Vertical deformation of the East Pilbara crust is not readily related to the horizontal deformation in the Northwest Pilbara during the De Grey Event. Instead, it is likely that the Mesoarchean doming was a response to the same crustal conditions that produced the Paleoarchean doming: namely, gravitational pressures on the crust (in this instance, resulting from deposition of the Gorge Creek Group) combined with crustal heating. Rising temperatures in the Mesoarchean East Pilbara crust are likely through an absence of volcanic heat loss between c. 3165 and 3000 Ma, and for reasons given in the 'conductive incubation' model of Sandiford et al. (2004). Conductive incubation involved the burial of radiogenic heat-producing elements beneath an accumulating supracrustal succession.

Deformation in the East Pilbara related to the northwest–southeast compression of the craton includes tight to isoclinal folding in the Wodgina, Pilbara Well, Cheearra, East Strelley, Pincunah, and Goldsworthy greenstone belts. It is notable that all this deformation, assigned to East Pilbara deformation D13 by Hickman (2021a), was confined to those parts of the EPT west of the Lalla Rookh–Western Shaw Structural Corridor where extension during the 3280–3165 Ma East Pilbara Terrane Rifting Event is interpreted to have substantially reduced the thickness of the crust. This distribution means that there is no obvious interference between these horizontal D13 structures and the Mesoarchean domes, although they appear to have formed at about the same time. The original orientation of the tight to isoclinal folds indicates north to south or north-northeast to south-southwest compression. Folds and faults that deform D13 structures in greenstone belts west of the Lalla Rookh–Western Shaw Structural Corridor are assigned to D14 (Hickman 2021a). The folds trend north-northeast and are upright, open to tight, and have subvertical axial planes (Van Kranendonk et al. 2010). A crenulation cleavage is locally developed where D14 deforms S13. Based on a structural correlation with D6 folds in the Mallina Basin, the age of D14 structures is interpreted to be about 2955 Ma, and part of the

North Pilbara Orogeny (Chap. 10). The difference between the trend of D13 and D14 indicates west-northwest to east-southeast compression.

References

Arndt NT, Bruzak G, Reischmann T (2001) The oldest continental and oceanic plateaus: geochemistry of basalts and komatiites of the Pilbara Craton, Australia. In: Ernst RE, Buchan K (eds) Mantle plumes: their identification through time. Geological Society of America, pp 359–387

Bagas L, Van Kranendonk MJ, Pawley MJ (2002) Split Rock, WA Sheet 2854. Geological Survey of Western Australia

Bagas L, Van Kranendonk MJ, Pawley MJ (2004) Geology of the Split Rock 1:100 000 sheet. 1: 100 000 Geological Series Explanatory Notes. Geological Survey of Western Australia

Barley ME (1987) The Archaean Whim Creek Belt, an ensialic fault-bounded basin in the Pilbara Block, Australia. Precambrian Res 37:199–215. https://doi.org/10.1016/0301-9268(87)90067-2

Barley ME, McNaughton NJ, Williams IS, Compston W (1994) Geological note: age of Archaean volcanism and sulphide mineralization in the Whim Creek Belt, West Pilbara. Aust J Earth Sci 41:175–177

Beintema KA (2003) Geodynamic evolution of the West and Central Pilbara Craton: A mid-Archean active continental margin. PhD, Geologica Ultraiectina. Utrecht University

Bekker A, Slack JF, Planavsky NJ, Krapež B, Hofmann AW, Konhauser KO, Rouxel OJ (2010) Iron formation: The sedimentary product of a complex interplay among mantle, tectonic, oceanic, and biospheric processes. Econ Geol 105:467–508

Bekker A, Planavsky NJ, Krapež B, Rasmussen B, Hofmann A, Slack JF, Rouxel OJ, Konhauser KO (2014) Iron formations: Their origins and implications for ancient seawater chemistry. In: Turekian KK (ed) Treatise on Geochemistry. Elsevier, Oxford, UK, pp 561–628

Blake TS, Barley ME (1992) Tectonic evolution of the Late Archaean to Early Proterozoic Mount Bruce megasequence set, Western Australia. Tectonics 11:1415–1425

Blewett RS (2002) Archaean tectonic processes: a case for horizontal shortening in the North Pilbara Granite–Greenstone Terrane, Western Australia. Precambrian Res 113:87–120

Dawes PR, Smithies RH, Centofanti J, Podmore DC (1995a) Sunrise Hill unconformity: a newly discovered regional hiatus between Archaean granites and greenstones in the Northeastern Pilbara Craton. Aust J Earth Sci 42:635–639

Dawes PR, Smithies RH, Centofanti J, Podmore DC (1995b) Unconformable contact relationships between the Muccan and Warrawagine batholiths and the Archaean Gorge Creek Group in the Yarrie mine area, Northeast Pilbara. Record, 1994/3. Geological Survey of Western Australia

Duuring P, Teitler Y, Hagemann SG (2017) Banded iron formation-hosted iron ore deposits of the Pilbara Craton. In: Phillips GN (ed) Australian Ore Deposits. Australasian Institute of Mining and Metallurgy, pp 345–350

Duuring P, Teitler Y, Hagemann S, Angerer T (2016) MRIWA Report Project M426: Exploration targeting for BIF-hosted iron deposits in the Pilbara Craton, Western Australia. Report, vol 163. Geological Survey of Western Australia

Eriksson KA, Kidd WSF, Krapez B (1988) Basin analysis in regionally metamorphosed and deformed early Archean terrains: examples from southern Africa and Western Australia. In: Kleinspehn K, Paola C (eds) New perspectives in basin analysis. Springer-Verlag, New York, pp 371–404

Fitton MJ, Horwitz RC, Sylvester G (1975) Stratigraphy of the early Precambrian in the West Pilbara. Western Australia, Report, FP, p 11

Gauger T, Konhauser K, Kappler A (2015) Protection of phototrophic iron(II)-oxidizing bacteria from UV irradiation by biogenic iron(III) minerals: implications for early Archean banded iron formation. Geology 43:1067–1170

Gross GA (1980) A classification of iron formations based on depositional environments. Can Mineral 18:215
Hickman AH (1977) Stratigraphic relations of rocks within the Whim Creek Belt. In: Annual report for the year 1976. Geological Survey of Western Australia, pp 53–56
Hickman AH (1980a) Excursion guide–Archaean geology of the Pilbara Block: Second International Archaean Symposium, Perth 1980. Geological Society of Australia, W.A. Division, Perth, Western Australia
Hickman AH (1980b) Lithological map and stratigraphic interpretation of the Pilbara Block. Geology if the Pilbara Block and its environs. Geological Survey of Western Australia
Hickman AH (1983) Geology of the Pilbara Block and its environs. Bulletin 127. Geological Survey of Western Australia
Hickman AH (1990) Excursion No 5: Pilbara and Hamersley Basin. In: Ho SE, Glover JE, Myers JS, Muhling JR (eds) Excursion guidebook: Third International Archaean Symposium, Perth 1990 (17–21 September). The University of Western Australia, Geology Department and University Extension, pp. 1–60
Hickman AH (1997) Dampier, WA Sheet 2256. Geological Survey of Western Australia
Hickman AH (2001a) East Pilbara diapirism: New evidence from mapping. In: Geological Survey of Western Australia (ed) GSWA 2001 extended abstracts: new geological data for WA explorers. Geological Survey of Western Australia, pp 23–25
Hickman AH (2001b) Geology of the Dampier 1:100 000 sheet. 1:100 000 Geological Series Explanatory Notes. Geological Survey of Western Australia
Hickman AH (2002) Geology of the Roebourne 1:100 000 sheet. 1:100 000 Geological Series Explanatory Notes. Geological Survey of Western Australia
Hickman AH (2004) Two contrasting Granite–Greenstones Terranes in the Pilbara Craton, Australia: Evidence for vertical and horizontal tectonic regimes prior to 2900 Ma. Precambrian Res 131:153–172
Hickman AH (2010) Marble Bar, WA Sheet SF50–8, 3rd edn. Geological Survey of Western Australia
Hickman AH (2012) Review of the Pilbara Craton and Fortescue Basin, Western Australia: Crustal evolution providing environments for early life. Island Arc 21:1–31
Hickman AH (2013) North Shaw, WA Sheet 2755, 2nd edn. Geological Survey of Western Australia
Hickman AH (2016) Northwest Pilbara Craton: a record of 450 million years in the growth of Archean continental crust. Report, 160. Geological Survey of Western Australia
Hickman AH (2021a) East Pilbara Craton: a record of one billion years in the growth of Archean continental crust. Report 143, Geological Survey of Western Australia
Hickman AH (2021b) Interpreted bedrock geology of the East Pilbara Craton (version 2). East Pilbara Craton: a record of one billion years in the growth of Archean continental crust. Geological Survey of Western Australia
Hickman AH, Gibson DL (1981) Port Hedland - Bedout Island, WA Sheet SF50–4 and part of Sheet SE50–16, 2nd edn. Geological Survey of Western Australia
Hickman AH, Lipple SL (1975) Explanatory notes on the Marble Bar 1:250 000 geological sheet, Western Australia. Record, 1974/20. Geological Survey of Western Australia
Hickman AH, Lipple SL (1978) Marble Bar, WA Sheet SF50–8, 2nd edn. Geological Survey of Western Australia
Hickman AH, Smithies RH, Huston DL (2000) Archaean geology of the West Pilbara Granite–Greenstone Terrane and the Mallina Basin, Western Australia–a field guide. Record, 2000/9. Geological Survey of Western Australia
Hickman AH, Smithies RH, Tyler IM (2010) Evolution of active plate margins: West Pilbara Superterrane, De Grey Superbasin, and the Fortescue and Hamersley Basins–a field guide. Record, 2010/3. Geological Survey of Western Australia
Hickman AH, Strong CA (2003) Dampier–Barrow Island, Western Australia. 1:250 000 Geological Series Explanatory Notes. Geological Survey of Western Australia

Hickman AH, Van Kranendonk MJ (2008) Archean crustal evolution and mineralization of the Northern Pilbara Craton–a field guide. Record, 2008/13. Geological Survey of Western Australia

Hickman AH, Van Kranendonk MJ (2012) Early Earth Evolution: evidence from the 3.5 – 1.8 Ga geological history of the Pilbara region of Western Australia. Episodes 35:283–297. https://doi.org/10.18814/epiiugs/2012/v35i1/028

Horwitz RC (1990) Palaeogeographic and tectonic evolution of the Pilbara Craton, Northwestern Australia. Precambrian Res 48:327–340

Huston DL, Smithies RH, Sun S-S (2000) Correlation of the Archaean Mallina–Whim Creek Basin: implications for base-metal potential of the central part of the Pilbara Granite-Greenstone Terrane. Aust J Earth Sci 47:217–230

Huston DL (2006) Mineralization and regional alteration at the Mons Cupri stratiform Cu-Zn-Pb deposit, Pilbara Craton, Western Australia. Mineral Deposita 41:17–32

Kiyokawa S, Ito T, Ikehara M, Yamaguchi KE, Koge S, Sakamoto R (2012) Lateral variations in the lithology and organic chemistry of a black shale sequence on the Mesoarchean seafloor affected by hydrothermal processes: the Dixon Island Formation of the coastal Pilbara Terrane, Western Australia. Island Arc 21:118–147

Konhauser KO, Planavsky NJ, Hardisty DS, Robbins LJ, Warchola TJ, Haugaard R, Lalonde SV, Partin CA, Oonk P, Tsikos H, Lyons TW, Bekker A, Johnson CM (2017) Iron formations: a global record of Neoarchaean to Palaeoproterozoic environmental history. Earth Sci Rev 172: 140–177. https://doi.org/10.1016/j.earscirev.2017.06.012

Krapež B (1984) Sedimentation in a small, fault-bounded basin: The Lalla Rookh sandstone, East Pilbara Block. In: Muhling JR, Groves DI, Blake TS (eds) Archaean and Proterozoic basins of the Pilbara, Western Australia: Evolution and mineralization potential. The University of Western Australia, Geology Department and University Extension, pp 89–110

Krapež B (1993) Sequence stratigraphy of the Archaean supracrustal belts of the Pilbara Block, Western Australia. Precambrian Res 60:1–45

Krapež B, Barley ME (1987) Archaean strike-slip faulting and related ensialic basins: evidence from the Pilbara Block. Australia Geol Mag 124:555–567. https://doi.org/10.1017/S0016756800017386

Krapež B, Barley ME, Pickard AL (2003) Hydrothermal and resedimented origins of the precursor sediments to banded iron formation: sedimentological evidence from the Early Palaeoproterozoic Brockman Supersequence of Western Australia. Sedimentology 50:979–1011

Krapež B, Eisenlohr B (1998) Tectonic settings of Archaean (3325–2775 Ma) crustal–supracrustal belts in the West Pilbara Block. Precambrian Res 88:173–205

Lipple SL (1975) Definitions of new and revised stratigraphic units of the Eastern Pilbara Region. In: Annual report for the year 1974. Geological Survey of Western Australia, pp. 58–63

Low GH (1963) Copper deposits of Western Australia. Geological Survey of Western Australia, Mineral Resources Bull, p 8

Martin DMB, Clendenin CW, Krapež B, NJ MN (1998a) Tectonic and geochronological constraints on late Archaean and Palaeoproterozoic stratigraphic correlation within and between the Kaapvaal and Pilbara Cratons. J Geol Soc 155:311–322

Martin DMB, Li ZX, Nemchin AA, Powell CMA (1998b) A pre-2.2 Ga age for giant hematite ores of the Hamersley Province, Australia? Econ Geol 93:1084–1090

Miller LJ, Gair HS (1975) Mons Cupri copper-lead-zinc-silver deposit. In: Knight CL (ed) Economic geology of Australia and Papua New Guinea-Metals, vol 5. Australasian Inst. Mining Metall., Mon, pp 195–202

Müller SG, Krapež B, Barley ME, Fletcher IR (2005) Giant iron-ore deposits of the Hamersley province related to the breakup of Paleoproterozoic Australia: new insights from in situ SHRIMP dating of baddeleyite from mafic intrusions. Geology 33:577–580

Nelson DR (1997) Compilation of SHRIMP U-Pb zircon geochronology data, 1996. Record, 1997/2. Geological Survey of Western Australia

Nelson DR (1998) Compilation of SHRIMP U-Pb zircon geochronology data, 1997. Record, 1998/2. Geological Survey of Western Australia
Nelson DR (1999) Compilation of geochronology data, 1998. Record, 1999/2. Geological Survey of Western Australia
Nelson DR (2000) Compilation of geochronology data, 1999. Record, 2000/2. Geological Survey of Western Australia
Nelson DR (2001) Compilation of geochronology data, 2000. Record, 2001/2. Geological Survey of Western Australia
Nelson DR (2002) Compilation of geochronology data, 2001. Record, 2002/2. Geological Survey of Western Australia
Nelson DR (2005a) 178019: coarse sandstone, Mount Grant; Geochronology Record 552. Geological Survey of Western Australia
Nelson DR (2005b) 178020: pebble conglomerate, Mount Grant; Geochronology Record 553. Geological Survey of Western Australia
Olivier N, Dromart G, Coltice N, Flament N, Rey P, Sauvestre R (2012) A deep subaqueous fan depositional model for the Palaeoarchean (3.46 Ga) Marble Bar Cherts, Warrawoona Group. Western Australia Geol Mag 149:743–749
Pickard AL, Barley ME, Krapež B (2004) Deep-marine depositional setting of banded iron formation: sedimentological evidence from interbedded clastic sedimentary rocks in the early Palaeoproterozoic Dales Gorge Member of Western Australia. Sediment Geol 170:37–62
Pike G (2001) Facies architecture of two contrasting volcano-sedimentary basin successions from the Archaean Whim Creek Belt, North Pilbara Terrain, Western Australia: the intra-continental arc-related Whim Creek Group and plume-related, continental rift-hosted Bookingarra Group: Monash University Victoria, Department of Earth Sciences, PhD thesis (unpublished)
Pike G, Cas R (2002) Stratigraphic evolution of Archaean volcanic rock-dominated rift basins from the Whim Creek Belt, West Pilbara Craton, Western Australia. In: Altermann W, Corcoran P (eds) Precambrian sedimentary environments: a modern approach to ancient depositional systems. Blackwell Science, Oxford, UK, pp 213–234
Pike G, Cas R, Smithies RH (2002) Geological constraints on base metal mineralization of the Whim Creek greenstone belt, Pilbara Craton, Western Australia. Econ Geol 97:827–845
Pike G, Cas RAF, Hickman AH (2006) Archean volcanic and sedimentary rocks of the Whim Creek greenstone belt, Pilbara Craton, Western Australia. Report, vol 101. Geological Survey of Western Australia
Rasmussen B, Krapež B, Meier DB (2014) Replacement origin for hematite in 2.5 Ga banded iron formation: evidence for post-depositional oxidation of iron-bearing minerals. Geol Soc Am Bull 126:438–446
Rasmussen B, Muhling JR, Suvorova A, Krapež B (2016) Dust to dust: Evidence for the formation of "primary" hematite dust in banded iron formations via oxidation of iron silicate nanoparticles. Precambrian Res 284:49–63. https://doi.org/10.1016/j.precamres.2016.07.003
Retallack GJ, Krinsley DH, Fischer R, Razink JJ, Langworthy KP (2016) Archean coastal-plain paleosols and life on land. Gondwana Res 40:1–20
Ryan GR, Kriewaldt M (1964) Facies changes in the Archaean of the West Pilbara Goldfield. In: Annual report for the year 1963. Geological Survey of Western Australia, pp. 28–30
Sandiford M, Van Kranendonk MJ, Bodorkos S (2004) Conductive incubation and the origin of dome-and-keel structure in Archean granite–greenstone terrains: A model based on the Eastern Pilbara Craton. Western Australia Tectonics 23. https://doi.org/10.1029/2002TC001452
Sheppard S, Krapež B, Zi J-W, Rasmussen B, Fletcher I (2017) SHRIMP U-Pb zircon geochronology establishes that banded iron formations are not chronostratigraphic markers across Archean greenstone belts of the Pilbara Craton. Precambrian Res 292:290–304
Smith JB (2003) The episodic development of intermediate to silicic volcano-plutonic suites in the Archaean West Pilbara, Australia. Chem Geol 194:275–295
Smith JB, Barley ME, Groves DI, Krapež B, McNaughton NJ, Bickle MJ, Chapman HJ (1998) The Sholl Shear Zone, West Pilbara; evidence for a domain boundary structure from integrated

tectonostratigraphic analysis, SHRIMP U–Pb dating and isotopic and geochemical data of granitoids. Precambrian Res 88:143–172

Smith AJB (2015) The life and times of banded iron formations. Geology 43:1111–1112. https://doi.org/10.1130/focus122015.1

Smithies RH (1998a) Geology of the Mount Wohler 1:100 000 sheet. 1:100 000 Geological Series Explanatory Notes. Geological Survey of Western Australia

Smithies RH (1998b) Geology of the Sherlock 1:100 000 sheet. 1:100 000 Geological Series Explanatory Notes. Geological Survey of Western Australia

Smithies RH (1999) Geology of the Yule 1:100 000 sheet. 1:100 000 Geological Series Explanatory Notes. Geological Survey of Western Australia

Smithies RH (2002) De Grey, WA Sheet 2757. Geological Survey of Western Australia

Smithies RH (2004) Geology of the De Grey and Pardoo 1:100 000 sheets. 1:100 000 Geological Series Explanatory Notes. Geological Survey of Western Australia

Smithies RH, Champion DC (2000) The Archaean high-Mg diorite suite: links to tonalite–trondhjemite–granodiorite magmatism and implications for early Archaean crustal growth. J Petrol 41:1653–1671

Smithies RH, Champion DC, Blewett RS (2001a) Wallaringa, WA Sheet 2656. Geological Survey of Western Australia

Smithies RH, Champion DC, Blewett RS (2002) Geology of the Wallaringa 1:100 000 sheet. 1:100 000 Geological Series Explanatory Notes. Geological Survey of Western Australia

Smithies RH, Champion DC, Sun S-S (2004) Evidence for early LREE-enriched mantle source regions: diverse magmas from the c. 3.0 Ga Mallina Basin, Pilbara Craton, NW Australia. J Petrol 45:1515–1537

Smithies RH, Champion DC, Van Kranendonk MJ, Hickman AH (2007a) Geochemistry of volcanic rocks of the Northern Pilbara Craton, Western Australia. Report, vol 104. Geological Survey of Western Australia

Smithies RH, Farrell TR (2000) Geology of the Satirist 1:100 000 sheet. 1:100 000 Geological Series Explanatory Notes. Geological Survey of Western Australia

Smithies RH, Hickman AH (2003) Archaean geology of the Mallina and Whim Creek Basins (1: 250 000 scale). Geological Survey of Western Australia

Smithies RH, Hickman AH (2004) Pyramid, WA Sheet SF 50–7, 2nd edn. Geological Survey of Western Australia

Smithies RH, Hickman AH, Nelson DR (1999) New constraints on the evolution of the Mallina Basin, and their bearing on relationships between the contrasting Eastern and Western Granite–Greenstone Terranes of the Archaean Pilbara Craton, Western Australia. Precambrian Res 94: 11–28

Smithies RH, Nelson DR, Pike G (2001b) Development of the Archaean Mallina Basin, Pilbara Craton, northwestern Australia; a study of detrital and inherited zircon ages. Sediment Geol 141–142:79–94

Smithies RH, Van Kranendonk MJ, Champion DC (2005) It started with a plume–early Archaean basaltic proto-continental crust. Earth Planet Sci Lett 238:284–297

Smithies RH, Van Kranendonk MJ, Champion DC (2007b) The Mesoarchean emergence of modern-style subduction. In: Maruyama S, Santosh M (eds) Island arcs: past and present, pp 50–68

Sugahara H, Sugitani K, Mimura K, Yamashita F, Yamamoto K (2010) A systematic rare-earth elements and yttrium study of Archean cherts at the Mount Goldsworthy greenstone belt in the Pilbara Craton: implications for the origin of microfossil-bearing black cherts. Precambrian Res 177:73–87

Sugitani K, Grey K, Allwood A, Nagaoka T, Mimura K, Minami M, Marshall CP, Van Kranendonk MJ, Walter MR (2007) Diverse microstructures from Archaean chert from the Mount Goldsworthy–Mount Grant area, Pilbara Craton, Western Australia: microfossils, dubiofossils, or pseudofossils? Precambrian Res 158:228–262

Sugitani K, Grey K, Nagaoka T, Mimura K (2009a) Three-dimensional morphological and textural complexity of Archean putative microfossils form the Northeastern Pilbara Craton: indications of biogenicity of large (>15 μm) spheroidal and spindle-like structures. Astrobiology 9:603–615

Sugitani K, Grey K, Nagaoka T, Mimura K, Walter MR (2009b) Taxonomy and biogenicity of spheroidal microfossils (c. 3.0 Ga) from the Mount Goldsworthy–Mount Grant area in the Northeastern Pilbara Craton, Western Australia. In: Schopf JW, Bottjer DJ (eds) World summit on ancient microfossils, 27 July–2 August 2008. IGPP Center for the Study of Evolution and the Origin of Life (CSEOL), University of California, pp. 50–59

Sugitani K, Mimura K, Suzuki K, Nagamine K, Sugisaki R (2003) Stratigraphy and sedimentary petrology of an Archean volcanic-sedimentary succession at Mt Goldsworthy in the Pilbara Block, Western Australia: implications of evaporite (nahcolite) and barite deposition. Precambrian Res 120:55–79

Sugitani K, Yamashita F, Nagaoka T, Minami M, Mimura K, Suzuki K (2006) Geochemistry and sedimentary petrology of Archean clastic sedimentary rocks at Mt. Goldsworthy, Pilbara Craton, Western Australia: evidence for the early evolution of continental crust and hydrothermal alteration. Precambrian Res 147:124–147

Trendall AF (1968) Three great basins of Precambrian banded iron-formation deposition: a systematic comparison. Geol Soc Am Bull 79:1527–1544

Trendall AF, Blockley JG (1970) The iron formations of the Precambrian Hamersley Group, Western Australia, with special reference to the associated crocidolite. Bulletin 119, Geological Survey of Western Australia

Van Kranendonk MJ (1999) North Shaw, WA Sheet 2755. Geological Survey of Western Australia

Van Kranendonk MJ (2000) Geology of the North Shaw 1:100 000 sheet. 1:100 000 Geological Series Explanatory Notes. Geological Survey of Western Australia

Van Kranendonk MJ (2003) Geology of the Tambourah 1:100 000 sheet. 1:100 000 Geological Series Explanatory Notes. Geological Survey of Western Australia

Van Kranendonk MJ (2004) Geology of the Carlindie 1:100 000 sheet. 1:100 000 Geological Series Explanatory Notes. Geological Survey of Western Australia

Van Kranendonk MJ (2010) Geology of the Coongan 1:100 000 sheet. 1:100 000 Geological Series Explanatory Notes. Geological Survey of Western Australia

Van Kranendonk MJ, Collins WJ (1998) Timing and tectonic significance of Late Archaean, sinistral strike-slip deformation in the Central Pilbara Structural Corridor, Pilbara Craton, Western Australia. Precambrian Res 88:207–232. https://doi.org/10.1016/s0301-9268(97)00069-7

Van Kranendonk MJ, Hickman AH, Smithies RH, Nelson DN, Pike G (2002) Geology and tectonic evolution of the Archaean North Pilbara Terrain, Pilbara Craton, Western Australia. Econ Geol 97:695–732. https://doi.org/10.2113/gsecongeo.97.4.695

Van Kranendonk MJ, Hickman AH, Smithies RH, Williams IR, Bagas L, Farrell TR (2006) Revised lithostratigraphy of Archean supracrustal and intrusive rocks in the Northern Pilbara Craton, Western Australia. Record, 2006/15. Geological Survey of Western Australia

Van Kranendonk MJ, Hickman AH, Williams IR, Nijman W (2001) Archaean geology of the East Pilbara Granite–Greenstone Terrane, Western Australia–a field guide. Record, 2001/9. Geological Survey of Western Australia

Van Kranendonk MJ, Smithies RH (2006) Port Hedland–Bedout Island, WA Sheet SF 50–4 and Part Sheet SE 50–16, 3rd Geological Survey of Western Australia

Van Kranendonk MJ, Smithies RH, Hickman AH, Champion DC (2007) Review: Secular tectonic evolution of Archean continental crust: interplay between horizontal and vertical processes in the formation of the Pilbara Craton, Australia. Terra Nova 19:1–38

Van Kranendonk MJ, Smithies RH, Hickman AH, Wingate MTD, Bodorkos S (2010) Evidence for Mesoarchean (~3.2 Ga) rifting of the Pilbara Craton: the missing link in an early Precambrian Wilson cycle. Precambrian Res 177:145–161

Wilhelmij HR, Dunlop JSR (1984) A genetic stratigraphic investigation of the Gorge Creek Group in the Pilgangoora syncline. In: Muhling JR, Groves DI, Blake TS (eds) Archaean and Proterozoic basins of the Pilbara, Western Australia: Evolution and mineralization potential. The University of Western Australia, Geology Department and University Extension, pp 68–88
Williams IR (1999) Geology of the Muccan 1:100 000 sheet. 1:100 000 Geological Series Explanatory Notes. Geological Survey of Western Australia
Williams IR (2003) Yarrie, Western Australia, 3rd edn. 1:250 000 Geological Series Explanatory Notes. Geological Survey of Western Australia
Williams IR (2000) Geology of the Cooragoora 1:100 000 sheet. 1:100 000 Geological Series Explanatory Notes. Geological Survey of Western Australia
Williams IR (2001) Geology of the Warrawagine 1:100 000 sheet. 1:100 000 Geological Series Explanatory Notes. Geological Survey of Western Australia
Williams IR (2004) Yarrie, WA Sheet SF 51–1, 3rd edn. Geological Survey of Western Australia
Williams IR, Bagas L (2004) Mount Edgar, WA Sheet 2955. Geological Survey of Western Australia
Williams IR (1998) Muccan, WA Sheet 2956. Geological Survey of Western Australia
Wingate MTD, Bodorkos S, Van Kranendonk MJ (2009a) 180048: felsic metavolcanic rock, Cattle Well; Geochronology Record 807. Geological Survey of Western Australia
Wingate MTD, Bodorkos S, Van Kranendonk MJ (2009b) 180056: granophyre, Eel Creek; Geochronology Record 808. Geological Survey of Western Australia

Chapter 10
Orogenies, Cratonization, and Post-Orogenic Granites

Abstract Two late Mesoarchean orogenic events completed the tectonic evolution of the Pilbara Craton: the 2955–2919 Ma North Pilbara Orogeny and the 2930–2900 Ma Mosquito Creek Orogeny. The deformation of these events compressed and closed the Mallina and Mosquito Creek Basins and effectively completed cratonization. From 2900 Ma onwards, the Pilbara Craton remained a stable unit of continental crust until uplift, crustal extension, and rifting began the breakup of the craton (Chap. 12).

The North Pilbara Orogeny was a major event of deformation, metamorphism, and granitic intrusion resulting from convergence between the Pilbara Craton and an exotic plate to the northwest (Chaps. 8 and 9). Effects of the event were concentrated in the Central Pilbara Tectonic Zone and Karratha Terrane, although compression also extended into the East Pilbara Terrane. Closure of the Mallina Basin at c. 2930 Ma resulted in deformation and metamorphism moving to the relatively thin crust of the Mosquito Creek Basin.

Keywords Plate convergence · Basin closure · Deformation · Metamorphism · North Pilbara Orogeny · Mosquito Creek Orogeny

10.1 Introduction

The Mesoarchean evolution of the Pilbara Craton included the development of three basins (Regal, Mallina, and Mosquito Creek) that were underlain by crust much thinner than in the adjacent continental microplates (East Pilbara, Karratha, and Kurrana Terranes). A metamorphic event at c. 3160 Ma ended plate separation of the East Pilbara Terrane (EPT) and Karratha Terrane (KT) and was followed by northwest–southeast convergence (Chap. 8). The change from separation to convergence is interpreted to have resulted from a collision between the Northwest Pilbara Craton and an exotic plate converging from the northwest (Hickman 2016).

The identity of the northwest plate is unknown (the northwest margin of the Pilbara Craton is concealed by the Indian Ocean), but based on evidence that it was progressively subducted beneath the Pilbara Craton over at least 140 million years, it

A. H. Hickman, *Archean Evolution of the Pilbara Craton and Fortescue Basin*, Modern Approaches in Solid Earth Sciences 24,
https://doi.org/10.1007/978-3-031-18007-1_10

was evidently a large plate. By 2920 Ma, melting of the subducted slab had reached the central part of the EPT, over 250 km from the northwest margin of the craton (Fig. 8.6).

Post-3160 Ma compression of the Regal Basin was taken up by the development of a c. 3130 Ma subduction zone within its thin basaltic crust (Chap. 8). Subduction of this crust allowed the EPT and KT to converge. Eventual collision of the EPT and KT at c. 3070 Ma, marked by the Prinsep Orogeny (Chap. 8), closed the Regal Basin and ended subduction within it. Ongoing compression of the Pilbara Craton resulted in the northwest plate being subducted along the northwest margin of the Pilbara Craton. This led to evolution of the Gorge Creek and Mallina Basins, inland from magmatic arcs on the KT, and above the closed Regal Basin. Initially, the crust of the Mallina Basin was extended (Smithies et al. 1999, 2001a, b; Van Kranendonk et al. 2002, 2006), but from 2955 Ma onwards, the Croydon Group within the basin became increasingly deformed. Intrusion of mafic and felsic intrusions of the Sisters Supersuite between 2954 and 2913 Ma indicates ongoing subduction of the northwest plate, and the Mallina Basin was compressed and metamorphosed at the same time. Deformation and metamorphism of the Mallina Basin and adjacent parts of the West Pilbara Superterrane (WPS) from 2955 to 2920 Ma are referred to as the North Pilbara Orogeny.

By about 2930 Ma, the Mallina Basin appears to have been compressed to the point that its tectonically thickened crust could not fully accommodate crustal shortening under the northwest–southeast compression. At this stage, the Mosquito Creek Basin in the Southeast Pilbara (Fig. 1.3) was still essentially a rift basin with a relatively thin underlying basaltic crust. At 2930 Ma, with all other parts of the Pilbara Craton composed of rigid continental crust, the regional northwest–southeast compression began to deform the Mosquito Creek Basin. Northwest–southeast deformation of the basin resulted in isoclinal folding and metamorphism of the Mosquito Creek Formation referred to as the Mosquito Creek Orogeny. Closure of the basin at about 2900 Ma effectively completed cratonization of the Pilbara Craton, although minor local deformation and isolated post-orogenic granitic intrusion continued sporadically until about 2830 Ma.

10.1.1 North Pilbara Orogeny

The North Pilbara Orogeny (Van Kranendonk et al. 2006) was a major event of deformation, metamorphism, and granitic intrusion. Although effects of the event were concentrated in the Central Pilbara Tectonic Zone (CPTZ) and KT, compression of the Pilbara Craton also extended into the EPT. The deformation and metamorphism of the orogeny were recognized from GSWA mapping in the 1970s (Hickman 1981), but the event was not named. The timing of the event was interpreted to be c. 2950 Ma based on the age of a widespread metamorphic event detected from Pb–Pb and Rb–Sr dating (Oversby 1976).

Detailed geological mapping and extensive U – Pb zircon geochronology between 1994 and 2005 have provided far more data on the orogeny. A study of the Mallina Basin (Smithies et al. 2001a, b) concluded that there were three main phases of deformation of the Croydon Group. All available geochronology indicates that igneous intrusions of the Sisters Supersuite were emplaced between c. 2954 and 2913 Ma. Because deformation and igneous intrusion occurred within the same tectonic setting, the age range of the supersuite closely approximates to the duration of the orogeny.

Because of the regional extent of the North Pilbara Orogeny, the thicknesses and rigidities of the crustal units affected were laterally variable, with the result that the timing and styles of deformation also varied. No single of history of deformation applies across the entire Northern Pilbara Craton. In the Northwest Pilbara, three main stages are recognized, whereas in the East Pilbara, only two stages are represented. Structural evidence, including changes in the sense of movement along major strike-slip faults, indicates that at some times convergence was oblique; regional stress fields changed during the orogeny.

The first event of the North Pilbara Orogeny occurred between c. 2955 Ma and c. 2948 Ma. Major folds in the Mallina Basin are illustrated on Fig. 10.1. Smithies et al. (2001a, b) classified folds formed at c. 2955 Ma as D2 structures, whereas they were assigned to D6 (Fig. 10.1) in a later regional interpretation (Hickman 2016). An associated steep axial plane schistosity (S6) cuts across refolded D5 folds between the Satirist Monzogranite and Nunyerry Gap. In the Whim Creek greenstone belt, Krapež and Eisenlohr (1998) described folds and thrusts affecting the lower part of the Bookingarra Group, and therefore post-2955 Ma, as 'Phase 3' structures, and Hickman (2016) assigned these to D7.

The second major deformation of the orogeny (D3, Smithies et al. 2001a, b; D8, Hickman 2016) was widespread across the Northwest Pilbara and produced major northeast-trending folds such as the Roebourne Synform, Prinsep Dome, Cleaverville Syncline, Bradley Syncline, and Whim Creek Anticline (Fig. 7.1). These folds are oriented oblique to major sinistral strike-slip faults and shear zones and are transpressional folds formed under compression from the north or north-northwest. Large folds of this generation in the East Pilbara Craton include the John Bull Syncline in the Pilbara Well greenstone belt and the Goldsworthy Syncline in the Goldsworthy greenstone belt. In the Mallina Basin, the age of these structures has been estimated at 2940 – 2930 Ma (Smithies et al. 2001a, b; Hickman 2016).

An apparently local third event of deformation is present in the Sholl Terrane east and west of the Maitland Shear Zone (Hickman 2001). Here, a steeply inclined tectonic foliation (S9, Hickman 2016) cuts northeast-trending D8 folds. The eastern side of the Cherratta granitic complex is dominated by S9 which strikes north–south. Some shear zones within the area deformed by D9 include granites containing zircons dated at c. 2944 and c. 2925 Ma. This suggests that the third event in the orogeny took place at c. 2930 Ma (Hickman 2016).

The final stage of the orogeny occurred at c. 2920 Ma and included dextral movement on pre-existing major shear zones (D10, Hickman 2016). U – Pb zircon dating close to the Sholl Shear Zone has revealed a metamorphic disturbance event at

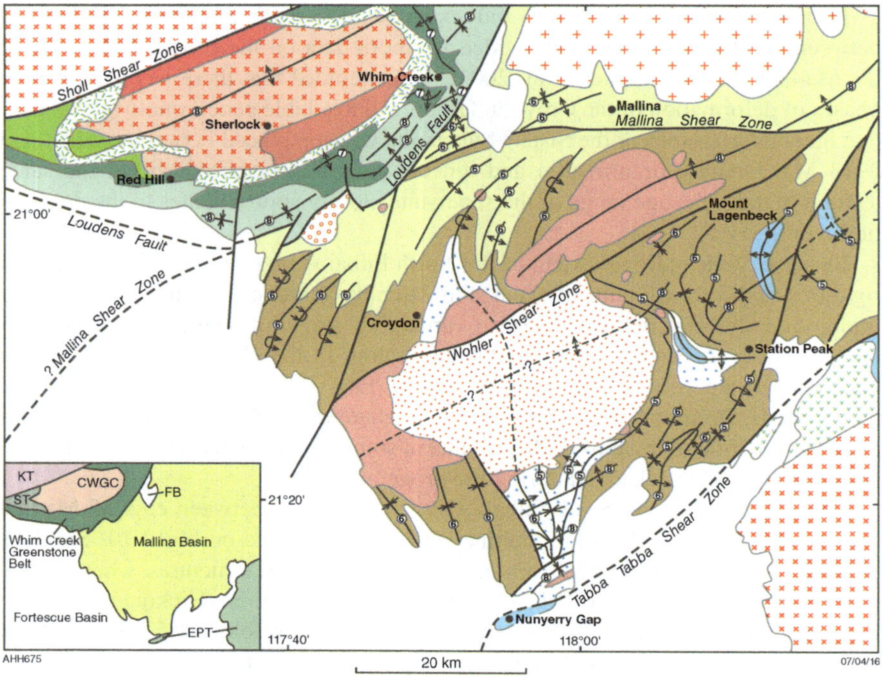

Fig. 10.1 Stratigraphic and structural differences between the central Mallina Basin and the Whim Creek greenstone belt. (**a**) Simplified geological map illustrating geological differences across the Loudens Fault. Note that the Gorge Creek Group, which underlies the Croydon Group in the Mallina Basin, is exposed within the cores of several anticlines. Fold axial traces simplified from Smithies (1998), Smithies and Farrell (2000), and Krapež and Eisenlohr (1998). (**a**) Reference to map, summarizing stratigraphy and deformation events (from Hickman 2016; with Geological Survey of Western Australia permission)

about 2920 Ma (Nelson 1997, GSWA 118976; Krapež and Eisenlohr 1998). Dextral strike-slip movement of 30 to 40 km is shown on geological maps of the Caines Well granitic complex and the Whim Creek greenstone belt (Fig. 10.2) (Hickman and Smithies 2000; Hickman 2001). Dextral movement on the east-northeast striking shear zone indicates northwest–southeast compression at c. 2920 Ma.

10.1.1.1 Lalla Rookh–Western Shaw Structural Corridor

Deformation related to the North Pilbara Orogeny has been described from the Lalla Rookh–Western Shaw Structural Corridor (Figs. 1.4 and 10.3) (Zegers 1996; Van Kranendonk and Collins 1998; Zegers et al. 1998; Van Kranendonk 2000, 2003, 2008; Van Kranendonk et al. 2002, 2006; Hickman 2021) and the Tabba Tabba Shear Zone (TTSZ) (Figs. 1.4 and 6.7) (Beintema et al. 2001; Smithies et al. 2002;

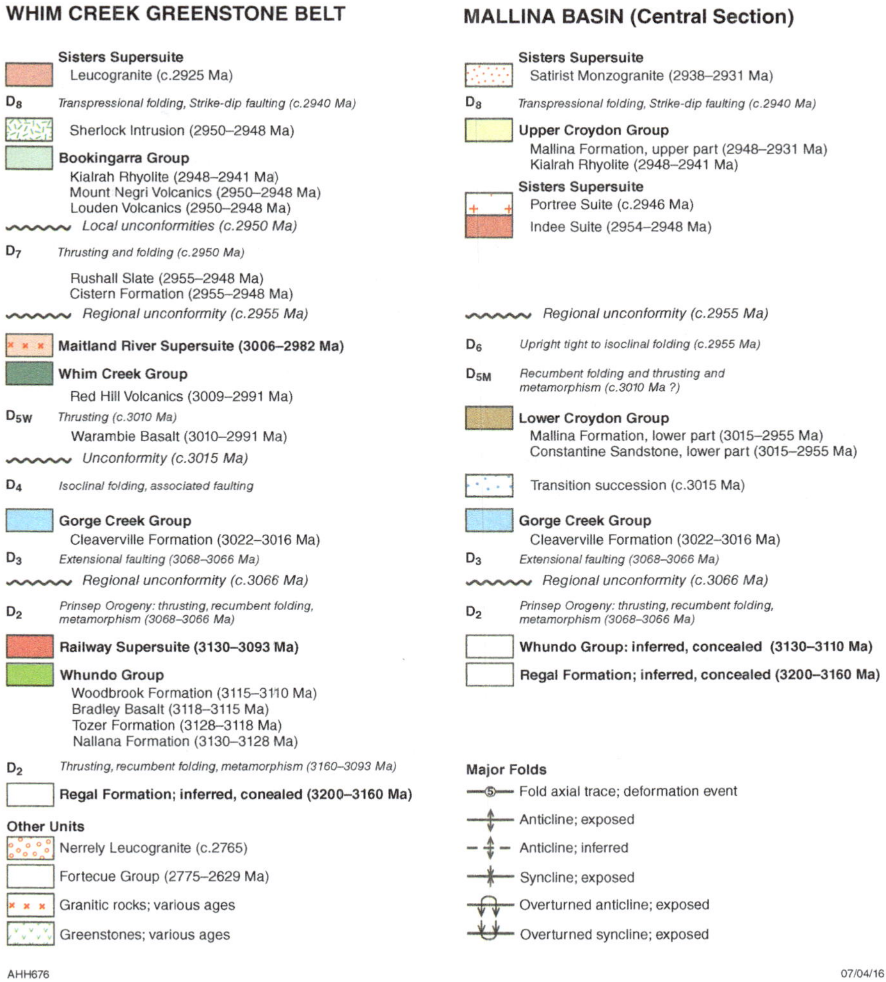

Fig. 10.1 (continued)

Beintema 2003; Hickman 2016, 2021). Both these major structures contain evidence of major sinistral strike-slip movement between 2940 and 2930 Ma, although earlier movement in the underlying crust is now interpreted to have occurred during the 3280 – 3165 Ma East Pilbara Terrane Rifting Event (Chap. 6).

Van Kranendonk and Collins (1998) identified the Lalla Rookh–Western Shaw Structural Corridor as a corridor of linked fold structures and shear zones, developed in response to c. 2950 Ma sinistral transpression. The timing of the compression and deformation was subsequently revised to between 2940 and 2930 Ma (Van Kranendonk et al. 2002; Van Kranendonk 2008). The structural corridor is exposed over a north–south strike length of about approximately for 150 km and varies in

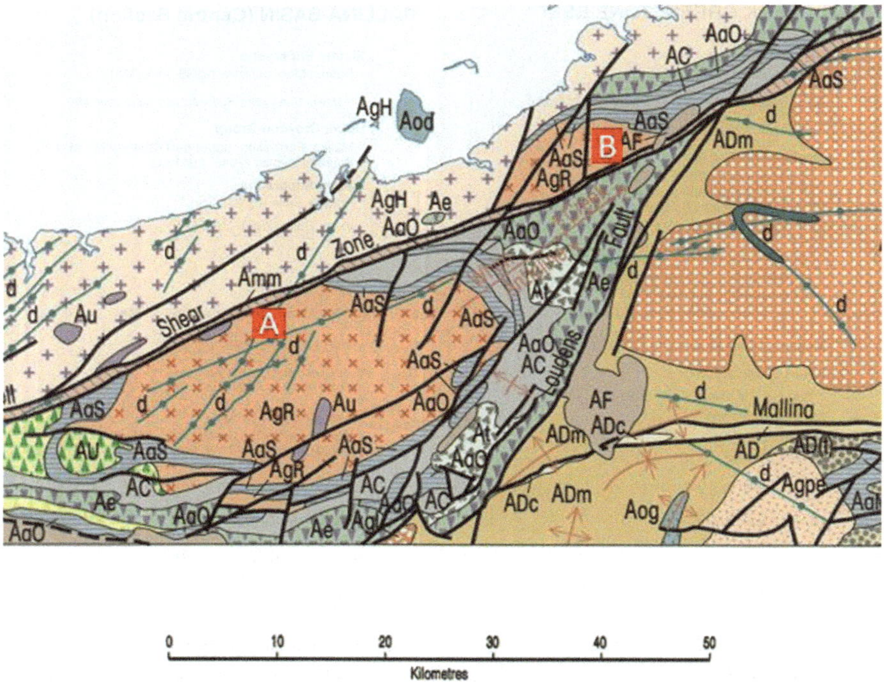

Fig. 10.2 Extract from an interpreted bedrock geology map accompanying the Roebourne 1: 250,000 map sheet, showing 30–40 km dextral displacement (A to B) of the Whim Creek and Bookingarra Groups, and of the Caines Well granitic complex, along the Sholl Shear Zone. Geochronology indicates that the dextral movement occurred at c. 2920 Ma. Maximum compression is interpreted to have been northwest–southeast (modified from Hickman and Smithies 2000; with Geological Survey of Western Australia permission)

width from a single shear zone between the Carlindi and Muccan Domes to 35 km between the Shaw and Yule Domes. The western boundary fault is the Lalla Rookh–Western Shaw Fault (LRWS), whereas along the western margin of the Shaw granitic complex, the eastern boundary fault is the Mulgandinnah Shear Zone (Zegers et al. 1998).

10.1.1.2 Tabba Tabba Shear Zone

The TTSZ is a 250-km-long zone of intense faulting and shearing that defines the southeast boundary of the CPTZ (Fig. 1.3). The TTSZ is up to 3 km wide and contains strongly foliated to mylonitic units derived from almost all lithologies in the EPT and the Mallina Basin. The origin and geological history of the TTSZ before the North Pilbara Orogeny are discussed in Chap. 6. However, the best-preserved structures in the shear zone were formed after c. 2955 Ma as northwest–southeast

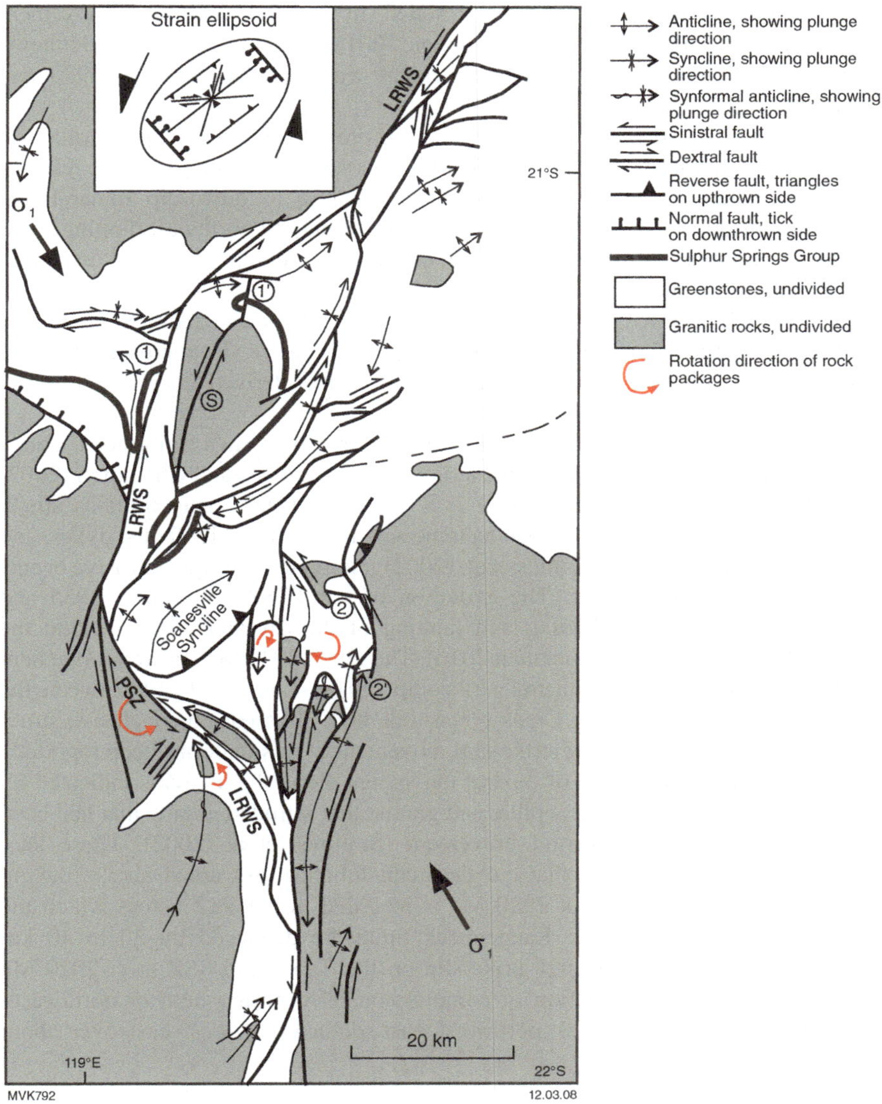

Fig. 10.3 Structural map of the Lalla Rookh Structural Corridor, showing major faults and senses of displacement, major folds, and rotation movements of rock panels. The σ1 direction is inferred from the orientation of the Soanesville Syncline and sinistral LRWS fault. 1–1′ and 2–2′ indicate points of measured offset across fault segments, as described by Van Kranendonk (2008). Circled S, Strelley Monzogranite; PSZ, Pulcunah Shear Zone. Strain ellipsoid in inset is oriented according to the σ1 direction in the map area and shows the major structures predicted from experimental studies. (From Van Kranendonk 2008; with Geological Survey of Western Australia permission)

compression of the Pilbara Craton resulted in sinistral strike-slip movement (Beintema et al. 2001, 2003; Hickman et al. 2001; Smithies et al. 2002; Beintema 2003). A later episode of dextral movement was recorded between 2930 and 2920 Ma.

Evidence of sinistral strike-slip movement is provided by mesoscopic structures, such as C – S fabrics and rotated feldspar phenocrysts. This movement also resulted in the formation of moderately northwest plunging to down-dip mineral and stretching lineations, suggesting that this steep, commonly northwest dipping shear zone had a normal, northwest-side-down, displacement component (Hickman et al. 2001; Smithies and Champion 2002; Smithies et al. 2002). Intrusion of 2955–2945 Ma gabbro and high-Mg diorite (Langenbeck and Indee Suites of the Sisters Supersuite, Chap. 8) along the TTSZ indicates a deep crustal structure controlling the migration and emplacement of mantle-derived magmas (Smithies and Champion 2000; Smithies et al. 2002).

In the central section of the TTSZ (Fig. 6.7), near Wallareenya Homestead, a bend in the shear zone resembles a jog-like structure that Smithies and Champion (2002) and Smithies et al. (2002) described as a 'releasing bend'. This contains small, northeast-oriented, diamond-shaped tectonic segments defined by faults, dykes, and sheared magma conduits. Smithies et al. (2002) interpreted multiple intrusive events over about 15 million years. The evolution of the TTSZ between c. 2955 and 2940 Ma coincides with folding and faulting of the Croydon Group within the Mallina Basin (D6 and D8, Hickman 2016). The 2955–2940 Ma sinistral movement might also correlate with sinistral strike-slip on the Loudens Fault between the Mallina Basin and the Whim Creek greenstone belt, although no conclusive structural evidence on the sense of strike-slip movement in this fault has been reported.

Evidence of a later phase of dextral movement along the TTSZ is indicated by dykes of unfoliated pegmatite, aplite, and granite that intruded granite that had been mylonitized during the sinistral movement (Smithies et al. 2002). These later intrusions fill dextral shears that cut the main foliation. As noted above, dextral strike-slip movement at about 2920 Ma is recorded in the SSZ across which the Whim Creek Group and the Langenbeck Suite are displaced by 30 to 40 km (Hickman 2001, 2016). Dextral strike-slip on the TTSZ and SSZ at c. 2920 Ma indicates a shift in the direction of compression from north–south or northwest–southeast at 2940 Ma to west-northwest–east-southeast or west–east over about 20 million years.

10.1.1.3 Sholl Shear Zone

As described in Chap. 6, the SSZ is a major sinistral strike-slip fault that separates Paleoarchean crust of the Karratha Terrane (KT) from Mesoarchean crust of the Sholl Terrane (ST) (Fig. 1.3). The fault originated during late Paleoarchean crustal extension of the East Pilbara Terrane Rifting Event and was reactivated during the evolution of the CPTZ, most notably during the Prinsep Orogeny. Subduction along the northwest margin of the Pilbara Craton between c. 2970 and 2955 Ma, resulting in emplacement of the Orpheus and Maitland River Supersuites northwest of the SSZ (Chap. 8), would have been accompanied by deformation along the SSZ, but the shear also provides evidence of movement during the North Pilbara Orogeny. One of the best exposures of the SSZ is in the rocky bed of the Nickol River 15 km south of Karratha (Fig. 6.8).

10.1.1.4 Loudens Fault

The Loudens Fault (LF) extends 120 km along the southeast and southwest margins of the Whim Creek greenstone belt, separating this belt from the Mallina Basin (Fig. 10.1). Aeromagnetic imagery indicates that the LF branches off the SSZ near Peawah Hill, northeast of Whim Creek, and rejoins the SSZ near Mount Ada, south of Roebourne (Fig. 10.4). Isotopic data indicate that Paleoarchean crust underlies the area north of the LF but is absent from the Mallina Basin to the south (Hickman 2016). This suggests that the zone of late Paleoarchean rifting that elsewhere underlies the SSZ (Chap. 6) is locally represented by the LF. There might have been two linked rift faults (SSZ and LF) along this northwest boundary of the Regal Basin, although it is possible that between Peawah Hill and Mount Ada the SSZ is a later structure formed by c. 2920 Ma dextral strike-slip movement.

Mesoarchean strike-slip movement on the LF occurred after deposition of the c. 3010–2990 Ma Whim Creek Group, and after deposition of all formations of the c. 2950 Ma Bookingarra Group except for the c. 2945 Ma Kialrah Rhyolite. Early stratigraphic correlations between the Whim Creek greenstone belt and the central trough of the Mallina Basin (Fitton et al. 1975; Horwitz 1990; Smithies et al. 1999, 2001a, b; Van Kranendonk et al. 2002, 2006) were discussed by Hickman (2016) who concluded that correlations had been made on inconclusive lithological or geochemical evidence. The only correlation supported by both lithology and geochronology involves the youngest formation of the Bookingarra Group, the c. 2945 Ma Kialrah Rhyolite which is present on both sides of the fault. Thus, the LF was a major focus of strike-slip movement until c. 2945 Ma, and the interval between 2945 and 2940 Ma coincides with a change in magmatic activity within the Mallina Basin.

Prior to c. 2945 Ma, the central trough of the Mallina Basin was intruded by mantle-derived, LREE-enriched suites of the Sisters Supersuite (Langenbeck, Indee and Portree Suites, Chap. 8), whereas after c 2940 Ma, intrusions such as the Satirist and Petermarer Monzogranites were K-rich granites derived from partial melting of older sialic crust. In contrast to the Langenbeck, Indee, and Portree Suites, the later K-rich granites of the Sisters Supersuite are distributed across most of the Northern Pilbara Craton. This suggests that the last major movement on the Loudens Fault coincided with final compression and closure of the Mallina Basin, which thereafter was mainly underlain by sialic crust of the amalgamated EPT, KT, and ST.

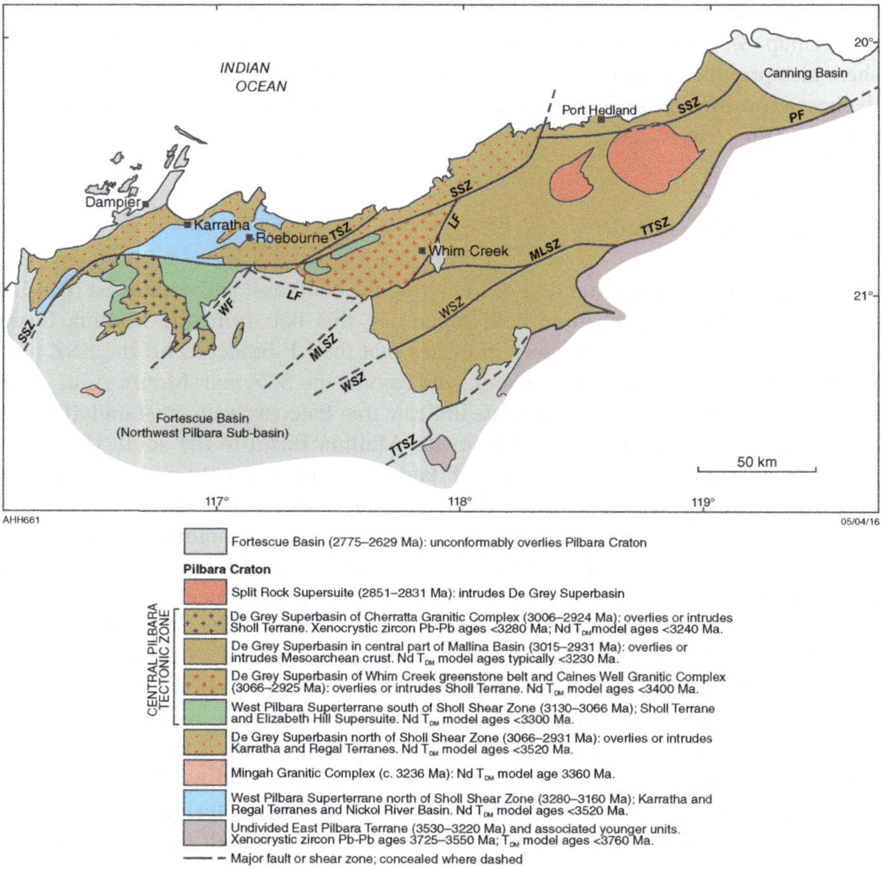

Fig. 10.4 Geological map showing the principal tectonostratigraphic divisions and structures of the Northwest Pilbara Craton, including an interpretation of underlying crustal ages. Note that units within the Central Pilbara Tectonic Zone are underlain by relatively young crust but that isotopic data indicates Paleoarchean crust between the Loudens Fault and Sholl Shear Zone. *LF* Loudens Fault; *MLSZ* Mallina Shear Zone; *PF* Pardoo Fault; *SSZ* Sholl Shear Zone; *TSZ* Terenar Shear Zone; *TTSZ* Tabba Tabba Shear Zone; *WF* Woodbrook Fault; *WSZ* Wohler Shear Zone (from Hickman 2016; with Geological Survey of Western Australia permission)

10.1.2 Mosquito Creek Orogeny

The Mosquito Creek Orogeny was the final orogeny in the Northern Pilbara Craton and occurred when the East Pilbara and Kurrana Terranes were forced together by north-northwest to south-southeast compression. The timing of the orogeny is indicated by a Pb–Pb model age of 2905 ± 9 Ma from gold mineralization within a shear zone (Thorpe et al. 1992). The orogeny commenced after c. 2930 Ma following deposition of the Mosquito Creek Formation (Bagas et al. 2004). At

2930 Ma, the Mosquito Creek Basin was probably at least 100 km wide, but compression during the orogeny converted it into a 30–35 km-wide synclinorium (Fig. 10.5). Unlike the earlier orogenic events in the Pilbara Craton, the Mosquito Creek Orogeny was not accompanied by granitic intrusion of the basin under compression. Granitic intrusions of the 2954–2913 Ma Sisters Supersuite were emplaced northwest of the Mosquito Creek Basin, but these were related to subduction of a plate northwest of the Pilbara Craton.

The first stage of the orogeny (D18 in the deformation history of the Northern Pilbara Craton, Hickman 2021) produced regional-scale, tight to isoclinal folds plunging shallowly east-northeast, accompanied by an east-northeast striking schistosity extending throughout the basin and into the northern margin of the Kurrana Terrane (KUT). The second stage (D19) resulted in the local development of north-trending and north-plunging folds, producing a crenulation cleavage superimposed on the regional schistosity. The third stage (D20) produced regional-scale, tight upright folds, trending east-northeast in the west and east in the east. These folds have an axial plane foliation, and the limbs of the folds are locally disrupted by major shear zones containing gold mineralization.

Compression of the Mosquito Creek Basin resulted in basin inversion, with the entire basin fill being tectonically thickened and uplifted. The marginal normal faults that originated during the c. 3200 Ma rift phase of basin development were reactivated as thrusts, and northern sections of the basin were thrust onto the EPT. On the southern margin of the basin, the Coondamar Formation was thrust onto the KUT. Much of the southern thrusting apparently occurred along the Kurrana Shear Zone (KSZ, Fig. 10.5).

Although the bulk of the Mosquito Creek Formation was deposited before the orogeny, parts of the succession were deposited during the deformation. Nijman et al. (2010) noted that conglomerate units in the North Dromedary and South Dromedary hills, 8 km south-southeast of Nullagine, unconformably overlie folded units of the Mosquito Creek Formation but are folded and sheared by late deformation of the orogeny.

10.1.2.1 Kurrana Shear Zone

The Kurrana Shear Zone (KSZ) (Fig. 10.5) is a major structure separating the Mosquito Creek Basin from the Kurrana Terrane (KUT) (Tyler et al. 1992; Bagas 2005). The evolution of the KSZ is interpreted to have occurred in several stages, commencing with its origin as a rift fault during the c. 3220 Ma breakup of the Pilbara Craton (Van Kranendonk et al. 2006; Hickman 2021). Tyler et al. (1992) interpreted the KSZ to be a suture zone joining the EPT and the KUT between 3000 and 2760 Ma. It is notable that no granitic rocks of similar age to the c. 3199–3178 Ma Golden Eagle Orthogneiss of the KUT have been identified in the EPT north of the Mosquito Creek Basin. One interpretation is that the TTG of the Golden Eagle Orthogneiss was derived by subduction of mafic crust of the Mosquito Creek Basin under the Paleoarchean crust of the KUT along the early KSZ fault

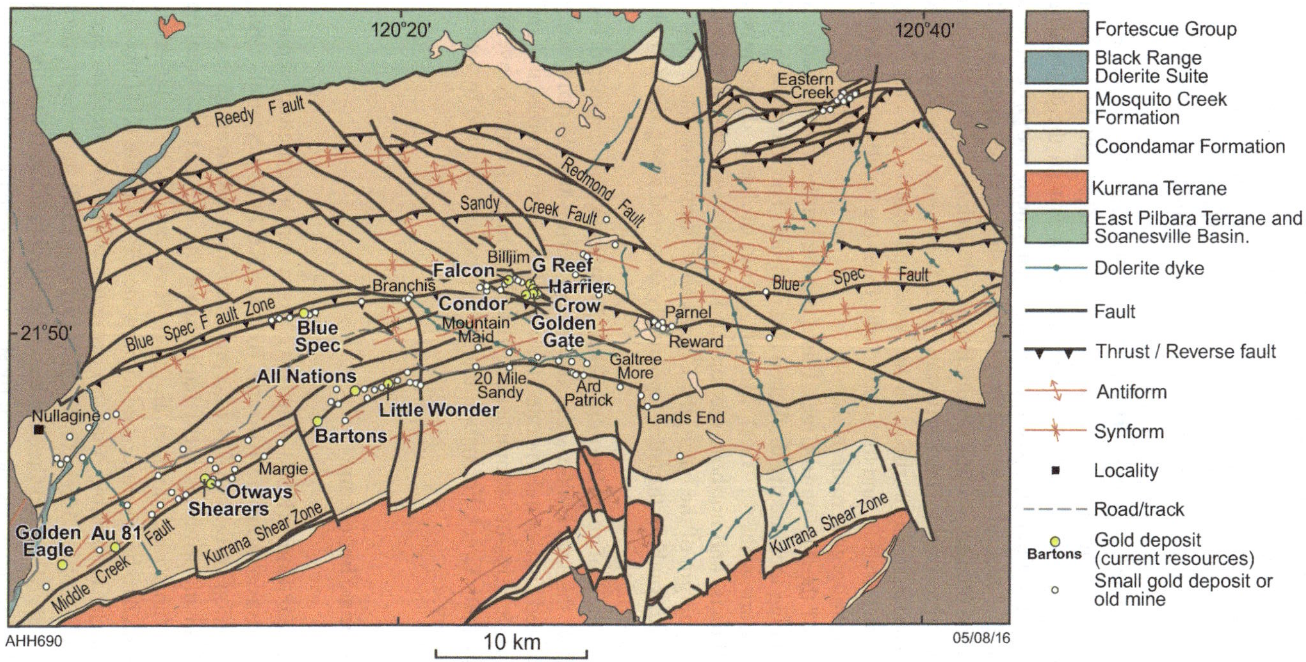

Fig. 10.5 Geological setting of the Mosquito Creek Basin showing major structures and areas of gold mineralization (from Hickman 2021; with Geological Survey of Western Australia permission)

(Hickman 2021). Other major rifting-related faults formed at c. 3220 Ma included the Tabba Tabba and Sholl Shear Zones of the Northwest Pilbara Craton. There is no evidence for further tectonic activity along the KSZ until the Mosquito Creek Orogeny between 2930 and 2890 Ma.

Regional-scale tight folding of the KSZ during the Mosquito Creek Orogeny was reported by Bagas (2005) and Farrell (2006). Thrusting of the Mosquito Creek Basin against the KUT is likely to have occurred during crustal shortening under north–south compression. As noted by Farrell (2006), the c. 2838 Ma Bonney Downs Monzogranite was strongly deformed immediately adjacent to the eastern section of the KSZ, indicating younger movement. Because the shear zone is unconformably overlain by undeformed units of the 2775–2629 Ma Fortescue Group, most deformation of the monzogranite is likely to have occurred at c. 2800 Ma. The KSZ might have been reactivated again during the Paleoproterozoic Capricorn Orogeny because pegmatite veining within the western exposures of the shear zone was dated at 1793 ± 17 Ma (GSWA 178232, Bodorkos et al. 2006).

Mafic and ultramafic rocks along the KSZ at the northern boundary of the KUT have not been dated but are likely to be either c. 3185 Ma intrusions of the Dalton Suite (Hickman and Van Kranendonk 2012) or c. 3200 Ma remnants of the Coondamar Formation. These mafic rocks are closely associated with thin units of chert and BIF, which might be parts of the Coondamar Formation. Some of these units might alternatively be recrystallized mylonites formed when the Mosquito Creek Formation was thrust against the KUT; no definitive studies have been reported.

10.2 Cutinduna Supersuite

The Cooninia Inlier of the KUT (Fig. 1.2) includes porphyritic biotite monzogranite and granodiorite dated at 2897–2896 Ma (Nelson 2005a, b, GSWA 178230, 178,231). Van Kranendonk et al. (2006) assigned these granitic rocks to the Cutinduna Supersuite. Information on this supersuite is limited because the Cooninia Inlier was not within the area remapped during the 1994–2005 Ma Pilbara Craton Mapping Project. Williams (1989) described the granites as post-orogenic, based mainly on composition and a lack of tectonic foliations. He recorded that they have been metamorphosed to lower greenschist facies and that all granitic rocks of the Cooninia Inlier and adjacent inliers of the KUT are unconformably overlain by the Fortescue Group. The published zircon U–Pb age of these granitic rocks suggests intrusion after the 2930–2900 Ma Mosquito Creek Orogeny. Both dated samples of Cutinduna Supersuite were altered and many zircons that were analysed showed effects of Pb loss. However, the least discordant zircon analyses gave $^{207}Pb/^{206}Pb$ ages close to 2900 Ma indicating that the intrusions are distinct from those of the 2851–2831 Ma Split Rock Supersuite.

Rare-metal pegmatites dated at 2890–2880 Ma (Kinny 2000; Sweetapple and Collins 2002) suggest that granitic intrusions of similar age to the Cutinduna

Supersuite might be present in the Carlindi and Yule Domes. Additionally, the timing of gold mineralization in the Pilgangoora area of the East Strelley greenstone belt, adjacent to some of the 2890–2880 Ma rare metal pegmatites, was c. 2890 Ma (Neumayr et al. 1998; Huston et al. 2001, 2002; Baker et al. 2002), indicating local deformation and hydrothermal activity at this time (Blewett and Champion 2005).

10.3 Split Rock Supersuite

The Split Rock Supersuite comprises multiple intrusions of highly fractionated, Sn–Ta–Li bearing, post-orogenic monzogranites that were emplaced in a broad, northwest trending linear belt across the KUT and EPT and into the northeast part of the Mallina Basin (Fig. 10.6) (Van Kranendonk et al. 2006, 2007). Dating of the Split Rock Supersuite has been limited to intrusions in the southeast EPT and in the KUT where ages of 2851–2831 Ma are recorded. However, as noted in the description of the Cutinduna Supersuite, there is evidence suggesting that some granitic intrusions in the Carlindi and Yule Domes might have been emplaced at c. 2880 Ma. Kinny (2000) dated tantalite in a Sn–Ta–Li pegmatite at Pilgangoora at 2879 ± 5 Ma. Intrusions assigned to the Split Rock Supersuite in the Carlindi and Yule Dome (Kadgewarrina, Kimmys Bore, Minnamonica, Poocatche, Gillam, and Numbana

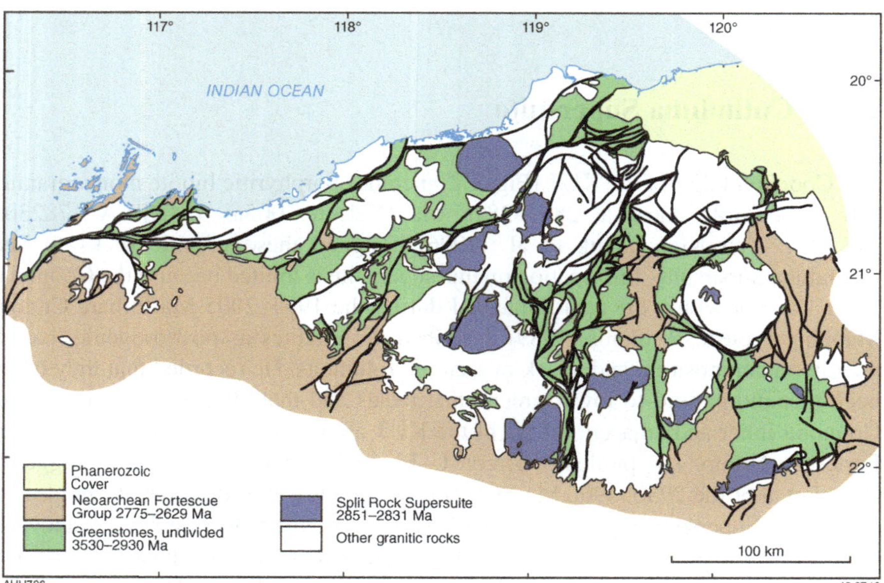

Fig. 10.6 Simplified geological map of the Northern Pilbara Craton showing the distribution of the Split Rock Supersuite (from Hickman 2021; with Geological Survey of Western Australia permission)

Monzogranites) have not been dated by the U–Pb zircon method. Some of these intrusions are weakly foliated, whereas others are non-foliated, suggesting different intrusive events. In the Carlindi Dome, the Minnamonica Monzogranite is interpreted to have been intruded by the Poocatche Monzogranite (Smithies et al. 2002).

In the Corunna Downs Dome, a sample from a thin wedge of the Mondana Monzogranite, which locally intrudes the Boobina Porphyry and the Carbana Monzogranite, was dated at 3317 ± 2 Ma (Barley and Pickard 1999). This age of intrusion conflicts with a previous interpretation that the Mondana Monzogranite is a post-orogenic intrusion (Hickman and Lipple 1975, 1978; Blockley 1980; Davy and Lewis 1981; Hickman 1983). Early attempts to date the Moolyella Monzogranite gave results similar to that for the Mondana Monzogranite, with an initial conclusion that the Moolyella Monzogranite was emplaced at 3313 ± 2 Ma (Nelson 2000, GSWA 142977). However, the highly fractionated chemical composition of the Moolyella Monzogranite, its massive unfoliated texture, and its geological relationship to nearby tin-bearing pegmatites with an interpreted age of 2830 Ma ± 30 Ma (Pidgeon 1978) questioned this result. Subsequent zircon U–Pb dating of the Moolyella Monzogranite indicated a date of 2831 ± 12 Ma (Nelson 2004, GSWA 169044). Consequently, the present interpretation is that the c. 3317 Ma date on the Mondana Monzogranite might represent a pooled age of xenocrystic zircons inherited from the adjacent c. 3315 Ma Boobina Porphyry and Carbana Monzogranite. Geochemical data (Bagas et al. 2003) indicate that the Mondana Monzogranite does not belong to the c. 3315 Ma Emu Pool Supersuite. Differences include its greater fractionation (higher SiO_2, K_2O, Rb, Y, Th), large negative Eu anomalies, and somewhat flatter chondrite-normalized REE patterns (Bagas et al. 2003). K:Th:U gamma-ray spectrometric imagery (Blewett et al. 2000) indicates that the Mondana Monzogranite has high contents of all three elements, which is a characteristic feature of established post-orogenic granites of the East Pilbara.

Geochemical features of the supersuite and Nd $T_{DM}{}^2$ model ages up to 3740 Ma (Table 2.2) indicate magma derivation from partial melting of much older granitic crust (Blockley 1980; Davy and Lewis 1986; Bickle et al. 1989; Champion and Smithies 2000, Smithies et al. 2003; Gardiner et al. 2017, 2018). Much younger Nd $T_{DM}{}^2$ model ages in the Carlindi and Yule Domes (Table 2.2) suggest far more juvenile sources, although ε_{Nd} values are moderately negative to chondritic. Champion and Smithies (2000) described the supersuite as consisting of silica-rich granitic rocks (>73% SiO_2), with high LILE, and moderate to large negative Eu anomalies, and depleted in Sr and undepleted in Y. They interpreted moderate to high Rb, Rb/Sr, Rb/Ba, Ca/Sr, and K_2O/Na_2O and low K/Rb ratios as being consistent with crystal fractionation. Fluorite is a common accessory mineral in most of the dated intrusions.

Field evidence indicates that some of the granites and pegmatites were intruded as sub-horizontal sheets into the earlier granites and greenstones of the craton. This is the situation in the Mount Edgar Dome where the Moolyella Monzogranite includes a sub-horizontal sheet of monzogranite underlain and overlain by banded orthogneiss of the Tambina Supersuite. Likewise, field exposures of the Cooglegong

Monzogranite in the Shaw Dome reveal this to be a sub-horizontal, sheet-like intrusion (Van Kranendonk et al. 2001). At Wodgina, on the boundary between the Yule and Carlindi Domes, almost horizontal Sn–Ta–Li bearing pegmatites cut across folded strata of the Soanesville Group (Van Kranendonk et al. 2006). Intrusion of horizontal sheets suggests reduced vertical pressures within the crust, possibly due to rapid erosion after the Mosquito Creek and North Pilbara Orogenies.

The distribution of the post-orogenic intrusions in a southeast to northwest trending zone across the East Pilbara suggests either some form of structural control or a process producing a succession of intrusions along this trend. One possible explanation is that intrusion of the supersuite occurred as the Pilbara crust drifted southeast to northwest across a hot spot (Hickman 2016). In this scenario, the northwest intrusions might have been emplaced at c. 2880 Ma, about 30 Ma earlier than the dated intrusions in the southeast. A drift rate of 150 km in 30 million years (c. 5 cm/yr) would be consistent with modern rates of plate movement.

References

Bagas L (2005) Geology of the Nullagine 1:100 000 sheet. 1:100 000 Geological Series Explanatory Notes. Geological Survey of Western Australia

Bagas L, Smithies RH, Champion DC (2003) Geochemistry of the Corunna Downs Granitoid Complex, East Pilbara Granite-Greenstone Terrane. In: Geological Survey of Western Australia Annual Review 2001–02. Geological Survey of Western Australia, pp 61–69

Bagas L, Farrell TR, Nelson DR (2004) The age and provenance of the Mosquito Creek Formation. In: Geological Survey of Western Australia Annual Review 2003–04, pp 62–70

Baker DEL, Seccombe PK, Collins WJ (2002) Structural history and timing of gold mineralization in the Northern East Strelley Belt, Pilbara Craton, Western Australia. Econ Geol 97:775–785

Barley ME, Pickard AL (1999) An extensive, crustally-derived, 3325 to 3310 Ma silicic volcanoplutonic suite in the Eastern Pilbara Craton: evidence from the Kelly Belt, McPhee Dome and Corunna Downs Batholith. Precambrian Res 96:41–62

Beintema KA (2003) Geodynamic evolution of the West and Central Pilbara Craton: A mid-Archean active continental margin. PhD, Geologica Ultraiectina, Utrecht University

Beintema KA, de Leeuw GAM, White SH, Hein KAA (2001) Tabba Tabba shear: a crustal-scale structure in the Pilbara Craton, WA. In: Cassidy KF, Dunphy JM, Van Kranendonk MJ (eds) Extended abstracts–4th international Archaean symposium, Perth, 24–28 September 2001. AGSO (Geoscience Australia), pp. 285–287

Beintema KA, Mason PRD, Nelson DR, White SH, Wijbrans JR (2003) New constraints on the timing of tectonic activity in the Archaean Central Pilbara Craton, Western Australia. J Virtual Explorer 13. https://doi.org/10.3809/jvirtex.vol.2003.013

Bickle MJ, Bettenay LF, Chapman HJ, Groves DI, McNaughton NJ, Campbell IH, de Laeter JR (1989) The age and origin of younger granitic plutons of the Shaw batholith in the Archean Pilbara Block, Western Australia. Contrib Mineral Petrol 101:361–376

Blewett RS (2000) A 'virtual' structural field trip of the North Pilbara. Record, 2000/45. Australian Geological Survey Organisation (Geoscience Australia)

Blewett RS, Champion DC (2005) Geology of the Wodgina 1:100 000 sheet. 1:100 000 geological series explanatory notes. Geological Survey of Western Australia

Blockley JG (1980) The tin deposits of Western Australia, with special reference to the associated granites. Mineral Resources Bulletin 12. Geological Survey of Western Australia

Bodorkos S, Love GJ, Nelson DR, Wingate MTD (2006) 178232: trondhjemitic pegmatite vein, Whatsamatta Well; Geochronology Record 654. Geological Survey of Western Australia

Champion DC, Smithies RH (2000) The geochemistry of the Yule Granitoid Complex, East Pilbara Granite–Greenstone Terrane; evidence for early felsic crust: Geological Survey of Western Australia, Annual Review 1999–2000, pp 42–48

Davy R, Lewis JD (1981) The geochemistry of the Mount Edgar Batholith, Pilbara area, Western Australia. In: Glover JE, Groves DI (eds) Archaean geology: second international symposium, Perth 1980. Geological Society of Australia, pp 373–383

Davy R, Lewis JD (1986) The Mount Edgar Batholith, Pilbara area, Western Australia: geochemistry and petrography. Report, vol 17. Geological Survey of Western Australia

Farrell TR (2006) Geology of the Eastern Creek 1:100 000 sheet. 1:100 000 Geological series explanatory notes. Geological Survey of Western Australia

Fitton MJ, Horwitz RC, Sylvester G (1975) Stratigraphy of the early Precambrian in the West Pilbara, Western Australia. Report, FP 11

Gardiner NJ, Hickman AH, Kirkland CL, Lu Y, Johnson T, Zhao J-X (2017) Processes of crust formation in the early earth imaged through Hf isotopes from the East Pilbara Terrane. Precambrian Res 297:56–76. https://doi.org/10.1016/j.precamres.2017.05.004

Gardiner NJ, Hickman AH, Kirkland CL, Lu Y, Johnson TE, Wingate MTD (2018) New Hf isotope insights into the Paleoarchean magmatic evolution of the Mount Edgar Dome, Pilbara Craton: implications for early Earth and crust formation processes. Report, vol 181. Geological Survey of Western Australia

Hickman AH (1981) Crustal evolution of the Pilbara Block. In: Glover JE, Groves DI (eds) Archaean geology: second International Symposium, Perth 1980. Geological Society of Australia, pp 57–69

Hickman AH (1983) Geology of the Pilbara Block and its environs. Bulletin, vol 127. Geological Survey of Western Australia

Hickman AH (2001) Geology of the Dampier 1:100 000 sheet. 1:100 000 Geological series explanatory notes. Geological Survey of Western Australia

Hickman AH (2016) Northwest Pilbara Craton: A record of 450 million years in the growth of Archean continental crust. Report, 160. Geological Survey of Western Australia

Hickman AH (2021) East Pilbara Craton: a record of one billion years in the growth of Archean continental crust. Report 143, Geological Survey of Western Australia

Hickman AH, Lipple SL (1975) Explanatory notes on the Marble Bar 1:250 000 geological sheet, Western Australia. Record, 1974/20. Geological Survey of Western Australia

Hickman AH, Lipple SL (1978) Marble Bar, WA sheet SF50-8, 2nd edn. Geological Survey of Western Australia

Hickman AH, Smithies RH (2000) Roebourne, W.A. Sheet SF 50-3, 2nd edn. Geological Survey of Western Australia

Hickman AH, Van Kranendonk MJ (2012) Early earth evolution: evidence from the 3.5–1.8 Ga geological history of the Pilbara region of Western Australia. Episodes 35:283–297. https://doi.org/10.18814/epiiugs/2012/v35i1/028

Hickman AH, Smithies RH, Pike G, Farrell TR, Beintema KA (2001) Evolution of the West Pilbara Granite–Greenstone Terrane and Mallina Basin, Western Australia–a field guide. Record, 2001/16. Geological Survey of Western Australia

Horwitz RC (1990) Palaeogeographic and tectonic evolution of the Pilbara Craton, Northwestern Australia. Precambrian Res 48:327–340

Huston DL, Blewett RS, Mernagh TP, Sun S-S, Kamprad J (2001) Gold deposits of the Pilbara Craton. Record, 2001/10. Australian Geological Survey Organisation (Geoscience Australia)

Huston DL, Sun S-S, Blewett R, Hickman AH, Van Kranendonk MJ, Phillips D, Baker D, Brauhart C (2002) The timing of mineralization in the Archean North Pilbara Terrain, Western Australia. Econ Geol 97:733–755

Kinny PD (2000) U–Pb dating of rare-metal (Sn–Ta–Li) mineralized pegmatites in Western Australia by SIMS analysis of tin and tantalum-bearing ore minerals. In: Abstracts and Proceedings: beyond 2000, New Frontiers in Isotope Geoscience, Lorne, Victoria, pp 113–116

Krapež B, Eisenlohr B (1998) Tectonic settings of Archaean (3325–2775 Ma) crustal–supracrustal belts in the West Pilbara Block. Precambrian Res 88:173–205

Nelson DR (1997) Compilation of SHRIMP U-Pb zircon geochronology data, 1996. Record, 1997/2. Geological Survey of Western Australia

Nelson DR (2000) Compilation of geochronology data, 1999. Record, 2000/2. Geological Survey of Western Australia

Nelson DR (2004) Compilation of geochronology data, 2003. Record, 2004/2. Geological Survey of Western Australia

Nelson DR (2005a) 178230: biotite granodiorite, Limestone Bore; Geochronology Record 574. Geological Survey of Western Australia

Nelson DR (2005b) 178231: biotite monzogranite, limestone bore; geochronology record 575. Geological Survey of Western Australia

Neumayr P, Ridley JR, McNaughton NJ, Kinny PD, Barley ME, Groves DI (1998) Timing of gold mineralization in the Mount York district, Pilgangoora greenstone belt and implications for the tectonic and metamorphic evolution of an area linking the Western and Eastern Pilbara Craton. Precambrian Res 88:249–265

Nijman W, Clevis Q, De Vries ST (2010) The waning stage of a greenstone belt: the Mesoarchean Mosquito Creek Basin of the East Pilbara, Western Australia. Precambrian Res 180:251–271

Oversby VM (1976) Isotopic ages and geochemistry of Archaean acid igneous rocks from the Pilbara, Western Australia. Geochim Cosmochim Acta 40:817–829

Pidgeon RT (1978) Geochronological investigation of granite batholiths of the Archaean Granite–Greenstone Terrain of the Pilbara Block, Western Australia. In: Smith IEM, Williams JG (eds) Proceedings of the 1978 Archaean geochemistry conference. University of Toronto, Ontario, pp 360–362

Smithies RH (1998) Geology of the Mount Wohler 1: 100 000 sheet. Geological Series Explanatory Notes. Geological Survey of Western Australia

Smithies RH, Champion DC (2000) The Archaean high-Mg diorite suite: links to tonalite–trondhjemite–granodiorite magmatism and implications for early Archaean crustal growth. J Petrol 41:1653–1671

Smithies RH, Champion DC (2002) Assembly of a composite granite intrusion at a releasing bend in an active Archean shear zone. In: Geological Survey of Western Australia Annual Review 2000–01. Geological Survey of Western Australia, Perth, pp 63–68

Smithies RH, Farrell TR (2000) Geology of the Satirist 1:100 000 sheet. 1:100 000 Geological Series Explanatory Notes. Geological Survey of Western Australia

Smithies RH, Hickman AH, Nelson DR (1999) New constraints on the evolution of the Mallina Basin, and their bearing on relationships between the contrasting Eastern and Western Granite–Greenstone Terranes of the Archaean Pilbara Craton, Western Australia. Precambrian Res 94: 11–28

Smithies RH, Champion DC, Blewett RS (2001a) Wallaringa, WA sheet 2656. Geological Survey of Western Australia

Smithies RH, Nelson DR, Pike G (2001b) Development of the Archaean Mallina Basin, Pilbara Craton, Northwestern Australia; a study of detrital and inherited zircon ages. Sedimentary Geol 141–142:79–94

Smithies RH, Champion DC, Blewett RS (2002) Geology of the Wallaringa 1:100 000 sheet. 1:100 000 geological series explanatory notes. Geological Survey of Western Australia

Smithies RH, Champion DC, Cassidy KF (2003) Formation of Earth's early Archaean continental crust. Precambrian Res 127:89–101

Sweetapple MT, Collins PLF (2002) Genetic framework for the classification and distribution of Archean rare metal pegmatites in the North Pilbara Craton, Western Australia. Econ Geol 97: 873–895

Thorpe R, Hickman AH, Davis DW, Mortensen JK, Trendall AF (1992) Constraints to models for Archaean lead evolution from precise U-Pb geochronology from the Marble Bar region, Pilbara Craton, Western Australia. In: Glover JE, Ho SE (eds) The Archaean: terrains, processes and metallogeny: proceedings for the third International Archaean Symposium, 17–21 September 1990. Geology Department and University Extension, The University of Western Australia, Perth, pp 395–408

Tyler IM, Fletcher IR, de Laeter JR, Williams IR, Libby WG (1992) Isotope and rare earth element evidence for a late Archaean terrane boundary in the Southeastern Pilbara Craton, Western Australia. Precambrian Res 54:211–229

Van Kranendonk MJ (2000) Geology of the North Shaw 1:100 000 sheet. 1:100 000 geological series explanatory notes. Geological Survey of Western Australia

Van Kranendonk MJ (2003) Geology of the Tambourah 1:100 000 sheet. 1:100 000 geological series explanatory notes. Geological Survey of Western Australia

Van Kranendonk MJ (2008) Structural geology of the central part of the Lalla Rookh–Western Shaw structural corridor, Pilbara Craton, Western Australia. Report 103. Geological Survey of Western Australia

Van Kranendonk MJ, Collins WJ (1998) Timing and tectonic significance of late Archaean, sinistral strike-slip deformation in the Central Pilbara structural corridor, Pilbara Craton, Western Australia. Precambrian Res 88:207–232. https://doi.org/10.1016/s0301-9268(97)00069-7

Van Kranendonk MJ, Hickman AH, Williams IR, Nijman W (2001) Archaean geology of the East Pilbara Granite–Greenstone Terrane, Western Australia–a field guide. Record, 2001/9. Geological Survey of Western Australia

Van Kranendonk MJ, Hickman AH, Smithies RH, Nelson DN, Pike G (2002) Geology and tectonic evolution of the Archaean North Pilbara Terrain, Pilbara Craton, Western Australia. Econ Geol 97:695–732. https://doi.org/10.2113/gsecongeo.97.4.695

Van Kranendonk MJ, Hickman AH, Smithies RH, Williams IR, Bagas L, Farrell TR (2006) Revised lithostratigraphy of Archean supracrustal and intrusive rocks in the Northern Pilbara Craton, Western Australia. Record, 2006/15. Geological Survey of Western Australia

Van Kranendonk MJ, Smithies RH, Hickman AH, Champion DC (2007) Review: secular tectonic evolution of Archean continental crust: interplay between horizontal and vertical processes in the formation of the Pilbara Craton, Australia. Terra Nova 19:1–38

Williams IR (1989) Balfour Downs, Western Australia, 2nd edn. 1:250 000 Geological Series Explanatory Notes. Geological Survey of Western Australia

Zegers TE (1996) Structural, kinematic, and metallogenic evolution of selected domains of the Pilbara Granite–Greenstone Terrain. PhD, Geologica Ultraiectina, Utrecht University

Zegers TE, de Keijzer M, Passchier CW, White SH (1998) The Mulgandinnah shear zone; an Archean crustal scale strike-slip zone, Eastern Pilbara, Western Australia. Precambrian Res 88: 233–248

Chapter 11
Mineralization in the Northern Pilbara

Abstract Styles of mineralization in the Northern Pilbara Craton, and the size of deposits, reflect the change in tectonic processes from the Paleoarchean to the Mesoarchean. Paleoarchean magmas and hydrothermal fluids were generated during vertical recycling of continental crust, and consequently the 'metal budget' was limited. None of the Paleoarchean gold and base metal deposits are large by international standards. In contrast, Mesoarchean mineralization occurred during plate-tectonic processes that allowed repeated influxes of metal-charged juvenile material, either above subduction zones or along major strike-slip faults.

The source of iron for the large iron ore deposits in Mesoarchean sedimentary basins of the Northern Pilbara was juvenile oceanic crust remote from the depositional sites. In contrast, Paleoarchean iron formations were deposited from local hydrothermal vents in volcanic environments and were formed as thin, low-grade, and laterally discontinuous units.

The most significant mineralization in the Neoarchean Fortescue Group is conglomerate-hosted gold, although sub-economic uranium mineralization is also present in some conglomerates.

Keywords Paleoarchean · Mesoarchean · Gold · Iron ore · Base metals

11.1 Paleoarchean Mineralization

Paleoarchean mineralization of the East Pilbara Terrane (EPT) occurred in an environment of crustal recycling by vertical tectonic processes (Hickman 2021). As a consequence, the scale of mineralization was limited by the amount of metal contained in the underlying crust. For example, almost all Paleoarchean gold mineralization occurred at c. 3315 Ma and is hosted in vertical shear zones and faults formed during doming. Although locally high-grade, Paleoarchean gold deposits in the EPT are small by international standards, almost all mines having recorded production of less than 30,000 oz. In contrast, Mesoarchean Pilbara gold deposits formed in plate-tectonic settings, such as those in the Mosquito Creek and

A. H. Hickman, *Archean Evolution of the Pilbara Craton and Fortescue Basin*, Modern Approaches in Solid Earth Sciences 24,
https://doi.org/10.1007/978-3-031-18007-1_11

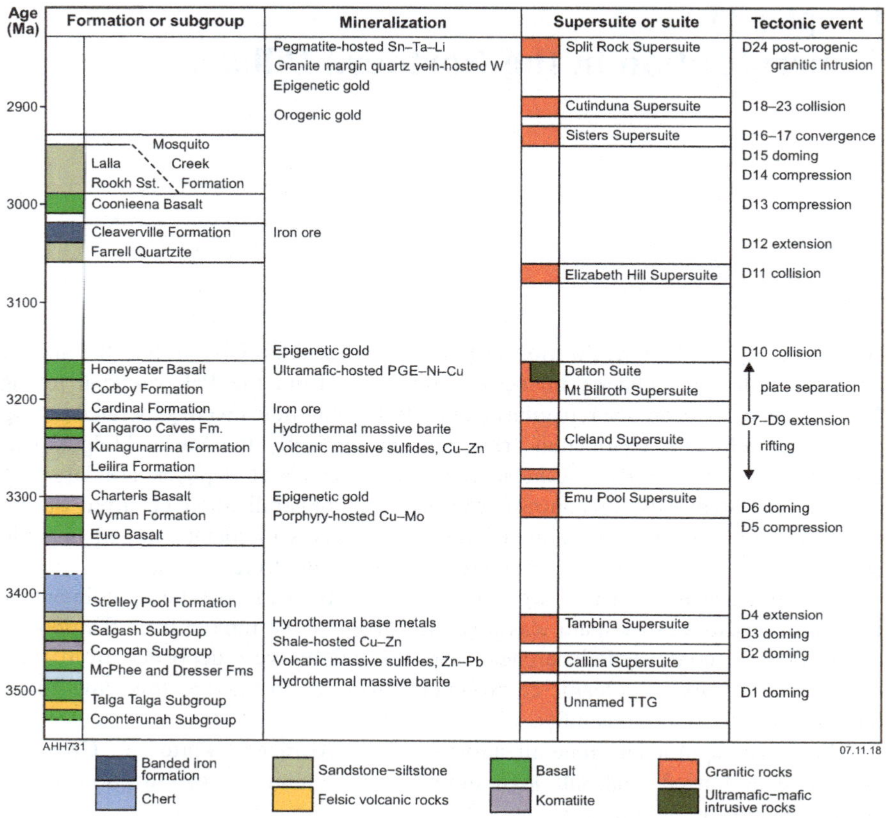

Fig. 11.1 Mineralization in the East Pilbara in relation to stratigraphy and tectonic events. (From Hickman 2021; with Geological Survey of Western Australia permission)

Mallina Basins, have been reported production and resources totalling well over one million oz.

Lithologies of the EPT host a variety of mineralization styles related to hydrothermal activity, volcanism, and magmatic intrusion (Fig. 11.1), including:

- Sediment-hosted, hydrothermal massive sulphates, Dresser Formation.
- Volcanogenic massive sulphides (VMS), Duffer Formation.
- Black shale-hosted Cu–Zn, Apex Basalt.
- Vein and hydrothermal base metals, Panorama Formation.
- Base and speciality metal mineralization at c. 3315 Ma.
- Precious metals, mainly c. 3315 Ma.

11.1.1 Sediment-Hosted, Hydrothermal Massive Sulphates

The oldest economic mineralization of the EPT is hydrothermal barite in the c. 3481 Ma Dresser Formation of the North Pole Dome (Fig. 11.2). The main barite deposits outcrop in an 8-km-long zone within and immediately underlying the Dresser Formation. Mineralization takes the form of stratabound layers and mounds, or veins of coarsely crystalline barite within hydrothermal chert–barite veins (Abeysinghe and Fetherston 1997). The veins intrude extremely altered metabasalt immediately underlying chert of the Dresser Formation. Barite forms coarsely crystalline mounds adjacent to growth faults (Nijman et al. 1998). The largest deposits are located where the Dresser Formation is thickest and consists of three stratigraphic horizons of chert–barite interlayered with metabasalt. The three mineralized beds are bound by an array of extensional normal growth faults, the main set of which passes just south of and through the Dresser mining centre. The geometry of the faults and their close relation to barite mineralization (Nijman et al. 1998) suggests that the barite was deposited during extensional faulting. It has been inferred that the deposits formed by exhalative chemical precipitation. Hickman (1973) interpreted the barite to have been mobilized from sedimentary deposits and concentrated in extensional fractures during doming. However, subsequent interpretations have been that the barite entered the depositional system in low-temperature hydrothermal volcanic emissions through the extensional faults (Van Kranendonk 2006). There is no evidence of c. 3481 Ma felsic volcanic activity at North Pole, but basalts underlying the barite deposits are extensively leached by acid-sulphate alteration (Van Kranendonk and Pirajno 2004) that might explain the source of barium.

11.1.2 Volcanogenic Massive Sulphides

Some of the world's oldest volcanogenic massive sulphide (VMS) Zn–Pb mineralization was discovered in 3474–3459 Ma felsic volcanic rocks of the Duffer Formation on the southwest and southeast sides of the Mount Edgar Dome. The best documented deposits are at the Big Stubby prospect, 6 km south of Marble Bar, and at Lennons Find, 50 km southeast of Marble Bar (Fig. 11.2). Reynolds et al. (1975) interpreted the VMS mineralization south of Marble Bar to be associated with seven rhyolite domes. The domes occupy a zone of stratabound mineralization 1 km long and several hundred metres thick in the upper Duffer Formation. Rhyolite within the domes was derived from a large underlying magma chamber that is now preserved as the c. 3466 Ma Homeward Bound Granite (Hickman and Van Kranendonk 2008). This alkali granite forms a large laccolith-like intrusion into the Warrawoona Group. The domes are capped by red, black, and white banded chert and lenses of massive pyrite.

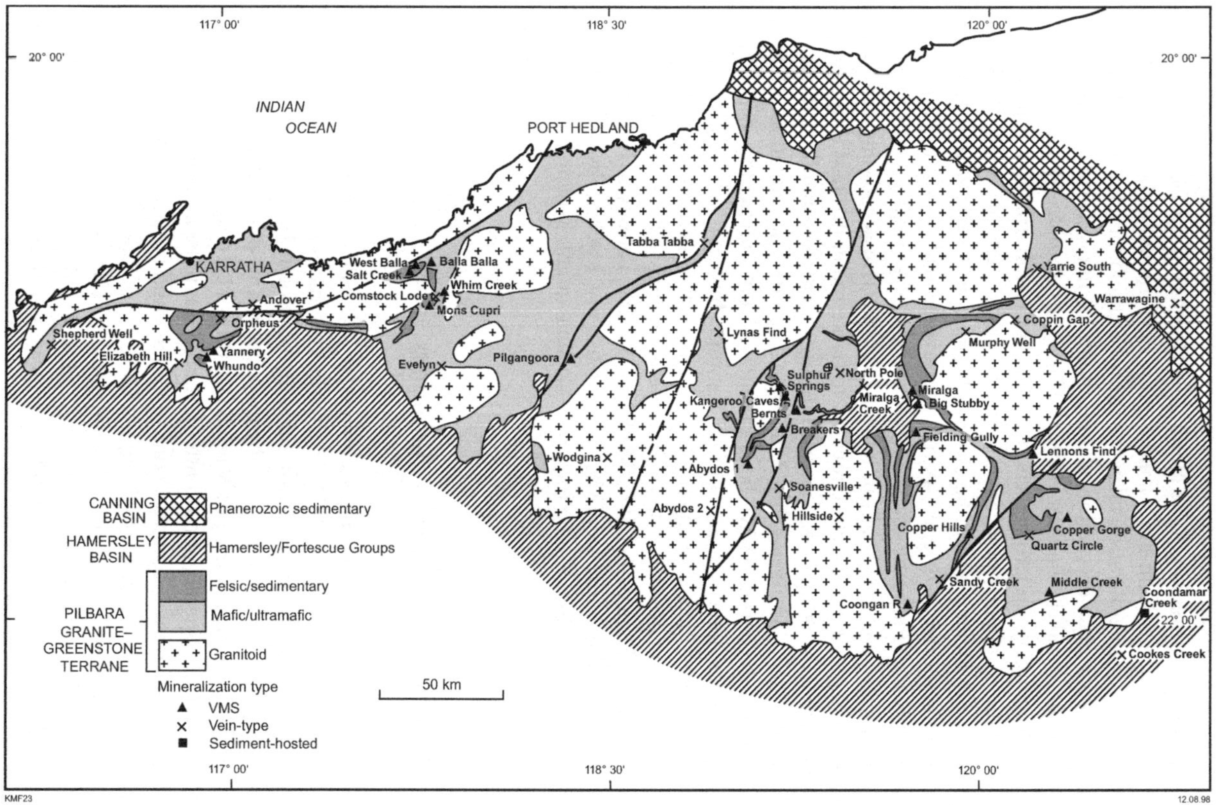

Fig. 11.2 Simplified geological map of the Northern Pilbara Craton, showing the distribution of important lead–zinc–silver occurrences. (From Ferguson 1999; with Geological Survey of Western Australia permission)

The flanks of the domes contain zinc, lead, silver, and barite mineralization. The Zn–Pb mineralization, which includes barite, stratigraphically overlies a copper-rich zone. Barite and gypsum are also located in the lateral extensions of the lenses. Sphalerite, galena, pyrite, and barite are the principal ore minerals. Silver occurs as acanthite. The mineralized zones are thin and discontinuous, with chlorite and carbonate alteration. Drilling intersections reported by Reynolds et al. (1975) included 4.7 m at 14.7% Zn, 3.2% Pb, 0.3% Cu, and 840 g/t Ag (DDH 21B) and 2.52 m at 21.0% Zn, 5.7% Pb, 0.1% Cu, 28 g/t Ag, and 10% Ba (DDH 16). Total resources (non-JORC) at Big Stubby have been estimated as 0.1 to 0.2 Mt. at 13.8% Zn, 4.5% Pb, 0.2% Cu, and 305 g/t Ag (Reynolds et al. 1975; Barley 1982). Reviews of the mineralization are provided by Van Kranendonk (2006), Huston et al. (2007), and Pirajno (2009).

The Lennons Find VMS mineralization occurs at the same stratigraphic level as the deposits at Big Stubby but is more deformed and metamorphosed. Lenses of Zn–Pb–Cu mineralization are located in five mineralized zones over a strike length of 4.3 km. The Duffer Formation is preserved as quartzofeldspathic schist overlain by clastic sedimentary rocks, which are in turn overlain by quartz–muscovite schist. Sulphide minerals are mainly sphalerite, chalcopyrite, galena, and pyrite. Associated minerals are barite, chlorite, carbonate, tourmaline, muscovite, and biotite. The mineralization is zoned, with a zinc-, lead-, and barium-rich zone at the top and a copper-rich zone at the base. The largest sulfidic zone (Hammerhead deposit) is 400 m long, 300 m wide, and 2 m thick, and massive barite occupies the contact with the overlying Apex Basalt.

Pb–Pb dating of galena in the mineralization has given ages of c. 3469 Ma at Big Stubby (Thorpe et al. 1992b) and c. 3472 Ma at Lennons Find [isotope data from Richards et al. (1981); recalculated by Thorpe et al. (1992b)]. Several U–Pb zircon dates on the Duffer Formation and Homeward Bound Granite in the Marble Bar area indicate an age for the host rocks at Big Stubby of c. 3464 Ma, consistent with a U–Pb zircon date of c. 3465 Ma (Thorpe et al. 1992a) for felsic schist at Yandicoogina on strike with the stratigraphy at Lennons Find.

11.1.3 Black Shale-Hosted Cu–Zn

About 15 km south of Marble Bar, black shale within a thin sedimentary member in the lower part of the Apex Basalt contains minor stratabound Cu–Zn mineralization. The sedimentary unit that outcrops intermittently over a strike length of 80 km on the western side of the Mount Edgar Dome, from Doolena Gap in the north, through the Marble Bar greenstone belt west of Marble Bar, southwards to Salgash, Warrawoona, and Yandicoogina. Midway along this strike length, the unit is exposed as the 'Apex chert' (informal name, Schopf 1993) west of Marble Bar Pool. South of Marble Bar at Salgash, it comprises metamorphosed carbonaceous shale and sandstone, mafic and ultramafic breccia, and ultramafic sills. The copper content of the shale is locally sufficiently high at Salgash to have supported small-scale mining. Drilling beneath old copper workings, referred to as the 'Salgash Copper Prospect' by Marston (1979), intersected wide zones with anomalous Cu values. Gossanous material on spoil tips includes malachite, chalcocite, chalcopyrite, and bornite (Marston 1979).

The Salgash mineralization is distinctly different from VMS mineralization in the underlying Duffer Formation (Cu–Zn in black shale as opposed to Zn–Pb in felsic volcanic rocks) and is similar to Besshi-type mineralization as described by Pirajno (2009). Where fully developed, the deposits are conformable, stratiform, blanket-like sheets of massive pyrrhotite or pyrite, or both, with variable contents of chalcopyrite, minor sphalerite, and rare galena. Host rocks are typically mafic volcanic rocks and associated marine sedimentary rocks including wacke, sandstone, and shale. Copper is the principal economic metal, and there is subordinate zinc, cobalt, silver, or gold (Peter and Scott 1997).

11.1.4 Vein and Hydrothermal Base Metals

Pb–Zn mineralization at Quartz Circle in the McPhee Dome, and at Breens Copper and Miralga Creek in the North Pole Dome, is associated with felsic volcanic rocks and related porphyritic granodiorite and monzogranite intrusions of the Panorama Formation. At Quartz Circle (Fig. 11.2), both massive and vein-type base metal mineralization are present, possibly with epithermal affiliations (Ferguson and Ruddock 2001). At Miralga Creek, gold, zinc, lead, and copper mineralization (Goellnicht et al. 1988) is associated with a felsic porphyry stock intruded into the Mount Ada Basalt on the southeast side of the North Pole Dome. Thorpe et al. (1992a) reported a U–Pb zircon crystallization age for this intrusion of 3449 ± 2 Ma. Galena in the Miralga Creek mineralization gave a range of Pb isotope model ages between c. 3447 and 3451 Ma (Groves 1987). The stock includes an outer zone of marginal intrusive breccia with disseminated, stringer, vein, and hydrothermal mineralization in association with porphyritic felsic dykes. Van Kranendonk (1999, 2000) related the mineralization to emplacement of the c. 3446 Ma North Pole Monzogranite in the core of the North Pole Dome. Elsewhere in the dome, the porphyritic felsic dykes radiate from the North Pole Monzogranite to feed into the Panorama Formation (Brown et al. 2006).

The Breens Copper deposit in the west-central part of the North Pole Dome is a complex, northeasterly striking belt of stratabound massive sulphide mineralization (pyrite, chalcopyrite, chalcocite, covellite, and neodiginite) and native copper, 500 m wide, consisting of stockworks and silicified breccia zones (Ferguson and Ruddock 2001). Mineralization is associated with a small dyke of porphyritic felsic rock related to the North Pole Monzogranite (Van Kranendonk 1999).

11.1.5 Copper and Molybdenum Mineralization

A porphyry copper–molybdenum system is developed at Spinifex Ridge where the northern tip of the c. 3315 Ma Coppin Gap Granodiorite intrudes the Marble Bar greenstone belt. Northeast of Nullagine, similar deposits are located at Gobbos, Lightning Ridge, and Reedies where stocks of the c. 3313 Ma Gobbos Granodiorite intrude the McPhee greenstone belt. Copper mineralization (without Mo) is associated with the c. 3315 Ma Boobina Porphyry at the Copper Hills and Kelly mining areas on the eastern margin of the Corunna Downs Dome (Barley and Pickard 1999).

The Spinifex Ridge Cu–Mo deposit, at Coppin Gap (Fig. 11.2), has been the subject of several investigations (Marston 1979; Barley 1982; De Laeter and Martyn 1986; Jones 1990; Van Kranendonk et al. 2006; Huston et al. 2007, 2017; Stein et al. 2007; Cummins and Cairns 2017), although no mining has been recorded. Combined measured and indicated resources are 652.3 Mt. at 0.05% Mo and 0.08% Cu (JORC 2004). Cummins and Cairns (2017) reported that the deposit also contains 40 Moz Ag. The higher grade mineralization (up to 0.18% Mo and 0.30% Cu) is located at the top of a long, narrow stock of porphyritic dacite and microgranodiorite that intruded the Apex Basalt, Panorama Formation, and Euro Basalt from the underlying Coppin Gap Granodiorite. The narrow stock is 1000 m long and in the main mineralized zone is 200 m thick and 50–80 m wide (Cummins and Cairns 2017). Drilling by Australian Anglo American Limited from 1970 to 1973 intersected a grade of 0.13% Mo and 0.23% Cu over 75.6 m (Marston 1979).

The c. 3315 Ma Coppin Gap Granodiorite represents one of several subvolcanic magma chambers contemporaneous with eruption of the 3325–3315 Ma Wyman Formation of the Kelly Group, and it is likely that the porphyritic microgranodiorite stock was emplaced as a late-stage vent structure. Prior to steep tilting of the greenstone succession on the northern side of the Mount Edgar Dome, the stock was approximately vertical. Mineralization within and around the porphyritic microgranodiorite is in the form of a stockwork of quartz and quartz–carbonate veins containing sulphide minerals, particularly molybdenite, chalcopyrite, pyrrhotite, pyrite, and scheelite. Scheelite is concentrated in a zone on the northeast side of the Cu–Mo mineralization (Cummins and Cairns 2017). Historic mining of tungsten is recorded from 'Talga Talga', which was the station property that included the area containing the Spinifex Ridge mineralization. Production of 162 kg of wolframite concentrate containing 95.2 kg WO_3 was attributed to 'Talga Talga' prior to 1977 (Hickman 1983). Potassic alteration introduced potassium feldspar into veins containing Mo and Cu and within the dacite has altered oligoclase to potassium feldspar (Huston et al. 2007). Biotite alteration is present in the porphyritic dacite, and intense sericite alteration surrounds and overprints the potassic alteration throughout the deposit.

Copper mineralization in the Copper Hills and Kelly mining areas on the east side of the Corunna Downs Dome is directly associated with the subvolcanic 3324–3307 Ma Boobina Porphyry. This intrusion, which outcrops over an area of 100 km^2, is essentially a laccolith that in its central outcrop, between the Copper

Hills (Fig. 11.2) and the Kelly mines, is 2 km thick, and was emplaced along the stratigraphic contact between the Euro Basalt and the Wyman Formation. North of Copper Hills, and in the southern area between Kelly and Ryan mines, the laccolith was intruded at lower stratigraphic levels. The northern area is almost devoid of mineralization, whereas close to the upper limit of its intrusion level, between Copper Hills and Kelly, there are numerous copper deposits. This central area of the intrusion includes by far the largest copper mine, Copper Hills; between discovery of the deposit in 1952 and mine closure in 1963, this produced 49.2 t of copper ore and concentrates averaging 35.08% Cu and 15455.67 t cupreous ore and concentrates averaging 12.68% Cu (Marston 1979). Most of the ore consisted of supergene malachite, azurite, chalcocite, and bornite. Features of the Cu mineralization at Copper Hills, including sericitization of feldspar, alteration of rare biotite phenocrysts to chlorite and sericite, and groundmass recrystallization to very fine-grained quartz, sericite, carbonate, chlorite, and rutile (Bagas et al. 2004), indicate a low-temperature, epithermal mineralization system.

At the southern end of its intrusion, the c. 3315 Ma Boobina Porphyry splits into several thin transgressive sills within the Euro Basalt. These southern sills are associated with relatively minor copper mineralization along faults. Minerals mined at Kelly included malachite, chrysocolla, azurite, bornite, cuprite, and chalcocite. In contrast to the Cu–Mo mineralization at Spinifex Ridge and Gobbos, quartz stockworks are absent at both Copper Hills and the Kelly mining area.

The stratigraphy and structure of the Corunna Downs Dome suggest that it was mainly formed at c. 3315 Ma. In this interpretation, the Euro Basalt would have been relatively flat-lying when the Boobina Porphyry was intruded, although south of Copper Hills there is a low-angle unconformity between the Euro Basalt and the Wyman Formation. The Boobina Porphyry, which intrudes volcanic rocks of the Wyman Formation between Copper Hills and Kelly, was therefore intruded very close to the c. 3315 Ma land surface, consistent with epithermal mineralization.

11.1.6 Precious Metals

Several gold and silver deposits in the EPT include minor galena that has been dated by the Pb–Pb method to indicate ages between c. 3430 and c. 3400 Ma (Thorpe et al. 1992b; Huston et al. 2001a, 2002; Zegers et al. 2002). However, very little EPT gold mineralization is now interpreted to be this old. For example, several >3400 Ma Pb isotope ages were obtained from the Bamboo Creek mining area on the northeast side of the Mount Edgar Dome (Fig. 3.18). Because the gold mineralization at this mining centre is hosted by a shear zone that cuts the 3350–3335 Ma Euro Basalt (Fig. 11.3), the Pb isotope ages indicate derivation of Pb from older crust. This interpretation is consistent with Sm–Nd and Lu–Hf isotope evidence that most Paleoarchean Pilbara rocks were derived from evolved crustal sources (Champion and Smithies 2007; Champion 2013; Champion and Huston 2016; Gardiner et al. 2017, 2018). The true age of the Bamboo Creek gold mineralization is here

Fig. 11.3 The pit and adit of the Bamboo Queen gold mine (looking northwest) on the Bamboo Creek Shear Zone. Gold mineralization is within sheared komatiitic rocks of the Euro Basalt. (From Ferguson and Ruddock 2001; with Geological Survey of Western Australia permission)

interpreted to be c. 3315 Ma, when gold-bearing hydrothermal fluids were introduced during major diapiric uplift of the Mount Edgar Dome.

Hydrothermal fluids were most likely derived by metamorphic devolatilization of greenstones (Phillips and Powell 2010) sagducted vertically during the c. 3315 Ma event. If so, this explains the relatively small size of the EPT gold deposits because limited volumes of source material would have been available compared to source-rock volumes possible during horizontal tectonic processes such as subduction. Precipitation of gold from silica-rich hydrothermal fluids was influenced by wallrock compositions, with ultramafic rocks and carbonaceous shale and chert being particularly favourable. In the Warrawoona Syncline, about 20 km southeast from Marble Bar (Fig. 11.4), the Warrawoona Group is exceptionally attenuated and sheared between the Mount Edgar and Corunna Downs Domes. Historically, gold mining at Warrawoona was concentrated along a few narrow shear zones, in particular the Klondyke and Copenhagen Shear Zones (Jones 1938; Hickman 1983; Huston et al. 2001a, 2002; Kloppenburg 2003).

11.1.6.1 Gold Mineralization during Doming

Gold mineralization in the EPT is mostly contained in hydrothermal quartz veins hosted by faults and shear zones. Because evolution of the EPT was dominated by doming and sagduction, gold mineralization coincided with doming events and was focused along dome-boundary faults or ring faults parallel to dome margins (Fig. 11.4). For example, the mineralized Bamboo Creek Shear Zone (BCSZ) on

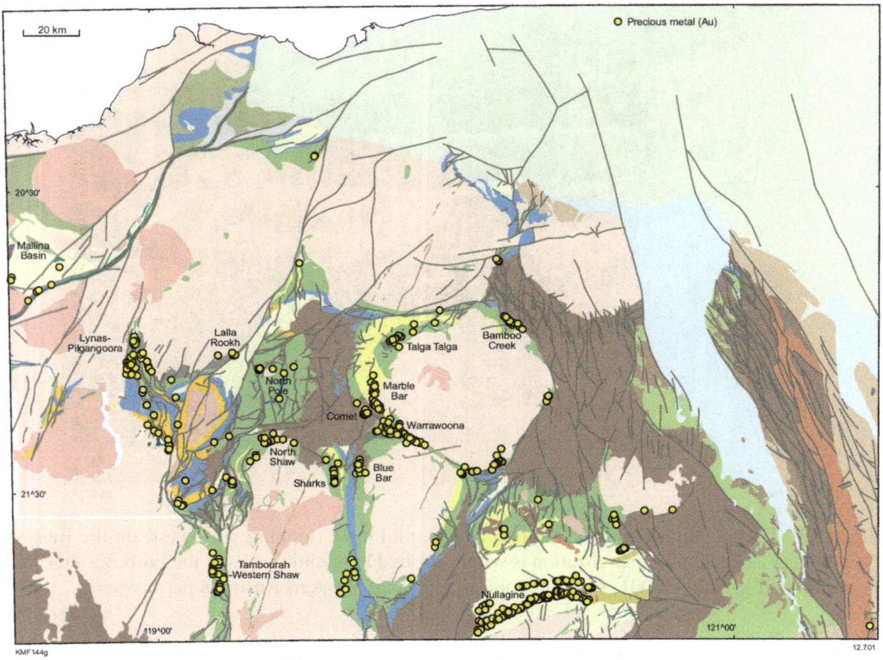

Fig. 11.4 Simplified geological map of the East Pilbara Terrane, showing the distribution of vein and hydrothermal gold deposits. Apart from gold mineralization within the Mosquito Creek Formation east of Nullagine, most of the deposits are located on shear zones within greenstones close to the granitic cores of the granite–greenstone domes. (From Ferguson and Ruddock 2001; with Geological Survey of Western Australia permission)

the northeast side of the Mount Edgar Dome (Fig. 3.18) developed on a major ring fault during D_6 at c. 3315 Ma. The hydrothermal fluids responsible for the gold mineralization along the BCSZ were most likely derived during the event that introduced felsic magma of the Emu Pool Supersuite through metamorphism of older crust.

The main mineralized structure at Warrawoona is the Klondyke Shear Zone (KLSZ, Fig. 3.18) that includes historically important mines such as Klondyke Boulder, Klondyke Queen, and Comet. Galena samples from the Klondyke group of workings yielded Pb isotope model ages of c. 3385, 3374, and 3050 Ma (Huston et al. 2001a, 2002). The c. 3385 and 3374 Ma model ages are unlikely to indicate the age of the mineralization because these ages fall within the c. 75 Ma year period of crustal stability during deposition of the Strelley Pool Formation. The present interpretation is that these ages are too old due to inclusion of Pb from recycling of older crust and that the true age of the KLSZ mineralization is c. 3315 Ma. The c. 3050 Ma Pb model age might be due to renewed hydrothermal activity on the KLSZ during the Prinsep Orogeny, or reactivation of the shear zone during the c. 2950 Ma North Pilbara Orogeny, again with introduction of older crustal Pb.

11.2 Mineralization during the EPTRE

The East Pilbara Terrane Rifting Event (EPTRE, Chap. 6) commenced during deposition of the Sulphur Springs Group and ended following deposition of passive margin successions such as the Soanesville Group. This stage in the evolution of the Pilbara Craton resulted in several different types of mineralization. For example, extensional faulting between c. 3280 and 3220 Ma resulted in syngenetic VMS Cu–Zn mineralization at Sulphur Springs and Kangaroo Caves (Fig. 11.2) (Vearncombe et al. 1995, 1998; Brauhart et al. 1998; Brauhart 1999; Van Kranendonk 2000, 2006; Huston et al. 2019; Pirajno and Huston 2019). Crustal extension in passive margin basins after the c. 3220 Ma breakup resulted in rift-related mafic intrusion and volcanism, providing potential for Ni–Cu and PGE mineralization.

11.2.1 Sulphur Springs Group

VMS Cu–Zn mineralization at the top of the Sulphur Springs Group is attributed to hydrothermal circulation through extensional faults above intrusions of the c. 3239 Ma Strelley Monzogranite (Brauhart et al. 1998; Van Kranendonk 2000, 2006; Huston et al. 2001b, 2007, 2019). Regional crustal extension across the EPT commenced at c. 3280 Ma and intrusion of the laccolith occurred within that tectonic environment. Sulphide mineralization is vertically zoned, from Cu rich at the base to Zn (± Pb) and barite at the top. Most of the mineralization is either immediately beneath or within the formerly named 'marker chert' at the top of the Kangaroo Caves Formation. Chlorite alteration of the Sulphur Springs Group 1000–2000 m beneath the VMS mineralization caused about 80% loss of ore-related metals, in particular Zn, Pb, Cu, and Mo, from which it is inferred that this chloritized zone was the source of the metal in the overlying mineralization (Huston et al. 2001b, 2007, 2019). Mineral resource estimates at Sulphur Springs have included 13.8 Mt. at 3.8% Zn, 1.5% Cu, and 18.0 g/t Ag and at Kangaroo Caves 3.6 Mt. at 6.0% Zn, 0.8% Cu, 0.32% Pb, and 15.2 g/t Ag (JORC 2012). Sulphide minerals include pyrite, chalcopyrite, sphalerite, and galena. Abundant barite is present at the top of several of the deposits (Vearncombe et al. 1995). Minor molybdenum, tin, and gold mineralization were also associated with emplacement of the intrusion.

11.2.1.1 Hydrothermal Barite Mineralization

In the western Warralong greenstone belt, about 110 km southeast of Port Hedland, the Sulphur Springs Group contains large hydrothermal veins of barite discovered during GSWA regional mapping. Based on surface exposures north of Cooke Bluff Hill, the larger veins contain about 10,400 t of massive barite to a depth of 10 m (Hickman 1977). The mineralization is adjacent to the faulted northwest margin of the LWSC which originated during extension and rifting of the EPT from c. 3280 Ma onwards. Hydrothermal veins of barite also intrude the western margin of the Warralong greenstone belt 20 km northeast of the Cooke Bluff Hill deposits (Van Kranendonk 2010), suggesting a zone of hydrothermal activity during early rifting. Abundant hydrothermal barite of the same age is also present in the rifting-related VMS deposits of the c. 3240 Ma Kangaroo Caves Formation in the Soanesville greenstone belt (Vearncombe et al. 1995). Analysis of the barite and gossans at Cooke Bluff Hill did not reveal any significant base metal or gold anomalies.

In view of the suggested correlation between the Sulphur Springs Group and the Fig Tree Group of South Africa (Chap. 6), it is notable that there are large sediment-hosted barite deposits in the 3260–3230 Ma Mapepe Formation in the Barberton Greenstone Belt (BGB). Descriptions of these deposits by Gutzmer et al. (2006) indicate that the BGB barite mineralization is similar to that in the sedimentary succession in the Warralong greenstone belt (Van Kranendonk 2004).

Chert–barite veins and sills at the top of the Kelly Group in the northern McPhee greenstone belt (Williams and Bagas 2007) might have been introduced at c. 3240 Ma rather than during the final stages of Wyman Formation volcanism at c. 3315 Ma. Approximately the same stratigraphic level is intruded by barite veins at Lionel North between the McPhee and Corunna Downs Domes (Thom et al. 1979; Bagas 2005). The host rocks (chert and sandstone) are undated but stratigraphically overlie the Wyman Formation.

11.2.2 Roebourne Group

11.2.2.1 Komatiite-Hosted Ni–Cu

The Ruth Well Ni–Cu deposits are hosted by serpentinized spinifex-textured peridotite flows in the c. 3280 Ma Ruth Well Formation. Detailed reports were provided by Tomich (1974) and Marston (1984). The mineralization consists of violaritized pentlandite, pyrrhotite, gersdorffite, niccolite, chalcopyrite, and magnetite, which is a similar mineral association to komatiite-hosted nickel deposits in the Kambalda area of the Yilgarn Craton. One diamond drill hole intersected 8.38 m of mineralization averaging about 3.52% Ni and 0.78% Cu (Marston 1984), but exploration indicated that the deposits are relatively small, with resources quoted at approximately 70,000 t at a grade of 3% Ni (Marston 1984).

11.2.3 VMS Cu–Zn Mineralization, Tabba Tabba Shear Zone

The Tabba Tabba Shear Zone (TTSZ) contains deposits of c. 3250 Ma VMS Zn–Pb mineralization associated with felsic volcanic and granitic rocks, although primary relationships have been destroyed by strong deformation. At Orchard Well (also known as Turner River VMS), exploration has suggested a total mineral resource (JORC 2012) of 3.5 Mt. at 3.2% Zn, 1.3% Pb, 0.8 g/t Au and 110 g/t Ag (De Grey Mining Ltd. 2020).

11.2.4 Soanesville Group

During breakup of the EPT, crustal extension established new tectonic settings including local continental rift systems followed by passive-margin basins (Chap. 8). Clastic sedimentary basins of the 3223–3165 Ma Soanesville Group included units of BIF up to 1 km thick, and in places these were subsequently enriched to form large iron ore deposits (e.g. Mount Webber and Iron Bridge).

11.2.4.1 Iron Ore: Supergene Enrichment of Banded Iron Formation

The passive margin setting of the lower Soanesville Group after breakup of the EPT included deposition of banded iron-formation (BIF) protoliths, in particular within the succession that became the c. 800 m-thick Pincunah Banded-Iron Member (PBIM) of the Cardinal Formation. Other clastic sedimentary formations of the Soanesville Group (Corboy, Paddy Market, and Pyramid Hill Formations) also contain BIF, although these are either much thinner than the PBIM or have not been sufficiently enriched to warrant mining. The depositional setting of the PBIM is interpreted to have been similar to that of Neoarchean–Paleoproterozoic BIF units of the Hamersley Group (Marra Mamba Iron Formation and Brockman Iron Formation). It is generally agreed the Hamersley Group was deposited on a continental margin platform during breakup of the Pilbara Craton (Morris and Horwitz 1983; Tyler and Thorne 1990; Blake and Barley 1992; Blake 1993; Martin et al. 1998; Thorne and Trendall 2001; Krapež et al. 2003; Bekker et al. 2010, 2014).

Mining of iron ore from BIF of the Soanesville Group has occurred at the Mt. Webber deposits in the Emerald Mine greenstone belt (Fig. 1.7). Published measured iron ore resources at Mt. Webber are 37.5 Mt. at 57.7% Fe. Duuring et al. (2017) reported that the Mt. Webber deposits were developed by supergene enrichment of quartz–magnetite BIF units where these are thickened in the hinge areas of tight synclines. In contrast to the steeply inclined BIFs, the supergene goethite ± martite ores are flat-lying, extend to depths of about 80 m, and are locally capped by nodular laterite.

Other potentially mineable iron ore deposits hosted by the PBIM are present in the Pincunah greenstone belt and are mainly unenriched BIF at the North Star, Eastern Limb, Glacier Valley, and West Star deposits, collectively referred to as the Iron Bridge Magnetite Project. Reported total resources of these deposits were 5448 Mt. at about 30.4% Fe. The mineralization extends along a strike length of >10 km on the eastern limb of the major syncline that forms the Pincunah greenstone belt. Drilling indicates that the mineralization extends to depths in excess of 600 m.

11.2.4.2 Ni–Cu and PGE Mineralization

Ultramafic–mafic intrusions of the c. 3185 Ma Dalton Suite (Chap. 6) were intruded during the 3280–3165 Ma EPTRE (Hickman 2012). The intrusions include layered sills in the Soanesville greenstone belt (Fig. 1.7), vertical intrusions between EPT domes, and sills between the Mosquito Creek Basin and the EPT and KUT.

Drilling of differentiated sills and layered intrusions at Soanesville intersected nickel sulphides in serpentinized peridotite. Marston (1979) reported that 11 diamond drill holes intersected low-grade disseminated sulphides and 2 adjacent drill holes intersected narrow widths (0.4 m) of high-tenor massive sulphides (20–26% nickel). Average intersections for two drill holes were 3.50 m at 2.55 wt.% Ni and 1.16 wt.% Cu and 3.66 m at 2.41 .wt.% Ni and 0.61 .wt.% Cu. Up to 0.55 wt.% Co was reported in these intersections (Ferguson and Ruddock 2001).

11.2.4.3 Gold Potential of the Nickol River Formation

The Nickol River Basin (Chap. 7) is likely to have received large quantities of detrital gold during post-breakup erosion of the EPT. The EPT contains numerous gold deposits older than 3220 Ma (Huston et al. 2002a, b), and erosion of this crust during rifting and plate separation must have shed detrital gold into the passive margin basins. Dark heavy mineral bands are locally present in the Nickol River Formation, and alluvial and colluvial deposits derived from weathering of the formation contain detrital PGEs including Os (Powell and Horwitz 1994) and Au at Lower Nickol. The PGEs were probably derived from erosion of peridotite in the Ruth Well Formation. There are no reported studies on background gold contents of quartzite and conglomerate in the Nickol River Formation, but these units could potentially host placer deposits. Thrusting and metamorphism of the Nickol River Formation are likely to have mobilized any placer gold and concentrated it in shear zones and brittle faults.

11.3 Mesoarchean Mineralization

11.3.1 Mineralization during Closure of the Regal Basin

Plate separation following breakup of the EPT was replaced by c. 3160 to 3070 Ma convergence. In the Northwest Pilbara Craton, the beginning of convergence was marked by a metamorphic event at 3160–3140 Ma (Karratha Event, Chap. 7). Compression of the Regal Basin resulted in subduction forming the Sholl Terrane, and obduction of part of the Regal Formation (c. 3200 Ma juvenile basaltic crust) onto the Paleoarchean Karratha Terrane to form the ophiolite succession of the Regal Terrane.

11.3.1.1 Karratha Event Mineralization

At c. 3165 Ma, plate divergence of the previous 60 million years was replaced by convergence, and TTG of the Mount Billroth Supersuite was intruded between the EPT and the Regal Basin. The influx of magma from juvenile sources at this time might have introduced convergent margin mineralization. Although no c. 3165 Ma East Pilbara mineral deposits have been identified, it is possible that gold mineralization in the Pilbara Well and East Strelley greenstone belts and copper mineralization in the Wodgina greenstone belt were introduced at this time. Using the method of Thorpe et al. (1992b), which compares $^{206}Pb/^{204}Pb$, $^{207}Pb/^{204}/Pb$, and $^{208}Pb/^{204}Pb$ ratios, a Pb–Pb model age of c. 3142 Ma was obtained from galena in an auriferous quartz vein at the McPhees gold deposit (Huston et al. 2001a, 2002a, b; Blewett and Champion 2005).

11.3.1.2 Sholl Terrane VMS

The Sholl Terrane contains volcanic-hosted massive sulphide (VMS) copper–zinc deposits at Whundo, West Whundo, Yannery Hill, and Ayshia. The closely spaced Whundo and West Whundo deposits (Fig. 11.2) are hosted by quartz–chlorite schist and quartz–sericite–chlorite schist within metabasalt close to the stratigraphic contact between the Nallana and Tozer Formations of the Whundo Group. Copper was mined intermittently at Whundo from 1911, and prior to renewed mining in 2006, total historical production stood at 12,000 tonnes of supergene ore grading 22.3% Cu (Collins and Marshall 1999a). Mining at West Whundo in 2006 produced over 5600 t of Cu-in-concentrate.

Roberts (1974) described two distinct types of primary sulphide mineralization at Whundo: fine- to medium-grained layered pyrite, sphalerite, and chalcopyrite and massive medium- to coarse-grained pyrite and pyrrhotite with minor sphalerite and chalcopyrite. At West Whundo, layered pyrite–sphalerite–chalcopyrite with disseminated magnetite is overlain by massive pyrrhotite and pyrite essentially devoid of sphalerite and chalcopyrite. Supergene mineralization is mainly massive chalcocite and goethite–limonite, with minor malachite, cuprite, and native copper.

The mineralization at Whundo and West Whundo is primarily VMS, but the thickness of the main deposits and their location as ore shoots was subsequently influenced by north-trending folding and north-northwest striking faults (Collins and Marshall 1999a). The tectonic foliation of the schist which hosts the deposits was probably produced by earlier shearing associated with D_2 thrusting. This would have occurred preferentially along the major lithological contact between the Nallana and Tozer Formations at Whundo. Evidence that the shearing at Whundo and West

Whundo was restricted to this narrow stratigraphic level is provided by the observation that overlying rocks of the Tozer Formation contain extremely well-preserved sedimentary structures.

The Cu–Zn mineralization at Yannery Hill, 1–2 km northeast of Whundo, is hosted by variably pyritic chloritic schist within metamorphosed volcanogenic sedimentary rock in the lower part of the Tozer Formation. Other local lithologies include quartz–sericite schist and chlorite–quartz–sericite schist with local andalusite porphyroblasts. Primary copper–zinc sulphide mineralization is concentrated in one stratabound unit that has been extensively folded within a northwest-plunging syncline. Thickening of the unit in fold cores controlled the geometry of most stopes. Ore minerals are secondary and include malachite, chalcocite, and cuprite within massive limonite (Marston 1979). Approximately 3045 t of high-grade oxidized and supergene ore averaging 15.9% Cu were produced prior to 1999.

11.3.1.3 Gold Mineralization

Gold mineralization in the area where the TTSZ is crossed by the Turner River has not been dated but, based on the interpreted 3160–3070 Ma compressional event in the TTSZ, might include mineralization related to convergence. A 3069 ± 41 Ma event in the shear zone is indicated by U–Pb zircon dating of Mesoarchean granitic rocks, and xenocrystic zircons in granitic rocks of the Sisters Supersuite were dated at 3123–3108 Ma (Beintema 2003; Beintema et al. 2003).

11.3.2 Gold and Copper North of the Sholl Shear Zone

The Regal Thrust (Chap. 7) extensively deformed basalt and komatiite at the base of the Regal Formation, the sedimentary Nickol River Formation, and basalt in the upper part of the Roebourne Group. Although the zone of shearing and faulting is only about 1 km wide, most of the known gold and copper mineralization in the Roebourne greenstone belt is situated within this zone (Hickman and Strong 2003). Accordingly, the distribution of known gold deposits within the Roebourne greenstone belt shows a strong correlation with the Regal Thrust (e.g. at Lower Nickol, Weerianna gold mine, Carlow Castle, and Sing Well) (Hickman 2002). This zone of gold mineralization was folded by the c. 2940 Ma Roebourne Synform and Prinsep Dome during the North Pilbara Orogeny. Gold mineralization is interpreted to have occurred between 3160 Ma and 3070 Ma, most likely during the Prinsep Orogeny at c. 3070 Ma. Sources of the gold might have included conglomerate and sandstone of the Nickol River Formation.

11.3.3 Mineralization in the De Grey Superbasin

Across the Northern Pilbara Craton, the c. 3067–2919 De Grey Superbasin comprises the Gorge Creek, Whim Creek, and Croydon Basins (Chap. 9), all of which are mineralized. Contemporaneous intrusive rocks emplaced within three magmatic arcs are represented by the Orpheus, Maitland River, and Sisters Supersuites (Chap. 8).

11.3.3.1 Iron Ore in the Gorge Creek Basin

In terms of production and resources, iron ore is by far the most important mineral commodity in the Pilbara Craton. Mining of 60% Fe hematite–goethite iron ore from deposits within the Cleaverville Formation commenced in the East Pilbara Craton in the mid-1960s and continues to the present. Important mines have been located at Mount Goldsworthy, Nimingarra, Shay Gap, Sunrise Hill, Yarrie (Fig. 11.5), Wodgina, and Pardoo. Deposits under development are located at McPhee Creek, Corunna Downs, and Abydos. The Pardoo iron project in the Ord Ranges includes several deposits (including Ridley, South Limb, Bobby, and Emma) and is within an inlier of the Gorge Creek Group in the Mallina Basin (Hickman 2016). Individual deposits were reviewed by Duuring et al. (2017).

It is notable that almost all of the mined iron ore deposits in the Northern Pilbara Craton are supergene enrichments of BIF that immediately underlie a Mesozoic–

Fig. 11.5 View of the Yarrie iron ore mine, showing the basal unconformity of the Gorge Creek Group on granitic rocks of the Warrawagine granitic complex. A thin sandstone of the Farrel Quartzite separates c. 3020 Ma BIF of the Cleaverville Formation from the underlying c. 3430 Ma granitic rocks. Kimberley Gap and terminal loop of Yarrie–Port Hedland Railway separate the mine from the Callawa Plateau to the south (from Ferguson and Ruddock 2001; with Geological Survey of Western Australia permission)

Cenozoic peneplain. North of the Fortescue River, this old land surface is preserved as isolated flat-topped hills capped by ferricrete (Hickman 1983). From the Neogene onwards, the surface was deeply dissected by northerly flowing river systems and is now best preserved above those lithologies most resistant to erosion, including chert, BIF, and quartzite. Duuring et al. (2017) commented that all iron ore deposits of the Northern Pilbara Craton display broad, near-surface, supergene goethite ± martite ores that constitute the bulk of the high-grade mineralization. In most deposits, the high-grade ores are limited to depths of between 50 and 200 m from the surface.

11.3.3.2 Mallina Basin Mineralization

The 2955–2940 Ma crustal evolution of the upper Mallina Basin succession in the Central Pilbara Tectonic Zone resulted in a greater diversity of mineralization than anywhere else in the Northern Pilbara Craton. Coinciding with intrusion of the Sisters Supersuite and mafic volcanism of the Bookingarra Group, 2955–2940 Ma mineralization included Au mineralization related to high-Mg diorites (sanukitoids) and shear zones produced by oblique convergence. In the Whim Creek Basin, there are base metal deposits of either VMS or epigenetic origin (Mons Cupri and Comstock), VMS Cu-Zn and Cu–Pb-Zn deposits (Whim Creek, Salt Creek, and Balla Balla), and V-Ti-Fe deposits in layered mafic intrusions (Balla Balla). The tectonic settings were mainly extensional, with the central part of the Mallina Basin being a retro-arc basin, whereas the Whim Creek Basin was on the southeast margin of a continental magmatic arc (Chap. 8).

11.3.3.2.1 Sanukitoid-Related Gold and PGE Mineralization

The Mallina Basin contains numerous high-Mg diorite (sanukitoid) intrusions (Chap. 8), the largest of which are shown on Fig. 11.6. First identified during the 1994–2005 Pilbara Craton Mapping Project (Chap. 1), the intrusions were emplaced into sedimentary rocks of the Croydon Group between 2954 and 2945 Ma. Mapping and geochemistry have revealed a northeast–southwest strike length of at least 150 km (Smithies and Champion 1999, 2000; Smithies 2002; Smithies and Hickman 2003; Hickman 2004; Smithies et al. 2004a, b).

Mineral exploration in the Mallina Basin has indicated a close spatial association between small high-Mg diorite intrusions and earlier ultramafic intrusions of the Langenbeck Suite, probably because both types of intrusion were emplaced along faults within the Mallina Basin. Drilling by De Grey Mining Limited has revealed extensive gold mineralization within small high-Mg diorite intrusions although, at the time of writing, mining has been limited to historic gold production at the Toweranna mine where gold was first discovered over 120 years ago. This old mining area is located 50 km west-southwest of the recent exploration activity.

High-Mg diorite intrusions have the potential to host platinum group element (PGE) mineralization, as at Lac des Iles in the Wabigoon Subprovince of Ontario

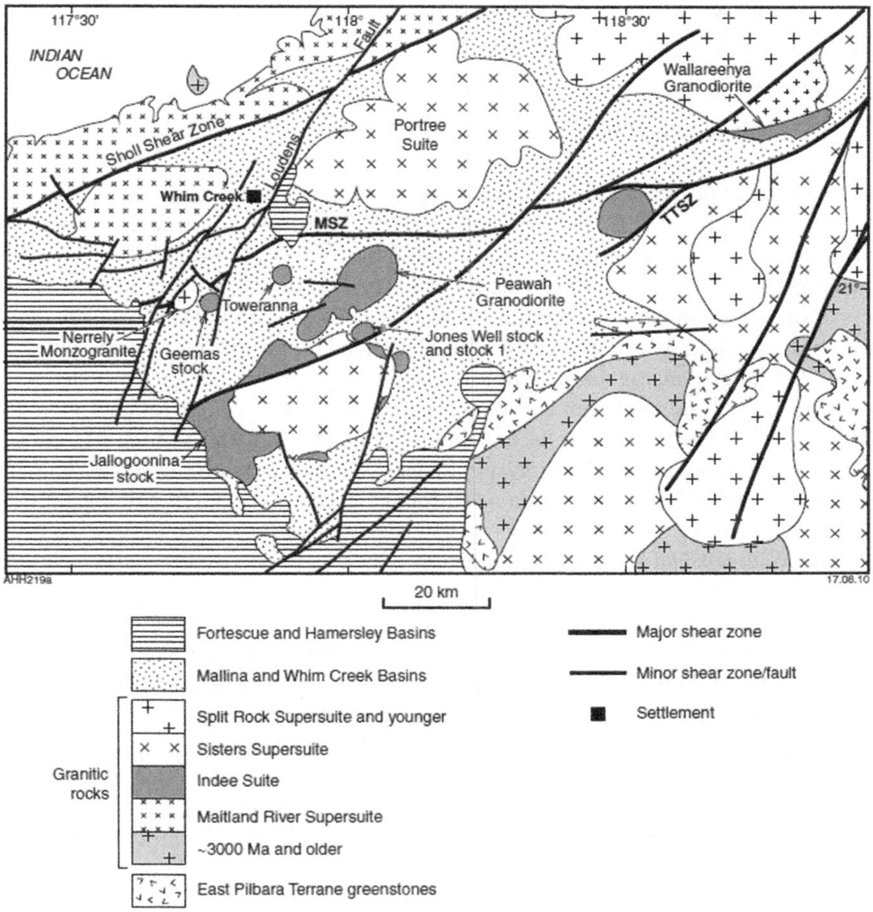

Fig. 11.6 Simplified interpreted bedrock geological map of the western part of the Mallina Basin, showing the largest high-Mg diorite (sanukitoid) intrusions of the Indee Suite. Extensive gold mineralization has been discovered under about 30 m of regolith north and west of the unnamed high-Mg diorite near Mount Dove between the Wallareenya and Peawah Granodiorites. MSZ, Mallina Shear Zone; TTSZ, Tabba Tabba Shear Zone. Star shows the location of the recently discovered Hemi gold mineralization (modified from Smithies and Champion 1999; with Geological Survey of Western Australia permission)

(Hickman 2004). Immediately north of the Tabba Tabba Shear Zone, between the Turner and Yule Rivers, PGE mineralization was discovered along a peridotite–pyroxenite contact next to a large high-Mg diorite intrusion north-northeast of Mount Dove. The best intersection reported from the prospect, known as Three Kings, was 6 m at 2.09 g/t PGE, which included 2 m at 3.18 g/t PGE and 1 m at 4.56 g/t PGE (De Grey Mining Limited 2005). Drilling 1 km to the east provided an intersection of 14 m at 1.32 PGE, whereas 3 km to the west an intersection of 6 m at

0.85 g/t was recorded. The Three Kings area is poorly exposed, and drilling was mainly focused on areas of outcrop.

11.3.3.2.2 Orogenic Lode Au and Au–Sb Deposits

Detailed descriptions of lode Au deposits in the Mallina Basin that were known prior to 2002 were provided by Huston et al. (2002b). The two main deposits recognized at that time were Withnell and Camel, both on the Mallina Shear Zone close to its junction with the Mount Wohler Shear Zone. The two shear zones are separated by a c. 2948 Ma high-Mg diorite intrusion, the Peawah Granodiorite which, as a pre-existing rigid body, probably caused deflection of the two shear zones at c. 2940 Ma. An analogous structural setting exists for another group of lode Au deposits in the Mallina Basin east of the Turner River, where the Tabba Tabba and Mallina shear zones were deflected around a large c. 2950 Ma high-Mg diorite intrusion at Mount Dove. The largest gold deposit in this area is Wingina Well on the TTSZ and others include Amanda and Mount Berghaus (De Grey Mining Limited 2012).

The Withnell and Camel mineralization is in quartz and quartz–carbonate vein systems on the northern margin of the Mallina Shear Zone. Both deposits contain pyrite and gold grades increase with pyrite content. Huston et al. (2002b) noted a difference between the two deposits in that Withnell contains sericite alteration and Camel has pyrophyllite alteration. They commented that the presence of pyrophyllite at Camel indicates involvement of magmatic-hydrothermal fluids. Sericite zones at Withnell are relatively high grade, although Au grades generally are not high compared to many other lode Au deposits in Western Australia. Prior to a decision to mine in 2005, combined resources of Withnell and Camel were reported as 10.46 Mt. at 1.6 g/t Au for 529,000 oz. Au (Range River Gold 2005). Mining during 2006 and 2007 produced approximately 30,000 oz. Au, after which the mines were placed on care and maintenance.

The Wingina Well deposit contains total measured, indicated, and inferred resources of 5.1 Mt. at 1.30 g/t for 268,000 oz. Au (Polymetals Mining Limited 2013). The host rocks are mainly sheared BIF and chert of the Cleaverville Formation, although sheared metabasalt is also present. Northeast plunging shoots of Au mineralization suggest combined structural and lithological controls of gold from hydrothermal fluids within the shear zone. Deflection of the TTSZ and Mallina Shear Zone around the high-Mg diorite intrusion at Mount Dove probably formed a triangular 'pressure-shadow'. Wingina Well is near the eastern apex of this triangular area, but other deposits and prospects on the shear zones either side of it include Lost Ark, Wingina Well 2, Last Crusade, Edkins, Amanda, Amanda East (all on the TTSZ), Brierly, and Mount Berghaus (on the Mallina Shear Zone). Published inferred resources at Amanda are 0.687 Mt. at 1.6 g/t for 35,000 oz. Au and at Mount Berghaus are 0.92 Mt. at 1.4 g/t for 43,000 oz. Au (De Grey Mining Limited 2012).

Another type of Au mineralization in the CPTZ is seen in a number of Au–Sb deposits. Mines with historic production of antimony are located at Balla Balla, Mallina Homestead, Peawah Hill, and Sherlock Homestead (Finucane and Telford 1939). With the possible exception of Balla Balla (old workings not relocated), the antimony was mined from quartz–stibnite(−cervantite) veins in metamorphosed shale or argillaceous greywacke of the Croydon Group. The Au–Sb association is common in orogenic gold deposits, and it is assumed that Sb was derived by metamorphic devolatilization of the host rocks (Hickman 1983).

11.3.3.2.3 VMS in the Whim Creek Greenstone Belt

The Whim Creek greenstone belt is composed of the c. 3000–2990 Ma Whim Creek Group and the c. 2955–2945 Ma Bookingarra Group (Chap. 9). Mineralization is restricted to isolated deposits within the basal volcaniclastic section of the Bookingarra Group.

At Mons Cupri, two levels of mineralization are present in conglomerate of the Cistern Formation: a lower, funnel-shaped, Cu-rich disseminated stockwork overlain by a stratiform, 5–10-m-thick Pb–Zn–Ag zone (Low 1963; Blockley 1971; Miller and Gair 1975; Marston 1979; Smith 1980; Collins and Marshall 1999b; Ferguson 1999; Hickman et al. 2000; Huston et al. 2000; Pike et al. 2002; Huston 2006). The copper-rich stockwork contains chalcopyrite with minor sphalerite and galena and is associated with intense chloritic alteration, whereas the stratiform Pb–Zn zone contains pyrite, sphalerite, and galena with minor chalcopyrite and tetrahedrite and has carbonate alteration. Total resources at Mons Cupri have been estimated at about 4.6 Mt. at 0.9% Cu and 1.3% Zn. The conglomerate host of the Mons Cupri mineralization is polymictic and poorly sorted with sub-angular to locally rounded clasts and blocks up to 10 m across. The clasts are mainly composed of rhyolite, granite, and lesser basalt. The matrix of the conglomerate contains rhyolite shards and has been intruded by felsic sills, dykes, and domal bodies (Huston et al. 2000) providing evidence of coeval volcanism during conglomerate deposition.

Although the Pb–Zn zone is stratiform, Huston et al. (2000) used textural evidence to conclude that the mineralization is epigenetic. Pb model ages of 2933–2920 Ma from galena in the Mons Cupri mineralization (Huston et al. 2000, 2002a) are approximately 20–30 million years younger than the depositional age of the Cistern Formation providing additional evidence that the mineralization might be epigenetic (Huston et al. 2002a; Pike et al. 2002; Huston 2006). However, the epigenetic interpretation is controversial, partly because there are no known 2933–2920 Ma potential source rocks, such as granites or felsic volcanic rocks in the Mallina Basin. Additionally, Pb model ages on apparently related deposits such as Whim Creek and Salt Creek are more in line with the interpreted depositional age of the Cistern Formation. At Whim Creek stratabound Cu–Zn deposit in the Rushall Slate, stratigraphically overlying or laterally equivalent to the Cistern Formation, has provided Pb model ages of 2948–2942 Ma (Huston et al. 2000, 2002a), and stratiform Pb-Zn mineralization in the Cistern Formation at Salt Creek has given

Pb model ages of 2960–2950 Ma (Richards et al. 1981). The same data were recalculated by Huston et al. (2000) to either 2962–2959 Ma or 2945–2942 Ma, depending on the model used. An alternative explanation of the textures observed in the Mons Cupri mineralization is that they were formed by metamorphism of the deposits, either by heat and hydrothermal fluids derived from c. 2950 Ma ultramafic, mafic, and granitic intrusions or during peak metamorphism (2940–2930 Ma) of the North Pilbara Orogeny. It is probable that quartz veins carrying Pb–Ag mineralization in high-Mg basalt or dolerite of the nearby epigenetic Comstock deposit were derived by metamorphism of the Mons Cupri VMS mineralization. Galena from the Comstock quartz veins, which immediately overlie the stratiform Pb–Zn–Ag mineralization at Mons Cupri, has provided Pb model ages of 2947–2921 Ma (Huston et al. 2000).

The Whim Creek, Cu–Zn VMS deposit is stratabound within a particular stratigraphic horizon of the Rushall Slate. Primary minerals are pyrite, pyrrhotite, chalcopyrite, and sphalerite with minor galena, magnetite, and arsenopyrite, whereas supergene alteration minerals are chalcocite, covellite, and malachite, with minor azurite and chrysocolla (Collins and Marshall 1999a). Collins et al. (2004) reported that the mineralization extends at least 5 km along strike, but except in the Whim Creek mine area, the sulphide zone is less than 0.5 m thick. In the mining area, the deposit is 10–15 m thick over a strike length of 600 m. Pb isotopic data were interpreted by Huston et al. (2002a, b) to provide a model age of c. 2948 Ma.

The Rushall Slate is a more distal sedimentary facies than the Cistern Formation, and the two formations are probably at least partly laterally equivalent. Between 1891 and 2009, total production from the Whim Creek mine was approximately 46,500 t Cu, 35,000 t of which was produced by Straits Resources between 2005 and 2009 (Hickman 2016).

The Pb–Zn (–Cu) deposits at Salt Creek and Balla Balla are situated 20 km northwest of Whim Creek, a short distance south of the Sholl Shear Zone. The local succession was described by Pike et al. (2002) and Pike et al. (2006). Massive sulphide lenses at Salt Creek contain sphalerite, pyrite, galena, and chalcopyrite within sandstone and shale overlying volcanic and volcaniclastic rocks (Huston et al. 2000). This is closely analogous to the stratigraphy of the Cistern Formation and Rushall Slate in the Mons Cupri–Whim Creek area, and Pb model ages (see Mons Cupri) also support a correlation.

The regional association of VMS mineralization with the Cistern Formation–Rushall Slate succession has been linked to the location of volcanic centres with intrusions of porphyritic dacite and rhyolite (Sylvester and de Laeter 1987; Collins and Marshall 1999a). Salt Creek and Whim Creek are located on opposite limbs of the Whim Creek Anticline, establishing that the succession of the Whim Creek greenstone belt is underlain by the Caines Well Granitic complex. Due to poor exposure, the detailed geology of this complex is uncertain, but it is likely that the uppermost intrusions of c. 2950 Ma granites within it were the source of the VMS mineralization in the Cistern Formation and Rushall Slate.

11.3.3.2.4 V–Ti in the Sherlock Intrusion

Approximately 10 km northwest of Whim Creek, the c. 2950 Ma Sherlock Intrusion includes a 20–30-m-thick layer of massive titanomagnetite with elevated vanadium. The ore zone extends over a strike length of almost 20 km from Balla Balla East to Don Well. Drilling has identified mineral reserves over the eastern 8 km of this strike length. The mineralization is tabular and positioned between layers of anorthositic gabbro and leucogabbro; other layers of the intrusion include pyroxenite, norite, anorthosite, and granophyre. At Balla Balla, the Sherlock Intrusion intrudes the axial region of the Whim Creek Anticline between the Caines Well Granitic complex and the Whim Creek Group. Layering is inclined 25° to the northeast in this area, confirming the sill was folded by the c. 2940 Ma D_8 anticline. At Don Well, southwest of Salt Creek, the intrusion lies on the northern limb of the anticline. Proposed products from mining are magnetite concentrate (containing V_2O_5) and ilmenite. A relatively minor occurrence of V–Ti mineralization in the Sherlock Intrusion has also been identified 35 km west-southwest of the deposits Balla Balla near Mount Fisher.

11.3.3.2.5 Ni–Cu in the Sherlock Intrusion

The Sherlock Bay Ni–Cu deposits occur in tectonic lenses within quartz–amphibole–magnetite–sulphide schist and metavolcanic and metasedimentary rocks on the southern margin of the Sholl Shear Zone. Deformed units of titaniferous magnetite outcrop in the same area (Miller and Smith 1975), and it is evident that both the titaniferous magnetite and Ni–Cu mineralization were derived from extreme deformation of the Sherlock Intrusion. This intrusion rims the Caines Well granitic complex in all areas except at Sherlock Bay where it is almost completely sheared out along the Sholl Shear Zone. The metamorphosed felsic volcanic and sedimentary units are almost certainly tectonic slices of the Red Hill Volcanics and Cistern Formation–Rushall Slate succession. The Ni–Cu mineralization has been drilled to a depth of 1000 m, and inferred resources are 16 Mt. at 0.75% Ni and 0.9% Cu (Ruddock 1999).

11.3.3.2.6 Radley Suite Ni–Cu and PGE

Some of the most significant mineralization in the CPTZ is contained in the large layered ultramafic–mafic intrusions of the 2930–2922 Ma Radley Suite (Sisters Supersuite). These intrusions, all of which were emplaced in a north-northeast-trending zone across the Sholl Terrane, host Ni–Cu and PGE deposits at Radio Hill, Munni Munni, and Mount Sholl. Detailed descriptions were provided by Hoatson et al. (1992) and Hoatson and Sun (2002), but three intrusions (Bullock Hide, Andover, and Sherlock) considered by these authors to be contemporaneous with the Radley Suite are now known to be older (Hickman 2016). Intrusions of the

Radley Suite typically comprise a lower section of ultramafic layers (dunite, peridotite, and pyroxenite) overlain by a layered mafic section of gabbro, leucogabbro, norite, and more rarely anorthosite and granophyre. Additionally, gabbroic units of the Radio Hill and North Whundo intrusions are overlain by granitic rocks.

Hunter Resources commenced a program of PGE exploration in the early 1980s. The exploration initially targeted all known ultramafic–mafic layered intrusions in the Northwest Pilbara but later focused on the ultramafic–mafic contact in the Munni Munni Intrusion. Hunter based its PGE exploration on the Bushveld and Stillwater models and in 1984 discovered the main layer of PGE mineralization in the Munni Munni Intrusion that became known as 'Hunter's reef'.

The Radio Hill Intrusion consists of a basal ultramafic zone comprised of lherzolite, dunite, and websterite and an overlying gabbroic zone, including quartz gabbro, gabbro, and gabbronorite. The basal layer of the ultramafic zone contains massive, disseminated, veined, and brecciated pyrrhotite, and pentlandite–chalcopyrite–magnetite at or near the contact with the Nallana Formation. Underground mining has been intermittent since the late 1980s and has mainly produced Ni and Cu with Co, Pd, Pt, and Ag as by-products. Between 1998 and 2002, the mine produced 21,000 t Ni and 16,000 t Cu (Hickman 2016). A second period of mining between early 2007 and mid-2008 resulted in production of an additional 2605 t Ni and 4120 t Cu (Hickman 2016). Cobalt production was 589 t, and Pd was 159 kg.

The c. 2925 Ma Munni Munni Intrusion, which outcrops over an area of 150 km^2 and is concealed by the Fortescue Group over a similar area, contains one of the most significant PGE deposits in Australia. The indicated and inferred resources are 23.6 Mt. at 2.9 g/t PGE + Au (Hickman 2016). These resources were calculated for one 7.5-km-long section of the mineralized layer, but its total known length is 22 km. Best intersections reported by Hoatson and Sun (2002) are 5–8 ppm combined Pt + Pd + Au over a vertical thickness of 0.5 m.

PGE mineralization in the Munni Munni Intrusion takes place near the top of a layer of porphyritic (large euhedral hypersthene grains) plagioclase websterite orthocumulate. The porphyritic websterite overlies a 1850 m thickness of ultramafic rocks and underlies a $>$ 3630-m-thick upper zone of gabbro (Hoatson 1986; Hoatson and Sun 2002). The ultramafic zone contains cyclic layers of dunite, lherzolite, wehrlite, olivine websterite, clinopyroxenite, and websterite with orthopyroxenite and norite. The base of the gabbro zone is marked by the first upward appearance of cumulus plagioclase and inverted pigeonite in the intrusion (Hoatson and Sun 2002). Hoatson and Sun (2002) interpreted the porphyritic websterite to be a product of mixing between the ultramafic and gabbroic zones and described an erosional contact with the underlying ultramafic zone. The total thickness of the mineralized orthocumulate layer is 30–80 m, and it extends 22 km around the western northern and eastern sides of the intrusion. The highest PGE grades are concentrated in a 7.5-km-long section of porphyritic websterite in the northern part of the intrusion. Mineralization consists of Pd and Pt accompanied by Au, Cu (0.3%), and Ni (0.2%). Disseminated sulphides are principally chalcopyrite, pentlandite, and pyrrhotite.

The Mount Sholl Intrusion contains Ni–Cu mineralization at a number of localities referred to as the A1, B1, and B2 zones. Abeysinghe and Flint (2008) referred to drilling data that indicate B2 as the most prospective zone, with indicated and inferred resources of 5.987 Mt. at 0.5% Ni and 0.6% Cu. The mineralization consists of aggregates of pentlandite, pyrrhotite, and chalcopyrite within a thin gabbroic marginal layer along the northwestern margin of the intrusion. This gabbroic layer underlies a thicker ultramafic layer of peridotite and pyroxenite that forms the main basal section of the intrusion. Marston (1984) provided additional information on the Mount Sholl deposits. The known mineralization at Mount Sholl is relatively low grade, but future mining may be possible if processing is by heap leaching.

11.3.3.2.7 Conglomerate-Hosted Gold Mineralization

The >2950 Ma Lalla Rookh Sandstone (Chap. 9) has sedimentological features in common with auriferous conglomerate–sandstone sequences in the Witwatersrand Basin of South Africa. These include braided-stream deposits, stacked and telescoped sequences above low-angle intraformational unconformities, fining-upwards sequences, heavy-mineral concentrations, and sedimentary reworking (Krapež and Groves 1984). Sub-economic gold and local uranium mineralization have been identified in three areas: the Lalla Rookh–Western Shaw Structural Corridor (Carter 1981; Krapež and Groves 1984; Hickman and Harrison 1986; Carter and Gee 1987, 1988; Krapež and Furnell 1987); the Goldsworthy greenstone belt (Guj et al. 1983, 1984; Taylor 1985); and the Shay Gap greenstone belt (Weir 1989a, b).

The Lalla Rookh Sandstone was deposited on a shallow water, fluviatile to marine shelf that had an area of at least 20,000 km^2. Until the North Pilbara Orogeny at c. 2950 Ma, the shelf was relatively stable with moderate vertical movements producing local erosional unconformities. Banks (1981) identified five erosional unconformities within the 3000-m-thick Lalla Rookh Sandstone succession preserved in the Lalla Rookh Syncline between the Carlindi and North Pole Domes. Intraformational unconformities were also reported by Hickman (1983) and Van Kranendonk (2000). These breaks in the succession provide evidence of a lengthy period of deposition interrupted by local uplift, erosion, and sedimentary reworking on the shelf.

Dating of the Lalla Rookh Syncline indicates that most detrital zircons have ages between c. 3330 and 3290 Ma (GSWA 142951, Nelson 2000). This age range corresponds to an important period of epigenetic gold mineralization in the EPT. In the Goldsworthy greenstone belt, visible detrital grains of gold in conglomerate, and small nuggets in recent alluvial deposits derived from its erosion (Guj et al. 1983, 1984; Taylor 1985), demonstrate gold sources during deposition. Exploration in the Lalla Rookh area revealed pyritic and auriferous conglomerate units in a 3–5-km-thick succession sedimentologically similar to units in the Witwatersrand Basin (Banks 1981; Furnell 1982). Banks (1981) recognized six vertically stacked conglomerate–sandstone sequences separated by five unconformities and disconformities, each of which is immediately overlain by a paleoplacer of oligomictic conglomerate. The majority of samples with anomalous gold contents assayed 0.05–0.5 g/t Au, with a maximum recorded value of 20 g/t Au (Banks 1981). However, of 472 surface rock samples analysed, only 2% contained more than 1 g/t Au.

The Lalla Rookh Sandstone of the Goldsworthy greenstone belt contains auriferous boulder and pebble conglomerate units in an area referred to as Amphitheatre (Hickman 2021). The conglomerate overlies basal and intraformational erosional unconformities (Guj et al. 1983, 1984; Taylor 1985). Guj et al. (1983) divided the formation into lower and upper associations, with the main auriferous conglomerate located at the base of the upper association, 300–500 m above the top of the underlying Cleaverville Formation. Most analyses of rock samples from this conglomerate unit contain approximately 0.1 g/t Au, although some intervals average more than 0.5 g/t Au. The maximum gold content reported by Guj et al. (1984) was 44.6 g/t Au. The conglomerate contains abundant detrital pyrite which locally forms up to 20% of the rock. Boulder conglomerate in the lower association was reported to contain up to 6.11 g/t Au, although contents were otherwise less than 0.5 g/t Au. Paleocurrent data from cross-bedded sandstone units indicate paleoslopes directed towards the east-northeast, which suggests higher potential for well-sorted mid-fan facies in that direction.

Based on limited exploration, auriferous pyritic conglomerate in the Lalla Rookh Sandstone of the Shay Gap greenstone belt is apparently concentrated in the central part of the formation. Analyses of surface samples from a synclinal structure 9.3 km northeast of Shay Gap revealed a weakly mineralized unit extending along a strike length of 1.3 km. The most anomalous gold content reported was 0.2 g/t Au (Weir 1989a, b), and subsequent drilling intersected no significant mineralization.

11.3.4 Gold in the Mosquito Creek Basin

Historical production from mines in the Mosquito Creek Basin (Fig. 10.5) has exceeded 12 t (about 40 million oz.) gold, with almost all of this production coming from mines along two parallel, east-trending shear zones: the Middle Creek Fault and Blue Spec Fault Zone (Huston et al. 2001a; Blewett et al. 2002; Bagas et al. 2008; Huston et al. 2017). Mineralization is hosted by sections of the Mosquito Creek Formation that contain unusually large amounts of carbonaceous pelitic schist (Hickman 1983). A structural interpretation by Nijman et al. (2010) suggests that the schist is stratigraphically high in the formation and might be a single stratigraphic unit duplicated by folding. Both shear zones are interpreted to have developed along major thrust structures formed during c. 2905 Ma convergence of the EPT and KUT (Chap. 10). Gold was most likely derived from metamorphism of mafic rocks of the Coondamar Formation that is interpreted to include oceanic-like mafic crust underlying the Mosquito Creek Formation.

The mine geology of the principal deposits on both shear zones was reviewed by Hickman (1983). Results of more recent exploration and mining have been described by Huston et al. (2001b, 2017), Blewett et al. (2002), and Bagas (2005).

11.3.5 Post-Orogenic Mineralization (2895–2830 Ma)

Mineralization in the Pilbara Craton following the Mosquito Creek Orogeny includes rare-metal pegmatite mineralization associated with post-orogenic granites of the Split Rock Supersuite. Mineral commodities include lithium, tin, tantalum, beryllium, and tungsten, all of which have been mined. Additionally, geochronology suggests that gold mineralization in the Mount York area, south of Pilgangoora, was introduced after the Mosquito Creek and North Pilbara Orogenies.

11.3.5.1 Lithium-Bearing Pegmatites

Spodumene is the principal lithium mineral in moderately dipping Sn–Ta–Li-bearing pegmatite sheets at Pilgangoora on the western margin of the East Strelley greenstone belt (Fig. 1.7). Available information indicates that the pegmatite sheets emanated from one or more intrusions of massive, non-foliated granite and syenogranite in the Carlindi Dome. Tantalite in a Pilgangoora pegmatite was dated at 2879 ± 5 Ma (Kinny 2000). This result suggests an intrusive age between intrusion of the Cutinduna and Split Rock Supersuites. However, more geochronology is required because the age of the Split Rock Supersuite in the Carlindi and Yule Domes has not been isotopically established. The lithium-bearing pegmatites at Pilgangoora are up to 70 m thick and form a north-trending swarm within mafic and ultramafic units correlated with the Euro Basalt and the Dalton Suite. Megacrystic spodumene and microcline are contained within a fine- to coarse-grained albite–quartz matrix. The strike length of the swarm is about 5 km, and some individual pegmatites are more than 1 km long (Huston et al. 2017; Sweetapple et al. 2017). Lithium resources (spodumene) at Pilgangoora (2021) exceeded 300 Mt. at >1.34% Li_2O.

11.3.5.2 Pegmatite of the 2851–2831 Ma Split Rock Supersuite

The Split Rock Supersuite (Fig. 10.6), intruded at 2851–2831 Ma, consists of post-tectonic, sheet-like granitic plutons distributed in a northwest-trending belt across the northern half of the Pilbara Craton. These highly fractionated intrusions are typically fringed by pegmatite veins containing tin–tantalum–lithium mineralization (Blockley 1980; Hickman 1983; Sweetapple and Collins 1998, 2002; Sweetapple 2000; Huston et al. 2001a; Sweetapple et al. 2001) and locally fluorite–barite mineralization (Bagas 2005). Three main types of pegmatite were described by

Hickman (1983): (1) simple pegmatite, with cassiterite and tantalite–columbite minerals, rarer lithium, and beryllium compounds; (2) layered albite pegmatites with a wider variety of mineral species and cassiterite concentrated at finer albitic margins; and (3) complex rare-earth pegmatites that mainly intrude greenstones (Ferguson and Ruddock 2001). A more detailed classification was provided by Sweetapple et al. (2001). Mining in the Shaw Dome recovered tin and tantalum mineralization from alluvial workings in recent drainage basins. In addition to alluvial mining, local underground mining took place in the Mount Edgar Dome where cassiterite occupies the margins of layered albite-pegmatite veins emplaced along the western side of the Moolyella Monzogranite. Pegmatite-hosted tantalite mineralization has been mined from the Wodgina area where there are thick, shallow-dipping pegmatite sheets (Ferguson and Ruddock 2001; Sweetapple et al. 2001). Fluorite is a common accessory mineral in most of the post-tectonic granites, and veins containing fluorite and barite fringe the southwestern margin of the Cookes Creek Monzogranite (Hickman 1983; Bagas 2005).

Tungsten mineralization is associated with the supersuite in several areas of the Northern Pilbara Craton, and small-scale mining has been recorded at Cookes Creek and Burrows Well and the Friendly Creek and Wodgina project areas (Baxter 1978; Hickman 1983). At all these localities, tungsten mineralization is present where intrusions of the Split Rock Supersuite have intruded greenstones, as opposed to older granitic rocks (Hickman 1975). Most of the recorded production came from Cookes Creek in 1951 and 1952 and was 26.6 t of wolframite and scheelite concentrate containing 17.6 t WO_3 (Hickman 1983). The workings were on quartz–pegmatite veins within the margin of the Cookes Creek Monzogranite, and in greenstones to the southwest and south of the intrusion (Hickman 1983). The largest undeveloped tungsten deposit is also at Cookes Creek and is hosted by metamorphosed pyroxenite and dolerite at the Big Hill deposit adjacent to the southern contact of the monzogranite. This mineralization comprises scheelite in veins of quartz, aplite, and pegmatite and is the same age as the monzogranite, locally dated at 2837 ± 16 Ma (GSWA 178011, Wingate et al. 2015). The Big Hill mineral resource has been reported as 11.5 Mt. at 0.15% WO_3.

11.3.5.3 Gold Mineralization

Gold mineralization in the Mount York area south of Pilgangoora was dated at 2888 ± 6 Ma (Pb–Pb isochron of alteration minerals; Neumayr et al. 1998; Huston et al. 2001a). The mineralization is hosted by shear zones interpreted to be related to late-stage strike-slip movement in the Pilgangoora and Wodgina areas. Baker et al. (2002) interpreted most of the hydrothermal gold mineralization at Pilgangoora to have been introduced at this time.

11.4 Neoarchean Mineralization

The Fortescue Group contains economically significant conglomerate-hosted gold mineralization in the Mount Roe Basalt and the Hardey Formation, and several deposits were mined during the latter part of the nineteenth century and the early twentieth century (Hickman 1983; Hickman and Harrison 1986; Carter and Gee 1987, 1988; Pirajno et al. 2018; Tyler et al. 2018). Conglomerate-hosted uranium mineralization has been explored in the Hardey Formation of the Northeast Pilbara Sub-basin and in the Pear Creek Formation in the southern part of the Marble Bar Sub-basin (Hickman and Harrison 1986; Carter and Gee 1987, 1988), but no mining has been recorded. Whereas the gold mineralization is concentrated immediately above the basal unconformity of the Fortescue Group, and was derived from erosion of hydrothermal gold deposits in the underlying Pilbara Craton, the uranium mineralization is apparently restricted to the post-2755 Ma stratigraphy when the Hardey Formation included large felsic intrusions and felsic volcanic units (Chap. 12).

Veins of fluorite have been mined from the Mount Roe Basalt at Meentheena (Hickman 1974), and abnormal concentrations of fluorite are present in stocks and dykes of porphyritic rhyodacite in the southwest section of the Warrawagine granitic complex (Hickman 1976). The fluorine content of the porphyritic rhyodacite is up to 3.25%, although contents below 1% are more typical. The intrusions have not been dated but are inferred to be related to the c. 2765 Ma Bamboo Creek Member of the Hardey Formation.

References

Abeysinghe PB, Fetherston JM (1997) Barite and fluorite in Western Australia. Mineral Resources Bulletin, vol 17. Geological Survey of Western Australia

Abeysinghe PB, Flint DJ (2008) Copper, lead and zinc in Western Australia: a commodity review for 2006-07: Geological Survey of Western Australia, Record 2008/4

Bagas L (2005) Geology of the Nullagine 1:100 000 sheet. 1:100 000 Geological series explanatory notes. Geological Survey of Western Australia

Bagas L, Van Kranendonk MJ, Pawley MJ (2004) Geology of the Split Rock 1:100 000 sheet. 1: 100 000 Geological series explanatory notes. Geological Survey of Western Australia

Bagas L, Bierlein FP, Bodorkos S, Nelson DR (2008) Tectonic setting, evolution, and orogenic gold potential of the late Mesoarchean Mosquito Creek Basin, North Pilbara Craton, Western Australia. Precambrian Res 160:237–244

Baker DEL, Seccombe PK, Collins WJ (2002) Structural history and timing of gold mineralization in the Northern East Strelley Belt, Pilbara Craton, Western Australia. Econ Geol 97:775–785

Banks MJ (1981) Progress report on temporary reserves 7437H and 7439H, Shaw River District, Western Australia: Anaconda Australia Inc. Geological Survey of Western Australia, Statutory mineral exploration report A9832

Barley ME (1982) Porphyry-style mineralization associated with early Archean calc-alkaline igneous activity, Eastern Pilbara, Western Australia. Econ Geol 77:1230–1236

Barley ME, Pickard AL (1999) An extensive, crustally-derived, 3325 to 3310 Ma silicic volcanoplutonic suite in the Eastern Pilbara Craton: evidence from the Kelly Belt, McPhee Dome and Corunna Downs Batholith. Precambrian Res 96:41–62

Baxter JL (1978) Molybdenum, tungsten, vanadium and chromium in Western Australia. Mineral Resources Bulletin, vol 11. Geological Survey of Western Australia

Beintema KA (2003) Geodynamic evolution of the West and Central Pilbara Craton: a mid-Archean active continental margin. PhD, Geologica Ultraiectina, Utrecht University

Beintema KA, Mason PRD, Nelson DR, White SH, Wijbrans JR (2003) New constraints on the timing of tectonic activity in the Archaean Central Pilbara Craton, Western Australia. J Virtual Explorer 13. https://doi.org/10.3809/jvirtex.vol.2003.013

Bekker A, Slack JF, Planavsky NJ, Krapež B, Hofmann AW, Konhauser KO, Rouxel OJ (2010) Iron formation: the sedimentary product of a complex interplay among mantle, tectonic, oceanic, and biospheric processes. Econ Geol 105:467–508

Bekker A, Planavsky NJ, Krapež B, Rasmussen B, Hofmann A, Slack JF, Rouxel OJ, Konhauser KO (2014) Iron formations: their origins and implications for ancient seawater chemistry. In: Turekian KK (ed) Treatise on geochemistry. Elsevier, Oxford, UK, pp 561–628

Blake TS (1993) Late Archaean crustal extension, sedimentary basin formation, flood basalt volcanism, and continental rifting. The Nullagine and mount Jope Supersequences, Western Australia. Precambrian Res 60:185–241

Blake TS, Barley ME (1992) Tectonic evolution of the late Archaean to early Proterozoic Mount Bruce megasequence set, Western Australia. Tectonics 11:1415–1425

Blewett RS, Champion DC (2005) Geology of the Wodgina 1:100 000 sheet. 1:100 000 Geological series explanatory notes. Geological Survey of Western Australia

Blewett RS, Huston DL, Mernagh TP, Kamprad J (2002) The diverse structure of Archean lode gold deposits of the southwest Mosquito Creek Belt, East Pilbara Craton, Western Australia. Econ Geol 97:787–800

Blockley JG (1971) The lead, zinc, and silver deposits of Western Australia. Mineral Resources Bulletin, vol 9. Geological Survey of Western Australia

Blockley JG (1980) The tin deposits of Western Australia, with special reference to the associated granites. Mineral Resources Bulletin, vol 12. Geological Survey of Western Australia

Brauhart C (1999) Regional alteration systems associated with Archean volcanogenic massive sulphide deposits at Panorama, Pilbara, Western Australia. PhD, The University of Western Australia

Brauhart C, Groves DI, Morant P (1998) Regional alteration systems associated with volcanogenic massive sulphide mineralization at Panorama, Pilbara, Western Australia. Econ Geol 93:292–302

Brown AJ, Cudahy TJ, Walter MR (2006) Hydrothermal alteration at the Panorama Formation, North Pole Dome, Pilbara Craton, Western Australia. Precambrian Res 151:211–223

Carter JD (1981) Exploration in Western Australia: A history of investigation and a guide to microfilm open-file information. Record, 1981/6. Geological Survey of Western Australia

Carter JD, Gee RD (1987) Geology and exploration history of Precambrian quartz-pebble conglomerates in Western Australia. In: Uranium deposits in quartz pebble conglomerates. International Atomic Energy Agency, Vienna, Austria, pp 387–425

Carter JD, Gee RD (1988) Geology and exploration history of uraniferous and auriferous pyritic conglomerates, Western Australia. In: Professional papers. Geological Survey of Western Australia, pp 17–36

Champion DC (2013) Neodymium depleted mantle model age map of Australia: explanatory notes and user guide. Record, 2013/44. Geoscience Australia

Champion DC, Huston DL (2016) Radiogenic isotopes, ore deposits and metallogenic terranes: novel approaches based on regional isotopic maps and the mineral systems concept. Ore Geol Rev 76:229–256. https://doi.org/10.1016/j.oregeorev.2015.09.025

Champion DC, Smithies RH (2007) Geochemistry of Paleoarchean granites of the East Pilbara Terrane, Pilbara Craton, Western Australia: implications for early Archean crustal growth. In: Van Kranendonk MJ, Bennett VC, Smithies RH (eds) Earth's oldest rocks. Elsevier B.V, Burlington, pp 369–410

Collins PLF, Marshall AE (1999a) Volcanic-hosted massive sulfide deposits at Whundo–Yannery in the Sholl belt. In: Ferguson KM, Lead, zinc and silver deposits of Western Australia. Geological Survey of Western Australia, Mineral Resources Bull. 15, pp 73–79

Collins PLF, Marshall AE (1999b) Volcanogenic base-metal deposits of the Whim Creek Belt. In: Ferguson KM, Lead, zinc and silver deposits of Western Australia. Geological Survey of Western Australia, Mineral Resources Bull. 15, pp 44–72

Collins P, Hooper B, Cornelius M (2004) Mons Cupri Cu–Zn–Pb Deposit, Pilbara, Western Australia. Cooperative Research Centre for Landscape Environments and Mineral Exploration, CSIRO Exploration and Mining, Bentley, Western Australia

Cummins B, Cairns B (2017) Spinifex Ridge molybdenum–copper deposit. In: Phillips GN (ed) Australian ore deposits. Australasian Institute of Mining and Metallurgy, pp 343–344

De Grey Mining Limited (2005) Quarterly operations report to Australian Stock Exchange, Three months ending September 2005

De Grey Mining Limited (2012) 40% increase in Turner River gold resources

De Grey Mining Limited (2020) Consolidated Annual Financial Report for the Year Ended 30 June 2020

De Laeter JR, Martyn JE (1986) Age of molybdenum copper–molybdenum mineralization at Coppin Gap, Western Australia. Austral J Earth Sci 33:65–71

Duuring P, Teitler Y, Hagemann S (2017) Tools for discovering BIF-hosted iron ore deposits in the Pilbara Craton. In: GSWA 2017 extended abstracts: promoting the prospectivity of Western Australia. Geological Survey of Western Australia, pp 1–3

Ferguson KM (1999) Lead, zinc, and silver deposits of Western Australia. Geological Survey of Western Australia, Mineral Resources Bull. 15

Ferguson KM, Ruddock I (2001) Mineral occurrences and exploration potential of the East Pilbara. Report 81. Geological Survey of Western Australia

Finucane KJ, Telford RJ (1939) The antimony deposits of the Pilbara Goldfield. Aerial Geological and Geophysical Survey of Northern Australia, Western Australia Report No. 47

Furnell R (1982) Anaconda Australia Inc. Annual Report for 1981: Lalla Rookh Temporary Reserve 7437H, Shaw River Temporary Reserves 7438H, George Range Temporary Reserve 7439H, Temporary Reserves for gold and other minerals 8119H–8152H, Marble Bar District–Western Australia. Geological Survey of Western Australia, Statutory mineral exploration report A11184

Gardiner NJ, Hickman AH, Kirkland CL, Lu Y, Johnson T, Zhao J-X (2017) Processes of crust formation in the early earth imaged through Hf isotopes from the East Pilbara Terrane. Precambrian Res 297:56–76. https://doi.org/10.1016/j.precamres.2017.05.004

Gardiner NJ, Hickman AH, Kirkland CL, Lu Y, Johnson TE, Wingate MTD (2018) New Hf isotope insights into the Paleoarchean magmatic evolution of the Mount Edgar Dome, Pilbara Craton: implications for early Earth and crust formation processes. Report, vol 181. Geological Survey of Western Australia

Geollnicht NM, Groves IM, Groves DI, Ho SE, McNaughton NJ (1988) A comparison between mesothermal gold deposits of the Yilgarn Block and gold mineralization at Telfer and Miralga Creek, Western Australia–indirect evidence for a non-magmatic origin for greenstone-hosted gold deposits. In: Ho SE, Groves DI (eds) Advances in understanding Precambrian gold deposits, vol 2. University of Western Australia, Geology Department and University Extension, Publication 12, pp 23–40

Groves IM (1987) Epithermal/porphyry-style base-metal and precious-metal mineralisation in the Miralga Creek area, Eastern Pilbara Block. University of Western Australia, B.Sc. (Hons) thesis (unpublished)

Guj P, McIntosh D, Stephenson P (1983) De Grey Gold Project, Mt Grant Prospect EL 45/50, Pilbara Mineral Field, W.A., First Progress Report to August 31, 1983: Technical Report

Guj P, McIntosh D, Stephenson P (1984) De Grey Gold Project, Mt Grant Prospect EL 45/50, Pilbara Mineral Field, W.A., Second Progress Report to December 31, 1983: Technical Report:

Carpentaria Exploration Company: Geological Survey of Western Australia, Statutory mineral exploration report A13800

Gutzmer J, Banks D, de Kock MO, McClung CR, Strauss H, Mezger K (2006) The origin and paleoenvironmental significance of stratabound barites from the Mesoarchean Fig Tree Group, Barberton Mountainland, South Africa. SSAGI–V, South America Symposium on Isotope Geology, 24–27 April 2006, Uruguay, Abstracts, pp 311–314

Hickman AH (1973) The North Pole barite deposits, Pilbara Goldfield. In: Annual report for the year 1972. Geological Survey of Western Australia, Perth, Western Australia, pp 57–60

Hickman AH (1974) The Meentheena fluorite deposits, Pilbara Goldfield. Geological Survey of Western Australia, Annual Report 1973:79–82

Hickman AH (1975) Precambrian structural geology of part of the Pilbara Region. In: Annual report for the year 1974. Geological Survey of Western Australia, pp 68–73

Hickman AH (1976) Fluorite porphyry at Ngarrin Creek, Yarrie. Geol Surv Western Australia, Record 3:68–73

Hickman AH (1977) Barite deposits near Cooke Bluff Hill, Port Hedland 1:250 000 sheet. Record, 1976/22. Geological Survey of Western Australia

Hickman AH (1983) Geology of the Pilbara Block and its environs. Bulletin 127. Geological Survey of Western Australia

Hickman AH (2002) Geology of the Roebourne 1:100 000 sheet. 1:100 000 Geological series explanatory notes. Geological Survey of Western Australia

Hickman AH (2004) Using new-generation geological maps of the Pilbara to guide mineral exploration. In: GSWA 2004 extended abstracts: promoting the prospectivity of Western Australia. Geological Survey of Western Australia, pp 1–4

Hickman AH (2012) Review of the Pilbara Craton and Fortescue Basin, Western Australia: crustal evolution providing environments for early life. Island Arc 21:1–31

Hickman AH (2016) Northwest Pilbara Craton: a record of 450 million years in the growth of Archean continental crust. Report, 160. Geological Survey of Western Australia

Hickman AH (2021) East Pilbara Craton: a record of one billion years in the growth of Archean continental crust. Report 143, Geological Survey of Western Australia

Hickman AH, Harrison PH (1986) A review of the occurrence and potential for Precambrian conglomerate-hosted gold mineralization within Western Australia. In: Geocongress '86: extended abstracts. Geological Society of South Africa, pp 301–319

Hickman AH, Strong CA (2003) Dampier–Barrow Island, Western Australia. 1:250 000 Geological series explanatory notes. Geological Survey of Western Australia

Hickman AH, Van Kranendonk MJ (2008) Archean crustal evolution and mineralization of the Northern Pilbara Craton–a field guide. Record, 2008/13. Geological Survey of Western Australia

Hickman AH, Smithies RH, Huston DL (2000) Archaean geology of the West Pilbara Granite–Greenstone Terrane and the Mallina Basin, Western Australia–a field guide: Geological Survey of Western Australia, Record 2000/9

Hoatson DM (1986) Geology of the Munni Munni layered intrusion: Preliminary Edition 1:20,000 geological map. Australia Bureau of Mineral Resources, Geology and Geophysics

Hoatson DM, Sun S-S (2002) Archean layered mafic-ultramafic intrusions in the West Pilbara Craton, Western Australia: a synthesis of some of the oldest orthomagmatic mineralizing systems in the world. Econ Geol 97:847–872

Hoatson DM, Wallace DA, Sun S-S, Macias LF, Simpson CJ, Keays RR (1992) Petrology and platinum-group element geochemistry of Archaean layered mafic–ultramafic intrusions, West Pilbara Block, Western Australia. Australian Geological Survey Organisation, Bulletin 242

Huston DL (2006) Mineralization and regional alteration at the Mons Cupri stratiform Cu-Zn-Pb deposit, Pilbara Craton, Western Australia. Mineral Deposita 41:17–32

Huston DL, Smithies RH, Sun S-S (2000) Correlation of the Archaean Mallina–Whim Creek Basin: implications for base-metal potential of the central part of the Pilbara Granite-Greenstone Terrane. Aust J Earth Sci 47:217–230

Huston DL, Blewett RS, Mernagh TP, Sun S-S, Kamprad J (2001a) Gold deposits of the Pilbara Craton: Australian Geological Survey Organisation (Geoscience Australia), Record 2001/10

Huston DL, Brauhart CW, Dreiberg SL, Davidson GJ, Groves DI (2001b) Metal leaching and inorganic sulphate reduction in volcanic-hosted massive sulphide mineral systems: evidence from the paleo-Archean Panorama district, Western Australia. Geology 29:687–690

Huston DL, Blewett RS, Keillor B, Standing J, Smithies RH, Marshall AE, Mernagh TP, Kamprad J (2002a) Lode gold and epithermal deposits of the Mallina Basin, North Pilbara Terrain, Western Australia. Econ Geol 97:801–818

Huston DL, Sun S-S, Blewett R, Hickman AH, Van Kranendonk MJ, Phillips D, Baker D, Brauhart C (2002b) The timing of mineralization in the Archean North Pilbara Terrain, Western Australia. Econ Geol 97:733–755

Huston DL, Morant P, Cummins B, Baker D, Mernaugh TP (2007) Paleoarchean mineral deposits of the Pilbara Craton: genesis, tectonic environment and comparison with younger deposits. In: Van Kranendonk MJ, Bennett VC, Smithies RH (eds) Earth's oldest rocks. Elsevier B.V, Burlington, Massachusetts, USA, pp 217–230

Huston DL, Wyche NL, Hickman AH, Norman M (2017) Pilbara Craton geology and metallogeny. In: Phillips GN Australian ore deposits. Australasian Institute of Mining and Metallurgy, pp. 325–330

Huston DL, Pirajno F, Morant P, Cummins B, Baker D, Mernagh P (2019) Paleoarchean mineral deposits of the Pilbara Craton: genesis, tectonic environment, and comparisons with younger deposits. In: Van Kranendonk MJ, Bennett VC, Hoffmann JE (eds) Earth's oldest rocks, 2nd edn. Elsevier B.V, pp 519–551

Jones FH (1938) Aerial, Geological and Geophysical Survey of Northern Australia: The Warrawoona area, Pilbara Gold-field, Western Australia. Report 20. Government Printer, Canberra, ACT

Jones CB (1990) Coppin Gap copper–molybdenum deposit. Austral Instit Mining Metall Monogr 14:141–144

Kinny PD (2000) U–Pb dating of rare-metal (Sn–Ta–Li) mineralized pegmatites in Western Australia by SIMS analysis of tin and tantalum-bearing ore minerals. In: Abstracts and Proceedings: Beyond 2000, New Frontiers in Isotope Geoscience, Lorne, Victoria, pp 113–116

Kloppenburg A (2003) Structural evolution of the Marble Bar Domain, Pilbara Granite-Greenstone Terrain, Australia: the role of Archaean mid-crustal detachments. PhD, Utrecht University

Krapez B, Furnell RG (1987) Sedimentology, origin and gold potential of the Late Archean Lalla Rookh Basin, East Pilbara Block, Western Australia. In: Uranium deposits in quartz pebble conglomerates. International Atomic Energy Agency, Vienna, Austria, pp 427–459

Krapež B, Groves DI (1984) Gold mineralization potential of Archaean clastic sequences in the East Pilbara Block. In: Muhling JR, Groves DI, Blake TS (eds) Archaean and Proterozoic basins of the Pilbara, Western Australia: evolution and mineralization potential. The University of Western Australia, Geology Department and University Extension, pp 111–122

Krapež B, Barley ME, Pickard AL (2003) Hydrothermal and re-sedimented origins of the precursor sediments to banded iron formation: sedimentological evidence from the early Palaeoproterozoic Brockman Supersequence of Western Australia. Sedimentology 50:979–1011

Low GH (1963) Copper deposits of Western Australia. Mineral Resources Bulletin, vol 8. Geological Survey of Western Australia

Marston RJ (1979) Copper mineralization in Western Australia. Mineral Resources Bulletin, vol 13. Geological Survey of Western Australia

Marston RJ (1984) Nickel mineralisation in Western Australia. Geological Survey of Western Australia, Mineral Resources Bulletin 14

Martin DMB, Li ZX, Nemchin AA, Powell CMA (1998) A pre-2.2 Ga age for giant hematite ores of the Hamersley Province, Australia? Econ Geol 93:1084–1090

Miller LJ, Gair HS (1975) Mons Cupri copper-lead-zinc-silver deposit. In: Knight CL (ed) Economic geology of Australia and Papua New Guinea-Metals, vol 5. Australasian Inst. Mining Metall., Mon, pp 195–202

Miller LJ, Smith ME (1975) Sherlock Bay nickel-copper deposit. In: Knight CL (ed) Economic geology of Australia and Papua New Guinea—Metals, vol 5. Australasian Inst. Mining and Metall., Mon, pp 168–174

Morris RC, Horwitz RC (1983) The origin of the iron formation-rich Hamersley Group of Western Australia–deposition on a platform. Precambrian Res 21:273–297. https://doi.org/10.1016/0301-9268(83)90044-X

Nelson DR (2000) Compilation of geochronology data, 1999. Record, 2000/2. Geological Survey of Western Australia Neumayr et al. 1998

Neumayr P, Ridley JR, McNaughton NJ, Kinny PD, Barley ME, Groves DI (1998) Timing of gold mineralization in the Mount York district, Pilgangoora greenstone belt and implications for the tectonic and metamorphic evolution of an area linking the Western and Eastern Pilbara Craton. Precambrian Res 88:249–265

Nijman W, de Bruijne KCH, Valkering ME (1998) Growth fault control of early Archaean cherts, barite mounds and chert–barite veins, North Pole Dome, Eastern Pilbara, Western Australia. Precambrian Res 88:25–52

Nijman W, Clevis Q, De Vries ST (2010) The waning stage of a greenstone belt: the Mesoarchean Mosquito Creek Basin of the East Pilbara, Western Australia. Precambrian Res 180:251–271

Peter JM, Scott SD (1997) Windy Craggy, Northwestern British Columbia: the world's largest Besshi-type deposit. In: Barrie CT, Hannington MD (eds) Volcanic-hosted massive sulfide deposits: processes and examples in modern and ancient settings, vol 8. Society of Economic Geologists, Reviews in Economic Geology, pp 261–295

Phillips GN, Powell R (2010) Formation of gold deposits: a metamorphic devolatilization model. J Metamorphic Geol 28:689–718

Pike G, Cas R, Smithies RH (2002) Geological constraints on base metal mineralization of the Whim Creek greenstone belt, Pilbara Craton, Western Australia. Econ Geol 97:827–845

Pike G, Cas RAF, Hickman AH (2006) Archean volcanic and sedimentary rocks of the Whim Creek greenstone belt, Pilbara Craton, Western Australia. Report, vol 101. Geological Survey of Western Australia

Pirajno F (2009) Hydrothermal processes and mineral systems. Springer, The Netherlands

Pirajno F, Huston DL (2019) Paleoarchean (3.6–3.2 Ga) mineral systems in the context of continental crust building and the role of mantle plumes. In: Van Kranendonk MJ, Bennett VC, Hoffmann JE (eds) Earth's oldest rocks, 2nd edn. Elsevier B.V, pp 187–209. https://doi.org/10.1016/B978-0-444-63901-1.00019-9

Pirajno F, Bagas L, Hickman AH (2018) Archaean palaeoplacer Au deposits of the Pilbara Craton and Fortescue Basin compared to palaeoplacer Au deposits in the Witwatersrand Basin. Australian Institute of Geoscientists, Gold18, August 2018, Perth, Western Australia, Abstracts, pp 48–52

Polymetals Mining Limited (2013) Turner River Gold Project: Mineral Resources increased to 345 koz Gold. Announcement to the Australian Securities Exchange, 13 March, 2013

Powell CMcA, Horwitz RC (1994) Late Archaean and Early Proterozoic tectonics and basin formation of the Hamersley Ranges. In: Excursion Guidebook 4. Geological Society of Australia

Range River Gold Limited (2005) Indee Gold Project to be developed: Report to Australian Stock Exchange, 16 September 2005

Reynolds DG, Brook WA, Marshall AE, Allchurch PD (1975) Volcanogenic copper–zinc deposits in the Pilbara and Yilgarn Archaean Blocks. In: Knight CL (ed) Economic geology of Australia and Papua New Guinea, 1. Metals. Australasian Institute of Mining and Metallurgy, pp 185–195

Roberts DL (1974) Geology and copper-zinc mineralization of the Whundo area, West Pilbara Goldfield, Western Australia. BSc, Hons, The University of Western Australia

Ruddock I (1999) Mineral occurrences and exploration potential of the West Pilbara. Report, vol 70. Geological Survey of Western Australia

Schopf JW (1993) Microfossils of the early Archaean Apex Chert: new evidence of the antiquity of life. Science 260:640–646

Smith ME (1980) Mons Cupri Cu–Pb–Zn–Ag deposit, Pilbara Block, W.A. J Geochem Exploration (Special Issue) 12:168–171

Smithies RH (2002) Archaean boninite-like rocks in an intracratonic setting. Earth Planet Sci Lett 197:19–34

Smithies RH, Champion DC (1999) High-Mg diorite from the Archaean Pilbara Craton; anorogenic magmas derived from a subduction-modified mantle. In: Geological Survey of Western Australia Annual Review 1998-99. Geological Survey of Western Australia, Perth, Western Australia, pp 45–50

Smithies RH, Champion DC (2000) The Archaean high-Mg diorite suite: links to tonalite–trondhjemite–granodiorite magmatism and implications for early Archaean crustal growth. J Petrol 41:1653–1671

Smithies RH, Hickman AH (2003) Archaean geology of the Mallina and Whim Creek Basins (1: 250 000 scale). Geological Survey of Western Australia

Smithies RH, Champion DC, Sun S-S (2004a) Evidence for early LREE-enriched mantle source regions: diverse magmas from the c. 3.0 Ga Mallina Basin, Pilbara Craton, NW Australia. J Petrol 45:1515–1537

Smithies RH, Champion DC, Sun S-S (2004b) The case for Archaean boninites. Contrib Mineral Petrol 147:705–725. https://doi.org/10.1007/s00410-004-0579-x

Stein HJ, Barley ME, Zimmerman A, Cummins B (2007) A 3.3 Ga Mo–Cu porphyry-style deposit at Spinifex Ridge, East Pilbara, Western Australia: Re–Os dating of Paleoarchean molybdenite. Geochimica et Cosmochimica Acta 71

Sweetapple MT (2000) Characteristics of Sn–Ta–Be–Li industrial mineral deposits of the Archaean Pilbara Craton, Western Australia. Record, 2000/044. AGSO, Canberra

Sweetapple MT, Collins PLF (1998) Tantalum–tin mineralized pegmatites at Wodgina and Mt. Cassiterite, Pilbara Craton, Western Australia. In: Geological Society of Australia (ed) Geoscience for the new millennium, p 435

Sweetapple MT, Collins PLF (2002) Genetic framework for the classification and distribution of Archean rare metal pegmatites in the North Pilbara Craton, Western Australia. Econ Geol 97: 873–895

Sweetapple MT, Collins PLF, Hickey RJ (2001) Classification, distribution, and petrogenetic affiliation of the Archaean rare metal pegmatites, Pilbara Craton, W.A. In: Cassidy KF, Dunphy JM, Van Kranendonk MJ (eds) Extended abstracts–4th international Archaean symposium, Perth, 24–28 September 2001. AGSO (geoscience Australia), pp. 475–477

Sweetapple MT, Holmes J, Yong J, Grigson MW, Barnes L, Till S (2017) Pilgangoora lithium–tantalum pegmatite deposit. In: Phillips GN (ed) Australian ore deposits. Australasian Institute of Mining and Metallurgy, pp 339–342

Taylor T (1985) De Grey Project–Mt Grant E.L. 45/50 Pilbara Mineral Field, W.A. Sedimentological Appraisal of the Amphitheatre Gold Prospect: Technical Report

Thom R, Hickman AH, Chin RJ (1979) Nullagine, WA sheet SF51-5, 2nd edn. Geological Survey of Western Australia

Thorne AM, Trendall AF (2001) Geology of the Fortescue Group, Pilbara Craton, Western Australia. Bulletin 144. Geological Survey of Western Australia

Thorpe R, Hickman AH, Davis D, Morton JGG, Trendall AF (1992a) U–Pb zircon geochronology of Archaean felsic units in the Marble Bar region, Pilbara Craton, Western Australia. Precambrian Res 56:169–189

Thorpe R, Hickman AH, Davis DW, Mortensen JK, Trendall AF (1992b) Constraints to models for Archaean lead evolution from precise U-Pb geochronology from the Marble Bar region, Pilbara Craton, Western Australia. In: Glover JE, Ho SE (eds) The Archaean: terrains, processes and metallogeny: proceedings for the third international Archaean symposium, 17–21 September

1990. Geology Department and University Extension, The University of Western Australia, Perth, Western Australia, pp 395–408

Tomich BNV (1974) The geology and nickel mineralization of the Ruth Well area, Western Australia. BSc (Hons), The University of Western Australia

Tyler IM, Thorne AM (1990) The northern margin of the Capricorn Orogen, Western Australia–an example of an early Proterozoic collision zone. J Struct Geol 12:685–701

Tyler IM, Smithies RH, Wingate MTD (2018) De Grey Superbasin and Fortescue Basin in the Pilbara: can they be compared with the Witwatersrand Basin? Geological Survey of Western Australia, GSWA 2018 Extended Abstracts, Record 2018/2, pp 8–13

Van Kranendonk MJ (1999) Two-stage degassing of the Archaean mantle: evidence from the 3.46 Ga Panorama volcano, Pilbara Craton, Western Australia. In: GSWA 99 extended abstracts: new geological data for WA explorers. Geological Survey of Western Australia, pp 1–3

Van Kranendonk MJ (2000) Geology of the North Shaw 1:100 000 sheet. 1:100 000 Geological series explanatory notes. Geological Survey of Western Australia

Van Kranendonk MJ (2004) Geology of the Carlindie 1:100 000 sheet. 1:100 000 geological series explanatory notes. Geological Survey of Western Australia

Van Kranendonk MJ (2006) Volcanic degassing, hydrothermal circulation, and the flourishing of early life on earth: a review of the evidence from c. 3490–3240 Ma rocks of the Pilbara Supergroup, Pilbara Craton, Western Australia. Earth-Sci Rev 74:197–240

Van Kranendonk MJ (2010) Geology of the Coongan 1:100 000 sheet. 1:100 000 Geological series explanatory notes. Geological Survey of Western Australia

Van Kranendonk MJ, Pirajno F (2004) Geochemistry of metabasalts and hydrothermal alteration zones associated with c. 3.45 Ga chert and barite deposits: implications for the geological setting of the Warrawoona Group, Pilbara Craton. Geochem Explor Environ Anal 4:253–278

Van Kranendonk MJ, Hickman AH, Huston DL (2006) Geology and mineralization of the East Pilbara–a field guide. Record, 2006/16. Geological Survey of Western Australia

Vearncombe S, Barley ME, Groves DI, McNaughton NJ, Mikucki EJ, Vearncombe JR (1995) 3.26 Ga black smoker-type mineralization in the Strelley Belt, Pilbara Craton, Western Australia. J Geol Soc Lond 152:587–590

Vearncombe S, Vearncombe JR, Barley ME (1998) Fault and stratigraphic controls on volcanogenic massive sulphide deposits in the Strelley belt, Pilbara Craton, Western Australia. Precambrian Res 88:67–82

Weir DJ (1989a) Annual Report for 1989 on C.E.C Joint Venture Area, Cattle Gorge, EL 45/430, Yarrie SF51-01, Western Australia

Weir DJ (1989b) Annual Report for 1989 on ELs 45/701 – 45/714 (SHAY 1–14), Yarrie SF51-01, Western Australia

Williams IR, Bagas L (2007) Geology of the Mount Edgar 1:100 000 sheet. 1:100 000 Geological series explanatory notes. Geological Survey of Western Australia

Wingate MTD, Kirkland CL, Hickman AH (2015) 178011: biotite granodiorite, Cookes Creek; geochronology record 1226. Geological Survey of Western Australia

Zegers TE, Barley ME, Groves DI, McNaughton NJ, White SH (2002) Oldest gold: deformation and hydrothermal alteration in the early Archean shear zone-hosted Bamboo Creek deposit, Pilbara, Western Australia. Econ Geol 97:757–773

Chapter 12
Fortescue Group: The Neoarchean Breakup of the Pilbara Craton

Abstract At 2775 Ma, the Neoarchean crust of the Pilbara Craton began to be extended and rifted resulting in the widespread eruption of basaltic lavas. Between c. 2775 and 2710 Ma, mafic–felsic volcanic and intrusive activity continued in stages that were separated by periods of uplift, folding, erosion, and sedimentation. The first basaltic formation deposited across the eroded surface of the craton was the Mount Roe Basalt, up to 2.44 km thick and fed from dolerite dykes intruded into extensional fractures; this was the first regionally extensive formation of the Fortescue Group. Deformation and erosion of the Mount Roe Basalt were followed by clastic deposition and felsic volcanism and intrusion of the 2766–2749 Ma Hardey Formation. The stratigraphic nomenclature of the Fortescue Group from 2749 Ma onwards differs between the North and South Pilbara. Even so, the same magmatic events affected both areas. Almost all volcanic activity ended at c. 2710 Ma following eruption of the Maddina Formation of the North Pilbara (correlated with the Bunjinah Formation in the south). Between c. 2710 and 2630 Ma, mainly clastic sedimentary rocks of the Jeerinah Formation, the upper formation of the Fortescue Group, were deposited in both areas. Because the stratigraphy and sedimentology of the Jeerinah Formation indicates passive margin basin deposition, it is interpreted that extension and rifting of the Pilbara Craton culminated in continental breakup and plate separation at c. 2710 Ma.

Most workers have interpreted the mainly volcanic 2775–2710 Ma lower Fortescue Group as a large igneous province formed by one or more mantle plumes. A mantle plume origin is consistent with the crustal extension and rifting of the Pilbara Craton, the continental breakup, and the stratigraphy of the volcanic succession that includes ultramafic–mafic–felsic volcanic cycles.

Keywords Mantle plume · Large igneous province · Crustal extension · Rifting · Continental breakup

A. H. Hickman, *Archean Evolution of the Pilbara Craton and Fortescue Basin*, Modern Approaches in Solid Earth Sciences 24,
https://doi.org/10.1007/978-3-031-18007-1_12

12.1 Introduction

The North Pilbara and Mosquito Creek Orogenies between 2955 and 2900 Ma (Chap. 10) effectively ended 300 million years of plate-tectonic activity following the 3220 Ma breakup of the Paleoarchean Pilbara Craton (Chap. 6). Closure of the Mosquito Creek Basin at 2900 Ma completed cratonization, forming a uniformly thick body of continental crust that remained tectonically stable until c. 2780 Ma. The only significant magmatic activity between 2900 and 2775 Ma was the local intrusion of post-orogenic granites, possibly related to continental drift across a hot spot (Chap. 10). At 2775 Ma, the crust of the Neoarchean Pilbara Craton began to be extended and rifted resulting in the widespread eruption of basaltic lavas. This marked the beginning of the Neoarchean breakup of the craton, an evolutionary stage referred to as the Fortescue Rifting Event (Hickman and Van Kranendonk 2008a, b; Hickman 2016, 2021).

The Fortescue Rifting Event lasted for about 60 million years during which a c. 6-km-thick succession of mainly volcanic formations was deposited to form the lower Fortescue Group. Due to substantial topographic relief on the surface of the craton, locally including granitic uplands rising over 2000 m above adjacent lowlands, the thickness of the group is regionally variable. Consequently, the lower formations were deposited on the lowlands and later formations progressively transgressed onto the uplands. In ascending stratigraphic order, the formations of the lower Fortescue Group are the Bellary Formation (sedimentary rocks and basalt), Mount Roe Basalt (basalt and basaltic andesite), Hardey Formation (sedimentary rocks with felsic volcanic and intrusive rocks), Kylena Formation (basalt, basaltic andesite, felsic volcanic rocks, and local stromatolitic carbonate rocks), Tumbiana Formation (sedimentary and pyroclastic rocks overlain by stromatolitic carbonate rocks), and Maddina Formation (basalt, basaltic andesite, and felsic volcanic rocks). The upper Fortescue Group, entirely assigned to the c. 2713–2630 Ma Jeerinah Formation, is mainly composed of shale, chert, and BIF intruded by dolerite sills. Local volcanic units (Baramine Volcanic Member and Isabella Member) form parts of the Jeerinah Formation in the Northeast Pilbara.

The stratigraphic succession of the group in the Southern Pilbara recognizes three formations correlated with the northern Pilbara succession: the Boongal Formation is correlated with the Kylena Formation; the Pyradie Formation with the Tumbiana Formation; and the Bunjinah Formation with the Maddina Formation (Thorne and Trendall 2001). The lithostratigraphy of the group was established in the 1960s and 1970s based on 1:250,000 scale mapping (reviewed by Hickman 1983). Following a detailed study of the group in the Northern Pilbara, Blake (1984) used the same stratigraphic nomenclature, although he subsequently used sequence stratigraphy to revise the nomenclature of the succession (Blake 1993). As explained by Thorne and Trendall (2001), this sequence stratigraphy has not been adopted during mapping by the Geological Survey of Western Australia.

The Fortescue Group was deposited in the Fortescue Basin that is preserved across at least 250,000 km^2 overlying the Pilbara Craton. For ease of reference to

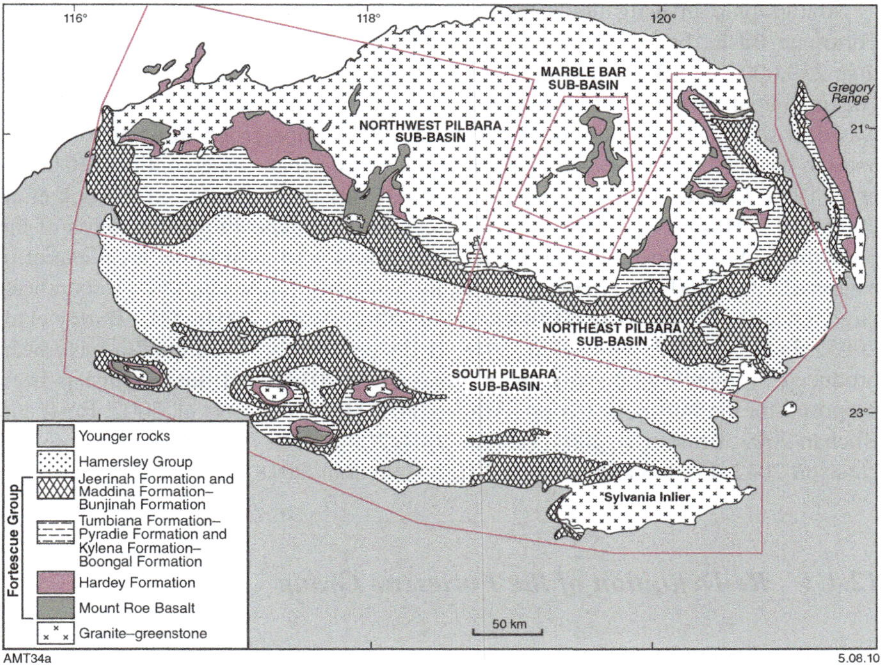

Fig. 12.1 Simplified geological map of the Fortescue and Hamersley Basins, showing sub-basins of the Fortescue Basin (Blake 1984). (Modified by Thorne and Trendall 2001; with Geological Survey of Western Australia permission)

depositional areas, the basin has been divided into four sub-basins (Fig. 12.1): Northwest Pilbara Sub-basin, Northeast Pilbara Sub-basin, Marble Bar Sub-basin, and South Pilbara Sub-basin (Thorne and Trendall 2001). The present interpretation of the lithostratigraphy of the Marble Bar Sub-basin is that the area shown as Hardey Formation on Fig. 12.1 is in reality composed of three formations: the Hardey Formation overlain by the Kylena Formation and the Pear Creek Formation. These three formations are separated by erosional unconformities.

The Fortescue Group succession of the Gregory Range area (Fig. 12.1) differs sufficiently from that of the adjacent Northeast Pilbara Sub-basin to be now assigned to the Gregory Range Sub-basin. Continuity of most elements of the basin stratigraphy between the sub-basins establishes that they were not separate depositional basins through most of the depositional history of the group. However, in detail the evolution of the basin was relatively complex and in the early stages deposition of the Mount Roe Basalt and Hardey Formation was restricted to low-lying areas on the uneven topography of the Pilbara Craton. Reactivation of domes and synclines in the underlying Pilbara Craton was an additional factor influencing the early evolution of the Fortescue Basin.

Stratigraphic investigations have concluded that the maximum extent of the Fortescue Basin, before the Neoarchean breakup of the craton, was much greater than 250,000 km^2 (Blake 1993; Martin et al. 1998a, b; Thorne and Trendall 2001; Hickman et al. 2010). There is geochronological evidence that the Fortescue Group was deposited at the same time as the Ventersdorp Supergroup of southern Africa and the basins might have been linked within the Vaalbara supercontinent (Cheney et al. 1988; Cheney 1996; Zegers et al. 1998; Eriksson et al. 2002; De Kock et al. 2009, 2012; Huston et al. 2012). In this scenario, the depositional thickness of the predominantly basaltic succession was up to 8 km, with an original areal extent in excess of 1,000,000 km^2; in other words, the volcanics formed part of a Neoarchean large igneous province (LIP) (Eriksson et al. 2002; Pirajno 2004, 2007; Barley et al. 2005; Hickman 2012; Mole et al. 2018). LIPs are mainly interpreted to have been produced by mantle plumes (Condie 2001; Ernst 2014), and this origin has been proposed for the Fortescue Group (Arndt et al. 2001; Eriksson et al. 2002; Ernst and Buchan 2003; Ernst et al. 2004; Pirajno 2004, 2007; Barley et al. 2005; Pirajno and Hoatson 2012; Pirajno and Santosh 2015; Mole et al. 2018).

12.1.1 Re-Definition of the Fortescue Group

A detailed description of the Fortescue Group was provided by Thorne and Trendall (2001). They interpreted the mainly volcanic and sedimentary parts of the succession (respectively, the Mount Roe Basalt–Maddina Formation succession and the overlying Jeerinah Formation) to have been deposited in a progressively deepening basin, with marine transgression from south to north. Consequently, despite a previous interpretation that the contact between the Maddina and Jeerinah Formations is locally unconformable in the Northern Pilbara (Williams 1968; Horwitz and Smith 1978; Simonson et al. 2000), and the long-term lithological change from volcanic to sedimentary rocks (Kriewaldt 1964; Hickman 1978; Seymour et al. 1988), Thorne and Trendall (2001) retained the Jeerinah Formation as part of the Fortescue Group.

The present interpretation is that the c. 2710 Ma contact between the Maddina and Jeerinah Formations marks the timing of the breakup of the Pilbara Craton. The c. 2775–2713 Ma volcanic succession was deposited during crustal extension and rifting, whereas the c. 2713–2630 Ma Jeerinah Formation is a post-breakup, sedimentary passive margin succession (Blake and Barley 1992; Blake 1993; Martin et al. 1998a, b; Blake 2001; Thorne and Trendall 2001; Hickman 2012). The interpretation that the volcanic succession constitutes a LIP, and was formed by mantle plume activity, adds to the difference between the lower and upper successions. Therefore, despite the limited field evidence for an unconformity between the Maddina and Jeerinah Formations, they evidently originated through different processes in different environments and should not be assigned to the same group. As previously concluded on purely lithological grounds, the Jeerinah Formation would be better assigned to the Hamersley Group (Kriewaldt 1964; Hickman 1978;

Seymour et al. 1988), and a formal stratigraphic revision to this effect is expected in the future.

12.2 Stratigraphy

The Fortescue Group unconformably overlies the Pilbara Craton across erosional unconformities. Because this erosion occurred after major deformation of the craton, and lasted more than 100 million years, the group overlies many different Paleoarchean–Mesoarchean groups, formations, and igneous suites. Numerous exposures of basal unconformities, some of which occur as high in the Fortescue Group stratigraphy as the Maddina Formation, reveal that the Neoarchean surface geology of the craton was very similar to that of today. In other words, many kilometres of the upper crust were eroded from the craton between 2900 and 2775 Ma. Many of the best exposures of the unconformity are in cliffs along the northern slopes of the Chichester Range (Fig. 12.2).

Fig. 12.2 The basal unconformity of the Fortescue Group in the Chichester Range 48 km north of the Auski Roadhouse. In this view, the Tumbiana Formation unconformably overlies granitic rocks of the Yule granitic complex. (From Van Kranendonk and Hickman 2012; with Geological Survey of Western Australia permission)

12.2.1 Tectono-Stratigraphic Sequences

Although the sequence stratigraphy of Blake (1993) has not been adopted (Thorne and Trendall 2001), the internal stratigraphic boundaries within the Fortescue Group, together with volcanic and sedimentary facies, geochemical compositions, and geochronological data, have been used to assign the eleven formations to five tectono-stratigraphic sequences, in descending stratigraphic order:

Sequence 5: Jeerinah Formation, 2713–2629 Ma; mainly sedimentary, but containing dolerite sills in the South Pilbara Sub-basin.

Sequence 4**:** Maddina Formation, 2719–2713 Ma; mafic–felsic volcanics in the northern sub-basins but including a thin sedimentary unit (Kuruna Member composed of volcaniclastic and stromatolitic carbonate rocks) in the Northeast Pilbara Sub-basin. In this review, Sequence 4 is interpreted to include the mainly sedimentary Pear Creek Formation (undated) in the Marble Bar Sub-basin. The Maddina Formation is correlated with the basaltic Bunjinah Formation in the South Pilbara Sub-basin.

Tumbiana Formation, 2727–2719 Ma; mainly sedimentary, including volcaniclastic units and stromatolitic carbonate units. Volcaniclastic and volcanic units of the formation are correlated with the ultramafic–mafic Pyradie Formation in the South Pilbara Sub-basin.

Sequence 3**:** Kylena Formation, 2749–2727 Ma; mafic–felsic volcanics in the northern sub-basins, but including a thin stromatolitic carbonate unit (Mopoke Member) in the Northeast Pilbara Sub-basin. The Kylena Formation is correlated with the basaltic Boongal Formation in the South Pilbara Sub-basin.

Sequence 2**:** Hardey Formation, 2766–2750 Ma; clastic sedimentary with felsic volcanic and subvolcanic intrusions in the northern sub-basins, but clastic sedimentary in the South Pilbara Sub-basin.

Sequence 1**:** Mount Roe Basalt, 2775–2766 Ma; thick successions of basalt and basaltic andesite in all sub-basins. In the Northeast Pilbara Sub-basin, includes locally thick basal clastic sedimentary units correlated with the undated Bellary Formation in the South Pilbara Sub-basin. The Bellary Formation might include sedimentary units significantly older than the Mount Roe Basalt.

12.2.1.1 Sequence 1

Crustal extension and rifting of the Pilbara Craton, commencing at c. 2775 Ma, led to the deposition of Sequence 1. The c. 2775–2766 Ma Mount Roe Basalt was extruded through a swarm of north-northeasterly trending fissures, now preserved as dolerite dykes (Fig. 12.3) (Black Range Dolerite). Dating of north-northeast trending dolerites at c. 2770 Ma in both the North and South Pilbara (Wingate 1999; Evans et al. 2017; Wingate et al. 2017) indicates west-northwest–east-southeast extension in both areas. In the context of the Fortescue Group volcanics constituting a LIP, it might be significant that a dolerite dyke swarm of the same age and trend is widespread across the Singhbhum Craton of eastern India (Kumar et al. 2017);

Fig. 12.3 View of the northern end of an exposure of the Black Range 40 km southwest of Marble Bar. The rocks forming the range are part of the Black Range Dolerite, a unit that forms a suite of c. 2770 Ma, north-northeast trending dolerite dykes across the Northern Pilbara. The dykes are interpreted to have been magma conduits for eruption of the Mount Roe Basalt. The Black Range exposes the largest dyke of the suite that has a length of 200 km. (From Van Kranendonk and Hickman 2012; with Geological Survey of Western Australia permission)

most supercontinent reconstructions place Western Australia in close proximity to India in the Neoarchean–Proterozoic.

Dykes of the Black Range Dolerite are intrusive into both greenstones and granites, but it appears that basalt flows were channelled from upland areas of granite into neighbouring valleys that had been preferentially developed on greenstone lithologies. There, the basalt was 'ponded' to locally build up volcanic piles over 2 km thick. Flows of the Mount Roe Basalt are only rarely preserved near dolerites in granitic areas. This is thought to be partly due to thinner basaltic accumulations over the uplands, and partly to result from erosion following uplift of the granitic areas during reactivation of the underlying Paleoarchean–Mesoarchean domes and synclines.

Within the deeper paleovalleys, the Mount Roe Basalt is separated from the Fortescue Group's basal unconformity by immature clastic sedimentary rocks that have been locally assigned to the Bellary Formation. The type area of the Bellary Formation is in the Bellary Dome of the South Pilbara Sub-basin where its basal contact with the Pilbara Craton is not exposed. In that area, the lowest exposed 100 m of the Bellary Formation is composed of basalt, and this raises the question if the sedimentary section of the Bellary Formation might be alternatively interpreted as a member within the Mount Roe Basalt. This possibility is consistent with observation that the sedimentary succession of the Bellary Formation is conformable with the Mount Roe Basalt. Correlation of the Bellary Formation with thick sedimentary units locally underlying the Mount Roe Basalt in the Marble Bar Sub-basin (Van Kranendonk 2004a, b, 2010; Hickman and Van Kranendonk 2008a, b) is problematic. Previously, all sedimentary units at the base of the Mount Roe Basalt in the sub-basins of the Northern Pilbara were interpreted to be basal conglomerates of the Mount Roe Basalt, and this remains the best interpretation for the majority of these units.

12.2.1.1.1 Bellary Formation

Locally thick sedimentary successions that conformably underlie the Mount Roe Basalt have been assigned to the Bellary Formation. In its type area (Bellary Dome), the formation has a minimum thickness of 400 m and consists of mudstone, siltstone, and sandstone, interbedded with smaller amounts of conglomerate, basalt, basaltic breccia, and tuff. No stratigraphic base is exposed. The lower part of the formation comprises over 100 m of subaqueous basalt flows and hyaloclastite overlain by 200 m of subaqueous fan-delta argillite and sandstone. In the eastern Bellary Dome, subaqueous facies are overlain by the lowest flows of the Mount Roe Basalt; farther west, they are capped by 100 m of subaerial fan-delta sandstone and conglomerate, interbedded with vesicular basalt flows and basaltic tuff.

In the Marble Bar Sub-basin, isolated thick units of conglomerate and sandstone at the base of the Mount Roe Basalt have been correlated with the Bellary Formation. Conglomerate and breccia form the base of the formation in the northern part of the Marble Bar Sub-basin. This clastic succession is locally over 500 m thick, but in most areas the thickness is far less than 50 m. Most of the conglomerate and sandstone units that separate the Mount Roe Basalt from underlying granite-greenstones of the Pilbara Craton fill early alluvial channels. These are interpreted as basal accumulations of detritus along the erosional unconformity and have been assigned to the Mount Roe Basalt. In areas where the Mount Roe Basalt is absent, similar channel deposits occur at the base of the Hardey Formation, as at Nullagine.

No definitive geochronology is available for the Bellary Formation. Hall (2002) provided U–Pb zircon data for a tuffaceous sandstone collected from the Bellary Dome north of Paraburdoo. Individual zircon ages ranged from c. 3592 to 2785 Ma, consistent with detrital zircon grains eroded from the underlying Pilbara Craton. Apart from a single grain dated at 2785 ± 15 Ma (analysis 99% concordant), no other analyses indicated crystallization ages younger than c. 2900 Ma. The c. 2785 Ma date is within error of dates obtained from the Mount Roe Basalt and dykes of the Black Range Dolerite.

12.2.1.1.2 Mount Roe Basalt

Except where sedimentary and volcanic rocks of the Bellary Formation unconformably overlie the Pilbara Craton, the Mount Roe Basalt is the oldest formation of the Fortescue Group. The formation has a maximum thickness of about 2.5 km (Blake 1993; Thorne and Trendall 2001; Van Kranendonk 2010).

The Mount Roe Basalt was erupted onto a Neoarchean landscape of low hills, ridges, and valleys. As a consequence, the basaltic lava flows of the formation were

Fig. 12.4 Ropy pahoehoe lava flow top in the Mount Roe Basalt. (From Van Kranendonk and Hickman 2012; with Geological Survey of Western Australia permission)

mainly confined to lowland areas, and overlying stratigraphic formations such as the Hardey and Kylena Formations onlapped onto increasingly higher ground. Flows of basalt and basaltic andesite, including local pillow lava and basalt displaying ropy texture on pahoehoe flows (Fig. 12.4), are interbedded with minor tuff, hyaloclastite, and epiclastic rocks. Pillow lavas form a relatively minor part of the formation, with most occurrences being recorded at or near the base of the succession in the North Pilbara. The pillows suggest local deposition in lakes that had formed either by volcanic damming of valleys in areas of rugged topography, or by local faulting. In the Marble Bar Sub-basin, a paleovalley including pillow basalt is exposed along a 15-km-long section of the Coongan River north and south of Blue Bar Pool (Hickman 1983; Hickman and Van Kranendonk 2008a, b). About 22 km north of Blue Bar Pool, at a locality where an old 'Flying Fox' crosses the Coongan River, there are excellent exposures of the basal Mount Roe Basalt on the eastern side of the same north–south paleovalley. Here, rocky exposures in the river bed show basaltic lava flows containing irregular sub-horizontal structures interpreted to be gas cavities deformed by the flow of the lava. Other unusual features include vertical lava pipes, typically 10 cm in diameter and up to 50 cm long that might represent late-stage volatile conduits.

A characteristic feature of the Mount Roe Basalt is the widespread occurrence of a glomeroporphyritic texture in which numerous clusters of radiating plagioclase phenocrysts, up to 10 mm long and variably sericitized, are set within a fine-grained basaltic groundmass (Fig. 12.5). Although this texture is not entirely

Fig. 12.5 Outcrop of the Mount Roe Basalt showing an example of the glomeroporphyritic texture characteristic of many parts of the formation. Clusters of 1-cm-long plagioclase phenocrysts are set in a basaltic matrix. (Modified from Van Kranendonk 2010; with Geological Survey of Western Australia permission)

confined to the Mount Roe Basalt, it is especially common in this formation (Hickman 1983).

Non-volcanogenic sedimentary rocks of the Mount Roe Basalt include clast- and matrix-supported conglomerate (Fig. 12.6), feldspathic quartz sandstone, and argillite. On the eastern side of the Marble Bar Sub-basin, the formation includes a 100-m-thick member of polymictic boulder conglomerate grading up into pebbly sandstone (Hickman and Van Kranendonk 2008a, b). The boulders include granite and chert in addition to clasts derived from erosion of the underlying units of the Mount Roe Basalt. Near Warrawagine in the Northeast Pilbara, the Mount Roe Basalt contains a thin unit of stromatolitic carbonate rocks. By analogy with similar carbonate units in the Kylena and Tumbiana Formations (this chapter), this was probably a lacustrine deposit.

In the Northwest Pilbara Sub-basin, two paleosol horizons have been recognized southeast of Whim Creek (Macfarlane et al. 1994; Yang et al. 2002; Nedachi et al. 2004; Neaman et al. 2005). These paleosols are up to 15 m thick and comprise an upper zone of massive, conchoidally fractured pale grey to white sericite-rich rock above a chloritic Fe-rich zone. The lower zones overlie fresh basalt, and in one

TRF57 07.06.06

Fig. 12.6 Poorly sorted, polymictic conglomerate at the base of the Mount Roe Basalt near Mount Elsie, Northeast Pilbara Sub-basin. Clasts are mainly composed of basaltic lithologies derived from the Euro Basalt (MGA Zone 51, 244360E 7604760N). (From Farrell 2006; with Geological Survey of Western Australia permission)

paleosol, the sericite-rich upper zone is overlain by shallow-water sandstone and siltstone, establishing a break in volcanism. The sericite-rich zones are depleted in Fe and other elements such as Mn, Mg, and Zn that are mobile under extreme weathering in anoxic conditions. Because Fe would not be depleted in the presence of free O_2 molecules, the Fe depletion of the sericite-rich zones has been interpreted as evidence for an anoxic atmosphere at c. 2770 Ma. However, the sericite-rich zones contain >10% K_2O which is far too high to be explained by residual enrichment (Thorne and Trendall 2001). An alternative explanation for the sericite-rich zones is alteration of the basalt by methane-rich hydrothermal fluids (Nedachi et al. 2002), although in this case it is unclear why this type of alteration would be restricted to such thin horizontal zones within the volcanic succussion. Macfarlane et al. (1994) attributed the high levels of K_2O and light REE to Paleoproterozoic metamorphism and metasomatism in addition to Neoarchean weathering.

A similar zone of massive white sericite and pyrophyllite is present along a 5-km-long contact between the Mount Roe Basalt and Hardey Formation south of the Comet gold mine on the southeast side of the Marble Bar Sub-basin (Hickman and Van Kranendonk 2008a, b). The same zone is repeated by faulting 5 km southwest of the Comet mine. These occurrences are consistent with the interpretation that the alteration was at depositional breaks where basalts were exposed to surface alteration over lengthy periods of time.

12.2.1.1.2.1 Geochronology

The maximum age of the Mount Roe Basalt is indicated by a U–Pb zircon date of 2775 ± 10 Ma from the lower part of the Mount Roe Basalt at the Wyloo Anticline (Arndt et al. 1991). Other geochronology has been obtained from the Black Range Dolerite which is interpreted to be a feeder to the Mount Roe Basalt (Hickman and Lipple 1975; Lewis et al. 1975; Hickman 1983; Blake 1993). A U–Pb baddeleyite date of 2772 ± 2 Ma was obtained on the main intrusion, the Black Range Dyke (Wingate 1999), and two other U–Pb baddeleyite dates on the Black Range Dolerite have been 2769 ± 1 Ma and 2764 ± 3 Ma (Evans et al. 2017). A dolerite dyke in the Rocklea Dome in the South Pilbara, correlated with the Black Range Dolerite, was dated at 2770 ± 4 Ma (Wingate et al. 2017, GSWA 205904). Based on the existence of three dates of c. 2766 Ma on the overlying Hardey Formation, the age range of the Mount Roe Basalt is interpreted to be 2775 to 2766 Ma.

12.2.1.2 Sequence *2*

In the Northwest Pilbara Sub-basin, the boundary between Sequences 1 and 2 is a disconformity, whereas in the Marble Bar and Northeast Pilbara Sub-basins, it varies locally between disconformities and angular unconformities (Blake 1993). Deposition of Sequence 2 commenced with reactivation of the major Paleoarchean–Mesoarchean domes and faults in the Northeast Pilbara, and with renewed movement on Mesoarchean faults and shear zones in the Northwest Pilbara. In the Northeast Pilbara Sub-basin, the Mount Roe Basalt was folded into synclines and formed graben that follow the trends of underlying synclines in the greenstone belts. Erosion of areas uplifted by the deformation led to rapid deposition of poorly sorted fluvial conglomerate and sandstone (Fig. 12.7) and subsequently lacustrine shale. A lithologically distinctive member of grey shale and mudstone, termed the ‘Glenn Herring Shale’ by Noldart and Wyatt (1962), appears to have occurred in both the Marble Bar and Northeast Pilbara Sub-basins. Shallow-water deposition of the shale is suggested by the widespread occurrence of ripple marks (Fig. 12.8).

In the Northwest Pilbara Sub-basin, deposition of non-volcanogenic sedimentary rocks of the Hardey Formation was concentrated within shallow rift basins. In the Northeast Pilbara Sub-basin, the Hardey Formation includes large sills of felsic intrusive rocks assigned to the Bamboo Creek Member. In many areas, these felsic sills are associated with felsic volcanic units. Close to the top of the Hardey Formation, there is an erosional unconformity above which clasts and boulder of the Bamboo Creek Member form an upper conglomerate; a similar conglomerate is present in the Hardey Formation of the Gregory Range. Large felsic sills are absent from the Northwest Pilbara Sub-basin, but felsic volcaniclastic rocks of the Lyre Creek Member are widespread.

In the South Pilbara Sub-basin, Sequence 2 evolved from braided fluvial to deltaic depositional environments (Thorne and Trendall 2001). As in the North Pilbara, braided fluvial sandstones are compositionally and texturally immature and are

Fig. 12.7 Basal conglomerate of the Hardey Formation near the Harding Dam. Polymictic conglomerate overlies the Mount Roe Basalt. Most of the boulders are vesicular or porphyritic basalt typical of lithologies in the underlying Mount Roe Basalt, but other lithologies including granite are also present. Scale provided by hammer (top centre). (From Hickman et al. 2010; with Geological Survey of Western Australia permission)

interpreted to have been derived from erosion Paleoarchean and Mesoarchean granitic rocks. Sediment transport directions in fluvial and deltaic systems were principally towards the southwest and west and indicate the presence of a basement high, referred to as the 'Yule–Sylvania High' (Thorne and Trendall (2001), to the north and northeast of the Milli Milli Dome. In the Southeast Pilbara, there is a marked northward thinning of the Hardey Formation across the Sylvania Inlier.

12.2.1.2.1 Hardey Formation

The Hardey Formation contains a wide range of sedimentary, volcanic, and intrusive rocks and locally forms successions up to 3 km thick. The sedimentary rocks include clast- and matrix-supported conglomerate, feldspathic quartz sandstone and pebbly sandstone, mudstone, and shale. Felsic volcanic and intrusive rocks form parts of the Hardey Formation in the four sub-basins of the North Pilbara but have not been recorded in the South Pilbara Sub-basin.

Fig. 12.8 Ripple marks in siltstone of the Hardey Formation, Northwest Pilbara Sub-basin 24 km south of Harding Dam (MGA Zone 50, 504250E 7655000N). (From Van Kranendonk and Hickman 2012; with Geological Survey of Western Australia permission)

12.2.1.2.1.1 Bamboo Creek Member

In the Northeast Pilbara Sub-basin, the c. 2760 Ma Bamboo Creek Member includes sills of porphyritic dacite up to 2 km thick and felsic lavas that include ignimbrites (Fig. 12.9). The sills are approximately the same age as A-type granites that occupy a 170-km-long belt in the Gregory Range of the East Pilbara and the same age as the Koongaling Volcanic Member that overlies the granites and genetically related granophyre (see Geochronology below). Numerous felsic dykes that intrude the Mount Edgar, Warrawagine, and Muccan granitic complexes have locally been dated between c. 2765 and 2758 Ma and are interpreted to be feeders to the Bamboo Creek Member (Williams and Bagas 2007).

12.2.1.2.1.2 Koongaling Volcanic Member

The Koongaling Volcanic Member is the lowest stratigraphic unit of the Fortescue Group in the Gregory Range Sub-basin and is up to 2 km thick. The main lithology is porphyritic rhyolite containing K-feldspar phenocrysts up to 5 mm in length, although the member also includes units of agglomerate and tuff (Hickman 1975, 1978). Unusual ovoid structures about 200 m across are visible on aerial photography north of Koongaling Hill (Hickman 1978). From the air, the structures display internal concentric layering resembling finger prints, and in several areas the

Fig. 12.9 Welded ignimbrite in the Hardey Formation, showing eutaxitic texture, with flamme, lithic fragments (dark), and compacted quartz and feldspar phenocrysts (MGA Zone 51, 224720E 7650705N). (From Van Kranendonk 2010; with Geological Survey of Western Australia permission)

layering in one ovoid structure is truncated by layering in one or more similar adjacent structures. Williams and Trendall (1996, 1998) interpreted the structures to be ogives, formed during non-explosive extrusion of highly viscous magma. Williams and Trendall (1998) used geochemical evidence to interpret a gradational and genetic relationship between the Koongaling Volcanic Member and underlying granophyre of the Gregory granitic complex.

12.2.1.2.1.3 Warri Warri and Tanguin Members

Two volcano-sedimentary members form parts of the Hardey Formation in the Gregory Range Sub-basin: the Warri Warri and Tanguin Members (Williams and Trendall 1996, 1998). The Warri Warri Member varies between 1 and 4 km thick and is mainly composed of tuffaceous sedimentary rocks with conglomerate, felsic agglomerate, and rhyolite. A 100-m-thick unit of massive felsic rock and porphyritic rhyolite is lithologically similar to the Bamboo Creek Member. The conglomerate contains boulders of porphyritic rhyolite that are likely to have been derived from erosion of the Bamboo Creek Member. The Tanguin Member is composed of felsic and mafic lavas with minor volcaniclastic sedimentary rocks. The lowest part of the succession is a 200-m-thick unit of volcaniclastic sandstone interbedded with quartz sandstone (Thorne and Trendall 2001). This is overlain by a 200-m-thick basaltic succession overlain by 100 m of rhyolite.

Fig. 12.10 Reworked volcaniclastic breccia of the Lyre Creek Member, northern face of Table Hill, Northwest Pilbara Sub-basin. Partly rounded boulders and pebbles of dacitic volcanic rocks are set in a poorly sorted dacitic tuffaceous matrix (MGA Zone 50, 510600E 7674200N). Scale card is 10 cm long. (From Hickman 2004a; with Geological Survey of Western Australia permission)

12.2.1.2.1.4 Lyre Creek Member

The Lyre Creek Member ('Lyre Creek Agglomerate Member', Williams 1968; renamed by Thorne and Trendall 2001) outcrops extensively in the Northwest Pilbara Sub-basin. The member comprises agglomerate, massive tuff, bedded lapilli and crystal tuff, quartzitic tuff, and local calcareous beds. Figure 12.10 shows an outcrop of the Lyre Creek Member on the northern slopes of Table Hill, with reworked volcaniclastic breccia containing rounded boulders and pebbles of dacitic volcanic rocks, and intercalated sandstone. Based on lateral thickness variations, the eruptive centre of the Lyre Creek Member is interpreted to have been in the Cooya Pooya area (Hickman 2004a, b). Thin, lenticular beds of sandstone or conglomerate are locally interstratified with the pyroclastic deposits and indicate local reworking in fluviatile channels. Excellent examples of convergent sandstone channels are exposed 1.5 km southeast of Mount Montagu (Fig. 12.11). The orientations of these channels indicate local paleocurrent directions to the east and northeast (Hickman 2004a, b).

The composition of the Lyre Creek Member is mainly andesitic to dacitic, but rhyolite and rhyolitic tuff outcrop about 1.5 km north of Mount Montagu where, due to complete erosion of the Kylena Formation, the member is unconformably overlain by the Tumbiana Formation.

Fig. 12.11 Convergent sandstone channels in felsic pyroclastic rocks of the Lyre Creek Member, 1.5 km southeast of Mount Montagu, Northwest Pilbara Sub-basin (MGA Zone 50, 534750E 7635250 N). (From Hickman 2004a; with Geological Survey of Western Australia permission)

12.2.1.2.1.5 Coolajacka Member

In the Northwest Pilbara Sub-basin, the Coolajacka Member (Hickman and Kojan 2003) is composed of komatiite and komatiitic basalt, although MgO contents do not exceed about 20 wt.%. The member is composed of fine-grained, silicified, pyroxene-rich, and olivine-bearing rock containing numerous quartz xenocrysts and blocks of quartzite. The xenocrysts and quartzite blocks were derived from underlying sandstone of the Hardey Formation. The Coolajacka Member is intruded by the Cooya Pooya Dolerite of the Kylena Formation (Hickman 2004a, b). Ultramafic clasts and boulders derived from the Coolajacka Member are locally present within pyroclastic units of the Lyre Creek Member.

12.2.1.2.1.6 Geochronology

Geochronology on various members of the Hardey Formation indicates depositional and intrusive events between c. 2766 and 2749 Ma. The most widely dated units have been dacite and rhyolite sills, lavas and tuffs, dykes, and granitic stocks of the Bamboo Creek Member. Sills of the member have been dated at 2768 ± 16 Ma (sample 77712, Pidgeon 1984), 2756 ± 8 Ma (sample 94761, Arndt et al. 1991), and 2766 ± 2 Ma (sample U938/1, Blake et al. 2004); lavas within the member have been dated at 2766 ± 3 Ma (sample U919/1, Blake et al. 2004) and 2766 ± 7 Ma (GSWA 169037, Nelson 2004); porphyritic felsic dykes have been dated at 2758 ± 4 Ma (GSWA 142875, Nelson 1999), 2765 ± 3 Ma (GSWA 169033, Nelson 2004), 2760 ± 4 Ma (GSWA 169043, Nelson 2004), and

2765 ± 2 Ma (sample 103281, RI Thorpe, in Thorne and Trendall 2001); and a granitic stock, interpreted to have originated in a pipe related to the sills and lavas, has been dated at 2757 ± 7 Ma (GSWA 142825, Nelson 1998). Felsic lavas and tuffaceous rocks, apparently significantly younger than the Bamboo Creek Member, have been dated at 2752 ± 4 Ma (sample U922/1, Blake et al. 2004) in the Northeast Pilbara Sub-basin and 2750 ± 4 Ma in the South Pilbara Sub-basin (sample 2944826/6, Hall 2002).

In the Gregory Range, a felsic volcanic rock of the Koongaling Volcanic Member was dated at 2764 ± 8 Ma (sample 94759, Arndt et al. 1991), and a rhyolite in the Warri Warri Member was dated at 2760 ± 10 Ma (sample 94760, Arndt et al. 1991). In the same area, a granophyre underlying the Koongaling Volcanic Member was dated at 2763 ± 8 Ma (GSWA 118923, Nelson 1996). Granitic rocks interpreted to be subvolcanic to felsic rocks of the Hardey Formation in the Gregory Range were dated at 2762 ± 4 Ma (GSWA 118920, Nelson 1996), 2761 ± 2 Ma (GSWA 118925, Nelson 1999), and 2757 ± 5 Ma (GSWA 118924, Nelson 1996).

In the Northwest Pilbara Sub-basin, Arndt et al. (1991) attempted to date the depositional age of the Lyre Creek Member, but the sample contained only inherited zircons dated between c. 3240 and 2900 Ma. The Nerrely Leucogranite, which intrudes the Mallina Basin south of the Whim Creek greenstone belt, was dated at 2765 ± 5 Ma (GSWA 117982, Nelson 1997) and might have been a subvolcanic intrusion feeding part of the Lyre Creek Member.

12.2.1.3 Sequence *3*

Sequence 3 is entirely composed of the Kylena Formation. In previous accounts of the Fortescue Group, the third sequence also included the Tumbiana and Maddina Formations (Thorne and Trendall 2001; Hickman et al. 2010). This grouping followed an early stratigraphic interpretation of the succession in which these three formations were assigned to the now obsolete 'Mount Jope Volcanics' (de la Hunty 1965; Kriewaldt and Ryan 1967). In the sequence stratigraphy of Blake (1993), the succession of the three formations was equated to the 'Mount Jope Supersequence'. However, an angular unconformity between the Kylena and Tumbiana Formations in the Northwest Pilbara Sub-basin (Fig. 12.1) (Hickman 2004a, b) has now led to separation of the Tumbiana and Maddina Formations into Sequence 4. Additional evidence of the stratigraphic break is provided by geochronology indicating of a time break of about ten million years.

In the Northeast Pilbara, Marble Bar, and Northwest Pilbara Sub-basins, boundaries between Sequences 2 and 3 are either angular unconformities or disconformities. Extreme thickness variations in the Hardey Formation are partly due to erosion prior to deposition of the Kylena Formation. For example, in the Eastern Creek area (Northeast Pilbara Sub-basin), Farrell (2006) reported that the Kylena Formation unconformably overlies the Hardey Formation; south of the Yilgalong granitic complex, the Hardey Formation is almost entirely represented by the felsic Bamboo Creek Member (intrusive and extrusive), and locally the Hardey Formation is entirely missing. In the Marble Bar Sub-basin, about 15 km north-northwest of Marble Bar, the Kylena Formation directly overlies the Mount Roe Basalt (Hickman and Van Kranendonk 2008a, b), and this is the situation throughout most of the Pear Creek area (Van Kranendonk 2003, 2004a, b, 2010). In the Gregory Range Sub-basin, a boulder conglomerate containing clasts of felsic rock near the top of the Warri Warri Member (Thorne and Trendall 2001) is evidence of erosion of the Hardey Formation prior to deposition of the Kylena Formation. A similar conglomerate is present close to the top of the Hardey Formation in the Taylor Creek area north of Nullagine.

12.2.1.3.1 Kylena Formation

In the Northeast Pilbara Sub-basin, the Kylena Formation comprises three stratigraphic divisions: the lower part of the formation comprises komatiitic basalt, basalt, and basaltic andesite; the central division is a thin but laterally extensive member of stromatolitic carbonate rocks, the Mopoke Member; and the upper part of the formation is composed of basaltic andesite, dacite, and minor rhyolite. The same vertical trend, but possibly lacking basal komatiitic basalt, is present in the Kylena Formation of the Northwest Pilbara Sub-basin where it has been confirmed by geochemical traverses (Fig. 12.12) (Kojan and Hickman 1998).

The upward lithological trend from ultramafic and mafic volcanics to felsic volcanics is similar to that of ultramafic–mafic–felsic volcanic cycles in the Paleoarchean Pilbara Supergroup (Chaps. 3 and 5). The Pilbara Supergroup is attributed to mantle plume activity (Van Kranendonk et al. 2002, 2006a, b, 2007; Smithies et al. 2005; Hickman 2011, 2021). In other parts of the Precambrian geological record, mantle plumes are interpreted to have produced many large igneous provinces (LIPs) (Condie 2001; Ernst 2014), and many authors now interpret that the lower volcanic succession of the Fortescue Group constitutes a LIP (Arndt et al. 2001; Eriksson et al. 2002; Bleeker 2003; Barley et al. 2005; Pirajno 2007; Hickman 2012; Pirajno and Hoatson 2012; Pirajno and Santosh 2015; Mole et al. 2018). More detailed stratigraphic and geochemical work is needed to examine the possibility that Sequences 1 and 2 represent a lower volcanic cycle in which the central section of basaltic andesite and andesite was removed by erosion during deposition of Sequence 2.

In the Gregory Range area, the Kylena Formation includes a unit of sandstone, rhyolite, and felsic pyroclastic rocks up to 800 m thick which outcrops along a strike length of 12 km (Williams and Trendall 1996 1998). The rocks include a dark, aphanitic, flinty rhyolite at the base, succeeded upwards by a sequence of well-stratified green tuffaceous siltstone and sandstone, locally with numerous beds of accretionary lapilli. These pyroclastic rocks are in turn overlain by rhyolite similar to that at the base but are locally associated with felsic agglomerate. A very coarse-grained felsic agglomerate caps the succession. This local succession of felsic and sedimentary rocks is interpreted to be at approximately the same stratigraphic level as sedimentary and felsic rocks immediately underlying the Mopoke Member east of

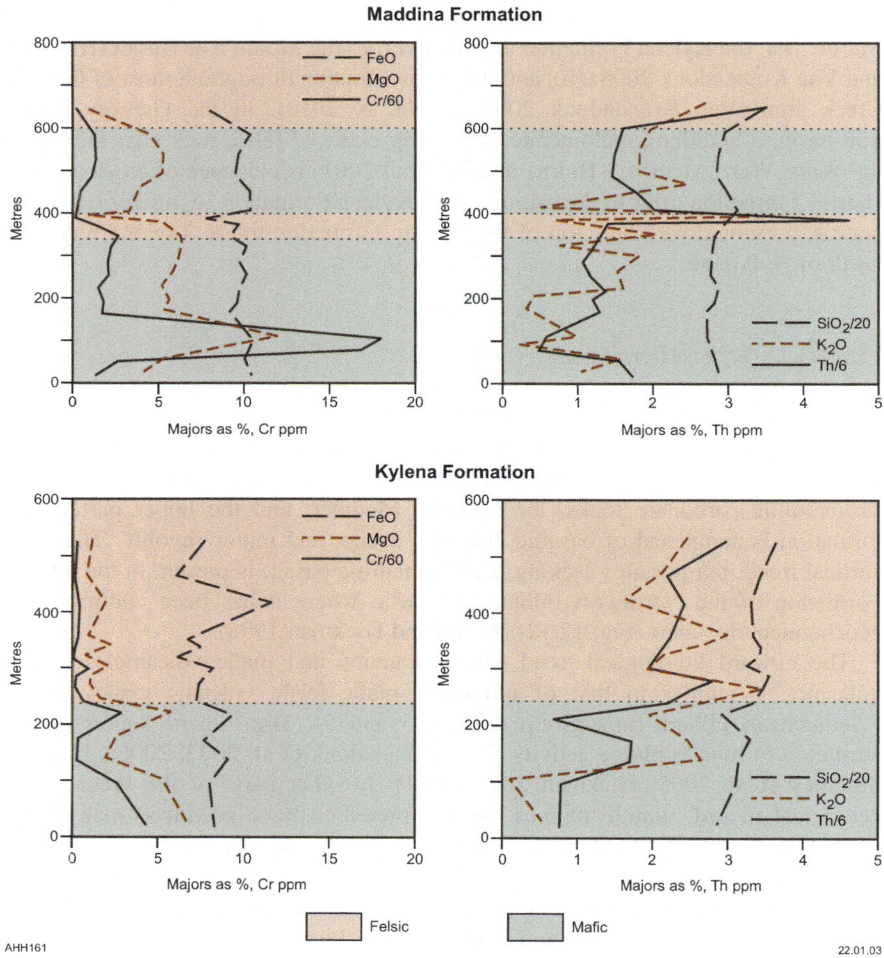

Fig. 12.12 Vertical geochemical trends in the Kylena and Maddina Formations of the Northwest Pilbara Sub-basin. (From Kojan and Hickman 1998; with Geological Survey of Western Australia permission)

the Yilgalong granitic complex in the Northeast Pilbara Sub-basin. The Gregory Range rocks have not been dated, but the possible lateral equivalents east of the Yilgalong granitic complex were dated at 2749 ± 5 Ma (Nelson et al. 2006, GSWA 178086) and 2735 ± 5 Ma (Bodorkos et al. 2006, GSWA 178090).

Fig. 12.13 Outcrop of silicified stromatolitic carbonate rock in the Mopoke Member (MGA Zone 51, 246377E 7641976N) 9 km east of Meentheena. (From Williams 2007; with Geological Survey of Western Australia permission)

12.2.1.3.1.1 Mopoke Member

The presence of a laterally extensive unit of stromatolitic carbonate rocks within the Kylena Formation was first recorded during GSWA geological mapping in the early 1970s (Thom et al. 1979; Hickman et al. 1983); it was formally named as the Mopoke Member by Williams (2007). The Mopoke Member is up to 10 m thick and outcrops intermittently across a strike length of about 170 km from the northwest part of the Oakover Syncline to the Balfour Downs area. The member is composed of ripple cross-laminated calcareous mudstone, cherty stromatolitic carbonate (Fig. 12.13), wavy-laminated and recrystallized stromatolitic carbonate, tabular clast edgewise conglomerate, and kerogenous siltstone and shale (Flannery et al. 2012). By analogy with the same lithologies in the Meentheena Member of the overlying Tumbiana Formation (Awramik and Buchheim 2009), the depositional environment was lacustrine.

Fig. 12.14 View of Table Hill from a quarry on the Robe River Railway, showing the upper sill of the Cooya Pooya Dolerite (top of hill) overlying volcaniclastic lithologies of the Lyre Creek Member. The dark rubbly outcrops and low hills in the middle distance are composed of the lower sill. (From Hickman (2004a; with Geological Survey of Western Australia permission)

12.2.1.3.1.2 Cooya Pooya Dolerite

In the Northwest Pilbara Sub-basin, the Hardey Formation is intruded by laccoliths, sills, and dykes of the Cooya Pooya Dolerite. This is a lenticular complex which intrudes two main stratigraphic levels: the lower part of the Hardey Formation beneath the Lyre Creek Member and the top of the Hardey Formation and lower levels of the Kylena Formation. The complex is of regional extent, with a strike length of about 130 km. At the lower stratigraphic level, it varies in vertical thickness up to 100 m and forms sills and small laccoliths. Networks of dolerite dykes transgress the stratigraphy and fragment the Coolajacka Member of the Hardey Formation. The dykes have no dominant orientation but outcrop in a polygonal pattern visible at map scale, suggesting either intrusion along pre-existing joints in the ultramafic lavas, or forceful injection above laccoliths. The upper sill of the Cooya Pooya Dolerite (Fig. 12.14) has an average thickness of about 50 m, and it ranges from being locally absent to 100 m thick. The lower and upper sills are separated by the Lyre Creek Member that has the same regional extent as the Cooya Pooya Dolerite.

It is interpreted that the Cooya Pooya Dolerite is an intrusive member of the Kylena Formation and that its emplacement was largely controlled by the local presence of the Lyre Creek Member within the Hardey Formation. Massive beds of the Lyre Creek Member are thought to have formed a barrier to upward intrusion of basaltic magma. This explains the laccolithic sill complex beneath the Lyre Creek Member, and local fragmentation of the Coolajacka Member, which directly underlies the Lyre Creek Member. The Kylena Formation is about 100 m thick where it is underlain by the Cooya Pooya Dolerite, whereas elsewhere it is 300 to 500 m thick. The lithological composition of the Cooya Pooya Dolerite is similar to that of the lower Kylena Formation.

12.2.1.3.1.3 Kylena Formation in the Marble Bar Sub-Basin

GSWA geological mapping of the Marble Bar Sub-basin interprets the Kylena Formation to overlie the Hardey Formation across an erosional unconformity (Hickman and Lipple 1975, 1978; Van Kranendonk 2004a, b, 2010; Hickman and Van Kranendonk 2008a, b). The formation is up to 1.1 km thick (Blake 1993, Thorne and Trendall 2001) and was deposited following folding, tilting, and erosion of the Hardey Formation and the Mount Roe Basalt (Van Kranendonk 2010). This is inconsistent with an interpretation by Blake (1993), later adopted by Thorne and Trendall (2001), that this thick basaltic succession is part of the Hardey Formation. It is also inconsistent with an almost complete absence of basalts elsewhere in the Hardey Formation; local thin basalt units, up to 15 m thick (Hall 2002), in some parts of the South Pilbara Sub-basin might have been related to early rifting.

12.2.1.3.1.4 Geochronology

Geochronology on the Kylena Formation in the Northeast Pilbara Sub-basin has provided dates between c. 2749 and 2735 Ma. Samples from a dacite and a tuffaceous sandstone immediately underlying Mopoke Member were dated at 2749 ± 5 Ma (GSWA 178086, Nelson et al. 2006) and 2735 ± 6 Ma (GSWA 178090, Bodorkos et al. 2006). A mafic tuffaceous unit from the near the base of the Kylena Formation was dated at 2741 ± 3 Ma (Blake et al. 2004). Dolerite dykes in the Sylvania Inlier, interpreted to be feeders to the Kylena Formation, were dated at 2747 ± 4 Ma (Wingate 1999). However, the youngest age of c. 2735 Ma was obtained from the top of the lower part of the formation, underlying the Mopoke Member. Accordingly, the minimum age of the Kylena Formation is constrained by the oldest date from the overlying Tumbiana Formation, 2727 ± 5 Ma (Blake et al. 2004).

The youngest date from the underlying Hardey Formation, 2750 ± 4 Ma (sample 2944826/6, Hall 2002), is inferred to constrain the maximum depositional age of the Kylena Formation.

12.2.1.4 Sequence 4

Sequence 4 comprises the Tumbiana Formation overlain by the Maddina Formation. In the Northwest Pilbara Sub-basin, the Tumbiana Formation overlies the Kylena Formation across an angular unconformity (Fig. 12.15) (Hickman 2004a, b). In most areas of the Northeast Pilbara Sub-basin and in the Gregory Range, the contact is disconformable to conformable (Blake et al. 2004; Williams 2007); geochronology

AHH190 29.04.04

Fig. 12.15 View of the angular unconformity between basalt flows of the Kylena Formation, dipping to the right (middle distance) and the horizontal Tumbiana Formation (far distance). Photograph taken 4 km southeast of Python Pool (MGA Zone 50, 524750E 7640900N). (From Hickman 2004a; with Geological Survey of Western Australia permission)

suggests a depositional time break of about ten million years, and Blake (2001) reported a major geochemical change in mafic rocks at this boundary. Sequence 4 is apparently absent from the Marble Bar Sub-basin where the Kylena Formation is unconformably overlain by the undated Pear Creek Formation. Blake (1993) interpreted the undated Pear Creek Formation to be part of the Hardey Formation, which would place it in Sequence 2, but this interpretation is inconsistent with significant erosional unconformities at the basal contacts of the Pear Creek and Kylena Formations (Van Kranendonk 2003, 2010).

The Maddina Formation conformably overlies the Tumbiana Formation, and geochronology indicates no significant break between the formations. Much of the Tumbiana Formation is composed of sandstone that was most likely deposited during erosion of the underlying Fortescue Group and the Pilbara Craton during uplift of the basin in advance of plume-related eruption of the Maddina Formation. Except in most of the Northwest Pilbara Sub-basin, the Maddina Formation contains a central unit of siltstone, accretionary lapilli tuff, and partly silicified stromatolitic carbonate rocks. This unit, renamed from 'Kuruna Siltstone' (MacLeod and de la Hunty 1966; Hickman 1983) to Kuruna Member (Thorne and Tyler 1997; Thorne and Trendall (2001), is lithologically very similar to the Tumbiana Formation.

In the Northeast Pilbara Sub-basin, there is thick unit of komatiitic basalt, basalt, and basaltic andesite between the Tumbiana Formation and the Kuruna Member. This lower unit of the Maddina Formation, once named as the 'Nymerina Basalt Member' (MacLeod and de la Hunty 1966) or 'Nymerina Basalt' (Hickman 1983), wedges out westwards in the Hooley area of the Chichester Range where the

Tumbiana Formation and the Kuruna Member come close together. This relationship suggests that eruption of the more mafic, lower Maddina Formation commenced in the Northeast Pilbara Sub-basin during northwest deposition of the Tumbiana Formation. In that situation, the Kuruna Member would not be recognized in the northwest because there it forms the upper part of the Tumbiana Formation.

12.2.1.4.1 Tumbiana Formation

In the Northeast Pilbara Sub-basin, the Tumbiana Formation is divided into the Mingah Member conformably overlain by the Meentheena Member. In the Northwest Pilbara Sub-basin, stromatolitic carbonate rocks characteristic of the Meentheena Member form laterally discontinuous units at several stratigraphic levels (Awramik and Buchheim 2009; Coffey et al. 2011).

The Mingah Member is composed of basaltic to andesitic volcaniclastic sandstone and siltstone, quartz sandstone, shale, and minor stromatolitic carbonate. Beds of accretionary lapilli (Fig. 12.16), forming graded or non-graded layers, were formed as pyroclastic fall deposits (Williams and Bagas 2007). Cross-bedded and ripple-marked beds, containing numerous broken lapilli, are interpreted to be reworked pyroclastic units, probably deposited in lakes or on flood plains. In the Meentheena area, a 200-m-thick unit of vesicular, amygdaloidal, and massive basalt overlies the tuffaceous deposits of the member.

Fig. 12.16 Accretion lapilli in tuff within the basal part of the Mingah Member, Tumbiana Formation, 19 km south of Nullagine, Northeast Pilbara Sub-basin (MGA Zone 51, 197430E 7567285N). (From Bagas 2005; with Geological Survey of Western Australia permission)

Fig. 12.17 Stromatolites in the Meentheena Member of the Tumbiana Formation: (**a**) natural cross-section exposure of columnar, umbellate, branching-style stromatolite bioherms covered by climbing rippled calcareous sandstone, about 15 km north of Meentheena, Northeast Pilbara Sub-basin. Columns are about 50 cm across. (From Williams and Bagas 2007; with Geological Survey of Western Australia permission). (**b**) Large domical stromatolites about 8 km west-southwest of Meentheena. Domes are over 1 m high. (Previously unpublished photograph by author). (**c**) Cross-sectional view of round-topped, branching, coniform columnar stromatolites truncated by rippled calcareous sandstone, Chichester Range (MGA Zone 50, 688400E 7565800N). Columns are 5 cm across. (From Van Kranendonk and Hickman 2012; with Geological Survey of Western Australia permission). (**d**) Bedding-plane view of the same stromatolites shown in c. (From Van Kranendonk and Hickman 2012; with Geological Survey of Western Australia permission)

The Meentheena Member comprises stromatolitic limestone and dolostone interbedded with volcaniclastic sandstone and shale. Exceptionally well-preserved stromatolites in this member (Figs 12.17a–d) have led to several detailed studies (Packer 1990; Van Kranendonk et al. 2006a, b; Bolhar and Van Kranendonk 2007; Awramik and Buchheim 2009; Coffey et al. 2011; Flannery and Walter 2012). In the type area around Meentheena, the main lithological facies have been described as ooidal flat-pebble conglomerate, ripple cross-laminated oolite and grainstone, planar to wavy-laminated calcilutite and calcisiltite, and shale and siltstone (Awramik and Buchheim 2009). Stromatolite morphologies include domes, columns, branched columns, and coniform types. Domal stromatolites are locally over 2 m in diameter and up to 1 m in height (Fig. 12.17b).

In most parts of the Gregory Range Sub-basin, the Tumbiana Formation is separated into the two members although the Mingah Member is the most extensive. The formation is apparently absent from the Marble Bar Sub-basin although sandstone of the undated Pear Creek Formation might be similar in age to the Mingah Member. The formation outcrops extensively along the northern slopes of the Chichester Range, extending into the Northwest Pilbara Sub-basin. Across much of this area, the formation is mainly represented by lithologies characteristic of the Mingah Member and is partly mapped as this member.

Interpretations of the depositional environment of the Tumbiana Formation have varied from marine (Packer 1990; Thorne and Trendall 2001; Sakurai et al. 2005) to fluvial and lacustrine (Kriewaldt and Ryan 1967; Walter 1983; Buick 1992; Blake et al. 2004; Bolhar and Van Kranendonk 2007; Awramik and Buchheim 2009; Coffey et al. 2011, 2013; Flannery and Walter 2012). A non-marine depositional environment is consistent with interpretations that the underlying Kylena Formation and overlying Maddina Formation are each composed of subaerial volcanics (Blake 1993; Thorne and Trendall 2001).

The depositional setting of the Fortescue Group, on thick continental crust of the North Pilbara (buoyant in relation to any adjacent oceanic crust), makes it unlikely that the craton was temporarily submerged during deposition of the Tumbiana Formation. Although there is evidence that deposition of the Fortescue Group in the South Pilbara Sub-basin changed from subaerial to subaqueous during eruption of the c. 2740 Ma Boongal Formation (Thorne and Trendall 2001), this might be explained by greater crustal extension, rifting, and subsidence in that sub-basin relative to the northern sub-basins.

12.2.1.4.2 Maddina Formation

The Maddina Formation is mainly composed of subaerial basalt and basaltic andesite with upper units of andesite, dacite, and rhyolite. In the Northeast Pilbara and Gregory Range Sub-basins, it includes a lower member of relatively mafic basalt, once named the 'Nymerina Basalt'. This lower member is separated from basaltic andesite, andesite, dacite, and rhyolite in the upper Maddina Formation by the sedimentary Kuruna Member.

The thickness of the formation varies from 2 km in the Balfour Downs area to about 350 m where the formation unconformably overlies granitic rocks of the Pilbara Craton 150 km to the west in the Chichester Range. The thickness is between 1 km and 700 m in the Gregory Range, and 550 km to the west, south of Cape Preston, the formation is about 1 km thick. Subaerial basaltic flows are the dominant volcanic facies of the Maddina Formation. Flows vary in thickness up to 50 m, and much of the basalt and basaltic andesite is more amygdaloidal than most lavas of the Fortescue Group. Amygdales are up to 10 cm in diameter and filled by variable amounts of quartz (including agate) and carbonate (Fig. 12.18).

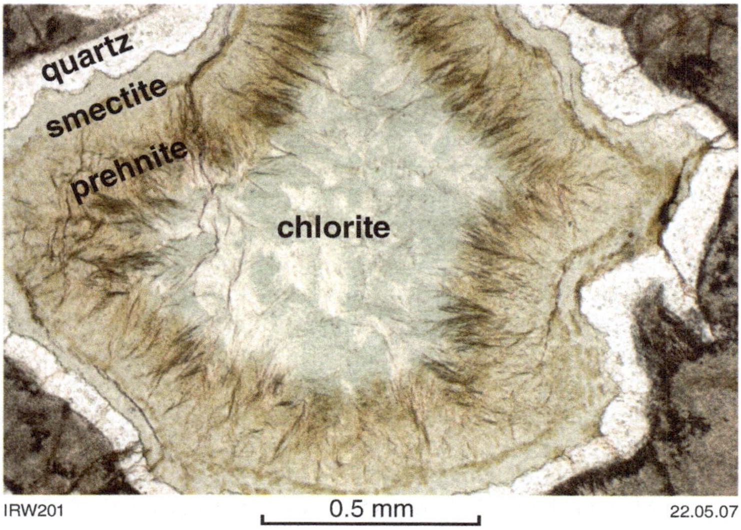

Fig. 12.18 Photomicrograph of a zoned amygdale in basalt of the Maddina Formation (MGA Zone 51, 281671E 7623044N), 15 km west of Carawine Pool. (From Williams 2007; with Geological Survey of Western Australia permission)

Radiometric ternary imagery (KTU) indicates relatively mafic compositions in the basal part of the lower succession in the Northwest Pilbara Sub-basin. The basal unit is composed of thick, massive flows of vesicular and amygdaloidal basalt that contains clinopyroxene phenocrysts up to 2 mm long. Four samples analysed by Glikson et al. (1986) contained more than 8 wt.% MgO; the average contents of Cr and Ni (relatively immobile elements) in these four samples were 845 ppm Cr and 279 ppm Ni, which is consistent with high-Mg compositions. However, the main part of the lower succession is more lithologically similar to most of the upper succession, although average compositions are more mafic (data in Thorne and Trendall 2001).

12.2.1.4.2.1 Kuruna Member

The Kuruna Member is composed of lithologies similar to those of the Tumbiana Formation, including pisolitic tuff (Fig. 12.19), sandstone, siltstone, shale, chert, and local stromatolitic carbonate rocks. The member is up to 150 m thick and forms a prominent stratigraphic marker horizon in the Maddina Formation of the Northwest Pilbara Sub-basin. Basaltic to andesitic volcaniclastic rocks include tuff containing beds of normally graded accretionary lapilli, and the bedding planes of siltstone and shale are commonly ripple marked. The stromatolitic carbonate units mainly lie towards the top of the member, and domal bioherms are up to 2 m across. In the Oakover Syncline, the Kuruna Member is intruded by several thick, coarse-grained dolerite sills. Geochemistry might indicate if these are subvolcanic to the overlying

Fig. 12.19 A graded bed of accretion lapilli in siltstone of the Kuruna Member (MGA Zone 51, 245467E 7664613N), 15 km north-northeast of the Nullagine River crossing on the Ripon Hills Road. (From Williams 2007; with Geological Survey of Western Australia permission)

upper Maddina Formation or were emplaced later, during rifting and deposition of the Jeerinah Formation.

12.2.1.4.3 Pear Creek Formation

The Pear Creek Formation is confined to the Marble Bar Sub-basin where it is up 1 km thick and composed of conglomerate, sandstone, siltstone, and shale with local thin units of basalt. The formation is exposed in two structural basins, or centroclines, separated by exposures of the underlying Kylena Formation, Hardey Formation, and Mount Roe Basalt. The southern exposure outcrops over 40 km^2 and is mainly composed of sandstone and thin beds of conglomerate. The northern exposure has an area of about 35 km^2 and has been described as occupying the 'Pear Creek Centrocline' (Blake 1993). The succession in this exposure is more complex than in the south, consisting of five mappable units. Conglomerates in the central section of the stratigraphy are mainly composed of basaltic boulders and pebbles derived from erosion of the underlying Kylena Formation and Mount Roe Basalt. Less common granite, quartzite, felsic volcanic, and chert clasts (Fig. 12.20) indicate erosion of the underlying Pilbara Craton. Due to a scarcity of lithologies suitable for dating, the age of the Pear Creek Formation remains to be determined. However, based on the erosional unconformity at the base of the formation (Van Kranendonk 2003) and the apparent lack of deformation elsewhere in the Northern Pilbara after deposition of Sequence 4, it is most likely of similar age to this

Fig. 12.20 Outcrops of conglomerate of the Pear Creek Formation, northern Marble Bar Sub-basin: (**a**) polymictic conglomerate near the base of the formation containing clasts of basalt, granite and vein quartz (MGA Zone 50, 768890E 7681956N); (**b**) conglomerate from the lower part of the formation (MGA Zone 50, 769800E 7680850N) composed of basalt clasts from the underlying Fortescue Group and clasts of granitic gneiss from adjacent outcrops of the Pilbara Craton. (From Van Kranendonk 2010; with Geological Survey of Western Australia permission)

sequence. Lithologies are similar to those in the Hardey Formation, but the Pear Creek Formation unconformably overlies the Kylena Formation that elsewhere in the Northern Pilbara disconformably to unconformably overlies the Hardey Formation.

The Marble Bar Sub-basin occupies a unique structural position in the Northern Pilbara, located over the widest area of greenstones and positioned between five Paleoarchean–Mesoarchean granite-greenstone domes (Muccan, Mount Edgar, Corunna Downs, Shaw and North Pole). The stratigraphy of the sub-basin (Van Kranendonk 2003) establishes reactivation of the adjacent domes during deposition of the Fortescue Group. An event of uplift and erosion of the domes after deposition of the Kylena Formation would explain the mixed lithological composition of the conglomerates in the Pear Creek Formation (Fig. 12.21).

12.2.1.4.4 Geochronology

In the Nullagine area of the Northeast Sub-basin, three units of the Tumbiana Formation were dated at 2727 ± 5 Ma, 2724 ± 4 Ma, and 2721 ± 4 Ma (Blake et al. 2004). In the Northwest Pilbara Sub-basin, units of the formation have been dated at 2719 ± 6 Ma (GSWA 168935, Nelson 2001) and 2715 ± 6 Ma (Arndt et al. 1991). Because c. 2715 Ma date is slightly inconsistent with dates on the overlying Maddina Formation, the depositional age range of the Tumbiana Formation is interpreted to be 2727 to 2719 Ma.

The Maddina Formation has been dated at four localities: in the Chichester Range of the Northwest Pilbara Craton, a sample of plagioclase-porphyritic dacite from the central part of the Maddina Formation was dated at 2717 ± 2 Ma (GSWA 144993, Nelson 1998); dates recorded from the central and upper parts of the formation in the Nullagine area of the Northeast Pilbara Sub-basin are 2718 ± 3 Ma, 2715 ± 2 Ma, and 2713 ± 3 Ma (Blake et al. 2004). Based on the interpreted minimum depositional

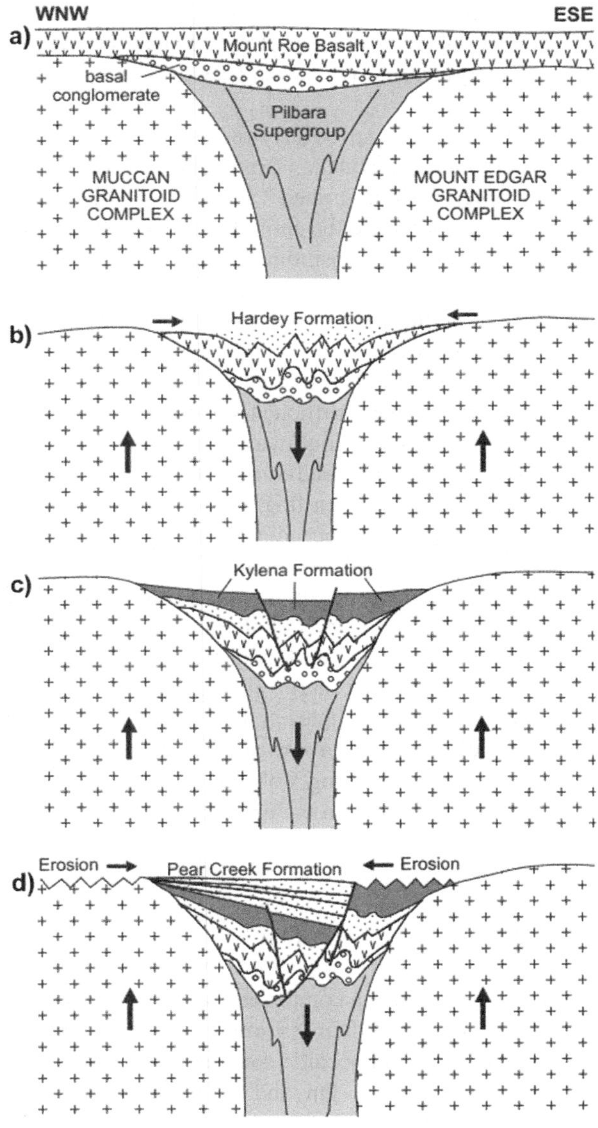

Fig. 12.21 Diagrammatic illustration of the evolution of the northern Marble Bar Sub-basin: (**a**) deposition of the Mount Roe Basalt and basal conglomerates on a greenstone basement with relict topography between domal granitic uplands; (**b**) deformation of the Mount Roe Basalt due to reactivation of the domes and intervening syncline in the underlying Pilbara Craton, and resulting erosion of uplifted areas with influx of detritus to deposit the unconformably overlying Hardey Formation; (**c**) with ongoing deformation, deposition of the Kylena Formation; (**d**) asymmetric downthrow across the Pear Creek Fault resulted in unconformable deposition of the Pear Creek Formation on the Kylena Formation. (Modified from Van Kranendonk 2003; with Geological Survey of Western Australia permission)

age of the underlying Tumbiana Formation, the age range of the Maddina Formation is interpreted to be c. 2719 to 2713 Ma.

12.2.1.5 Sequence 5

Rifting during the Fortescue Rifting Event is interpreted to have evolved into continental breakup and plate separation prior to the c. 2713–2629 Ma deposition of the Jeerinah Formation in Sequence 5. The main evidence for the breakup is that the depositional environment of the Jeerinah Formation is interpreted to be that of a passive margin basin (Blake and Barley 1992; Blake 1993; Thorne and Trendall 2001).

Thorne and Trendall (2001) referred to this sequence as 'Sequence 4' because they interpreted the Kylena, Tumbiana, and Maddina Formations to be essentially conformable. They described 'Sequence 4' (here Sequence 5) as comprising the Jeerinah Formation and the entire Hamersley Group, with deposition spanning about 267 million years. Depositional environments were interpreted as ranging from a nearshore shelf to 'deeper' offshore shelf facies in the Jeerinah Formation changing to 'deeper' marine shelf facies in the Hamersley Group.

The Jeerinah Formation is exposed over wide areas of the Southern Pilbara within and south of the Chichester Range. Water depths in the area now occupied by the Chichester Range were sufficiently shallow to allow the growth of stromatolites, but depths increased to the southwest. Following the Fortescue Rifting Event, ongoing rifting on the shelf occurred during plate separation of the Pilbara Craton from the inferred detached southern fragment of the craton in the area now occupied by the Capricorn Orogen. An oceanic basin is interpreted to have opened up south of preserved exposures of the Pilbara Craton, and a similar oceanic basin is likely to have formed northeast of the craton. Rifting on the passive margins of the craton resulted extensive intrusion of dolerite dykes and sills into the Jeerinah Formation. In the Northeast Pilbara, the Jeerinah Formation is exposed on both sides of the Oakover Valley where dolerite sills make up over 50% of the succession. Additionally, the formation includes a 1-km-thick basaltic unit, the Baramine Volcanic Member. Felsic volcanic rocks are present in the Isabella Member in the Northeast Pilbara and the Nallanaring Volcanic Member in the Northwest Pilbara. Thin layers of felsic tuff are present in the upper part of the Jeerinah Formation (Rasmussen and Fletcher 2010).

12.2.1.5.1 Jeerinah Formation

The Jeerinah Formation comprises a shallow- to deep-water succession of sandstone, siltstone, shale, jaspilitic iron formation, chert, and carbonate rocks widely intruded by dolerite sills, and locally containing volcanic rocks. The formation varies in thickness between 1.2 km and 100 m and outcrops in the South Pilbara, the Chichester Range sections of the Northwest and Northeast Pilbara, and the Gregory Range area. Thin successions are also present in the western part of the Northwest Pilbara and the eastern part of the Northeast Pilbara. There is no evidence that the Jeerinah Formation was deposited over that part of the Northern Pilbara Craton now exposing large granitic complexes.

The formation has a laterally variable stratigraphy that includes six members. In areas along the Chichester Range, the formation is divided into four members, in ascending order: the basal Woodiana Member, mainly composed of sandstone and conglomerate but locally including wacke, mudstone, and silicified clastic and carbonate rocks; the Warrie Member composed of mudstone, shale, ferruginous shale and chert, jaspilitic iron formation, and dolomite; the Nallanaring Volcanic Member comprising a lower unit of spherulitic felsic lava and pyroclastic rocks overlain by an upper unit of locally pillowed basaltic lava; and the Roy Hill Member consisting of black carbonaceous shale in which rounded limonite nodules have replaced pyrite.

Lithologies characteristic of the Warrie Member are widespread in the south Pilbara, although mapping in this area of the Pilbara has not subdivided the Jeerinah Formation into members. In the Northeast Pilbara, there are two volcanic members in the Jeerinah Formation: the Baramine Volcanic Member composed of massive, amygdaloidal, and vesicular dark-grey to grey-green basalt and basaltic andesite, interlayered with basaltic volcanic sandstone and siltstone, and local beds of accretionary lapilli, and the Isabella Member comprising extrusive and intrusive rhyolite and felsic volcaniclastic rocks.

In the Balfour Downs area, the Woodiana Member is separated from the Warrie Member by an unnamed, <300-m-thick unit of carbonated pillow basalt, tuff, and dolerite (Williams 1989; Farrell and Smithies 2005). Dolerite sills are most common in deep-water facies of the formation and are thought to have been intruded along extensional rifts in outer sections of the basin.

Sedimentary lithologies of the Jeerinah Formation in the South Pilbara mainly consist of argillite with minor units of volcanic arenite, chert, and dolostone, whereas volcanic rocks include massive and pillow basalt and basaltic hyaloclastite (Thorne and Trendall 2001).

12.2.1.5.1.1 Baramine Volcanic Member

The Baramine Volcanic Member is composed of massive, amygdaloidal, and vesicular dark-grey to grey-green basalt and basaltic andesite, interlayered with basaltic volcanic sandstone and siltstone, and local beds of accretionary lapilli. Pillow basalt forms part of the succession in the Coonanbunna Creek area, but elsewhere the member is almost entirely pyroclastic rocks thought to include surge and debris-flow deposits. Thick pyroclastic units are composed of angular and rounded clasts of basaltic lava between 5 and 20 cm across, although clasts up to 1 metre across are locally present.

The stratigraphic position of the member within the regional stratigraphy of the Jeerinah Formation has not been established. In the Isabella Range, there are locally only thin sedimentary units of the Jeerinah Formation separating the member from the underlying Maddina Formation, whereas in the Coonanbunna Creek area, the member is apparently at least 1 km above the base of the Jeerinah Formation. This might be due to southwest to northeast changes of thickness and facies within the passive margin basin succession of the Jeerinah Formation. In both areas, the Jeerinah Formation is intruded by thick sills of dolerite, and some of these might be subvolcanic to the Baramine Volcanic Member.

12.2.1.5.1.2 Isabella Member

The Isabella Member is restricted to outcrops in the Isabella Range in the Northeast Pilbara. The member is composed of extrusive and intrusive rhyolite, felsic volcaniclastic rocks, and interbedded shale. Rhyolite of the member is feldspar-porphyritic and locally contains small spherical amygdales filled by quartz, chlorite, and carbonate. The phenocrysts are composed of potash feldspar, quartz, and, less commonly, plagioclase in a fine-grained granular to spherulitic matrix. The rhyolite is overlain by coarse agglomerate containing of clasts of felsic lava and minor stromatolitic dolomite. Wacke, shale, chert, and dolomite underlying the rhyolite are interpreted to be part of the main succession of the Jeerinah Formation. The thickness of the member averages about 50 m, although if the sedimentary rocks are included the total thickness is about 250 m. Leucocratic microgranite intruding the Jeerinah Formation 6.5 km north of the site of Baramine Homestead is interpreted to be part of the Isabella Member.

12.2.1.5.1.3 Woodiana Member

The Woodiana Member is 15–60 m thick and conformably to unconformably overlies the Maddina Formation along a west–east strike length of 550 km, from Cape Preston in the west to the Gregory Range and Balfour Downs in the east. As mentioned in the Introduction, the contact marks a major depositional change from subaerial deposition on the 2775–2713 Ma volcanic plateau of the Fortescue Group to 2713–2629 Ma marine deposition of the mainly sedimentary Jeerinah Formation. The contact is evidence of a marine transgression (Thorne and Trendall 2001), and the present interpretation is that the Jeerinah Formation was deposited in a subsiding passive margin basin following c. 2710 Ma breakup of the Pilbara Craton. Regions of the breakup were located hundreds of kilometres south, east, and probably west of the present exposures of the Maddina Formation–Woodiana Member contact, in areas now overlying the concealed margins of the Pilbara Craton (Chap. 1).

The member is mainly composed of shallow-water sandstone, silicified stromatolitic carbonate rocks, and chert breccia. Sandstone at the base of the Jeerinah Formation is almost continuous along the Chichester Range and east into the Balfour Downs area. The sandstone is fine to very coarse grained, variably silicified, and forms beds between 0.05 and 1.5 m thick. Pebbly sandstone and conglomerate are locally included in the member. Sedimentary structures include various types of cross-stratification, symmetrical and asymmetrical ripples, and graded bedding (Thorne and Trendall 2001). In several areas, the sandstone is underlain by a laterally continuous unit that has been described as 'black siliceous mudstone' (Williams 1968, 1989) that might be tuffaceous. In places, this is separated from the underlying Maddina Formation by lenticular units of coarse-grained sandstone and breccia up to 2.5 m thick (Williams 1968).

Chert within the Woodiana Member is interpreted to have been formed by the silicification of carbonate rocks and fine-grained clastic sedimentary rock. Almost all the information from the chert has been derived from local investigations near

Tambrey in the central section of the Chichester Range (Packer 1990). In this area, the chert is preserved in 50–400-m-diameter, saucer-shaped remnants overlying the Maddina Formation. Lithologies include black, irregularly laminated chert containing bulbous, columnar, and branching stromatolites (Fig. 12.22) in addition to more tabular biostromes (Thorne and Trendall 2001). Other silicified lithologies included volcaniclastic and epiclastic sedimentary rock containing low-angle cross-stratification.

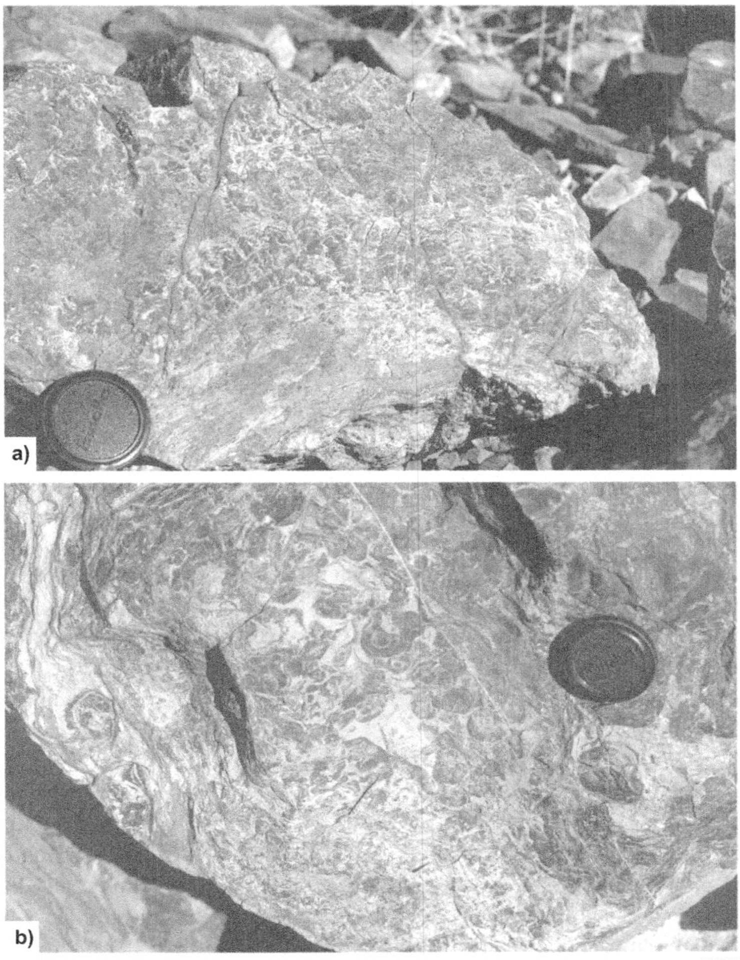

Fig. 12.22 Stromatolites in the Woodiana Member at Tambrey, Northwest Pilbara Sub-basin: (**a**) cross section of cumulate microcolumnar stromatolites in chert (silicified sedimentary carbonate rock); (**b**) transverse section of the same type of stromatolites. (From Thorne and Trendall 2001; with Geological Survey of Western Australia permission)

12.2.1.5.1.4 Warrie Member

Like the conformably underlying Woodiana Member, the Warrie Member has been mapped along the Chichester Range and farther east in the Balfour Downs area. The member is about 100 m thick and is composed of mudstone and siltstone, with subordinate chert, jaspilitic chert, shale, carbonate rocks, and sandstone. Turbidites form part of the succession. Argillaceous lithologies, which include thin beds and pods of chert and chert–carbonate layers, are mainly parallel-laminated with local ripple cross-stratification. Thicker beds of dolostone are tabular to nodular and locally contain domical stromatolites. Felsic volcaniclastic sandstone forms thin layers less than 20 mm thick but is locally present in thickly bedded tabular units up to 10 m thick. In the eastern part of the Chichester Range, the upper 20 m of the Warrie Member is composed of thinly bedded dolostone that has been used as a marker to separate the member from the conformably overlying Roy Hill Member. However, west of the Roy Hill area, the upper dolostone is absent, and the contact is defined by the base of the distinctive thick black shale succession of the Roy Hill Member.

The lithologies characteristic of the Warrie Member in the Chichester Range also make up most of the Jeerinah Formation in the South Pilbara and the Northeast Pilbara north of the Balfour Downs area. In both areas the Jeerinah Formation is extensively intruded by dolerite dykes and sills, and pillow basalts make up significant parts of the succession (Thorne and Trendall 2001). These mafic lithologies suggest deep rifting, which would be consistent greater extension of the Pilbara crust closer to zones of breakup and plate separation.

12.2.1.5.1.5 Nallanaring Volcanic Member

Conformably overlying the Warrie Member in the Northwest Pilbara, the Nallanaring Volcanic Member is about 75 m thick and outcrops over a strike length of 80 km. The member comprises a lower unit of spherulitic felsic lava and pyroclastic rocks overlain by basaltic lavas that include pillow structures. The felsic volcanic rocks are inferred to have been derived from partial melting of underlying granitic crust of the Pilbara Craton, and an evolved ε_{Nd} value (Nelson et al. 1992) is consistent with this interpretation.

12.2.1.5.1.6 Roy Hill Member

The Roy Hill Member is a distinctive unit of black shale, between 30 and 120 m thick, outcropping along a west–east strike length of about 550 km in the Chichester Range and in the Balfour Downs area. Intersections of the black shale in drilling establish that in most areas the Roy Hill Member is thicker than 100 m; thin layers of felsic tuff, chert, and sedimentary carbonate rocks are very minor constituents. The shale is carbonaceous and contains numerous spherical pyrite concretions altered to limonite. Surface exposures of the shale are limited and almost white due to deep weathering. However, the succession has been intersected in several diamond drill holes, including WRL-1, FVG-1, DDH-186, AIDP-2, AIDP-3, RHD0296, ECD-0001, and ECD-0002 (Hurst et al. 2013; Glikson and Hickman 2014).

Carbonaceous black shale such as that forming the Roy Hill Member indicates anoxic deposition, commonly but not invariably in relatively deep water. The shale lacks sedimentary structures such as cross-stratification which is consistent with a very low-energy depositional environment. Geochronology (Rasmussen and Fletcher 2010) indicates that deposition of the 100-m-thick shale occurred over about 50 Ma, indicating very slow deposition with little detrital influx; Trendall et al. (2004) also drew this conclusion. The depositional setting of the Roy Hill Member might have been an outer shelf or slope environment. However, this might not apply in the Northeast Pilbara where the Roy Hill Member is directly overlain by the Carawine Dolomite, the upper part of which includes stromatolites and evaporites (Williams 2007).

About 3 m beneath the top of the Roy Hill Member, there is an asteroid impact ejecta layer containing impact spherules and tsunami deposits. This layer, of variable total thickness up to about 20 m, has been named as the 'Jeerinah Impact Layer' (JIL). The JIL has been recorded in surface exposures at Hesta Siding (Zone 50 K, MGA 709000E 7544300 N) and in diamond drill holes over a strike length of about 200 km in the Fortescue Valley: WRL-1, FVG-1, DDH-186, AIDP-2, AIDP-3, ECD-0001, and ECD-0002 (Hurst et al. 2013; Glikson and Hickman 2014).

12.2.1.5.2 Geochronology

Ash-fall felsic tuff layers within black shale of the Jeerinah Formation were dated by Rasmussen and Fletcher (2010). The samples were collected from cores of four drill holes located between Ripon Hills in the Northeast Pilbara and an area north Tom Price in the South Pilbara: hole RHD2A at Ripon Hill provided dates of 2643 ± 13 Ma and 2636 ± 10 Ma; hole FVG-1 in the Fortescue Valley included a tuff dated at 2662 ± 7 Ma; hole WRL-1 near Wittenoom provided dates of 2632 ± 7 Ma and 2657 ± 13 Ma; and DDH 186 north of Tom Price gave dates of 2676 ± 11 Ma and 2664 ± 7 Ma. All these dated samples were from thin felsic tuff layers within the upper 70 m of the Jeerinah Formation, and based on regional stratigraphy, these tuffs are parts of the Roy Hill Member. Significantly, this indicates that the 100-m-thick Roy Hill Member was deposited over about 50 million years.

There have been four additional dates on the Jeerinah Formation: a date of 2690 ± 16 Ma was obtained on a tuffaceous sandstone in the South Pilbara (Arndt et al. 1991), possibly from the stratigraphic level of the Warrie Member; an andesitic ignimbrite in the Nallanaring Volcanic Member was dated at 2684 ± 6 Ma (Arndt et al. 1991); another andesitic ignimbrite at the top of the formation was dated at 2629 ± 5 Ma (sample 120048, Nelson et al. 1999); and dating of a sandstone (Woodiana Member) at the base of the formation indicated a maximum depositional age of 2715 ± 2 Ma (Blake et al. 2004). Because the minimum depositional age of the underlying Maddina Formation is interpreted to be c. 2713 Ma, the age range of the Jeerinah Formation is inferred to be c. 2713 to 2629 Ma.

References

Arndt N, Bruzak G, Reischmann T (2001) The oldest continental and oceanic plateaus: geochemistry of basalts and komatiites of the Pilbara Craton, Australia. In: Ernst RE, Buchan K (eds) Mantle plumes: their identification through time. Geological Society of America, pp 359–387

Arndt NT, Nelson DR, Compston W, Trendall AF, Thorne AM (1991) The age of the Fortescue Group, Hamersley Basin, Western Australia, from ion microprobe zircon U–Pb results. Austral J Earth Sci 38:261–281

Awramik SM, Buchheim HP (2009) A giant late Archean lake system: the Meentheena member (Tumbiana Formation; Fortescue Group), Western Australia. Precambrian Res 174:215–240

Bagas L (2005) Geology of the Nullagine 1:100 000 sheet. 1:100 000 geological series explanatory notes. Geological Survey of Western Australia

Barley M, Bekker A, Krapež B (2005) Late Archean to early Paleoproterozoic global tectonics, environmental change and the rise of atmospheric oxygen. Earth Planet Sci Lett 238:156–171. https://doi.org/10.1016/j.epsl.2005.06.062

Blake TS (1984) The lower Fortescue Group of the Northern Pilbara Craton–stratigraphy and paleogeography. In: Muhling JR, Groves DI, Blake TS (eds) Archaean and Proterozoic basins of the Pilbara, Western Australia: evolution and mineralization potential. The University of Western Australia, Geology Department and University Extension, pp 123–143

Blake TS (1993) Late Archaean crustal extension, sedimentary basin formation, flood basalt volcanism, and continental rifting. The Nullagine and Mount Jope Supersequences, Western Australia. Precambrian Res 60:185–241

Blake TS (2001) Cyclic continental mafic tuff and flood basalt volcanism in the late Archaean Nullagine and mount Jope Supersequences in the Eastern Pilbara, Western Australia. Precambrian Res 107:139–177

Blake TS, Barley ME (1992) Tectonic evolution of the late Archaean to early Proterozoic Mount Bruce megasequence set, Western Australia. Tectonics 11:1415–1425

Blake TS, Buick R, Brown SJA, Barley ME (2004) Geochronology of a late Archaean flood basalt province in the Pilbara Craton, Australia: constraints on basin evolutioncanic and sedimentary accumulation, and continental drift rates. Precambrian Res 133:143–173. https://doi.org/10.1016/j.precamres.2004.03.012

Bleeker W (2003) The late Archean record: a puzzle in ca. 35 pieces. Lithos 71:99–134

Bodorkos S, Nelson DR, Love GJ, Wingate MTD (2006) 178090: tuffaceous volcaniclastic sandstone, Dingo Well; Geochronology Record 633: Geological Survey of Western Australia, 4 p

Bolhar R, Van Kranendonk MJ (2007) A non-marine depositional setting for the northern Fortescue Group, Pilbara Craton, inferred from trace element geochemistry of stromatolitic carbonates. Precambrian Res 155:229–250. https://doi.org/10.1016/j.precamres.2007.02.002

Buick R (1992) The antiquity of oxygenic photosynthesis: evidence from stromatolites in sulphate-deficient Archaean lakes. Science 255:74–77. https://doi.org/10.1126/science.11536492

Cheney ES (1996) Sequence stratigraphy and plate tectonic significance of the Transvaal succession of southern Africa and its equivalent in Western Australia. Precambrian Res 79:3–24

Cheney ES, Roering C, Stettler E (1988) Valbara. In: Geocongress '88: extended abstracts. Geological Society of South Africa, pp 85–88

Coffey JM, Walter MR, George SC, Hill AC (2011) Paleontology and field observations of the ~2720 Ma Tumbiana Formation in the Northwest Pilbara, Western Australia. Record, 2011/8. Geological Survey of Western Australia

Coffey JM, Flannery DT, Walter MR, George SC (2013) Sedimentology, stratigraphy and geochemistry of a stromatolite biofacies in the 2.72 Ga Tumbiana Formation, Fortescue Group, Western Australia. Precambrian Res 236:282–296

Condie KC (2001) Mantle plumes and their record in earth history. Cambridge University Press, Cambridge

de Kock MO, Evans DAD, Beukes NJ (2009) Validating the existence of Vaalbara in the Neoarchean. Precambrian Res 174:145–154
de Kock MO, Beukes NJ, Armstrong RA (2012) New SHRIMP U–Pb zircon ages from the Hartswater Group, South Africa: implications for correlations of the Neoarchean Ventersdorp Supergroup on the Kaapvaal Craton and with the Fortescue Group on the Pilbara Craton. Precambrian Res 204–205:66–74
de la Hunty LE (1965) Mount Bruce, W.A. Geological Survey of Western Australia, 1:250 000 Geological Series Explanatory Notes
Eriksson PG, Condie KC, van der Westhuizen W, van der Merwe R, de Bruiyn H, Nelson DR, Altermann W, Catuneanu O, Bumby AJ, Lindsay J (2002) Late Archaean superplume events: a Kaapvaal-Pilbara perspective. J Geodyn 34:207–247
Ernst RE (2014) Large igneous provinces. Cambridge University Press, Cambridge
Ernst RE, Buchan KL (2003) Recognizing mantle plumes in the geological record. Annu Rev Earth Planet Sci 31:469–523
Ernst RE, Buchan KL, Prokoph A (2004) Large Igneous Province record through time. In: Eriksson PG, Altermann W, Nelson DR, Mueller WU, Catuneanu O (eds) The Precambrian earth: tempos and events in Precambrian time, vol 12. Elsevier, Amsterdam, The Netherlands, pp 173–-180Developments in Precambrian geology
Evans DAD, Smirnov AV, Gumsley AP (2017) Paleomagnetism and U–Pb geochronology of the black range dykes, Pilbara Craton, Western Australia: a Neoarchean crossing of the polar circle. Austral J Earth Sci 64:225–237. https://doi.org/10.1080/08120099.2017.1289981
Farrell TR (2006) Geology of the Eastern Creek 1:100 000 sheet. 1:100 000 geological series explanatory notes. Geological Survey of Western Australia
Farrell TR, Smithies RH (2005) Noreena Downs, WA Sheet 2953. Geological Survey of Western Australia
Flannery DT, Walter MR (2012) Archean tufted microbial mats and the great oxidation event: new insights into an ancient problem. Austral J Earth Sci 59:1–11
Flannery DT, Hoshino Y, George SC, Walter MR (2012) Field observations relating to the c. 2740 Ma Mopoke Member, Kylena Formation, Fortescue Group, Pilbara region, Western Australia. Record, 2012/8. Geological Survey of Western Australia
Glikson AY, Hickman AH (2014) Coupled asteroid impacts and banded iron formations, Fortescue and Hamersley Groups, Pilbara, Western Australia. Austral J Earth Sci 61:689–701
Glikson AY, Davy R, Hickman AH (1986) Geochemical data files of Archaean volcanic rocks, Pilbara Block, Western Australia. BMR (Geoscience Australia) Record, 1986/014
Hall C (2002) Sedimentology, geochemistry and palaeogeography of the lower Fortescue Group, Hamersley Province, Western Australia. PhD, The University of Western Australia
Hickman AH (1975) Explanatory notes on the Nullagine 1:250 000 geological sheet. Record, 1975/5. Geological Survey of Western Australia
Hickman AH (1978) Nullagine, Western Australia (2nd edition): Geological Survey of Western Australia, 1:250 000 Geological Series Explanatory Notes
Hickman AH (1983) Geology of the Pilbara Block and its environs. Bulletin 127. Geological Survey of Western Australia
Hickman AH (2004a) Geology of the Cooya Pooya 1:100 000 sheet. 1:100 000 Geological series explanatory notes. Geological Survey of Western Australia
Hickman AH (2004b) Two contrasting Granite–Greenstones Terranes in the Pilbara Craton, Australia: evidence for vertical and horizontal tectonic regimes prior to 2900 Ma. Precambrian Res 131:153–172
Hickman AH (2011) Pilbara Supergroup of the East Pilbara Terrane, Pilbara Craton: updated lithostratigraphy and comments on the influence of vertical tectonics. Geological Survey of Western Australia, Annual Review 2009–10, pp 50–59
Hickman AH (2012) Review of the Pilbara Craton and Fortescue Basin, Western Australia: crustal evolution providing environments for early life. Island Arc 21:1–31

Hickman AH (2016) Northwest Pilbara Craton: A record of 450 million years in the growth of Archean continental crust. Report 160. Geological Survey of Western Australia
Hickman AH (2021) East Pilbara Craton: a record of one billion years in the growth of Archean continental crust. Report 143. Geological Survey of Western Australia
Hickman AH, Kojan CJ (2003) Geology of the Pinderi Hills 1:100 000 sheet. 1:100 000 Geological series explanatory notes. Geological Survey of Western Australia
Hickman AH, Lipple SL (1975) Explanatory notes on the Marble Bar 1:250 000 geological sheet, Western Australia. Record, 1974/20. Geological Survey of Western Australia
Hickman AH, Lipple SL (1978) Marble Bar, WA sheet SF50–8, 2nd edn. Geological Survey of Western Australia
Hickman AH, Van Kranendonk MJ (2008a) Marble Bar, WA Sheet 2855. Geological Survey of Western Australia
Hickman AH, Van Kranendonk MJ (2008b) Archean crustal evolution and mineralization of the Northern Pilbara Craton–a field guide. Record, 2008/13. Geological Survey of Western Australia
Hickman AH, Chin RJ, Gibson DL (1983) Yarrie, Western Australia: Geological Survey of Western Australia, 1:250,000 Geological Series Explanatory Notes
Hickman AH, Smithies RH, Tyler IM (2010) Evolution of active plate margins: West Pilbara Superterrane, De Grey Superbasin, and the Fortescue and Hamersley Basins–a field guide. Record, 2010/3. Geological Survey of Western Australia
Horwitz RC, Smith RE (1978) Bridging the Yilgarn and Pilbara Blocks. Precambrian Res 6:293–322
Hurst J, Krapež B, Hawke P (2013) Stratigraphy of the Marra Mamba iron formation within the Chichester Range and its implications for iron ore genesis at Roy Hill–evidence from deep diamond drill holes within the Fortescue Valley. In: Australian Institute of Mining and Metallurgy, pp 91–106
Huston DL, Blewett RS, Champion DC (2012) Australia through time: a summary of its tectonic and metallogenic evolution. Episodes 35:23–43
Kojan CJ, Hickman AH (1998) Late Archaean volcanism in the Kylena and Maddina Formations, Fortescue Group, West Pilbara. In: Geological Survey of Western Australia Annual Review 1997–98. Geological Survey of Western Australia, pp 43–54
Kriewaldt M (1964) The Fortescue Group of the Roebourne Region, north-west division. In: Annual report for the year 1963. Geological Survey of Western Australia, pp 30–34
Kriewaldt MJB, Ryan GR (1967) Pyramid, Western Australia. 1:250 000 geological series explanatory notes. Geological Survey of Western Australia
Kumar A, Parashuramula V, Shankar R, Besse J (2017) Evidence for a Neoarchean LIP in the Singhbhum Craton, Eastern India: implications to Vaalbara supercontinent. Precambrian Res 292:163–174
Lewis JD, Rosman KJR, de Laeter JR (1975) The age and metamorphic effects of the Black Range dolerite dyke. In: Annual report for the year 1974. Geological Survey of Western Australia, pp 80–88
Macfarlane AW, Danielson A, Holland HD, Jacobsen SB (1994) REE chemistry and Sm-Nd systematics of late Archean weathering profiles in the Fortescue Group, Western Australia. Geochim Cosmochim Acta 58:1777–1794
MacLeod WN, de la Hunty LE (1966) Roy Hill, Western Australia. 1:250 000 geological series explanatory notes. Geological Survey of Western Australia
Martin DMB, Clendenin CW, Krapež B, McNaughton NJ (1998a) Tectonic and geochronological constraints on late Archaean and Palaeoproterozoic stratigraphic correlation within and between the Kaapvaal and Pilbara Cratons. J Geol Soc 155:311–322
Martin DMB, Li ZX, Nemchin AA, Powell CMA (1998b) A pre-2.2 Ga age for giant hematite ores of the Hamersley Province, Australia? Econ Geol 93:1084–1090

Mole DR, Barnes SJ, Yao Z, White AJ, Maas R, Kirkland CL (2018) The Archean Fortescue large igneous province: a result of komatiite contamination by a distinct Eo-Paleoarchean crust. Precambrian Res 310:365–390. https://doi.org/10.1016/j.precamres.2018.02.017

Neaman A, Chorover J, Brantley SL (2005) Element mobility patterns record organic ligands in soils on early earth. Geology 33:117–120

Nedachi M, Nozaki J, Hoashi M, Nedachi Y, Hidaka H, Ohmoto H (2002) The Loss of Iron from the 2.7 Ga Mt. Roe Paleosol (Pilbara, Australia) by Methane-richHydrothermal Fluids. Lunar and Planetary Science XXXIII

Nedachi Y, Nedachi M, Taguchi S, Ohmoto H (2004) Role of hydrothermal system at shallower depth in 2.77 Ga alteration of Mt. Roe Basalt, Pilbara, Western Australia. AGU Fall Meeting, Abstract B53B-0998

Nelson DR (1996) Compilation of SHRIMP U-Pb zircon geochronology data, 1995. Record, 1996/5. Geological Survey of Western Australia

Nelson DR (1997) Compilation of SHRIMP U-Pb zircon geochronology data, 1996. Record, 1997/2. Geological Survey of Western Australia

Nelson DR (1998) Compilation of SHRIMP U-Pb zircon geochronology data, 1997. Record, 1998/2. Geological Survey of Western Australia

Nelson DR (1999) Compilation of geochronology data, 1998. Record, 1999/2. Geological Survey of Western Australia

Nelson DR (2001) Compilation of geochronology data, 2000. Record, 2001/2. Geological Survey of Western Australia

Nelson DR (2004) Compilation of geochronlogy data, 2003. Record, 2004/2. Geological Survey of Western Australia

Nelson DR, Trendall AF, de Laeter JR, Grobler NJ, Fletcher IR (1992) A comparative study of the geochemical and isotopic systematics of late Archaean flood basalts from the Pilbara and Kaapvaal Cratons. Precambrian Res 54:231–256

Nelson DR, Trendall AF, Altermann W (1999) Chronological correlations between the Pilbara and Kaapvaal Cratons. Precambrian Res 97:165–189

Nelson DR, Love GJ, Bodorkos S, Wingate MTD (2006) Compilation of geochronology data, June 2006 update. Geological Survey of Western Australia

Noldart AJ, Wyatt JD (1962) The geology of portion of the Pilbara goldfield, covering the Marble Bar and Nullagine 4 mile map sheets. Bulletin 115. Geological Survey of Western Australia

Packer BM (1990) Sedimentology, paleontology and stable-isotope geochemistry of selected formations in the 2.7 billion-year-old Fortescue Group, Western Australia. University of California, Los Angeles, PhD thesis (unpublished)

Pidgeon RT (1984) Geochronological constraints on early volcanic evolution of the Pilbara Block, Western Australia. Austr J Earth Sci 31:237–242

Pirajno F (2004) The Fortescue Group, Western Australia; An Archaean (2.7 Ga) large igneous province. International Association of Volcanology and Chemistry of the Earth's Interior, December 2004 LIP of the Month

Pirajno F (2007) Ancient to modern earth: the role of mantle plumes in the making of continental crust. In: Van Kranendonk MJ, Bennett VC, Smithies RH (eds) Earth's oldest rocks. Elsevier B. V, Burlington, pp 1037–1064

Pirajno F, Hoatson DM (2012) A review of Australia's large igneous provinces and associated mineral systems: implications for mantle dynamics through geological time. Ore Geol Rev 48: 2–54

Pirajno F, Santosh M (2015) Mantle plumes, supercontinents, intracontinental rifting and mineral systems. Precambrian Res 259:243–261. https://doi.org/10.1016/j.precamres.2014.12.016

Rasmussen B, Fletcher IR (2010) Dating sedimentary rocks using in situ U-Pb geochronology of syneruptive zircon in ash-fall tuffs <1mm thick. Geology 38:299–302

Sakurai R, Ito M, Ueno Y, Kitajima K, Maruyama S (2005) Facies architecture and sequence-stratigraphic features of the Tumbiana Formation in the Pilbara Craton, Northwestern Australia:

implications for depositional environments of oxygenic stromatolites during the late Archean. Precambrian Res 138:255–273
Seymour DB, Thorne AM, Blight DF (1988) Explanatory notes on the Wyloo 1:250 000 geological sheet, Western Australia, 2nd edn. Record, 1987/3. Geological Survey of Western Australia
Simonson BM, Davies D, Hassler SW (2000) Discovery of a layer of probable impact melt spherules in the late Archaean Jeerinah Formation, Fortescue Group, Western Australia. Austr J Earth Sci 47:315–325. https://doi.org/10.1046/j.1440-0952.2000.00784.x
Smithies RH, Van Kranendonk MJ, Champion DC (2005) It started with a plume–early Archaean basaltic proto-continental crust. Earth Planet Sci Lett 238:284–297
Thom R, Hickman AH, Chin RJ (1979) Nullagine, WA sheet SF51–5, 2nd edn. Geological Survey of Western Australia
Thorne AM, Trendall AF (2001) Geology of the Fortescue Group, Pilbara Craton, Western Australia. Bulletin 144. Geological Survey of Western Australia
Thorne AM, Tyler IM (1997) Mount Bruce, Western Australia. Geological Survey of Western Australia, 1:250 000 Geological Series Explanatory Notes
Trendall AF, Compston W, Nelson DR, de Laeter JR, Bennett VC (2004) SHRIMP zircon ages constraining the depositional chronology of the Hamersley Group, Western Australia. Austr J Earth Sci 51:621–644
Van Kranendonk MJ (2003) Stratigraphic and tectonic significance of eight local unconformities in the Fortescue Group, Pear Creek Centrocline, Pilbara Craton, Western Australia. In: Geological Survey of Western Australia Annual Review 2001–02
Van Kranendonk MJ (2004a) Coongan, WA Sheet 2856. Geological Survey of Western Australia
Van Kranendonk MJ (2004b) Geology of the Carlindie 1:100 000 sheet. 1:100 000 Geological series explanatory notes. Geological Survey of Western Australia
Van Kranendonk MJ (2010) Geology of the Coongan 1:100 000 sheet. 1:100 000 Geological series explanatory notes. Geological Survey of Western Australia
Van Kranendonk MJ, Hickman AH (2012) A billion years of Earth history: A geological transect through the Pilbara Craton and Mount Bruce Supergroup–a field guide to accompany 34th IGC Excursion WA-2. Record, 2012/10. Geological Survey of Western Australia
Van Kranendonk MJ, Hickman AH, Smithies RH, Nelson DN, Pike G (2002) Geology and tectonic evolution of the Archaean North Pilbara Terrain, Pilbara Craton, Western Australia. Econ Geol 97:695–732. https://doi.org/10.2113/gsecongeo.97.4.695
Van Kranendonk MJ, Hickman AH, Smithies RH, Williams IR, Bagas L, Farrell TR (2006a) Revised lithostratigraphy of Archean supracrustal and intrusive rocks in the Northern Pilbara Craton, Western Australia. Record, 2006/15, Geological Survey of Western Australia
Van Kranendonk MJ, Philippot P, Lepot K (2006b) The Pilbara drilling project: c. 2.72 Ga Tumbiana Formation and c. 3.49 Ga Dresser Formation, Pilbara Craton, Western Australia. Record, 2006/14. Geological Survey of Western Australia
Van Kranendonk MJ, Smithies RH, Hickman AH, Champion DC (2007) Review: secular tectonic evolution of Archean continental crust: interplay between horizontal and vertical processes in the formation of the Pilbara Craton, Australia. Terra Nova 19:1–38
Walter MR (1983) Archean Stromatolites: evidence of the Earth's earliest benthos. In: Schopf JW (ed) Earth's earliest biosphere: its origin and evolution. Princeton University Press, Princeton, pp 187–213
Williams IR (1968) Yarraloola, Western Australia. 1:250 000 Geological series explanatory notes. Geological Survey of Western Australia
Williams IR (1989) Explanatory notes on the Balfour Downs 1:250 000 geological sheet, Western Australia. Record, 1989/2. Geological Survey of Western Australia
Williams IR (2007) Geology of the Yilgalong 1:100 000 sheet. 1:100 000 Geological series explanatory notes. Geological Survey of Western Australia
Williams IR, Bagas L (2007) Geology of the Mount Edgar 1:100 000 sheet. 1:100 000 Geological series explanatory notes. Geological Survey of Western Australia

Williams IR, Trendall AF (1996) Braeside, W.A. Sheet 3155, Geological Survey of Western Australia, 1:100 000 Geological Series

Williams IR, Trendall AF (1998) Geology of the Braeside 1:100 000 sheet. 1:100 000 Geological series explanatory notes. Geological Survey of Western Australia

Wingate MTD (1999) Ion microprobe baddeleyite and zircon ages for late Archaean mafic dykes of the Pilbara Craton, Western Australia. Austr J Earth Sci 46:493–500

Wingate MTD, Lu Y, Johnson SP (2017) 205904: metagabbro dyke, Black Hill Bore: Geochronology Record 1365. Geological Survey of Western Australia

Yang WB, Holland HD, Rye R (2002) Evidence for low or no oxygen in the late Archean atmosphere from the ~2.76 Ga Mount Roe #2 paleosol, Western Australia. Geochim Cosmochim Acta 66:3707–3718

Zegers TE, de Wit MJ, Dann J, White SH (1998) Vaalbara, Earth's oldest assembled continent? A combined structural, geochronological, and palaeomagnetic test. Terra Nova 10:250–259

The manufacturer's authorised representative in the EU is Springer Nature Customer Service Centre GmbH, Europaplatz 3, 69115 Heidelberg, Germany. If you have any concerns regarding our products, please contact ProductSafety@springernature.com

Printed and bound by CPI Group (UK) Ltd, Croydon, CR0 4YY
28/09/2026
02224056-0015